DISCRETE MATHEMATICS

S.K. CHAKRABORTY
Associate Professor
Department of Mathematics
BIT Mesra, Ranchi

B.K. SARKAR
Assistant Professor
Department of Information Technology
BIT Mesra, Ranchi

OXFORD
UNIVERSITY PRESS

Oxford University Press is a department of the University of Oxford.
It furthers the University's objective of excellence in research, scholarship,
and education by publishing worldwide. Oxford is a registered trademark of
Oxford University Press in the UK and in certain other countries.

Published in India by
Oxford University Press
YMCA Library Building, 1, Jai Singh Road, New Delhi 110001, India

© Oxford University Press 2011

The moral rights of the authors have been asserted.

First Edition published in 2011
Second impression 2011

All rights reserved. No part of this publication may be reproduced, stored in
a retrieval system, or transmitted, in any form or by any means, without the
prior permission in writing of Oxford University Press, or as expressly permitted
by law, or under terms agreed with the appropriate reprographics
rights organization. Enquiries concerning reproduction outside the scope of the
above should be sent to the Rights Department, Oxford University Press, at the
address above.

You must not circulate this work in any other form
and you must impose this same condition on any acquirer.

ISBN-13: 978-0-19-806543-2
ISBN-10: 0-19-806543-4

Typeset in Times New Roman
by Dash Publishing Solutions, Delhi 110002
Printed in India by Ram Printograph, Delhi 110051

To
The Almighty
and
My Loving Parents
(Late) Sudhir Krishna Chakraborty and Smt. Suroma Chakraborty

S.K. Chakraborty

To
The Almighty
and
My Beloved Parents
(Late) Haripada Sarkar and (Late) Kanak Lata Sarkar

B.K. Sarkar

Preface

Mathematics, in general, is a vast, all encompassing subject. There are many facets to this subject, each having its own distinct flavour—calculus on one hand and discrete mathematics on the other. While the fundamentals of the former lie in continuity, the latter is almost the opposite of it. Discrete mathematics represents discrete objects, i.e., distinct and unconnected elements. These elements are characterized by integers. This subject is applied in cases where the objects are countable, where relationships between finite (or countable) sets are studied, and in the analysis of processes involving a finite number of steps.

Since the advent of computers and their increasing demand in areas such as applied mathematics and computer science, the importance of discrete mathematical structures has been steadily increasing. This subject is the gateway to several advanced courses in mathematics. It also provides the mathematical foundations for many computer science disciplines such as data structures, algorithms, database theory, formal languages, computer theory, computer security, and operating systems. It also provides the necessary mathematical background to solve problems of operations research, chemistry, engineering, biology, etc. Its application lies in programming, coding, and computing.

The study of this subject serves several purposes, the most important of which is to help students gain an intuitive understanding of mathematics and thereby in developing the skill of mathematical reasoning, as well as enhancing the student's capability for abstract imagination. The second and equally important skill developed through the study of this subject is that of problem solving.

Discrete mathematics is a core subject in the undergraduate and postgraduate degree courses in computer science and engineering, and information technology. The aim of this course is to enable students to develop logical and mathematical reasoning abilities and apply them in practice.

ABOUT THE BOOK

This book provides an introduction to the subject for beginners and serves as a useful reference for senior students. It provides a comprehensive and rigorous explanation of the key mathematical concepts and seeks to motivate students through a wealth of relevant examples that are imbued with a practical essence.

This book explains the fundamental concepts in modern mathematics (sets, relations, functions, and logic) and explores them in the context of three particular topics, namely, combinatorics (the mathematics of counting), probability (the mathematics of chance), and graph theory (the mathematics of connections in network). The application of discrete mathematical structures to the study of automata theory has also been dealt with and is a key feature of this book.

The book provides numerous solved examples and end-chapter exercises. Many of the worked-out examples are deliberately straight forward, but some are quite challenging. The chapters have also been interspersed throughout with numerous interesting and instructional notes.

CONTENT AND STRUCTURE

The contents of the book are divided into eleven chapters and an appendix. A brief description of each chapter is as follows:

Chapter 1 deals with sets, sequences, number systems, relations, and functions. The subjects such as computer science and engineering make use of these fundamental concepts. The idea of set, relation, and function is fundamental in the study of mathematical structures, a sole part of discrete mathematics.

Chapter 2 covers combinatorics, in which finite probability spaces, random variables, recursion, recurrence relation, generating function, and counting (combinatorial) method have been discussed.

Chapter 3 focuses on the study of methods of reasoning, which includes propositions and truth tables, tautologies and contradictions, rules of inference, predicate calculus, and introduction to proofs.

Chapter 4 is an integral part of discrete mathematics and deals with the study of algebraic structures such as groups, rings, and fields. These topics have many applications in computer science, such as formal languages and automata theory, and coding theory.

Chapter 5 provides a thorough coverage of matrices. They are used in discrete mathematics to represent a variety of discrete structures, like relations and graphs.

Chapter 6 discusses partially ordered sets. Important applications of partial order relation are PERT (Program Evaluation and Review Technique) and CPM (Critical Path Method). These techniques deal with complexities of scheduling the individual activities needed to complete large projects.

Chapter 7 deals with Boolean algebra. An important application of Boolean algebra exists in the analysis and design of digital computers, especially of the combinational or gating circuits. It also discusses the Quine McCluskey algorithm as well as the Stone's representation theorem as an annexure.

Chapter 8 introduces complexity which treats several data structures, basic techniques for calculating time complexity of algorithms, and algorithms on common types of searching and sorting at the end.

Chapter 9 covers graph theory, which is one of the most important branches of discrete mathematics and a source for algorithms of practical importance. It is used in areas such as syntactic analysis, fault detection and diagnosis in computers, and minimal path problems. Indeed, many real world problems deal with discrete objects and binary relation, and graphical representations of them is very convenient.

Chapter 10 deals with trees which is another variation of data structure. They are used to organize information in database systems and to represent the syntactic structure of source programs in compiler. The file directory system of operating systems and the symbol table part of compilers also use tree structures.

Chapter 11 includes formal language and automata. Automata theory is a mathematical discipline concerned with the invention and the study of mathematically abstract, idealized machines and their capabilities. It also gives a mathematical description of language construction for any new computer language. Designing of DFA for the automata theory is also described at the end of the chapter. The book concludes with an appendix on number theory.

ACKNOWLEDGEMENTS

Dr S.K. Chakraborty would like to thank Mr Abdul Kadir Zilani, CAD Laboratory, BIT, Mesra for his strenuous attempt in drafting the figures in the text. He also thanks Mr Pramod Kumar, CAD Lab, BIT, Mesra for his support. He would further like to thank Mr S.P. Bhattacharya, faculty of Architecture Department, BIT, Mesra, for his constant efforts. Dr Chakraborty takes immense pleasure in thanking his wife Smt. Rina Chakraborty and loving son Sayantan for their boundless inspiration and encouragement.

Mr B.K. Sarkar would like to acknowledge the infusion and fruitful discussions of Mr Ambuj Kumar, former student of the Department of Computer Science and Engineering, BIT, Mesra. Mr Sarkar would like his beloved son Sudipta to be motivated in future by this joint venture.

The authors thank the anonymous reviewers of the text. These reviewers have provided much helpful criticism and encouragement. They hope that this text lives up to their high expectations. Finally, the authors would like to thank the editorial team of Oxford University Press for their cooperation.

S.K. Chakraborty
B.K. Sarkar

Brief Contents

Preface *(v)*

Chapter 1 Sets, Relations, and Functions 1
Chapter 2 Combinatorics 69
Chapter 3 Mathematical Logic 121
Chapter 4 Algebraic Structures 161
Chapter 5 Matrix Algebra 219
Chapter 6 Order, Relation, and Lattices 277
Chapter 7 Boolean Algebra 319
Chapter 8 Complexity 347
Chapter 9 Graph Theory 383
Chapter 10 Tree 438
Chapter 11 Formal Language and Automata 480

Appendix: Number System *546*
Answers to Exercises *555*
Bibliography *562*
Index *563*

Detailed Contents

Preface   *v*

Chapter 1   Sets, Relations, and Functions   1

1.1 Introduction *1*
1.2 Sets *2*
   1.2.1 Representation of a Set *2*
   1.2.2 Sets of Special Status *2*
   1.2.3 Universal Set and Empty Set *3*
   1.2.4 Subsets *3*
   1.2.5 Power Set *4*
   1.2.6 Cardinality of a Set *4*
1.3 Ordered Pairs *5*
   1.3.1 Cartesian Product of Sets *5*
   1.3.2 Properties of Cartesian Product *6*
1.4 Venn Diagrams *8*
1.5 Operations on Sets *9*
   1.5.1 Union of Sets *9*
   1.5.2 Intersection of Sets *9*
   1.5.3 Complements *10*
   1.5.4 Symmetric Difference *11*
1.6 Countable and Uncountable Sets *12*
1.7 Algebra of Sets *14*
1.8 Multiset *17*
   1.8.1 Operations on Multisets *17*
1.9 Fuzzy Set *18*
   1.9.1 Operations on Fuzzy Set *18*
1.10 Computer Representation of Sets *19*
1.11 Introduction to Relations *20*
1.12 Binary Relation *21*
1.13 Classification of Relations *22*
   1.13.1 Reflexive Relation *22*
   1.13.2 Symmetric Relation *22*
   1.13.3 Antisymmetric Relation *23*
   1.13.4 Transitive Relation *23*
   1.13.5 Equivalence Relation *23*
   1.13.6 Associative Relation *24*
1.14 Composition of Relations *24*
1.15 Inverse of a Relation *25*
1.16 Representation of Relations on a Set *25*
1.17 Closure Operation on Relations *26*
   1.17.1 Reflexive Closure *26*
   1.17.2 Symmetric Closure *27*
1.18 Matrix Representation of Relation *27*
1.19 Digraphs *29*
   1.19.1 Transitive Closure *34*
   1.19.2 Warshall's Algorithm *36*
1.20 Partial Ordering Relation *40*
1.21 *n*-Ary Relations and Their Applications *40*
1.22 Relational Model for Databases *41*
1.23 Introduction to Functions *42*
1.24 Addition and Multiplication of Functions *45*
1.25 Classification of Functions *46*
   1.25.1 One-to-One (Injective) Function *46*
   1.25.2 Onto (Surjective) Function *47*
   1.25.3 One-to-One and Onto (Bijective) *47*
   1.25.4 Identity Function *48*
   1.25.5 Constant Function *48*
1.26 Composition of Function *48*
   1.26.1 Associativity of Composition of Functions *50*
1.27 Inverse Function *51*
   1.27.1 Invertible Function *54*
   1.27.2 Image of a Subset *54*
1.28 Hash Function *54*
1.29 Recursively Defined Functions *55*
1.30 Some Special Functions *55*
   1.30.1 Floor and Ceiling Functions *56*

    1.30.2 Integer and Absolute Value
          Functions 58
    1.30.3 Remainder Function 58
1.31 Functions of Computer Science 59
    1.31.1 Partial and Total Functions 59
    1.31.2 Primitive Recursive Function 60
    1.31.3 Ackermann Function 60
1.32 Growth of Function 61
1.33 The Inclusion–Exclusion Principle 61
    1.33.1 Applications of Inclusion–Exclusion
          Principle 63
1.34 Sequence and Summation 65
    1.34.1 Sequence 65
    1.34.2 Summation 66

Chapter 2   Combinatorics   69

2.1 Introduction 69
2.2 Basic Principles of Counting 69
    2.2.1 Multiplication Principle (the
          Principle of Sequential Counting) 70
    2.2.2 Addition Rule (the Principle of
          Disjunctive Counting) 70
2.3 Factorial Notation 71
2.4 Binomial Theorem 72
    2.4.1 Pascal's Triangle 73
    2.4.2 Multinomial Theorem 73
2.5 Permutations (Arrangements of
    Objects) 75
    2.5.1 Permutations with Repetitions 77
    2.5.2 Circular Permutations 78
2.6 Combinations (Selection of Objects) 79
    2.6.1 Combinations of n Different
          Objects 80
    2.6.2 Combinations with Repetitions 82
2.7 Discrete Probability 83
    2.7.1 Terminology (Basic Concepts) 84
2.8 Finite Probability Spaces 85
2.9 Probability of an Event 86
    2.9.1 Axioms of Probability 86
    2.9.2 Odds in Favour and Odds
          Against an Event 86
    2.9.3 Addition Principle 87
2.10 Conditional Probability 87
    2.10.1 Multiplication Rule 88
2.11 Independent Repeated Trials and
    Binomial Distribution 90

    2.11.1 Repeated Trials with Two
          Outcomes and Bernoulli Trials 91
2.12 Random Variables 92
    2.12.1 Probability Distribution of a
          Random Variable 93
    2.12.2 Expectation of a Random Variable 94
    2.12.3 Variance and Standard Deviation
          of a Random Variable 95
    2.12.4 Binomial Distribution 96
2.13 Recursion 96
    2.13.1 Recursively Defined Sequences 96
    2.13.2 Recursive Definition 97
    2.13.3 Recursively Defined Sets 98
    2.13.4 Recursively Defined Functions 98
2.14 Recurrence Relation 99
    2.14.1 Order and Degree of
          Recurrence Relation 100
    2.14.2 Linear Homogeneous and
          Non-Homogeneous Recurrence
          Relations 100
    2.14.3 Solution of Linear Recurrence
          Relation with Constant
          Coefficients 100
    2.14.4 Homogeneous Solution 101
    2.14.5 Particular Solution 104
2.15 Generating Functions 109
2.16 Counting (Combinatorial) Method 117
2.17 Pigeonhole Principle 118
    2.17.1 Generalized Pigeonhole
          Principle 118

Chapter 3   Mathematical Logic   121

3.1 Introduction 121
3.2 Statement (Propositions) 122
3.3 Laws of Formal Logic 123
3.4 Basic Set of Logical Operators/
    Operations 123
    3.4.1 Conjunction (AND, $p \wedge q$) 123
    3.4.2 Disjunction (OR, $p \vee q$) 124
    3.4.3 Negation (NOT, $\sim p$) 125
3.5 Propositions and Truth Tables 125
    3.5.1 Connectives 126
    3.5.2 Compound Propositions 127
    3.5.3 Conditional Statement 128
    3.5.4 Converse, Contrapositive, and Inverse 128
    3.5.5 Biconditional Statement 129

3.6 Algebra of Propositions *130*
3.7 Propositional Functions *130*
3.8 Tautologies and Contradictions *131*
   3.8.1 System Specifications (Consistency) *132*
   3.8.2 Principle of Substitution *132*
3.9 Logical Equivalence *133*
   3.9.1 De Morgan's Laws *134*
3.10 Logical Implication *136*
3.11 Normal Forms *136*
   3.11.1 Disjunctive Normal Form *137*
   3.11.2 Conjunctive Normal Form *138*
3.12 Arguments *138*
3.13 Rules of Inference *141*
   3.13.1 Law of Detachment (or Modus Pones) *141*
   3.13.2 Law of Contraposition (Modus Tollens) *142*
   3.13.3 Disjunctive Syllogism *143*
   3.13.4 Hypothetical Syllogism *143*
3.14 Well-Formed Formulae *145*
3.15 Predicate Calculus *146*
3.16 Quantifier *148*
   3.16.1 Universal Quantifier *148*
   3.16.2 Existential Quantifier *149*
   3.16.3 Bound Variables *150*
3.17 Introduction to Proofs *150*
   3.17.1 Brief Status of Terminology *151*
   3.17.2 Methods of Proof *151*
   3.17.3 Direct Proof *151*
   3.17.4 Consistency *151*
   3.17.5 Contraposition *152*
   3.17.6 Contradiction (reductio ad absurdum) *152*
   3.17.7 Mathematical Induction *153*
   3.17.8 Proof by Cases *157*

Chapter 4 Algebraic Structures 161

4.1 Introduction *161*
4.2 Binary Operations *162*
   4.2.1 Properties of Binary Operations *163*
4.3 Groups *168*
   4.3.1 Abelian Group *168*
   4.3.2 Properties of Groups *171*
   4.3.3 Products and Quotients of Groups *174*
4.4 Semigroups *176*
   4.4.1 Isomorphism and Homomorphism *178*
   4.4.2 Products and Quotients of Semigroups *181*
4.5 Subgroup *183*
4.6 Cyclic Group *185*
4.7 Permutation Groups *187*
   4.7.1 Equality of Permutations *188*
   4.7.2 Permutation Identity *188*
   4.7.3 Composition of Permutations (or, Product of Permutations) *189*
   4.7.4 Inverse Permutation *190*
   4.7.5 Cyclic Permutations *191*
   4.7.6 Transposition *192*
   4.7.7 Even and Odd Permutations *193*
4.8 Symmetric Group *194*
4.9 Cosets *196*
   4.9.1 Properties of Cosets *197*
4.10 Normal Subgroup *198*
4.11 Lagrange's Theorem *200*
4.12 Group Codes *202*
   4.12.1 Coding of Binary Information *203*
   4.12.2 Parity and Generator Matrices *204*
   4.12.3 Decoding and Error Correction *207*
4.13 Algebraic Systems with Two Binary Operations *208*
   4.13.1 Rings *208*
   4.13.2 Elementary Properties of a Ring *208*
   4.13.3 Special Kinds of Rings *210*
   4.13.4 Integral Domain *210*
   4.13.5 Field *211*
4.14 Subring *212*
   4.14.1 Ideal *213*
   4.14.2 Quotient Ring *215*
   4.14.3 Morphisms of Rings *216*
   4.14.4 Properties of Homomorphism of Ring *216*

Chapter 5 Matrix Algebra 219

5.1 Introduction *219*
5.2 Definition of a Matrix *220*
5.3 Types of Matrices *220*
   5.3.1 Rectangular and Square Matrices *220*
   5.3.2 Row Matrix or a Row Vector *221*
   5.3.3 Column Matrix or a Column Vector *221*
   5.3.4 Zero or Null Matrix *222*
   5.3.5 Diagonal Elements of a Matrix *222*
   5.3.6 Diagonal Matrix *222*

- 5.3.7 Scalar Matrix *223*
- 5.3.8 Unit Matrix or Identity Matrix *223*
- 5.3.9 Comparable Matrices *223*
- 5.3.10 Equal Matrices *223*
- 5.3.11 Upper Triangular Matrix *224*
- 5.3.12 Lower Triangular Matrix *224*
- 5.4 Operations on Matrices *225*
  - 5.4.1 Addition of Matrices *225*
  - 5.4.2 Subtraction of Matrices *225*
  - 5.4.3 Scalar Multiple of a Matrix *226*
  - 5.4.4 Multiplication of Matrices *226*
  - 5.4.5 Properties of Matrix Multiplication *227*
  - 5.4.6 Positive Integral Powers of Matrices *231*
  - 5.4.7 Sub-Matrix *231*
  - 5.4.8 Partition of Matrices *233*
- 5.5 Related Matrices *233*
  - 5.5.1 Transpose of a Matrix *233*
  - 5.5.2 Symmetric and Skew-symmetric Matrices *234*
  - 5.5.3 Complex Matrix *234*
  - 5.5.4 Conjugate of a Matrix *234*
  - 5.5.5 Conjugate Transpose of a Matrix *235*
  - 5.5.6 Hermitian and Skew-Hermitian Matrices *235*
- 5.6 Determinant of a Matrix *238*
  - 5.6.1 Minor and Co-Factor *239*
  - 5.6.2 Expansion of the Determinant (Δ) *239*
  - 5.6.3 Difference between a Matrix and a Determinant *240*
- 5.7 Typical Square Matrices *241*
  - 5.7.1 Orthogonal Matrix *241*
  - 5.7.2 Unitary Matrix *242*
  - 5.7.3 Involutory Matrix *243*
  - 5.7.4 Idempotent Matrix *243*
  - 5.7.5 Nilpotent Matrix *244*
- 5.8 Adjoint and Inverse of a Matrix *244*
  - 5.8.1 Singular and Non-singular Matrices *244*
  - 5.8.2 Adjoint of a Square Matrix *244*
  - 5.8.3 Properties of Adjoint of a Matrix *245*
- 5.9 Inverse of a Matrix *247*
  - 5.9.1 Properties of Inverse of a Matrix *247*
- 5.10 Rank of a Matrix *253*
  - 5.10.1 Elementary Transformations (Operations) of a Matrix *254*
- 5.11 Boolean Matrix or a Zero-One Matrix *256*
  - 5.11.1 Operations on Zero-One Matrices *256*
  - 5.11.2 Boolean Product of Matrices *257*
- 5.12 Elementary Row Operation on a Matrix *257*
  - 5.12.1 Echelon Matrix (Row-Reduced Echelon Form) *257*
  - 5.12.2 Normal Form of a Matrix *259*
  - 5.12.3 Procedure of Reduction of a Matrix A to Its Normal Form *259*
- 5.13 Solution of Linear Algebraic Equations *260*
  - 5.13.1 Linear Homogenous Equations ($Ax = 0$) *261*
  - 5.13.2 Linear Non-homogenous Equations ($Ax = b$) *261*
  - 5.13.3 Consistent and Inconsistent Equations *262*
  - 5.13.4 Gaussian Elimination (Direct Method) *265*
- 5.14 Eigenvalues and Eigenvectors *267*
  - 5.14.1 Determination of Eigenvalues and Eigenvectors *268*
  - 5.14.2 Linear Transformations *269*
  - 5.14.3 Properties of Eigenvalues and Eigenvectors *270*
- 5.15 Cayley–Hamilton Theorem *273*
  - 5.15.1 Inverse of a Matrix *273*

Chapter 6   Order, Relation, and Lattices   277

- 6.1 Introduction *277*
- 6.2 Partially Ordered Set *278*
  - 6.2.1 Comparability of Elements *278*
  - 6.2.2 Linearly Ordered Set *279*
  - 6.2.3 Cover of an Element *281*
- 6.3 Hasse Diagram *281*
  - 6.3.1 Topological Sorting *284*
  - 6.3.2 Chain *285*
  - 6.3.3 Antichain *285*
- 6.4 Isomorphism *286*
  - 6.4.1 Isomorphic Ordered Sets *287*
- 6.5 Lexicographic Ordering *288*

6.6 Extremal Elements of Posets *290*
  6.6.1 Greatest and Least Elements *291*
  6.6.2 Upper and Lower Bounds *292*
  6.6.3 Least Upper Bound (Supremum) *292*
  6.6.4 Greatest Lower Bound (Infimum) *293*
6.7 Well-Ordered Set *294*
6.8 Consistent Enumerations *295*
6.9 Lattices *296*
  6.9.1 Principle of Duality *300*
  6.9.2 Isotonocity Property *302*
6.10 Sublattices *303*
6.11 Direct Product of Lattices *304*
6.12 Some Special Class of Lattices *305*
  6.12.1 Complete Lattice *306*
  6.12.2 Bounded Lattice *306*
  6.12.3 Properties of Bounded Lattices *306*
  6.12.4 Distributive Lattice *306*
  6.12.5 Modular Lattice *309*
  6.12.6 Complemented Lattices *310*
  6.12.7 Isomorphic Lattices *313*
  6.12.8 Join-Irreducible *313*
  6.12.9 Meet-Irreducible *314*
6.13 Lattice Homomorphism *315*

Chapter 7 Boolean Algebra 319

7.1 Introduction *319*
7.2 Laws on Boolean Algebra *320*
7.3 Truth Tables of Boolean Operations *321*
7.4 Unique Features of Boolean Algebra *325*
7.5 Minterm and Maxterm *325*
  7.5.1 Boolean Expression in Sum of Products (SOP) and Product of Sums (POS) Form or Normal Form *325*
7.6 Boolean Function *326*
7.7 Switching Network from Boolean Expression Using Logic Gates *329*
7.8 Karnaugh Map *332*
  7.8.1 Rules Used by K-Map for Simplification *333*
  7.8.2 Labelling of K-Map Squares *336*
A.1 Annexure 1 *341*
  A.1.1 Quine–McCluskey Algorithm *341*
  A.1.2 Quine–McCluskey Method *342*
A.2 Annexure 2 *345*
  A.2.1 Stone's Representation Theorem for Boolean Algebras *345*
  A.2.2 Stone Spaces *345*
  A.2.3 Representation Theorem *345*

Chapter 8 Complexity 347

8.1 Introduction *347*
8.2 Algorithm *348*
  8.2.1 Basic Criteria of Algorithm *348*
8.3 Data Structure *348*
  8.3.1 Operations on Data Structure *348*
  8.3.2 Categorizations of Data Structure *349*
  8.3.3 Abstract Data Type *350*
  8.3.4 Linear and Non-linear Data Structure *350*
8.4 Complexity *350*
  8.4.1 Idea on Complexity Function of any Algorithm *351*
  8.4.2 Asymptote and Its Behaviour *351*
  8.4.3 Why Asymptotic Notations to Express Inexact Running Time? *352*
  8.4.4 Counting Strategy of Operations in Algorithm *353*
  8.4.5 Discussion on Order of Complexity *354*
  8.4.6 Mathematical Definitions of Some Useful Asymptotic Notations *354*
  8.4.7 Standard Cases *359*
  8.4.8 Some Properties of Time Complexity Functions *360*
  8.4.9 Complexity of Recursive Procedure *361*
  8.4.10 Solving Recurrence Relation: $T(n) = aT(n/b) + f(n)$, $a \geq 1, b > 0$ *364*
  8.4.11 Comparison of Complexity *368*
8.5 Searching and Sorting *370*
  8.5.1 Searching *370*
  8.5.2 Sorting *374*

Chapter 9 Graph Theory 383

9.1 Introduction *383*
9.2 Graphs and Basic Terminologies *384*
  9.2.1 Undirected and Directed Graphs *384*
  9.2.2 Weighted Graph *385*
  9.2.3 Self-Edge or Self-Loop *385*
  9.2.4 Multiple or Parallel Edges *386*
9.3 Types of Graphs *397*
  9.3.1 Null Graph *397*
  9.3.2 Complete Graph *397*

9.3.3 Regular Graph *398*
9.3.4 Bipartite Graph *399*
9.3.5 Complete Bipartite Graph *401*
9.4 Subgraph and Isomorphic Graph *401*
  9.4.1 Subgraph *401*
  9.4.2 Isomorphic Graph *403*
9.5 Operations on Graphs *404*
9.6 Representation of Graph *405*
  9.6.1 Matrix (Adjacency Matrix) Representation *405*
  9.6.2 Linked List (Adjacency List) Representation *406*
  9.6.3 Advantages and Disadvantages of Matrix and Linked List Representations *408*
  9.6.4 Incidence Matrix Representation of Graph *409*
9.7 Graph Algorithms *409*
  9.7.1 Breadth First Search *410*
  9.7.2 Depth First Search *410*
  9.7.3 Single-Source Shortest Path Problem *412*
9.8 Euler Graph *418*
  9.8.1 Some Useful Results on Euler Graph *419*
9.9 Hamiltonian Graph *419*
  9.9.1 Useful Hints on Hamiltonian Circuit *421*
9.10 Planar Graph *422*
  9.10.1 Properties of Planar Graph *423*
9.11 Colouring of Graph *424*
9.12 Component *426*
9.13 Cut Vertex *427*
9.14 Flow Network *428*
  9.14.1 Ford–Fulkerson Algorithm *431*

Chapter 10 Tree — 438

10.1 Introduction *438*
10.2 Tree *439*
  10.2.1 Common Terminologies of Tree *439*
  10.2.2 Labelled Tree *440*
  10.2.3 Some Diagrams of Directed and Undirected Trees *441*
  10.2.4 Review of the Basic Properties of a Tree *443*
  10.2.5 m-Ary Tree *443*
  10.2.6 Why Are Skewed Trees Considered as Binary Trees? *444*
10.3 Some Important Results on Tree *445*
10.4 Sequential Representation of a Binary Tree *450*
10.5 Operations on Tree *451*
  10.5.1 Tree Traversal *451*
  10.5.2 More Discussions on Tree Traversals *452*
  10.5.3 Construction of a Unique Binary Tree When the Pre-Order and the In-Order Traversal Sequences Are Given *453*
  10.5.4 Algorithm to Construct a Unique Binary Tree Using the Pre-Order and the In-Order Sequences *455*
10.6 Binary Search Tree *455*
  10.6.1 Linked List Representation of a Binary Tree *456*
  10.6.2 Construction of Binary Search Tree *456*
  10.6.3 Useful Results from Binary Search Tree *457*
10.7 Recursive Prodecure for Binary Tree Traversal *458*
  10.7.1 Analysis of Time Complexities for Some Operations on Binary Tree *460*
10.8 Predecessor and Successor Node *460*
10.9 Expression Tree *460*
10.10 AVL Tree *461*
10.11 Spanning Tree *462*
  10.11.1 Minimum Spanning Tree *463*
10.12 General Tree *473*
  10.12.1 Conversion of a General Tree to a Binary Tree *473*
  10.12.2 Pre-Order Traversal for General Tree *474*
10.13 Some Important Applications of Tree *474*

Chapter 11 Formal Language and Automata — 480

11.1 Introduction *480*
11.2 Mathematical Preliminaries *481*
  11.2.1 Symbol *481*
  11.2.2 Alphabet *481*
  11.2.3 String *481*
  11.2.4 Language *482*

11.3 Automata *483*
   11.3.1 Basic Categories of Automata *484*
   11.3.2 Finite Automaton and Its Types *484*
   11.3.3 Importance of NDFA *488*
   11.3.4 Graphical Notations Used in Drawing Finite Automata *488*
   11.3.5 Discussion on Designing of Some Basic FAs *489*
   11.3.6 Some Basic Tips to Design FA *496*
   11.3.7 Conversion Strategy from NDFA to DFA *498*
   11.3.8 Finite Automaton with Output *599*
11.4 Regular Expression *503*
   11.4.1 Minimization of FA *503*
   11.4.2 Brief Discussion to Derive REs *506*
   11.4.3 Solved Problems on RE *507*
   11.4.4 The Identities on Regular Expresssion *509*
   11.4.5 Rules for Constructing NDFA from Regular Expression *510*
   11.4.6 Tips to Get Quick Answer of Some Special Problems on FA and RE *512*
   11.4.7 Pumping Lemma for Regular Language *514*
   11.4.8 Applications of FA and Regular Expression *516*
11.5 Grammar *516*
   11.5.1 Formai Definition of Grammar *517*
   11.5.2 The Chomsky Hierarchy *518*
   11.5.3 Derivation (Parsing) *524*
   11.5.4 Parsing Techniques *526*
   11.5.5 Ambiguous Grammar *526*
11.6 Pushdown Automaton *528*
   11.6.1 Formal Definition of PDA *529*
   11.6.2 Types of PDA *532*
11.7 Turing Machine *532*
   11.7.1 Formal Definition of TM *533*
   11.7.2 Improvement in TM *534*
   11.7.3 Variations of TMs *534*
   11.7.4 Halting Problem *536*
   11.7.5 Turing Acceptable Language *536*
   11.7.6 Turing Decidable Language *537*
   11.7.7 Properties of Recursive and Recursively Enumerable Languages *537*
   11.7.8 Church Thesis *537*
11.8 Post-Correspondence Problem *537*
11.9 Classes of Problems *537*
11.10 Cellular Automaton *539*
11.11 Fuzzy Sets and Fuzzy Logic *540*
11.12 Russell's Paradox *541*
   11.12.1 History of the Paradox *541*
A.1 Annexure *544*

Appendix: Number System   546
Answers to Exercises   555
Bibliography   562
Index   563

1 Sets, Relations, and Functions

LEARNING OBJECTIVES

After reading this chapter you will be able to understand the following:
- Sets and their components drawn from set-theoretic approach
- Ordered pairs that are indeed useful in computer science
- Cartesian product of sets on the basis of which many of the discrete structures are discussed in later chapters
- Computer representation of sets which leads to store elements
- Relationships between elements of sets
- Definition of function and its basic properties
- How functions can be combined, using the idea of a composition of functions
- The several different ways of visualizing the action of function
- How sequences, an important data structure in computer science, are ordered in terms of lists of elements

1.1 INTRODUCTION

Set theory provides an important foundation for contemporary mathematics. Its essence has spilled over into a number of related disciplines. Discrete structures are one of them, which are built using sets and are a collection of objects. Among the discrete structures that emerged from sets are combinations, unordered collection of objects used in counting, relations, sets of ordered pairs that represent relationships between objects; graphs, sets of vertices and edges that connect vertices; and finite set machines used to model computing machines. Moreover, the correct basis for a study of computer science is to tackle programming languages that are capable of handling sets and in a sense everything one needs can be constructed from sets.

In this chapter, we study the fundamental discrete structure on which all other discrete structures are built, namely, the set. Here, we have discussed extensively the set theory starting from its definition via cartesian product of sets, Venn diagram, operation on sets to representation of set in a computer.

1.2 SETS

A set is any *well-defined* collection of objects, known as *elements* or *members* of the set. We generally use capital letters, A, B, X, Y, to denote sets and lowercase letters, a, b, x, y, ..., to denote elements of sets. Here, the term *well-defined* implies that there exists certain rule by which it can be decided that whether a given object or element belongs to the collection or not. The statement 'x is an element of A' or, equivalently, 'x belongs to A' is written as $x \in A$. The statement 'x is not an element of A' is written as $x \notin A$.

1.2.1 Representation of a Set

There are mainly two types of representation of a set:
  I. Roster or tabular form
  II. Set builder form or rule method

  I. *Roster or tabular form* All the members of the set, here, are listed and being separated by commas, and are enclosed within braces.

  For example The notation $\{a, b, c, d\}$ represents the set with the four elements $a, b, c,$ and d. Moreover,
  (i) The set V of all vowels in the English alphabet can be written as $V = \{a, e, i, o, u\}$.
  (ii) The set O of odd positive integers less than 8 can be represented by $O = \{1, 3, 5, 7\}$.

  II. *Set builder or rule method* In this method, we define certain property which recognises the elements in the set.

  For example
  (i) Let $B = \{x : x \text{ is an even integer}, x > 0\}$, which implies that '$B$ is the set of x such that x is an even integer and x is greater than 0'. Here, colon (:) is read as 'such that' and comma (,) as 'and'. A vertical bar is also used in place of colon.
  (ii) The set $X = \{x : x \text{ is a real number}, 1 < x < 8\}$ is the set of real numbers that lie strictly between 1 and 8.

EXAMPLE 1.1 Redefine each of the following sets using set-builder notation:
  (i) $\{2, 3, 4, 6, 8, 9, 10, \ldots\}$
  (ii) $\{1, \frac{1}{2}, \frac{1}{3}, \frac{1}{4}, \frac{1}{5}, \ldots\}$
  (iii) $\{0, 1, 2, \ldots, 50\}$

Solution
  (i) $\{x : x = 2n \text{ or } x = 3n \text{ for } n \in N\}$
  (ii) $\{x : x = \frac{1}{n} \text{ for } n \in N\}$
  (iii) $\{x : 0 \leq x \leq 50 \text{ and } x \in Z\}$

1.2.2 Sets of Special Status

There are some important sets which can be represented by special name:

Z, the set of all integers $= \{\ldots, -2, -1, 0, 1, 2, \ldots,\}$

Z^+, the set of positive integers = $\{1, 2, 3, 4, \ldots,\}$
N, the set of natural numbers = $\{0, 1, 2, 3, 4, \ldots,\}$
R, the set of real numbers
Q, the set of rational numbers = $\{p/q : p \in Z, q \in Z, \text{ and } q \neq 0\}$
C, the set of complex numbers

1.2.3 Universal Set and Empty Set

If there are some sets under consideration, then there happens to be a fixed set which contains each one of the given sets. Such a fixed set is known as the *universal set*, and is denoted by U or ξ.

For example
(i) In human population studies, the universal set consists of all the people in the world.
(ii) If $A = \{1, 2, 3, 4\}$, $B = \{2, 3, 5, 7\}$, and $C = \{2, 4, 6, 8\}$, then $U = \{1, 2, 3, 4, 5, 6, 7, 8\}$ is the universal set.

A set consisting of no element at all is called an *empty set* or a *null set* or a *void set* and is denoted by ϕ. In roster form, ϕ is denoted by $\{\}$.

For example
(i) $\{x : x \in N, 2 < x < 3\} = \phi$
(ii) $\{x : x \in R, x^2 = -1\} = \phi$

A set which possesses at least one element is called a *non-empty set*. A set consisting of a single element is called a *singleton set*.

For example The set $\{0\}$ is a singleton set, whose only member is 0.

1.2.4 Subsets

If A and B are two sets such that every element in a set A is also an element of a set B, then A is called a *subset* of B. We also say that A is contained in B or that B contains A. This relationship is expressed as

$$A \subseteq B \quad \text{or} \quad B \supseteq A$$

If there exists even a single element in A, which is not in B, then A is not in B; hence A is not a subset of B and we write

$$A \nsubseteq B \quad \text{or} \quad B \nsupseteq A$$

If $A \subseteq B$, then it is possible that $A = B$. When $A \subseteq B$, but $A \neq B$, we say that A is a *proper subset* of B. Again, if $A \subset B$, then A is a proper subset of B.

For example Let $A = \{1, 3\}$, $B = \{1, 2, 3\}$, and $C = \{1, 3, 2\}$. Then A and B are both subsets of C, but A is a proper subset of C. On the other hand, since $B = C$, we cannot say that B is a proper subset of C.

The following properties of set are very important:
  I. Every set is a subset of itself.
 II. The empty set is a subset of every set.
III. The total number of all possible subsets of a given set containing n elements is 2^n.

Proof (Property III) Let A be any set containing n elements. Then, one of its subsets is the empty set. Apart from this,

the number of singleton subsets of $A = n = {}^nC_1$,
the number of subsets of A, each containing 2 elements $= {}^nC_2$,
the number of subsets of A, each containing 3 elements $= {}^nC_3$,
... ...
the number of subsets of A, each containing n elements $= {}^nC_n$.

Hence, the total number of all possible subsets of A

$$= (1 + {}^nC_1 + {}^nC_2 + {}^nC_3 + \cdots + {}^nC_n)$$
$$= (1 + 1)^n = 2^n$$

using binomial theorem.

For example If the set $A = \{1, 2, 3, 4\}$ contains 4 elements, then the total number of its subsets is $2^4 = 16$.

Note A set A with n elements has 2^n subsets.

1.2.5 Power Set

Consider the set $A = \{a, b\}$. The subsets of A are ϕ, $\{a\}$, $\{b\}$, and $\{a, b\}$. Then, the family of all the subsets of A is called the *power set* of A, which is denoted by $P(A)$. Thus, $P(A) = \{\phi, \{a\}, \{b\}, \{a, b\}\}$. Symbolically, $P(A) = \{x : x \text{ is a subset of } A\}$.

It may be noted here that both the empty set and the set A themselves are members of the power set.

For example Consider $A = \phi$. Then $P(A) = \{\phi\}$ and contains one element, the empty set itself.

EXAMPLE 1.2 List all the members of the power set of each of the following sets:
(i) $A = \{1, 2, 3\}$ (ii) $B = \{\{a\}, \{b\}\}$ (iii) $C = \{\phi\}$

Solution

(i) $P(A) = \{\phi, \{1\}, \{2\}, \{3\}, \{1, 2\}, \{1, 3\}, \{2, 3\}, A\}$
(ii) $P(B) = \{\phi, \{a\}, \{b\}, B\}$
(iii) $P(C) = \{\phi, C\}$

Note If the set A is finite and contains n elements, then the power set of A will have 2^n elements.

1.2.6 Cardinality of a Set

The number of distinct elements contained in a finite set is called the *cardinality* or the *cardinal number* of the set. The cardinality of a set is denoted by various notations like, $n(A)$ or, card (A), $|A|$, and A.

For example Cardinality of empty set, ϕ, is 0 and is denoted by $n(\phi) = 0$. Let $A = \{2, 3, 5, 7, 11\}$, then $n(A) = 5$.

1.3 ORDERED PAIRS

An *ordered pair* of objects is a pair of objects arranged in some order. Thus, in the set $\{a, b\}$ of two objects, a is the first one and b is the second object of a pair. Hence, (a, b) and (b, a) are two different ordered pairs. Moreover, the two objects in ordered pair need not be distinct. So, (a, a) is a well-defined ordered pair.

An ordered triple is an ordered triple of objects $\{a, b, c\}$ where a is the first object, b is the second one, and c is the third element of triple. An ordered triple can also be written in terms of ordered pair as $\{(a, b), c\}$. Similarly, an ordered quadruple is an ordered pair $\{((a, b), c), d\}$ with first element as ordered triple. Thus, an ordered n-tuple is an ordered pair whose first component is an ordered $(n-1)$-tuple.

Let us now define an *ordered set* and simultaneously redefine an ordered pair. Actually, an ordered set is defined as an ordered collection of distinct objects. In general, an ordered n-tuple is an ordered set with n elements. First component of this n-tuple is an ordered $(n-1)$-tuple and the nth element is the second component. Also, an ordered set of n elements is an ordered pair of $(n-1)$-tuple and an element.

For example An ordered set of 5 elements $\{a, b, c, d, e\}$ can be expressed as $\{(((a, b), c), d), e\}$.

Many of the discrete structures, we will study in later chapters, are based on the notion of the *cartesian product of sets*.

1.3.1 Cartesian Product of Sets

Cartesian product is a mode by which two or more sets can be combined to obtain another one. If A and B are two non-empty sets, the *cartesian product* (or cross product or direct product) of A and B is the set,

$$A \times B = \{(a, b) : a \in A \text{ and } b \in B\}$$

If $A = \phi$ or $B = \phi$, we define $A \times B = \phi$.

> **Note**
>
> (i) If A and B are finite sets, then $n(A \times B) = n(A) \cdot n(B)$.
> (ii) If either A or B is an infinite set, then $A \times B$ is an infinite set.

The cartesian product of $A \times A$ is denoted by A_2. More generally,

$$A_n = A \times A \times \cdots \times A \,(n \text{ times}) = \{(a_1, a_2, \ldots, a_n) : a_i \in A, i = 1, 2, \ldots, n\}$$

Let us redefine the cartesian product in terms of ordered pairs.

The cartesian product of A and B is denoted by $A \times B$ and is the set of all ordered pairs of the form (a, b), where $a \in A$ and $b \in B$.

For example Let $A = \{a, b\}$ and $B = \{a, c, d\}$. Then,

$$A \times B = \{a, b\} \times \{a, c, d\} = \{(a, a), (a, c), (a, d), (b, a), (b, c), (b, d)\}$$

and

$$B \times A = \{a, c, d\} \times \{a, b\} = \{(a, a), (a, b), (c, a), (c, b), (d, a), (d, b)\}$$

The elements of $A \times B$ are ordered pairs of the order count, i.e., $(a, b) \neq (b, a)$ unless $a = b$.

For example A software firm used to provide the following three characteristics for each program that it sold:

Language FORTRAN (f); PASCAL (p); LISP (l)
Memory 2 meg (2); 4 meg (4); 8 meg (8)
Operating system UNIX (u); DOS (d)

Let $L = \{f, p, l\}$, $M = \{2, 4, 8\}$, and $O = \{u, d\}$. Then the cartesian product $L \times M \times O$ provides all the categories that describe a program. There are $3 \cdot 3 \cdot 2$ or 18 categories in this classification scheme.

EXAMPLE 1.3 If $A = \{1, 2, 3\}$, $B = \{3, 4\}$, and $C = \{4, 5, 6\}$, determine
 (i) $A \times (B \cup C)$ (ii) $A \times (B \cap C)$ (iii) $(A \times B) \cap (B \times C)$

Solution

(i) $B \cup C = \{3, 4\} \cup \{4, 5, 6\} = \{3, 4, 5, 6\}$

Thus,

$$A \times (B \cup C) = \{1, 2, 3\} \times \{3, 4, 5, 6\}$$
$$= \{(1, 3), (1, 4), (1, 5), (1, 6), (2, 3), (2, 4), (2, 5), (2, 6), (3, 3), (3, 4), (3, 5), (3, 6)\}$$

(ii) $B \cap C = \{3, 4\} \cap \{4, 5, 6\} = \{4\}$

Thus,

$$A \times (B \cap C) = \{1, 2, 3\} \times \{4\} = \{(1, 4), (2, 4), (3, 4)\}$$

(iii) $A \times B = \{(1, 3), (1, 4), (2, 3), (2, 4), (3, 3), (3, 4)\}$

and

$$B \times C = \{(3, 4), (3, 5), (3, 6), (4, 4), (4, 5), (4, 6)\}$$

Thus,

$$(A \times B) \cap (B \times C) = \{3, 4\}$$

1.3.2 Properties of Cartesian Product

I. For any three sets A, B, C
  (i) $A \times (B \cup C) = (A \times B) \cup (A \times C)$
  (ii) $A \times (B \cap C) = (A \times B) \cap (A \times C)$

Proof

(i) Let a and b be arbitrary elements of $A \times (B \cup C)$. Then,

$(a, b) \in A \times (B \cup C)$
$\Rightarrow a \in A$ and $b \in (B \cup C)$
$\Rightarrow a \in A$ and $(b \in B$ or $b \in C)$
$\Rightarrow (a \in A$ and $b \in B)$ or $(a \in A$ and $b \in C)$
$\Rightarrow (a, b) \in (A \times B)$ or $(a, b) \in (A \times C)$
$\Rightarrow (a, b) \in (A \times B) \cup (A \times C)$

Thus,
$$A \times (B \cup C) \subseteq (A \times B) \cup (A \times C)$$

Similarly,
$$(A \times B) \cup (A \times C) \subseteq A \times (B \cup C)$$

Hence,
$$A \times (B \cup C) = (A \times B) \cup (A \times C)$$

(ii) This may be proved in a similar manner and is left to the reader.

II. For any three sets A, B, C, $A \times (B - C) = (A \times B) - (A \times C)$. The proof is left to the reader.

III. For any sets A, B, C, D, $(A \times B) \cap (C \times D) = (A \cap C) \times (B \cap D)$.

Proof Let $(a, b) \in (A \times B) \cap (C \times D)$, i.e., $(a, b) \in (A \times B)$ and $(a, b) \in (C \times D)$

$\Rightarrow (a \in A$ and $b \in B)$ and $(a \in C$ and $b \in D)$
$\Rightarrow (a \in A$ and $a \in C)$ and $(b \in B$ and $b \in D)$
$\Rightarrow a \in (A \cap C)$ and $b \in (B \cap D)$
$\Rightarrow (a, b) \in (A \cap C) \times (B \cap D)$.

Thus,
$$(A \times B) \cap (C \times D) \subseteq (A \cap C) \times (B \cap D)$$

Similarly,
$$(A \cap C) \times (B \cap D) \subseteq (A \times B) \cap (C \times D)$$

Hence,
$$(A \times B) \cap (C \times D) = (A \cap C) \times (B \cap D)$$

EXAMPLE 1.4 For any non-empty sets A and B prove that $A \times B = B \times A \Leftrightarrow A = B$.

Solution Assume $A = B$. Then, $A \times B = A \times A$ and $B \times A = A \times A$. Thus,
$$A \times B = B \times A$$

Conversely, let $A \times B = B \times A$ and let $x \in A$. Then,

$x \in A \Rightarrow (x, b) \in A \times B$ for $b \in B$
$\Rightarrow (x, b) \in B \times A$ [since $A \times B = B \times A$]
$\Rightarrow x \in B$

Thus,
$$A \subseteq B$$

Similarly,
$$B \subseteq A$$

Hence,
$$A = B$$

EXAMPLE 1.5 If A and B are two non-empty sets having n elements in common, then prove that $A \times B$ has n^2 elements in common.

Solution Assume $C = A \cap B$. Then,

$C \times C = (A \times B) \cap (B \times A)$, since $(a, b) \in C \times C$
$\Leftrightarrow a \in C$ and $b \in c \Leftrightarrow a \in A \cap B$ and $b \in A \cap B$ [since $C = A \cap B$]
$\Leftrightarrow (a \in A$ and $a \in B)$ and $(b \in A$ and $b \in B)$
$\Leftrightarrow (a \in A$ and $b \in B)$ and $(a \in B$ and $b \in A)$
$\Leftrightarrow (a, b) \in A \times B$ and $(a, b) \in B \times A$
$\Leftrightarrow (a, b) \in (A \times B) \cap (B \times A)$

Thus,
$$C \times C = (A \times B) \cap (B \times A)$$

Again, since $C \times C$ has n^2 elements, so $(A \times B) \cap (B \times A)$ has n^2 elements, i.e., $(A \times B)$ and $(B \times A)$ have n^2 elements in common.

EXAMPLE 1.6 Assume A and B as two sets having two elements in common. If $n(A) = 5$ and $n(B) = 3$, find $n(A \times B)$ and $n\{(A \times B) \cap (B \times A)\}$.

Solution Here, $n(A \times B) = n(A) \cdot n(B) = 5 \cdot 3 = 15$. Since A and B have 2 elements in common, hence $A \times B$ and $B \times A$ have 2^2 elements, i.e., 4 elements in common. Hence, $n\{(A \times B) \cap (B \times A)\} = 4$.

1.4 VENN DIAGRAMS

To express the relationship among sets in a perspective way, we represent them (sets) pictorially by means of diagrams, known as *Venn diagrams*.

The universal set is usually represented by a rectangular region and its subsets by closed bounded regions (circles) inside this rectangular region.

If $A \subseteq B$, then the circle representing A will be entirely within the circle representing B as shown in Figure 1.1(a). If A and B are disjoint, that is if they have no element in common, then the circle representing A will be separated from the circle representing B as in Figure 1.1(b). However, if A and B are two arbitrary sets, it may be possible that some elements are present in A but absent in B, some are present in B but absent in A, some may be neither present in A nor in B. Hence, in general we can represent A and B as shown in Figure 1.1(c).

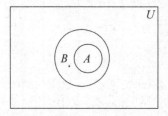
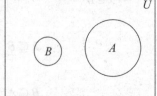
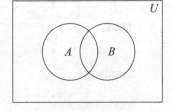

(a) $A \subseteq B$ (b) A and B are disjoint   (c) A and B

FIGURE 1.1

1.5 OPERATIONS ON SETS

This section presents a number of key operations on sets.

1.5.1 Union of Sets

The union of two sets A and B, denoted by $A \cup B$ (read as 'A union B'), is the set of all those elements, each one of which belongs to A or to B or belongs to both A and B. Symbolically,

$$A \cup B = \{x : x \in A \text{ or } x \in B\}$$

Here, 'or' is used in the sense of and/or. Figure 1.2 shows a Venn diagram in which $A \cup B$ is shaded.

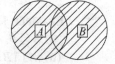

FIGURE 1.2 $A \cup B$

For example

(i) If $A = \{5, 7, 8\}$ and $B = \{2, 7, 9, 10, 11\}$, then
$A \cup B = \{2, 5, 7, 9, 10, 11\}$.
(ii) If $A = \{x : x \in Z, x \geq 3\}$ and $B = \{x : x \in Z, x \geq 8\}$, then
$A \cup B = \{x : x \in Z, x \geq 3\}$, where Z denotes the set of integers.

In general, if $A_1, A_2, A_3, \ldots, A_n$ represent sets, then the union of these sets, denoted by $\bigcup_{i=1}^{n} A_i$, is expressed as

$$\bigcup_{i=1}^{n} A_i = \{x : x \in A_i \text{ for at least one set } A_i\}$$

Also, $x \in A \cup B \Rightarrow x \in A$ or $x \in B$; and $x \notin A \cup B \Rightarrow x \notin A$ and $x \notin B$.

1.5.2 Intersection of Sets

The intersection of two sets A and B, denoted by $A \cap B$, is the set of all elements, common to both A and B. Thus, $A \cap B = \{x : x \in A \text{ and } x \in B\}$. Figure 1.3 shows a Venn diagram in which $A \cap B$ is shaded.
Also,

$x \in A \cap B \Rightarrow x \in A$ and $x \in B$ and $x \notin A \cap B \Rightarrow x \notin A$ or $x \notin B$

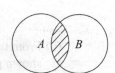

FIGURE 1.3 $A \cap B$

For example

(i) Let $A = \{1, 2, 3, 4\}$ and $B = \{2, 4, 6\}$. Then, $A \cap B = \{2, 4\}$.
(ii) Let $A = \{x : x = 3n, n \in N\}$ and $B = \{x : x = 4n, n \in N\}$.

Then, $A \cap B = \{x : x = 12n, n \in N\}$.

If $A \cap B = \phi$, i.e., if A and B do not have any common elements, then A and B are said to be *disjoint* or *non-intersecting sets*.

EXAMPLE 1.7 Let A and B be two sets such that $(A \cap B) \subseteq B$ and $B \not\subset A$. Draw the corresponding Venn diagram.

Solution Since $(A \cap B) \subseteq B$, it means that every member of A is also a member of B, i.e., $A \subseteq B$. Also, $B \not\subset A$ implies that A and B are not equal. Hence, Figure 1.4 shows a Venn diagram in which $A \subseteq B$ is shown.

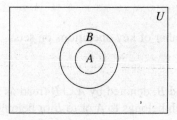

FIGURE 1.4 Venn diagram of $A \subseteq B$

 EXAMPLE 1.8 Let A, B, and C be sets such that
(i) $A \subseteq B, A \subseteq C, (B \cap C) \subseteq A$, and $A \subseteq (B \cap C)$
(ii) $(A \cap B \cap C) = \phi, A \cap B = \phi, B \cap C = \phi$, and $A \cap C = \phi$

Draw the corresponding Venn diagrams.

Solution

(i) Here, $(B \cap C) \subseteq A$ and $A \subseteq (B \cap C)$. This implies that $A = B \cap C$.

Hence, Figure 1.5 shows a Venn diagram in which $A \subseteq B$ and $A \subseteq C$ are shown.

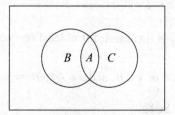

FIGURE 1.5 Venn diagram of $A \subseteq B$ and $A \subseteq C$

(ii) From the given conditions it can be said that A, B, and C are disjoint sets. This can be shown pictorially using Venn diagram (Figure 1.6).

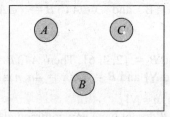

FIGURE 1.6 Sets A, B, and C are disjoint

1.5.3 Complements

Let U be the universal set and let $A \subset U$. Then the *complement* of A, denoted by A', is the set of elements which belong to U but do not belong to A, i.e.,

$$A' = \{x : x \in U, x \notin A\}$$

Clearly, $x \in A' \Leftrightarrow x \notin A$. Figure 1.7(a) is a Venn diagram in which A' is shaded.

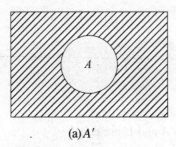

(a) A' (b) $A \sim B$

FIGURE 1.7

For example
(i) If $U = \{1, 2, 3, 4, 5\}$ and $A = \{2, 4\}$, then $A' = \{1, 3, 5\}$.
(ii) Let $U = \{x : x$ is a letter in English alphabet$\}$ and $A = \{x : x$ is a vowel$\}$, then $A' = \{x : x$ is a consonant$\}$.

The *relative complement* of a set B with respect to a set A, or simply, the *difference* of A and B, denoted by $A \sim B$, is the set of elements which belong to A, but which do not belong to B, i.e.,

$$A \sim B = \{x : x \in A, x \notin B\}$$

The set $A \sim B$ is read as 'A minus B'. Figure 1.7(b) shows a Venn diagram in which $A \sim B$ is shaded.

1.5.4 Symmetric Difference

The *symmetric difference* of sets A and B, denoted by $A \ominus B$, or $A \Delta B$, is the set of elements which belong to A or B but not to both, i.e.,

$$A \ominus B = (A \cup B) \sim (A \cap B)$$

This may also be expressed as

$$A \ominus B = (A \sim B) \cup (B \sim A)$$

For example If $A = \{-2, 0, 1, 2\}$ and $B = \{1, 2, 3, 4\}$, then

$$A \sim B = \{-2, 0\} \text{ and } B \sim A = \{3, 4\}$$

Thus,

$$A \ominus B = (A \sim B) \cup (B \sim A) = \{-2, 0\} \cup \{3, 4\} = \{-2, 0, 3, 4\}$$

EXAMPLE 1.9 Express the relation which must be valid between sets under the following conditions. Also, draw Venn diagrams.
(i) $A' \cap U = \phi$
(ii) $(A \cap B)' = B'$
(iii) $A \cap B = A \cap C$ and $A' \cap B = A' \cap C$

Solution
(i) The given condition shows that A' and U are disjoint sets. Thus, Figure 1.8 shows a Venn diagram in which A and U are equal.

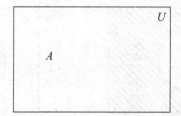

FIGURE 1.8 Sets A and U are equal

(ii) The left-hand side of the given condition can be rewritten as

$$(A \cap B)' = A' \cup B', \text{ i.e., } A' \cup B' = B'$$

$\Rightarrow B$ must be a subset of A and the Venn diagram for this is shown in Figure 1.9.

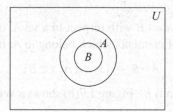

FIGURE 1.9 $B \subseteq A$

(iii) The given conditions imply that the sets B and C are equal and its Venn diagram is depicted in Figure 1.10.

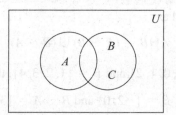

FIGURE 1.10 Sets B and C are equal

1.6 COUNTABLE AND UNCOUNTABLE SETS

If a rule is such that it associates with each element $a \in A$, one and only one element $b \in B$, then this rule is called *one-to-one*. Also, if for each element $b \in B$, there exists exactly one element $a \in A$, then this rule is called *one-to-one and onto*, which is also known as *one-to-one correspondence*.

A set which is not finite is called an *infinite* set. A set is said to be *countably infinite*, if there exists one-to-one correspondence between the elements in the set and elements in N; that is, if the set is countably infinite, we can make a list of its members in such a way that each one corresponds uniquely to a natural number. A countably infinite set is also termed as *denumerable*. The set of non-negative even integers is countably infinite.

A set which is either finite or denumerable is called *countable*. A set which is not countable is called as *uncountable*. A set which is not countably infinite is called *uncountably infinite* set or *non-denumerable* set or simply *uncountable* set.

For example The most accomplished example of an uncountable set is the set of all real numbers that can be represented by an infinite decimal of the form $0 \cdot a_1 \cdot a_2 \cdot a_3 \cdot \ldots$, where a_i is an integer satisfying the inequality $0 \le a_i \le 9$.

EXAMPLE 1.10 Investigate on finite, countably infinite, or uncountable sets with respect to one-to-one correspondence between the set of N and the following sets:

(i) The set of even integers
(ii) The set of negative integers
(iii) The set of integers that of multiples of 5
(iv) The set of real numbers between 0 and 0.6
(v) The set of all trees on earth

Solution

(i) The set of even integers is countably infinite as we can form one-to-one correspondence with N as,

$$\begin{array}{ccccccc} 1 & 2 & 3 & 4 & 5 & 6 & 7 & \ldots \\ \updownarrow & \updownarrow & \updownarrow & \updownarrow & \updownarrow & \updownarrow & \updownarrow \\ 0 & -2 & 2 & -4 & 4 & -6 & 6 & \ldots \end{array}$$

(ii) The set of negative integers is countably infinite, as we can form one to one correspondence with N as,

$$\begin{array}{ccccccc} 1 & 2 & 3 & 4 & 5 & 6 & 7 & \ldots \\ \updownarrow & \updownarrow & \updownarrow & \updownarrow & \updownarrow & \updownarrow & \updownarrow \\ -1 & -2 & -3 & -4 & -5 & -6 & -7 & \ldots \end{array}$$

(iii) The set of integers that of multiples of 5 is countably infinite, as the elements of above-mentioned set has one-to-one correspondence with elements in N,

$$\begin{array}{cccccccc} 1 & 2 & 3 & 4 & 5 & 6 & 7 & 8 & \ldots \\ \updownarrow & \updownarrow & \updownarrow & \updownarrow & \updownarrow & \updownarrow & \updownarrow & \updownarrow \\ 0 & 5 & 10 & 15 & 20 & 25 & 30 & 35 & \ldots \end{array}$$

(iv) The set of real numbers between 0 and 0.6 is uncountable.
(v) The set of all trees on earth, even though the number is large, is finite.

EXAMPLE 1.11 State whether the following sets are finite, countably finite, or uncountably finite.

(i) Set of all prime numbers
(ii) Set of real numbers between 0 and 1
(iii) Number of fish in Indian ocean

Solution
(i) Countably infinite
(ii) Uncountably infinite
(iii) Finite, even though the count is too large

 EXAMPLE 1.12 Illustrating with an example, show that the intersection of countably infinite number of infinite sets may be finite.

Solution

Suppose A = Set of all integers less than or equal to 2
$A = \{2, 1, 0, -1, -2, -3, \ldots,\}$
$N = \{\text{Set of natural numbers}\}$
$= \{1, 2, 3, \ldots,\}$

Thus, $A \cap N = \{1, 2\}$ is finite. Hence, intersection of two countably infinite sets could be finite.

1.7 ALGEBRA OF SETS

There exists a number of relationships among set operations. These operations satisfying the laws or identities form the basis of the algebraic laws of set operations. Sets under the operations of union, intersection, and complement also satisfy various laws or identities which are listed in Table 1.1.

There are two ways of tackling the proofs of the identities involving set operations. One way to show that the two sets are equal is to illustrate that each is a subset of the other.

Recall that to show that one set is a subset of a second set, we can prove that if an element belongs to the first set, then it must also belong to the second set. We will use here a direct proof. On the other way we can use Venn diagrams. The proofs of some of the laws or identities given in Table 1.1 will be presented here and the remaining is left to the reader.

TABLE 1.1 Set identities

Name	Identity	
I. Identity laws	$A \cup \phi = A$	$A \cap U = A$
II. Domination laws	$A \cup U = U$	$A \cap \phi = \phi$
III. Idempotent laws	$A \cup A = A$	$A \cap A = A$
IV. Complement laws	$A \cup A' = U$	$A \cap A' = \phi$
V. Involution law or complementation laws	$(A')' = A$	
VI. Commutative laws	$A \cup B = B \cup A$	$A \cap B = B \cap A$
VII. Associative laws	$(A \cup B) \cup C = A \cup (B \cup C)$	
	$(A \cap B) \cap C = A \cap (B \cap C)$	
VIII. Distributive laws	$A \cup (B \cap C) = (A \cup B) \cap (A \cup C)$	
	$A \cap (B \cap C) = (A \cap B) \cup (A \cap C)$	
IX. De Morgan's laws	$(A \cup B)' = A' \cap B'$	$(A \cap B)' = A' \cup B'$
X. Absorption laws	$A \cup (A \cap B) = A$	$A \cap (A \cup B) = A$

III. *Idempotent law* $A \cap A = A$

Proof Here, $A \cap A \subseteq A$. Let x be a member of A, i.e., $x \in A$

$$\Rightarrow x \in A \text{ and } x \in A \quad \text{[by the repetition of the statement]} \quad (1.1)$$
$$\Rightarrow x \in A \cap A$$

So,
$$A \subseteq A \cap A \quad (1.2)$$

Combining Eqs (1.1) and (1.2), we get

$$A \cap A = A$$

VI. *Commutative law* $A \cap B = B \cap A$

Proof Let $x \in A \cap B$. Then, $x \in A$ and $x \in B$, i.e., $x \in B$ and $x \in A \Rightarrow x \in B \cap A$

Thus,
$$A \cap B \subseteq B \cap A \quad (1.3)$$

Conversely, let $x \in B \cap A$. Then, $x \in B$ and $x \in A$, i.e., $x \in A$ and $x \in B \Rightarrow x \in A \cap B$

Thus,
$$B \cap A \subseteq A \cap B \quad (1.4)$$

Combining Eqs (1.3) and (1.4),

$$A \cap B = B \cap A$$

VII. *Associative law* $(A \cap B) \cap C = A \cap (B \cap C)$

Proof Let $x \in (A \cap B) \cap C$. Then, $x \in (A \cap B)$ and $x \in C$, i.e., $(x \in A$ and $x \in B)$ and $x \in C$

$$\Rightarrow x \in A \text{ and } (x \in B \text{ and } x \in C)$$
$$\Rightarrow x \in A \text{ and } x \in B \cap C$$
$$\Rightarrow x \in A \cap (B \cap C)$$

Thus,
$$(A \cap B) \cap C \subseteq A \cap (B \cap C) \quad (1.5)$$

Conversely, let $x \in A \cap (B \cap C)$. Then,

$$x \in A \text{ and } (x \in B \cap C)$$
$$\Rightarrow x \in A \text{ and } (x \in B \text{ and } x \in C)$$
$$\Rightarrow (x \in A \text{ and } x \in B) \text{ and } x \in C$$
$$\Rightarrow x \in (A \cap B) \text{ and } x \in C$$
$$\Rightarrow x \in (A \cap B) \cap C$$

Thus,
$$A \cap (B \cap C) \subseteq (A \cap B) \cap C \qquad (1.6)$$

Combining Eqs (1.5) and (1.6),
$$(A \cap B) \cap C = A \cap (B \cap C)$$

VIII. *Distributive law* $A \cap (B \cup C) = (A \cap B) \cup (A \cap C)$

Proof We will prove this identity in a similar fashion as we have done for the previous identities, i.e., by showing that each side is a subset of the other side.

Suppose that $x \in A \cap (B \cap C)$. Then, $x \in A$ and $x \in B \cap C$. By the definition of union, $x \in A$ and $x \in B$ or $x \in C$ (or both). Consequently, we know that $x \in A$ and $x \in B$ or that $x \in A$ and $x \in C$. By the definition of intersection, it follows that $x \in A \cap B$ or $x \in A \cap C$. Using the definition of union, we conclude that
$$x \in (A \cap B) \cup (A \cap C)$$

Thus,
$$A \cap (B \cap C) \subseteq (A \cap B) \cup (A \cap C) \qquad (1.7)$$

Again, suppose that $x \in (A \cap B) \cup (A \cap C)$. Then, by the definition of union, $x \in A \cap B$ or $x \in A \cap C$. By the definition of intersection, $x \in A$ and $x \in B$ or that $x \in A$ and $x \in C$. From this, we can see that $x \in A$ and $x \in B$ or $x \in C$. Consequently, by the definition of union, we see that $x \in A$ and $x \in B \cup C$. Furthermore, by the definition of intersection, $x \in A \cap (B \cup C)$. Thus,
$$(A \cap B) \cup (A \cap C) \subseteq A \cap (B \cup C) \qquad (1.8)$$

Combining Eqs (1.7) and (1.8),
$$A \cap (B \cap C) = (A \cap B) \cup (A \cap C)$$

IX. *De Morgan's law* $(A \cap B)' = A' \cup B'$

Proof Let x be any arbitrary element of the set $(A \cap B)'$. Then,
$$x \in (A \cap B)' \Rightarrow x \notin (A \cap B)$$
$$\Rightarrow x \notin A \text{ and } x \notin B$$
$$\Rightarrow x \in A' \text{ or } x \in B'$$
$$\Rightarrow x \in A' \cup B' \qquad (1.9)$$

Conversely, let x be any arbitrary element of the set $A' \cup B'$. Then,
$$x \in A' \cup B' \Rightarrow x \in A' \text{ or } x \in B'$$
$$\Rightarrow x \notin A \text{ or } x \notin B$$
$$\Rightarrow x \notin (A \cap B)$$
$$\Rightarrow x \in (A \cap B)'$$

i.e.,
$$A' \cup B' \subseteq (A \cap B)' \qquad (1.10)$$

From Eqs (1.9) and (1.10),
$$(A \cap B)' = A' \cup B'$$

1.8 MULTISET

A set is inherently an unordered collection of distinct elements. This unordered collection matters in case of repetitive occurrence of elements. For instance, the repetition occurs when we count each mark obtained by the number of students in an examination, i.e., in case of counting of frequency of numbers. This typical situation can be tackled by the concept like multiset.

Multisets are unordered collection of elements in which an element can occur as a member more than once (or repeatedly). Hence, multiset is a set in which elements are not necessarily distinct.

For example Let $A = \{1, 1, 1, 2, 2, 3\}$. Here, 1 appears three times, 2 appears twice, and 3 appears once in the set. Also, $\{3, 3, 3\}$, $\{x, y, z\}$, and $\{a, a, a, b, b, c, c, d\}$ are the multisets.

The notation used to represent multiset is given by

$$S = \{m_1 \cdot a_1, m_2 \cdot a_2, \ldots, m_n \cdot a_n\}$$

in which the element a_1 appears m_1 times, element a_2 appears m_2 times, and so on. The numbers m_i, $i = 1, 2, \ldots, n$ are called the *multiplicities* of the elements a_i. Actually, the *multiplicity* of an element in a multiset is defined to be the number of items the element appears in the multiset.

Note A set is a special case of multiset in which multiplicity of each element is either 0 or 1.

1.8.1 Operations on Multisets

Union of Multisets

Let A and B be two multisets. Then, the *union* of the multisets A and B is the multiset where the multiplicity of an element is the maximum of its multiplicities in A and B.

Let $A = \{a, a, b, b, c\}$ and $B = \{a, a, b, d\}$. Then,

$$A \cup B = \{a, a, b, b, c, d\}$$

Intersection of Multisets

The *intersection* of A and B is the multiset in which the multiplicity of an element is the minimum of its multiplicities in A and B, i.e.,

$$A \cap B = \{a, a, b\}$$

Sum of Multisets

The *sum* of A and B, denoted by $A + B$, is the multiset in which the multiplicity of an element is the sum of multiplicities in set A and set B.

Let $A = \{a, a, b, c, c\}$ and $B = \{a, b, b, d\}$. Then,

$$A + B = \{a, a, a, b, b, b, c, c, d\}$$

Difference of Multisets

The *difference* of A and B, denoted by $A - B$, is the multiset in which the multiplicity of an element equals to the multiplicity of element in A minus the multiplicity of element in B if the difference is positive, and is equal to zero if the difference is zero and negative.

Let $A = \{a, a, b, c, c\}$ and $B = \{a, b, b, d\}$. Then,

$$A - B = \{a, c, c\}$$

EXAMPLE 1.13 Let A and B be two multisets defined as $A = \{2 \cdot a, 3 \cdot b, 1 \cdot c\}$ and $B = \{1 \cdot a, 4 \cdot b, 1 \cdot d\}$, respectively. Then, determine

(i) $A \cup B$ (ii) $A \cap B$ (iii) $A + B$ (iv) $A - B$ (v) $B - A$

Solution

(i) $A \cup B = \{2 \cdot a, 4 \cdot b, 1 \cdot c, 1 \cdot d\}$
(ii) $A \cap B = \{2 \cdot a, 3 \cdot b\}$
(iii) $A + B = \{3 \cdot a, 7 \cdot b, 1 \cdot c, 1 \cdot d\}$
(iv) $A - B = \{1 \cdot a, 1 \cdot c\}$
(v) $B - A = \{1 \cdot b, 1 \cdot d\}$

1.9 FUZZY SET

A *fuzzy set* is defined as the set of pairs of elements alongwith their *degree of membership* (a real number lies between 0 and 1, including 0 and 1).

For example Let a set F (of famous people) be defined as

$$F = \{0.7 \text{ Gora}, 0.9 \text{ Somnath}, 0.3 \text{ Gautam}, 0.2 \text{ Rana}, 0.5 \text{ Monu}\}$$

Here, the set indicates that Gora has a 0.7 degree of membership in F, Somnath has a 0.9 degree of membership in F, Gautam has a 0.3 degree of membership in F, Rana has a 0.2 degree of membership in F, and Monu has a 0.5 degree of membership in F, so that Somnath is the most famous and Rana is the least famous of these people. Also, assume that R is the set of rich people with

$$R = \{0.5 \text{ Gora}, 0.8 \text{ Somnath}, 0.2 \text{ Gautam}, 0.7 \text{ Rana}, 0.6 \text{ Monu}\}$$

1.9.1 Operations on Fuzzy Set

Complement

The *complement* of a fuzzy set S is the set S', with the degree of membership of an element in S' equal to 1 minus the degree of membership of this element in S.

EXAMPLE 1.14 Find F' (the fuzzy set of people who are not famous) and R' (the fuzzy set of people who are not rich).

Solution

$$F' = \{(1 - 0.7) \text{ Gora}, (1 - 0.9) \text{ Somnath}, (1 - 0.3) \text{ Gautam}, (1 - 0.2) \text{ Rana}, (1 - 0.5) \text{ Monu}\}$$

i.e.,

$$F' = \{0.3 \text{ Gora}, 0.1 \text{ Somnath}, 0.7 \text{ Gautam}, 0.8 \text{ Rana}, 0.5 \text{ Monu}\}$$

Similarly,

$$R' = \{0.5 \text{ Gora}, 0.2 \text{ Somnath}, 0.8 \text{ Gautam}, 0.3 \text{ Rana}, 0.4 \text{ Monu}\}$$

Union

The *union* of two fuzzy sets S and T is the fuzzy set $S \cup T$, where the degree of membership of an element in $S \cup T$ is the maximum of the degrees of membership of this element in S and in T.

EXAMPLE 1.15 Find the fuzzy set $F \cup R$ of rich or famous people.

Solution Using the sets F (of famous people) and R (of rich people) defined under Section 1.9, we can write

$$F \cup R = \{0.7 \text{ Gora}, 0.9 \text{ Somnath}, 0.3 \text{ Gautam}, 0.7 \text{ Rana}, 0.6 \text{ Monu}\}$$

Intersection

The *intersection* of two fuzzy sets S and T is the fuzzy set $S \cap T$, in which the degree of membership of an element in $S \cap T$ is the minimum of the degrees of membership of this element in S and in T.

EXAMPLE 1.16 Find the fuzzy set $F \cap R$ of rich and famous people.

Solution Using the sets F (of famous people) and R (of rich people) defined under Section 1.9, we can write

$$F \cap R = \{0.5 \text{ Gora}, 0.8 \text{ Somnath}, 0.2 \text{ Gautam}, 0.2 \text{ Rana}, 0.5 \text{ Monu}\}$$

1.10 COMPUTER REPRESENTATION OF SETS

There are a number of ways to represent sets using computer. One of the ways is to store the elements of the set in an unordered fashion. Here, the operations of computing the union, intersection, or difference of two sets would be time-consuming, since each of these operations would require a large amount of searching for elements. This constraint can be overcome by storing elements using an arbitrary ordering of the elements of the universal set.

Assume that the universal set U is finite (and of reasonable size so that the number of elements of U is not larger than the memory size of the computer being used). At first, let us specify an arbitrary ordering of the elements of U, like $a_1, a_2, \ldots, a_n$. Then, represent a subset A of U with the *bit string* (a string over the alphabet $\{0, 1\}$) of length n, where the ith bit in this string is 1 if a_i belongs to A and is 0 if a_i does not belong to A. This technique can be explained by the following illustration.

For example Let $U = \{1, 2, 3, 4, 5, 6, 7, 8\}$ and the ordering of elements of U contains the elements in increasing order, i.e., $a_i = i$. Then the question arises that how do the bit strings

represent the following:
(i) The subset of all odd integers in U
(ii) The subset of all even integers in U
(iii) The subset of integers not exceeding 5, say, in U

For case (i), the bit string that represents the set of odd integers in U, namely, $\{1, 3, 5, 7\} = A$, say, possesses a length of 8 and has a value of 1 when x is 1 or 3 or 5 or 7 and a zero elsewhere. The form of the string can be expressed as

$$10\ 10\ 10\ 10$$

Similarly, in Case (ii), we represent the subset of all even integers in U, namely, $\{2, 4, 6, 8\}$,

$$01\ 01\ 01\ 01$$

Lastly, for Case (iii), the set of all integers in U that do not exceed 5, namely, $\{1, 2, 3, 4, 5\}$ is represented by the string

$$11\ 11\ 10\ 00$$

Using bit strings to represent sets, one can find complements of sets, unions, and intersection of sets in the following fashion:

I. The complement of a set from the bit string can be obtained by simply changing each 1 to 0 and 0 to 1.
II. The bit string for the union is the bitwise OR (Boolean operation) of the bit strings for the two sets.
III. The bit string for the intersection is the bitwise AND (Boolean operation) of the bit strings for the two sets.

EXAMPLE 1.17 If $U = \{1, 2, 3, 4, 5, 6\}$, $A = \{1, 2, 3, 4\}$, and $B = \{3, 4, 5, 6\}$, find the bit string for the set A and B. Using bit string, determine the complement of the set A, i.e., A'. Also, find the union and intersection of the sets A and B.

Solution Here, the bit string is of length 6. The representation of the bit string for the sets A and B, respectively, are

$$11\ 11\ 00 \qquad 00\ 11\ 11$$

The bit string for the complement of the set A, i.e., A' is

$$00\ 00\ 11 \quad \text{[which corresponds to the set } \{5, 6\}\text{]}$$

The bit string for the union of the sets A and B is

$$11\ 11\ 00 \cup 00\ 11\ 11 = 11\ 11\ 11 \text{ [which corresponds to the set } \{1, 2, 3, 4, 5, 6\}\text{]}$$

The bit string for the intersection of the sets A and B is

$$11\ 11\ 00 \cap 00\ 11\ 11 = 00\ 11\ 00 \quad \text{[which corresponds to the set } \{3, 4\}\text{]}$$

1.11 INTRODUCTION TO RELATIONS

A *relation* may be thought of a family tie between, such as, 'is the son of', 'is the daughter of', 'is the brother of', and 'is the sister of', etc. A relation may also involve equality or inequality. The mathematical concept of a relation deals with the way the variables are related or paired.

In mathematics, the expressions such as 'is less than', 'is greater than', 'is parallel to', 'is perpendicular to' are relations.

Formally, relations in mathematics describe connections between different elements of the same set, whereas functions describe connections between two different sets. As an illustration, imagine a set A of students,

$$A = \{\text{Kavita, Sonal, Babu, Puneet, Pappu}\}$$

and a set of ages,

$$B = \{5, 7, 9, 11, 13, 15\}$$

The phrase 'is younger than' defines a relation between any two elements of A. It is seen that Kavita is younger than Babu, but Pappu is not younger than Sonal. On the other hand, if we actually want to know how old the students in A are, we need to make a connection between elements of A and elements of B; so, 'age' is a function which operates between the sets A and B.

In the present assignment, relations will be considered as *binary relations* only.

1.12 BINARY RELATION

Let A and B be two non-empty sets. Then any subset of R of the cartesian product $A \times B$ is called a *binary relation R* from A to B. If $(a, b) \in R$, one can say that a is related to b and is written as aRb.

The set $\{a \in A : (a, b) \in R \text{ for some } b \in B\}$ is called the *domain* of R and is denoted by $D_R(R)$. The set $\{b \in B : (a, b) \in R \text{ for some } a \in A\}$ is called the *range* of R and is denoted by $R_R(R)$.

For example

(i) Let $A = \{3, 6, 9\}$, $B = \{4, 8, 12\}$. Then $R = \{(3, 4), (3, 8), (3, 12)\}$ is a relation from A to B.

(ii) Let $A = \{2, 3, 4\}$, $B = \{a, b\}$. Then, $A \times B = \{(2, a), (2, b), (3, a), (3, b), (4, a), (4, b)\}$. If $R = \{(2, a), (3, b)\}$, then $R \subseteq A \times B$ and R is a relation from A to B.

(iii) Let $A = \{2, 3, 4\}$ and $B = \{3, 4, 5, 6, 7\}$. If a relation R from A to B is defined by $(a, b) \in R$ such that a divides b (with zero remainder), then $R = \{(2, 4), (2, 6), (3, 3), (3, 6), (4, 4)\}$. Here, the domain of R is the set $\{2, 3, 4\}$ and the range of R is the set $\{3, 4, 6\}$.

For example Let A be the set of all possible inputs to a given computer, and let B be the set of all possible outputs from the same program. Then the relation R from A to B can be defined as aRb if and only if b is the output of the program when input a is used.

If R_1 and R_2 are two relations with same domain D and same range R, then one can also define the relation $R_1 \cup R_2$ and $R_1 \cap R_2$ with the same domain D and the same range R.

EXAMPLE 1.18 Consider $A = \{1, 2, 3, 4\}$ and $B = \{3, 4, 5, 6\}$. Find the elements of each relation R stated below. Also, find the domain and range of R.

(i) $a \in A$ is related to $b \in B$, i.e., aRb, if and only if $a < b$.

(ii) $a \in A$ is related to $b \in B$, i.e., aRb, if and only if a and b are both odd numbers.

Solution

(i) Here, $R = \{(1, 3), (2, 3), (1, 4), (1, 5), (1, 6), (2, 4), (2, 5), (2, 6), (3, 4), (3, 5), (3, 6), (4, 5), (4, 6)\}$.

Also, $D_R(R) = \{1, 2, 3, 4\}$ and $R_R(R) = \{3, 4, 5, 6\}$.

(ii) Here, $R = \{(1, 3), (1,5), (3, 3), (3, 5)\}$; $D_R(R) = \{1, 3\}$, and $R_R(R) = \{3, 5\}$.

EXAMPLE 1.19 Find the number of distinct relations from a set A to a set B.

Solution Let m and n be the number of elements contained in A and B, respectively. Then the number of elements of $A \times B$ is mn. So, the number of elements of power set of $A \times B$ is 2^{mn}, i.e., $A \times B$ has 2^{mn} distinct subsets. However, every subset of $A \times B$ is a relation from A to B. Thus, the number of distinct relations from A to B is 2^{mn}.

Note With respect to Example 1.19, we can say that there exists 2^{n^2} binary relations on a set A, for $A \times A$ possesses n^2 elements.

1.13 CLASSIFICATION OF RELATIONS

In several applications of computer science and applied mathematics, we generally treat relations on a set A rather than relations from A to B. Furthermore, these relations often satisfy certain properties which we will study in the next section.

1.13.1 Reflexive Relation

Let R be a relation defined on a set A. Then R is *reflexive* if aRa holds for all $a \in A$, i.e., if $(a, a) \in R$ for all $a \in A$.

For example

(i) Let R be the relation on $A = \{1, 2, 3, 4\}$. Then the relation $R = \{(1, 1), (2, 2), (3, 3), (4, 4)\}$ is reflexive, since for each element $a \in A$, $(a, a) \in R$.
(ii) Let $A = \{a, b, c\}$ and $R = \{(a, a), (b, b), (c, c)\}$. Then R is a reflexive relation in A.

Note The relation 'equality' is a reflexive relation, since an element equals itself.

1.13.2 Symmetric Relation

A relation R defined on a set A is said to be *symmetric* if bRa holds whenever aRb holds for $b \in A$, i.e., R is symmetric on A if $(a, b) \in R \Rightarrow (b, a) \in R$.

For example

(i) Let $A = \{1, 2, 3\}$ and $R = \{(2, 2), (2, 3), (3, 2)\}$. Then R is symmetric, since both $(2, 3)$ and $(3, 2)$ are in R.
(ii) Let R be a relation defined by 'is perpendicular to' in set of all straight lines, then R is a symmetric relation.

1.13.3 Antisymmetric Relation

A relation R on a set A is called *antisymmetric* if for all $a, b \in A$, if $(a, b) \in R$ and $a \neq b$, then $(b, a) \notin R$.

For example Let R be a relation on $A = \{1, 2, 3\}$ defined by $(a, b) \in R$ if $a \leq b$, $a, b \in A$, then $R = \{(1, 1), (1, 2), (1, 3), (2, 2), (2, 3), (3, 3)\}$. Here, $(1, 2) \in R$, but $(2, 1) \notin R$. So, the relation R is antisymmetric.

1.13.4 Transitive Relation

A relation R on a set A is said to be *transitive*, if $(a, b) \in R$, $(b, c) \in R \Rightarrow (a, c) \in R$, i.e., if aRb and $bRc \Rightarrow aRc$, $a, b, c \in A$.

For example

(i) Let $A = \{1, 2, 3\}$ and $R = \{(1, 1), (2, 2), (2, 3), (3, 2), (3, 3)\}$. Then R is transitive, since $2R2$ and $2R3 \Rightarrow 2R3 \in R$; also, $(2, 3)$ and $(3, 2) \Rightarrow 2R2 \in R$.
(ii) Let R be a relation on $A = \{a, b, c, d\}$, given by $R = \{(a, a), (b, c), (c, b), (d, d)\}$. Here, (b, c) and (c, b) are in R, but (b, b) is not in R. So, the relation R is not transitive.
(iii) Let A denote the set of straight lines in a plane and R be a relation in A defined by 'is parallel to', then R is a transitive relation in A.

EXAMPLE 1.20 Let $A = \{1, 2, 3\}$. Then, investigate the following relations for reflexive, symmetric, antisymmetric, or transitive.

(i) $R = \{(1, 1), (2, 2), (3, 3), (1, 3), (1, 2)\}$
(ii) $R = \{(1, 1), (2, 2), (1, 3), (3, 1)\}$
(iii) $R = \{(1, 1), (2, 2), (3, 3), (1, 2), (2, 1), (2, 3), (3, 2)\}$

Solution

(i) The given relation is reflexive and transitive but not symmetric, for $(1, 3)$ and $(1, 2) \in R$, but $(3, 1)$ and $(2, 1) \notin R$.
(ii) The given relation is symmetric and transitive but not reflexive, since $(3, 3) \notin R$.
(iii) The given relation is reflexive and symmetric but not transitive, for $(1, 3) \notin R$.

EXAMPLE 1.21 Let Z^* be the set of all non-zero integers and R be the relation on Z^* such as $(a, b) \in R$, if a is the factor of b, i.e., a/b. Investigate R for reflexive, symmetric, antisymmetric, or transitive.

Solution Since a/a for all $a \in Z^*$, hence R is reflexive. If a/b and b/c then a/c which implies that R is transitive. However, a/b does not imply that b/a is true, for $2/16$ but $16/2$ is not true, i.e., R is not symmetric.

Again, if $4/-4$ and $-4/4$ are true, then $4 \neq -4$, i.e., R is not antisymmetric.

1.13.5 Equivalence Relation

A relation R on a set A is said to be an *equivalence relation*, if R is reflexive, symmetric, and transitive.

For example Let $A = \{a, b, c\}$ and $R = \{(a, a), (a, b), (b, a), (b, b), (b, c), (a, c), (c, a), (c, b), (c, c)\}$. Then R is an equivalence relation in A.

EXAMPLE 1.22 Let Z denote the set of integers and the relation R in Z be defined by aRb iff $a - b$ is an even integer. Then, show that R is an equivalence relation.

Solution The relation R is reflexive, since $a - a$ is even, hence aRa for every $a \in Z$.

The relation R is symmetric, if $a - b$ is even, then $b - a = -(a - b)$ is also even, i.e., $aRb \Rightarrow bRa$.

The relation R is transitive, for if aRb and bRc, then both $a - b$ and $b - c$ are even. Consequently, $a - c = (a - b) + (b - c)$ is also even. So, aRb and $bRc \Rightarrow aRc$.

Thus, R is an equivalence relation.

EXAMPLE 1.23 Let A be the set of triangles in the Euclidean plane and R is the relation in A defined by 'a is similar to b' then show that R is an equivalence relation in A.

Solution Every triangle is similar to itself, i.e., the relation R is reflexive.

If a is similar to b, then b is similar to a, i.e., $(a, b) \Rightarrow (b, a) \in R$. So, R is symmetric.

If a is similar to b and b is similar to c, then a is similar to c, i.e., the relation R is transitive. Thus, the relation R is reflexive, symmetric, and transitive. Hence, R is an equivalence relation.

1.13.6 Associative Relation

It is known that relations from A to B are the subsets of $A \times B$; so, two relations from A to B can be associated in any way as two sets can be associated.

For example Let $A = \{a, b, c\}$ and $B = \{a, b, c, d\}$. Then, the associative relations of $R_1 = \{(a, a), (b, b), (c, c)\}$ and $R_2 = \{(a, a), (a, b), (a, c), (a, d)\}$ are represented notationwise as, $R_\cup$, $R_\cap$, R_{12}, and R_{21}, respectively

$$R_\cup = R_1 \cup R_2 = \{(a, a), (a, b), (a, c), (a, d), (b, b), (c, c)\}$$
$$R_\cap = R_1 \cap R_2 = \{(a, a)\}$$
$$R_{12} = R_1 - R_2 = (b, b), (c, c)\}$$
$$R_{21} = R_2 - R_1 = \{(a, b), (a, c), (a, d)\}$$

1.14 COMPOSITION OF RELATIONS

Let R_1 be a relation from A to B and R_2 be a relation from B to C. The *composition* of R_1 and R_2, denoted by $R_2 \circ R_1$, the relation from A to C, is defined as

$$R_2 \circ R_1 = \{(a, c) : (a, b) \in R_1 \text{ and } (b, c) \in R_2 \text{ for some } b \in B\}$$

EXAMPLE 1.24 Find the composition of the relations

$$R_1 = \{(1, 2), (1, 6), (2, 4), (3, 4), (3, 6), (3, 8)\}$$

and
$$R_2 = \{(2, x), (4, y), (4, z), (6, z), (8, x)\}$$

Solution Here, $R_2 \circ R_1 = \{(1, x), (1, z), (2, y), (2, z), (3, y), (3, z), (3, x)\}$.

Note Composition of parent relation with itself. Let R be the relation on the set of all people such that $(a, b) \in R$ if person a is a parent of person b. Then, $(a, c) \in R \circ R$ if and only if there is a person b such that $(a, b) \in R$ and $(b, c) \in R$, i.e., if and only if there is a person b such that a is a parent of b and b is a parent of c. In other words, $(a, c) \in R \circ R$ if and only if a is a grandparent of c.

1.15 INVERSE OF A RELATION

Let R be a relation from A to B. The inverse of R, denoted by R^{-1}, is the relation from B to A defined by

$$R^{-1} = \{(b, a): (a, b) \in R\}$$

For example

Let $A = \{2, 3, 4\}$, $B = \{3, 4, 5, 6, 7\}$, and $R = \{(2, 4), (2, 6), (3, 3), (3, 6), (4, 4)\}$. The inverse of the relation R is

$$R^{-1} = \{(4, 2), (6, 2), (3, 3), (6, 3), (4, 4)\}$$

This relation, in words, can be mentioned as 'is divisible by'.

 EXAMPLE 1.25 If a relation R is transitive, then prove that its inverse relation R^{-1} is also transitive.

Solution Let (a, b) and $(b, c) \in R^{-1}$. Then, $(b, a) \in R$ and $(c, b) \in R$. Again, $(c, b) \in R$ and $(b, a) \in R \Rightarrow (c, a) \in R \Rightarrow (a, c) \in R^{-1}$.

1.16 REPRESENTATION OF RELATIONS ON A SET

Relations from a set A to itself possess practical significance. A *relation on a set A* is a relation from A to A, i.e., a relation on a set A can be treated as a subset of $A \times A$.

 EXAMPLE 1.26 Let A be the set consisting of elements such as 1, 2, 3, and 4, i.e., $A = \{1, 2, 3, 4\}$. Also, assume the relation $R = \{(a, b) : a \text{ divides } b\}$. Then, find the ordered pairs which exist in R.

Solution From the given relation R, it is seen that (a, b) is in R if and only if a and b are positive integers not exceeding such that a divides b, and the following ordered pairs exist in R as

$$R = \{(1, 1), (1, 2), (1, 3), (1, 4), (2, 2), (2, 4), (3, 3), (4, 4)\}$$

The ordered pairs in R are shown graphically in Figure 1.11.

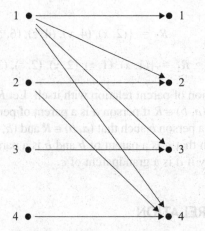

FIGURE 1.11 The ordered pairs in R

1.17 CLOSURE OPERATION ON RELATIONS

Let R be a relation on a set A. It may be often possible that R may or may not have the relational properties such as reflexivity, symmetry, and transitivity. If R does not possess any property, what is to be done next? To fulfil R with a property, we should add new pairs to R to get a relation having the required property. For this, we wish to add as few new pairs as possible, so what we need to find is the *smallest* relation R_1 on A which contains R and possesses the necessary property. However, R_1 does not exist always. If a relation like R_1 does exist, we call it the *closure* of the relation R with respect to the property we desire. Several types of closures will be discussed as follows.

1.17.1 Reflexive Closure

Consider a relation R on a set A. Relation R_R is called *reflexive closure* of R if R_R is the smallest relation containing R, having the reflexive property, i.e., $R_R = R \cup \Delta_A$, where $\Delta_A = \{(a, a) : a \in A\}$ is the *diagonal* or *equality* relation on A. In other words, the reflexive closure of R can be obtained by adding to R all pairs of the form (a, a), $a \in A$, not already in R. The addition of these pairs produces a new relation that is reflexive, contains R, and is contained within any reflexive relation containing R.

For example Consider the relation $R = \{(1, 1), (1, 2), (2, 1), (3, 2)\}$ on the set $A = \{1, 2, 3\}$. This relation is not reflexive. The relation can be made reflexive, by adding $(2, 2)$ and $(3, 3)$ to R, since these are the only pairs of the form (a, a), not in R. This new relation contains R. Moreover, any reflexive relation that contains R must also contain $(2, 2)$ and $(3, 3)$.

 EXAMPLE 1.27 Let $A = \{a, b, c\}$ and R be a relation on A defined by $R = \{(a, a), (a, b), (b, c), (c, a)\}$. Find the reflexive closure of R.

Solution Here, $R \cup \Delta_A$ is the smallest relation having the reflexive property. Thus, the reflexive closure of R is given by

$$R_R = R \cup \Delta_A = \{(a, a), (a, b), (b, b), (b, c), (c, c), (c, a)\}$$

1.17.2 Symmetric Closure

Consider a relation R on a set A. The relation R_S is called the *symmetric closure* of R if R_S is the smallest relation containing R, having the symmetric property. The relation $R_S = R \cup R^{-1}$ is the smallest symmetric relation containing R, i.e., $R \cup R^{-1}$ is the symmetric closure of R. In other words, the symmetric closure of R is obtained by adding all ordered pairs of the form (b, a), whenever (a, b), belongs to the relation, that are not already present in R. Addition of these pairs produces a relation, that is symmetric, contains R, and is contained within any symmetric relation containing R.

For example The relation $R = \{(1, 1), (1, 2), (2, 2), (2, 3), (3, 1), (3, 2)\}$ on the set $A = \{1, 2, 3\}$ is not symmetric. The relation will be a symmetric one if we add $(2, 1)$ and $(1, 3)$ to R, because these are the only pairs of the form (b, a) with $(a, b) \in R$ that are not in R. This new relation is symmetric, contained in R.

EXAMPLE 1.28 Let R be the relation on the set $A = \{4, 5, 6, 7\}$ defined by

$$R = \{(4, 5), (5, 5), (5, 6), (6, 7), (7, 4), (7, 7)\}$$

Find the symmetric closure of R.

Solution The smallest relation containing R, having the symmetric property, is $R \cup R^{-1}$, i.e.,

$$R_S = R \cup R^{-1} = \{(4, 5), (5, 4), (5, 5), (5, 6), (6, 5), (6, 7), (7, 6), (7, 4), (4, 7), (7, 7)\}$$

The *transitive closure* of a relation R is the *smallest* transitive relation containing R. We will study the transitive closure in Section 1.19.1.

1.18 MATRIX REPRESENTATION OF RELATION

Let $A = \{a_1, a_2, \ldots, a_m\}$ and $B = \{b_1, b_2, \ldots, b_n\}$ are two finite sets, containing m and n elements, respectively. Let R be a relation from A to B. Then, we can represent the *relation matrix* of R, denoted by M_R, by the $m \times n$ matrix, i.e.,

$$M_R = [m_{ij}]_{m \times n}$$

which is defined by

$$m_{ij} = \begin{cases} 0, & \text{if } (a_i, b_j) \notin R \\ 1, & \text{if } (a_i, b_j) \in R \end{cases}$$

where m_{ij} is the element in the ith row and the jth column. Moreover, M_R can be described both in the tabular form and in the matrix form.

For example
(i) Let $A = \{1, 2, 3\}$ and $R = \{(1, 2), (1, 3), (2, 3)\}$. Then we can describe M_R in the following fashion:

	1	2	3
1	0	1	1
2	0	0	1
3	0	0	0

$$\begin{bmatrix} 0 & 1 & 1 \\ 0 & 0 & 1 \\ 0 & 0 & 0 \end{bmatrix}$$

(ii) Let $A = \{1, 2, 3\}$, $B = \{a, b\}$, and $R = \{(1, a), (2, b), (3, a)\}$. Then M_R can be represented both in tabular and in matrix forms as shown below.

	a	b
1	1	0
2	0	1
3	1	0

$$M_R = \begin{bmatrix} 1 & 0 \\ 0 & 1 \\ 1 & 0 \end{bmatrix}$$

EXAMPLE 1.29 Let $A = \{1, 4, 5\}$ and $R = \{(1, 4), (1, 5), (4, 1), (4, 4), (5, 5)\}$. Determine M_R.

Solution Given that $R = \{(1, 4), (1, 5), (4, 1), (4, 4), (5, 5)\}$. Then the relation matrix of R, i.e., M_R is given by

$$M_R = \begin{bmatrix} 0 & 1 & 1 \\ 1 & 1 & 0 \\ 0 & 0 & 1 \end{bmatrix}$$

EXAMPLE 1.30 Let $A = \{1, 2, 3, 4\}$, $B = \{p, q, r, s\}$, and $R = \{(1, p), (1, q), (1, r), (2, q), (2, r), (2, s)\}$. Find M_R.

Solution The matrix representation of relation R, i.e., M_R is given by

$$M_R = \begin{array}{c} \\ 1 \\ 2 \\ 3 \\ 4 \end{array} \begin{array}{cccc} p & q & r & s \end{array} \\ \begin{bmatrix} 1 & 1 & 1 & 0 \\ 0 & 1 & 1 & 1 \\ 0 & 0 & 0 & 0 \\ 0 & 0 & 0 & 0 \end{bmatrix}$$

EXAMPLE 1.31 Let

$$A = \{a, b, c\} \quad \text{and} \quad M_R = \begin{bmatrix} 1 & 1 & 0 \\ 0 & 0 & 1 \\ 0 & 0 & 0 \end{bmatrix}$$

Find the relation R defined on A.

Solution Here, M_R can be re-written as

$$M_R = \begin{array}{c} \\ a \\ b \\ c \end{array} \begin{array}{ccc} a & b & c \end{array} \\ \begin{bmatrix} 1 & 1 & 0 \\ 0 & 0 & 1 \\ 0 & 0 & 0 \end{bmatrix}$$

Thus, $R = \{(a, a), (a, b), (b, c)\}$.

1.19 DIGRAPHS

Let R be a relation on the set $A = \{a_i, a_2, \ldots, a_n\}$. The element a_i of A are represented by points (or circles) called *nodes* (or *vertices*). If $(a_i, a_j) \in R$, then we connect the vertices a_i and a_j by means of an arc and place an arrow in the direction from a_i to a_j. If $(a_i, a_j) \in R$ and $(a_j, a_i) \in R$, then we draw two arcs between a_i and a_j, on one hand and vice versa (i.e., one arc starts from node a_i and ends at a_j, and vice versa for the other arc). When all the nodes corresponding to the ordered pairs in R are connected by arcs with proper arrows, we get a graph of the relation R. This diagram or graph is called the *directed graph*, or in precise *digraph* of the relation R.

If R is a relation on a set A, a path of *length* (the number of edges in a path) n in R from a_i to a_j is a finite sequence P, such as, $a_i, a_1, a_2, \ldots, a_{n-1}, a_j$, beginning with a_i and ending with a_j such that $a_i R a_1, a_1 R a_2, \ldots, a_{n-1} R a_j$.

If n is a positive integer then the relation R^n on the set A can be defined as that there is a path of length n from a_i to a_j in R, i.e., $(a_i, a_j) \in R^n$. The relation R^∞ is defined on A, by letting $(a_i, a_j) \in R^\infty$ means, that there is some path in R from a_i to a_j. Also, a *cycle* in a digraph can be defined as a path of length $n \geq 1$ from a vertex to itself (or, in other words, a path that begins and ends at the same vertex is called a *cycle*).

Paths in a relation R can be used to define new relations that are quite useful. The relation R^∞ is sometimes called the *connectivity relation* for R.

Note The $R^n(x)$ consists of all vertices that can be reached from x by means of a path in R of length n. The set $R^\infty(x)$ consists of all vertices that can be reached from x by some path in R.

If R is reflexive, then there must exist a *loop* at each node in the digraph of R. If R is symmetric, then $(a_i, a_j) \in R$ implies $(a_j, a_i) \in R$ and the nodes a_i and a_j will be connected by two arcs (edges), i.e., one from a_i to a_j and the other from a_j to a_i.

For example The directed graph or digraph of the following relation R on the set $A = \{1, 2, 3, 4\}$ is shown in Figure 1.12.

$$R = \{(1, 2), (2, 2), (2, 4), (3, 2), (3, 4), (4, 1), (4, 3)\}$$

It may be observed from Figure 1.12 that there exists an arrow from 2 to itself, since 2 is related to 2 under R.

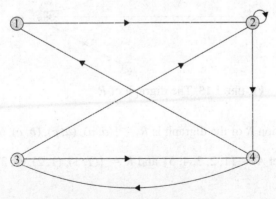

FIGURE 1.12 The directed graph of R

EXAMPLE 1.32 Let $A = \{a, b, d\}$ and $R = \{(a, b), (a, d), (b, d), (d, a), (d, d)\}$ be a relation on A. Construct the digraph of R.

Solution The digraph of R is shown in Figure 1.13.

 EXAMPLE 1.33 Let $A = \{1, 2, 3, 4\}$ and $R = \{(1, 1), (1, 2), (2, 1), (2, 2), (2, 3), (2, 4), (3, 4), (4, 1), (4, 4)\}$. Construct the digraph of R.

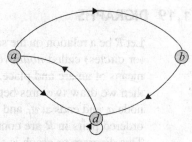

FIGURE 1.13 The digraph of R

Solution The digraph of R is shown in Figure 1.14.

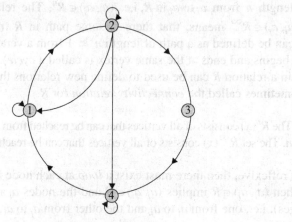

FIGURE 1.14 The digraph of R

 EXAMPLE 1.34 Find the relation R from the digraph of Figure 1.15.

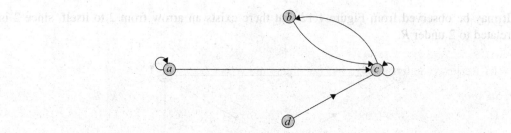

FIGURE 1.15 The digraph of R

Solution The relation R of the digraph is $R = \{(a, a), (a, c), (b, c), (c, b), (c, c), (d, c)\}$.

 EXAMPLE 1.35 Let $A = \{1, 2, 3, 4, 5\}$ and $R = \{(1, 1), (1, 2), (2, 3), (3, 5), (3, 4), (4, 5)\}$. Determine
  (i) R^2 (ii) R^∞

Solution The digraph of R is shown in Figure 1.16.

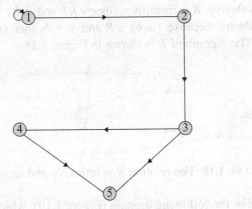

FIGURE 1.16 The digraph of R

(i) Here, $(1, 1) \in R$ and $(1, 1) \in R \Rightarrow (1, 1) \in R^2$
Again,
$$(1, 1) \in R \text{ and } (1, 2) \in R \Rightarrow (1, 2) \in R^2$$
$$(1, 2) \in R \text{ and } (2, 3) \in R \Rightarrow (1, 3) \in R^2$$
$$(2, 3) \in R \text{ and } (3, 5) \in R \Rightarrow (2, 5) \in R^2$$
$$(2, 3) \in R \text{ and } (3, 4) \in R \Rightarrow (2, 4) \in R^2$$
$$(3, 4) \in R \text{ and } (4, 5) \in R \Rightarrow (3, 5) \in R^2$$

Thus,
$$R^2 = \{(1, 1), (1, 2), (1, 3), (2, 5), (2, 4), (3, 5)\}$$

(ii) There is a path from 1 to $4 \Rightarrow (1, 4) \in R^\infty$, whose length is 3 and there is a path from 1 to $5 \Rightarrow (1, 5) \in R^\infty$, whose length is 3.
Thus,
$$R^\infty = \{(1, 1), (1, 2), (1, 3), (1, 4), (1, 5), (2, 3), (2, 4), (2, 5), (3, 4), (3, 5), (4, 5)\}$$

 EXAMPLE 1.36 Find a non-empty set and a relation on the set that satisfy each of the following combinations of properties. Simultaneously, draw a digraph of each relation.
(i) Reflexive and symmetric but not transitive.
(ii) Reflexive and transitive but not antisymmetric.

Solution
(i) Let $A = \{a, b, c\}$ and $R = \{(a, a), (b, b), (c, c), (a, b), (a, c), (b, a), (c, b)\}$. Here, R is reflexive, since for each element $a \in A$, $(a, a) \in R$. R is symmetric, since both (a, b) and (b, a) are in R. R is not transitive, because (b, a) and (a, c) are in R, but $(b, c) \notin R$. The digraph of R is given in Figure 1.17.

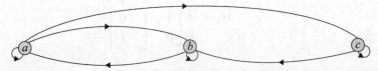

FIGURE 1.17 The relation R is reflexive and symmetric but not transitive

(ii) Let $A = \{a, b, c\}$ and $R = \{(a, a), (b, b), (c, c), (a, b), (a, c), (b, c), (b, a), (c, b), (c, a)\}$. Here, R is reflexive. R is transitive, since aRb and bRc implies $aRc \in R$. However, R is not antisymmetric because $(a, b) \in R$ and $a \neq b$, then $(b, a) \in R$ [R to be antisymmetric, $(b, a) \notin R$]. The digraph of R is shown in Figure 1.18.

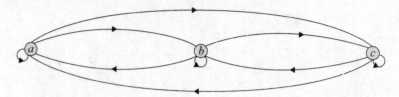

FIGURE 1.18 The relation R is reflexive and transitive but not antisymmetric

EXAMPLE 1.37 For the following digraph (Figure 1.19), which of the special properties is satisfied by digraph's relation?

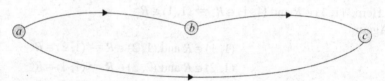

FIGURE 1.19 The digraph's relation R is both transitive and antisymmetric

Solution Here, $R = \{(a, b), (b, c), (a, c)\}$. R is both transitive and antisymmetric on $A = \{a, b, c\}$.

EXAMPLE 1.38 Let $A = \{1, 4, 5\}$ and let R be given by the digraph shown in Figure 1.20.

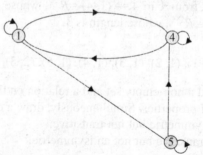

FIGURE 1.20 The digraph of R

Determine M_R and R.

Solution Here,

$$M_R = \begin{array}{c} \\ 1 \\ 4 \\ 5 \end{array} \begin{array}{c} 1 \quad 4 \quad 5 \end{array} \left[\begin{array}{ccc} 1 & 1 & 1 \\ 1 & 1 & 0 \\ 0 & 1 & 1 \end{array} \right]$$

and $R = \{(1, 1), (1, 4), (1, 5), (4, 1), (4, 4), (5, 4), (5, 5)\}$.

EXAMPLE 1.39 Let

$$A = \{a, b, c, d, e\} \text{ and } M_R = \begin{bmatrix} 1 & 1 & 0 & 0 & 0 \\ 0 & 0 & 1 & 1 & 0 \\ 0 & 0 & 0 & 1 & 1 \\ 0 & 1 & 1 & 0 & 0 \\ 1 & 0 & 0 & 0 & 0 \end{bmatrix}$$

Find the relation R defined on A.

Solution Here, $R = \{(a, a), (a, b), (b, c), (b, d), (c, d), (c, e), (d, b), (d, c), (e, a)\}$. The digraph of R is shown in Figure 1.21.

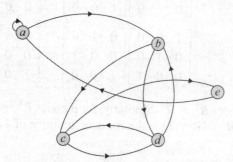

FIGURE 1.21 The digraph of R

EXAMPLE 1.40 Let R be a binary relation on $A = \{a, b, c, d, e, f, g, h\}$ represented by the following two-component digraphs. Find the smallest integers m and n such that $m < n$ and $R^m = R^n$.

Solution According to the given condition, the following relation R is formed as

$$R = \{(a, b), (b, c), (c, a), (d, e), (e, f), (f, g), (g, h), (h, d)\}$$

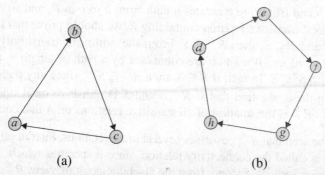

(a)   (b)

FIGURE 1.22 The component digraphs of the binary relation R

Adjacency matrix of Figure 1.22(a) is

$$R = \begin{array}{c} \\ a \\ b \\ c \end{array}\begin{array}{c} a \ \ b \ \ c \end{array} \\ \left[\begin{array}{ccc} 0 & 1 & 0 \\ 0 & 0 & 1 \\ 1 & 0 & 0 \end{array}\right]$$

$$R^2 = \left[\begin{array}{ccc} 0 & 1 & 0 \\ 0 & 0 & 1 \\ 1 & 0 & 0 \end{array}\right]\left[\begin{array}{ccc} 0 & 1 & 0 \\ 0 & 0 & 1 \\ 1 & 0 & 0 \end{array}\right] = \left[\begin{array}{ccc} 0 & 0 & 1 \\ 1 & 0 & 0 \\ 0 & 1 & 0 \end{array}\right]$$

$$R^3 = \left[\begin{array}{ccc} 0 & 0 & 1 \\ 1 & 0 & 0 \\ 0 & 1 & 0 \end{array}\right]\left[\begin{array}{ccc} 0 & 1 & 0 \\ 0 & 0 & 1 \\ 1 & 0 & 0 \end{array}\right] = \left[\begin{array}{ccc} 1 & 0 & 0 \\ 0 & 1 & 0 \\ 0 & 0 & 1 \end{array}\right]$$

$$R^4 = \left[\begin{array}{ccc} 1 & 0 & 0 \\ 0 & 1 & 0 \\ 0 & 0 & 1 \end{array}\right]\left[\begin{array}{ccc} 0 & 1 & 0 \\ 0 & 0 & 1 \\ 1 & 0 & 0 \end{array}\right] = \left[\begin{array}{ccc} 0 & 1 & 0 \\ 0 & 0 & 1 \\ 1 & 0 & 0 \end{array}\right]$$

Therefore, for 3 nodes $R^1 = R^4$. Similarly, for 5 nodes $R^1 = R^6$. From these two relations, we can conclude that $R^1 = R^{12}$ (12 is the common multiple of 4 and 6). Thus, $m = 1$ and $n = 12$.

1.19.1 Transitive Closure

In this section, we form the relational-based set, which possesses several interpretations and variety of important applications to computer science. Assume that R is a relation on a set A and that R is not transitive. Here, we will present the *transitive closure* of R which is just the *connectivity relation* R^∞ [also represented as transitive (R)], mentioned in Section 1.19, and is the smallest transitive relation containing R.

Theorem 1.1 Let R be a relation on a set A. Then R^∞ is the transitive closure of R.

Proof It is known that if a and b belong to the set A, then $aR^\infty b$ iff there exists a path in R from a to b. Here, R^∞ is transitive, since, if $aR^\infty b$ and $bR^\infty c$, the composition of the paths from a to b and from b to c creates a path from a to c in R, and so $aR^\infty c$. To show that R^∞ is the smallest transitive relation containing R, we should prove that if S is any transitive relation on A and $R \subseteq S$, then $R^\infty \subseteq S$. From the notion of transitivity, if S is transitive, then $S^n \subseteq S$ for all n; i.e., if a and b are connected by a path of length n, then aSb. It follows that $S^\infty = \bigcup_{n=1}^\infty S^n \subseteq S$. In fact, if $R \subseteq S$, then $R^\infty \subseteq S^\infty$, since any path in R is also a path in S. Combining these, we find that if $R \subseteq S$ and S is transitive on A, then $R^\infty \subseteq S^\infty \subseteq S$, which implies that R^∞ is the smallest of all transitive relations on A that contain R.

It may be seen that R^∞ possesses several interpretations. First of all, from the graphical point of view, it is called the connectivity relation, since it specifies which vertices are connected (by paths) to other vertices. Second, from the algebraic point of view, R^∞ is transitive closure of R (from the above theorem). In this form, it plays key roles in the theory of equivalence relations and in the theory of certain languages, namely BASIC, FORTRAN, JAVA, PASCAL, C^{++}, etc.

EXAMPLE 1.41 Let $A = \{1, 2, 3, 4\}$ and $R = \{(1, 2), (2, 3), (3, 4), (2, 1)\}$. Find the transitive closure of R.

Solution

Method 1 Graphical representation: The digraph of R is shown in Figure 1.23. Since R^∞ is the transitive closure, we can proceed graphically by computing all the paths. It is observed that from vertex 1 the paths are formed by joining to vertices 2, 3, and 4. Moreover, the path from 1 to 1 proceeds from 1 to 2 to 1. Thus, the ordered pairs $(1, 1)$, $(1, 2)$, $(1, 3)$, and $(1, 4)$ exist in R^∞.

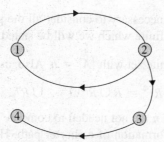

FIGURE 1.23 The graphical representation of R

Again, beginning with vertex 2 and the paths to vertices 2, 1, 3, and 4 are constructed; so, the ordered pairs $(2, 1)$, $(2, 2)$, $(2, 3)$, and $(2, 4)$ exist in R^∞. The only other path formed is from vertex 3 to vertex 4. Thus,

$$R^\infty = \{(1, 1), (1, 2), (1, 3), (1, 4), (2, 1), (2, 2), (2, 3), (2, 4), (3, 4)\}$$

Method 2 Matrix representation: The matrix of R is

$$M_R = \begin{bmatrix} 0 & 1 & 0 & 0 \\ 1 & 0 & 1 & 0 \\ 0 & 0 & 0 & 1 \\ 0 & 0 & 0 & 0 \end{bmatrix}$$

Now, we compute the powers of M_R, i.e.,

$$(M_R)^2 = M_R \circ M_R = \begin{bmatrix} 1 & 0 & 1 & 0 \\ 0 & 1 & 0 & 1 \\ 0 & 0 & 0 & 0 \\ 0 & 0 & 0 & 0 \end{bmatrix}, \quad (M_R)^3 = (M_R)^2 \circ M_R = \begin{bmatrix} 0 & 1 & 0 & 1 \\ 1 & 0 & 1 & 0 \\ 0 & 0 & 0 & 0 \\ 0 & 0 & 0 & 0 \end{bmatrix}$$

and

$$(M_R)^4 = (M_R)^3 + M_R = \begin{bmatrix} 1 & 0 & 1 & 0 \\ 0 & 1 & 0 & 1 \\ 0 & 0 & 0 & 0 \\ 0 & 0 & 0 & 0 \end{bmatrix}$$

Continuing in this way, it is observed that $(M_R)^n$ equals $(M_R)^2$, if n is even and equals $(M_R)^3$ if n is odd and greater than 1. Thus,

$$M_{R^\infty} = M_R \cup (M_R)^2 \cup (M_R)^3 = \begin{bmatrix} 1 & 1 & 1 & 1 \\ 1 & 1 & 1 & 1 \\ 0 & 0 & 0 & 0 \\ 0 & 0 & 0 & 0 \end{bmatrix}$$

which gives the same relation as Method 1.

In Example 1.41, it is not necessary to consider all the powers of R^n to find R^∞. This idea is true whenever the set A is finite which we will be stated in the following theorem.

Theorem 1.2 Let A be a finite set with $|A| = n$. Also, assume R as a relation on A. Then,

$$R^\infty = R \cup R^2 \cup \cdots \cup R^n$$

i.e., powers of R greater than n are not needed to compute R^∞.

The theorem reveals the formation of a shorter path. Here, we prove the theorem using a simple example of transitive closure in which a set has only four/three elements.

EXAMPLE 1.42 Let the relation R be defined on the set $A = \{1, 2, 3, 4\}$ as $R = \{(1, 2), (2, 3), (2, 4)\}$. Compute transitive (R).

Solution Here, $R^2 = R \circ R = \{(1, 2), (2, 3), (2, 4)\} \circ \{(1, 2), (2, 3), (2, 4)\} = \{(1, 3), (1, 4)\}$ (We unite (a, b) and (b, c) in R to obtain (a, c) in R^2) and $R^3 = R^2 \circ R = \{(1, 3), (1, 4)\} \circ \{(1, 2), (2, 3), (2, 4)\} \neq \phi$. (Here, no pair (a, b) in R^2 can be united with any pair in R). Furthermore, $R^4 = R^5 = \cdots = \phi$. Consequently, transitive $(R) = R \cup R^2 = \{(1, 2), (2, 3), (2, 4), (1, 3), (1, 4)\}$.

EXAMPLE 1.43 Let the relation R be defined on the set $A = \{1, 2, 3\}$ as $R = \{(1, 2), (2, 3), (3, 3)\}$. Compute transitive (R).

Solution Here, $R^2 = R \circ R = \{(1, 3), (2, 3), (3, 3)\}$ and $R^3 = R^2 \circ R = \{(1, 3), (2, 3), (3, 3)\}$. Consequently, transitive $(R) = R \cup R^2 \cup R^3 = \{(1, 2), (2, 3), (3, 3), (1, 3)\}$.

In Examples 1.41–1.43, the methods used are not systematic. The graphical method is impractical for large sets and relations. The matrix method can be used in general and more or less systematic enough to be programmed for a computer; however, it is costly enough. A significant and a more efficient algorithm for computing the transitive closure of a relation are given by Warshall as *Warshall's algorithm*.

1.19.2 Warshall's Algorithm

The aim of the approach of algorithm is to generate a sequence of matrices $M^0, M^1, \ldots, M^k, \ldots, M^n$ for a graph of n vertices with $M^0 = M$ (path matrix). Initially, $M^0 = A$ (adjacency matrix).

The first iteration consists of searching the existence of paths from a vertex to another one directly (via an edge) or indirectly (through the intermediate vertex or a pivotal one, say, v_1). Then M^1 represents the resulting matrix with its elements $m_{ij}^{(1)}$, given by.

$$m_{ij}^{(1)} = \begin{cases} 1 \\ 0 \end{cases}$$

where '1' corresponds to a path (of length 2) from v_i to v_1 and v_1 to v_j and '0' otherwise.

The second iteration is to search a path from a vertex to another one with v_1 or v_2 or both as pivots. Then we compute $m_{ij}^{(2)}$ and its elements are

$$m_{ij}^{(2)} = \begin{cases} 1 \\ 0 \end{cases}$$

where '1' corresponds to a path from v_i to v_j using only pivots (intermediate vertices) from (v_1, v_2) and '0' otherwise.

Finally, in the kth iteration the pivots are to be fetched from the set $V = \{v_1, v_2, \ldots, v_k\}$. Then we compute $m_{ij}^{(k)}$, where

$$m_{ij}^{(k)} = \begin{cases} 1 \\ 0 \end{cases}$$

and '1' corresponds to a path from v_i to v_j using only pivots from $(v_1, v_2, \ldots, v_k)$ and '0' otherwise. This is the basis for Warshall's algorithm. We can also compute $m_{ij}^{(k)}$ from the previous iteration as follows:

$$m_{ij}^{(k)} = m_{ij}^{(k-1)} \vee (m_{ij}^{(k-1)} \wedge m_{ij}^{(k-1)})$$

where $\vee$ and $\wedge$ denote 'or' and 'and', respectively, which means that

$$m_{ij}^{(k)} = 1 \text{ if } m_{ij}^{(k-1)} = 1 \quad \text{or both} \quad m_{ij}^{(k-1)} = 1 \quad \text{and} \quad m_{ij}^{(k-1)} = 1$$

So, the only way the value of $m_{ij}^{(k)}$ can modify 0 is to search a path through v_k, i.e., there exist a path from v_i to v_k and a path from v_k to v_j. Finally, the Warshall's algorithm is presented below.

Warshall's algorithm:

Procedure $M_R : n \times n$ zero-one matrix
$m^{(0)} = M_R$
for $k = 1$ to n
begin
    for $i = 1$ to n
    begin
        for $j = 1$ to n
            $m_{ij}^{(k)} = m_{ij}^{(k-1)} \vee (m_{ij}^{(k-1)} \wedge m_{ij}^{(k-1)})$
    end
end $\{m_{ij}^{(k)}\}$

EXAMPLE 1.44 Let $R = \{(1, 2), (2, 3), (3, 3)\}$ be a relation on the set $A = \{1, 2, 3\}$. Using Warshall's algorithm, compute transitive closure of R.

Solution The matrix of R is

$$M_R = \begin{bmatrix} 0 & 1 & 0 \\ 0 & 0 & 1 \\ 0 & 0 & 1 \end{bmatrix} = M^{(0)}$$

Then, $M^{(1)}$ takes value of 1 in its (i, j)th entry, if there is a path (edge) from v_i to v_j that has only $v_1 = 1$ as an immediate vertex. Since the first column of $M^{(0)}$ has value of 1, there exist no edges that possess the value of 1 as terminal vertex. So, no new path is formed and, therefore, no new value as 1 can be inserted in $M^{(1)}$. Thus,

$$M^{(1)} = M^{(0)} = \begin{bmatrix} 0 & 1 & 0 \\ 0 & 0 & 1 \\ 0 & 0 & 1 \end{bmatrix}$$

Again, $M^{(2)}$ possesses the value of 1 in its (i, j)th entry, if there is a path from v_i to v_j that has only $v_1 = 1$ and/or $v_2 = 2$ as its intermediate vertices. Then, the new entry in $M^{(2)}$ occurs, given by $M^{(2)}(1, 3) = 1$ because $M^{(1)}(1, 2) = 1$ and $M^{(1)}(2, 3) = 1$. Thus,

$$M^{(2)} = \begin{bmatrix} 0 & 1 & 1 \\ 0 & 0 & 1 \\ 0 & 0 & 1 \end{bmatrix}$$

Finally, $M^{(3)}$ accepts the value 1 in its (i, j)th entry, if there exist a path from v_i to v_j that has only $v_1 = 1, v_2 = 2$, and/or $v_3 = 2$ as its path. It is also seen that there is no edges that possess 3 as intermediate vertex. So, no new path can be created and, therefore, no new value of 1 is inserted in $M^{(3)}$. Thus,

$$M^{(3)} = M^{(2)} = \begin{bmatrix} 0 & 1 & 1 \\ 0 & 0 & 1 \\ 0 & 0 & 1 \end{bmatrix}$$

is a matrix of transitive closure of R. Hence, the transitive closure of R is given by

$$R^\infty = \{(1, 2), (1, 3), (2, 3), (3, 3)\}$$

EXAMPLE 1.45 Let $A = \{4, 6, 8, 10\}$ and $R = \{(4, 4), (4, 10), (6, 6), (6, 8), (8, 10)\}$ be a relation defined on a set A. Using Warshall's algorithm, find the transitive closure of R.

Solution The diagraph of the relation R and its corresponding matrix M_R is depicted in Figure 1.24.

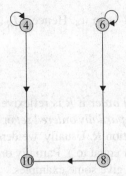

$$M_R = \begin{array}{c} \\ 4 \\ 6 \\ 8 \\ 10 \end{array} \begin{array}{c} 46810 \\ \begin{bmatrix} 1 & 0 & 0 & 1 \\ 0 & 1 & 1 & 0 \\ 0 & 0 & 0 & 1 \\ 0 & 0 & 0 & 0 \end{bmatrix} \end{array}$$

FIGURE 1.24 The digraph of the relation R

The value of $n = |A| = 4$. Thus, we have to find the Warshall's sets w_0, w_1, w_2, w_3, and w_4. The first set w_0 is identical to M_R and is given below.

$$w_0 = M_R = \begin{bmatrix} 1 & 0 & 0 & 1 \\ 0 & 1 & 1 & 0 \\ 0 & 0 & 0 & 1 \\ 0 & 0 & 0 & 0 \end{bmatrix}$$

Now, to find w_1 from w_0 we have row number 1 for column 1 in w_0 and column numbers 1 and 4 for row 1 in w_0. Thus, new entries in w_1 are (4, 4) and (4, 10), which are already 1. Thus, w_1 is same as w_0, which is

$$w_1 = \begin{bmatrix} 1 & 0 & 0 & 1 \\ 0 & 1 & 1 & 0 \\ 0 & 0 & 0 & 1 \\ 0 & 0 & 0 & 0 \end{bmatrix}$$

Next, to find w_2 from w_1, we have row numbers 2 for column 2 in w_1 and column numbers 2 and 3 for row 2 in w_1. Thus, the new entries in w_2 are (6, 6) and (6, 8) which are already 1. Thus, w_2 is same as w_1, which is

$$w_2 = \begin{bmatrix} 1 & 0 & 0 & 1 \\ 0 & 1 & 1 & 0 \\ 0 & 0 & 0 & 1 \\ 0 & 0 & 0 & 0 \end{bmatrix}$$

Similarly, w_3 is obtained from w_2. Here, we have row number 2 for column 3 and column number 4 in row 3. Thus, the new entries in w_2 are (6, 10). So, w_3 is

$$w_3 = \begin{bmatrix} 1 & 0 & 0 & 1 \\ 0 & 1 & 1 & 1 \\ 0 & 0 & 0 & 1 \\ 0 & 0 & 0 & 0 \end{bmatrix}$$

Similarly, w_4 is obtained from w_3. But there are no entries of 1s in w_4. Hence, $M_{R^\infty} = w_4 = w_3$, which is the transitive closure of R.

1.20 PARTIAL ORDERING RELATION

A relation R on a set S is called a *partial ordering* or a *partial order* if R is reflexive, antisymmetric, and transitive. A set S with partial order R on it is called *partially ordered set* or *n-ordered set* or a *poset*. We write (S, R) to specify the partial-order relation R. Usually, we denote a partial-order relation by symbol '$\leq$' (not necessarily 'less than or equal to'). Partially ordered sets will be discussed in more detail in Chapter 6. Here, simply we give some examples.

For example

(i) Let S be a non-empty set and $P(S)$ denote the power set of S. Then, the relation set inclusion denoted by '$\leq$' in $P(S)$ is a partial ordering.
(ii) Let $R = \{(1, 2), (1, 3), (2, 1), (2, 2), (2, 4), (3, 3)\}$. The relation '$a$ divides b' is a partial ordering on the set Z^+ of positive integers. However, 'a divides b' is not a partial ordering on the set Z of integers, since $a|b$ and $b|a$ do not imply $a = b$, since $2|-2$ and $-2|2$ but $2 \neq -2$.

1.21 n-ARY RELATIONS AND THEIR APPLICATIONS

The relationships among elements from more than two sets are called *n-ary relations*. Clearly, the basic definition of *n*-ary relations can be presented as a relation R is called an *n*-ary relation if it is composed of a set of ordered *n*-tuples, i.e., if $R \subseteq A_1 \times A_2 \times \cdots \times A_n$, where $A_1, A_2, \ldots, A_n$ *n*-tuples are sets. These sets are called as *domains* of the relation R and n is called its *degree*. If $n = 2$, then R is a *binary relation*. If $n = 3$, then R is a *ternary relation* and $n = 4$ yields a *quaternary relation*, and so on.

The relations are used to represent computer databases and the representations, on the other hand, help us to answer queries regarding the information stored in databases, such as: Which flights land at Netaji International Airport, Kolkata, between 3 p.m. and 4 p.m.? Which employees of a company have joined before 23.01.09?

For example Table 1.2 represents a 4-ary relation. This table expresses the relationships among roll numbers, names, branch, and grade point average (GPA).

TABLE 1.2 Student

Roll number	Name	Branch	GPA
1241	Sayantan	Civil	8.58
1425	Vikas	Mechanical	9.01
1351	Anita	Electrical	7.51
1561	Vijay	Mining	6.50

The *n*-ary relation can also be expressed as a collection of *n*-tuples.

The above table can be expressed in the following set notation: {(1241, Sayantan, Civil, 8.58), (1425, Vikas, Mechanical, 9.01), (1315, Anita, Electrical, 7.51), (1561, Vijay, Mining, 6.50)} of 4-tuples.

For example

(i) Let R be a relation on $N \times N \times N$ consisting of triples (a, b, c), where a, b, and c are integers with $a < b < c$. Then $(1, 2, 3) \in R$, but $(2, 4, 3) \notin R$. The degree of this relation is 3 and its domains are the set of natural numbers.

(ii) Let R be the relation on $Z \times Z \times Z$ consisting of all triples of integers (a, b, c) in which a, b, and c form an arithmetic progression; i.e. $(a, b, c) \in R$ if and only if there exists an integer k such that $b - a = k$ and $c - b = k$. It may be noted here that $(1, 3, 5) \in R$ because $3 - 1 = 2$ and $5 - 3 = 2$, but $(1, 3, 6) \notin R$, $3 - 1 = 2$ while $6 - 3 = 3$. The degree of this relation is 3 and its domains are the set of integers.

(iii) Assume that the equation of a sphere S with radius 1, centre at the origin $(0, 0, 0)$ as $x^2 + y^2 + z^2 = 1$. This equation determines a ternary relation T on the set R of real numbers; i.e., a triple $(x, y, z) \in T$ if (x, y, z) satisfies the equation, which means (x, y, z) is the coordinates of a point in R^3 on the sphere S.

1.22 RELATIONAL MODEL FOR DATABASES

The operations of adding and deleting records, updating records, and searching for records are performed millions of times each day in a *database* (a collection of records or data usually stored in a computer). Because of the importance of these operations, various options for representing databases have been developed. We will discuss one of these methods, called *relational model for databases*, or *relational data model*, based on the concept of a relation.

A database, having records of *n*-tuples, is made up of fields. The fields are the entries of the *n*-tuples. For instance, a database of student records may be made up of fields containing the name, roll number, branch, and GPA of the student. The relational data model represents a database of records as an *n*-ary relation. Thus, student records are represented as 4-tuples of the form (student's roll number, name, branch, and GPA). A sample database of four such records is already shown in Table 1.2.

Each column of Table 1.2 corresponds to an *attribute* of the database. For instance, the attributes of this database are student's roll number, name, branch, and GPA.

A single attribute or a combination of attributes for a relation is a *key* if the values of the attributes uniquely define an *n*-tuple. For instance, in the above table, we can take the attribute 'roll number' as a key (it is supposed to be that each student possesses a unique roll number). The attribute 'name' is not a key because different students can have the same name. For the same reason, we cannot take the attribute 'branch' or 'GPA' as a key. However, name and branch, in combination, and GPA solely could be used as a key for the above table, since in our example a student is uniquely defined by a name and a branch.

A domain of an *n*-ary relation is called a *primary key* when the value of the *n*-tuple from this domain determines the *n*-tuple. That is, a domain is a primary key if there exists no two *n*-tuples in the relation which have the same value from this domain.

The current collection of *n*-tuples in a relation is called the *extension* of the relation. The more salient part of a database, including the name and attributes of the database, is called its *intension*. When selecting a primary key, the goal should be to choose that key which can serve as a primary key for all possible extensions of the database. To do the same, it is necessary to examine the intension of the database for understanding the set of possible *n*-tuples that can occur in an extension.

A *database management system* is a program that helps users to access the information in databases. This system responds to *queries*. A query is a request for information from the database, for instance 'find the students who acquire GPA above ...' is a meaningful query for the relation given by the above table.

There is variety of operations on *n*-ary relations. If the operations are implemented together, then these operations can reply user queries. Some of such operations are 'SELECT', 'PROJECT', and 'JOIN'.

EXAMPLE 1.46 Let R be the relation consisting of 5-tuples (A, N, S, D, T) representing airplane flights, where A is the airline, N is the flight number, S is the starting point, D is the destination, and T is the departure time. If Air India, number 961, has flight from Netaji Airport, Kolkata, to Indira Gandhi Airport, New Delhi, at 6:15 a.m., then Air India, number 961, Netaji Airport, Indira Gandhi Airport, 6:15 a.m. belong to R. Find the degree of R and its domains.

Solution The degree of the relation is 5, since there are 5-tuples. Here, the domains are the set of all airlines, the set of flight numbers, the set of cities, the set of cities (repeated), and the set of times.

1.23 INTRODUCTION TO FUNCTIONS

The concept of a function is extremely important in discrete mathematics. Functions are used in defining sequences and strings concretely. Functions also express the time duration taken by a computer to solve problems of a given size. Functions defined in terms of themselves are the recursive functions, which are needed in computer science.

The present section reviews the basic concepts involving functions needed in discrete mathematics.

Definition Let A and B be any two sets. A *function* f from A to B is defined if for every element $a \in A$, there exists a unique element $b \in B$, such that $f(a) = b$, or $(a, b) \in f$. A function f from A to B is also defined having the following properties:
 (i) Domain of $f = A$
 (ii) If $(a, b) \in f$ and $(a, c) \in f$, then $b = c$

A function from A to B is denoted by $f: A \rightarrow B$. Functions are also called *mappings* or *transformations*.

Given a function f from A to B, according to the above definition, for each element a of the domain A, there exists exactly one $b \in B$ with $(a, b) \in f$. The unique value b is denoted by $f(a)$. In other words, $b = f(a)$ is another approach of expressing the function f, with $(a, b) \in f$. In general, the function f can be represented as $y = f(x)$, $(x, y) \in f$.

Definition If f is a function from A to B, then it can be said that A is the *domain* of f and B is the *codomain* of f. If $f(a) = b$, then b is the *image* of a and a is a pre-image of b. The *range* of f is the set of all images of elements of A. Also, if f is a function from A to B, one can express that f maps A to B, i.e., $f : A \to B$, mathematically.

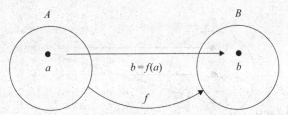

FIGURE 1.25 The function f maps A to B

Figure 1.25 represents a function f from A to B. Notice that the domain of a function is uniquely defined because the function must do something to every element of the domain. However, the codomain is not uniquely defined. If B is a codomain of a function f, then so is any other set C for which $B \subseteq C$. This asymmetry is the formal expression of the fact that, in mathematics, we must always be prepared to specify precisely what questions we are asking, but we may not know in advance precisely what answers we expect to get! If we can specify precisely what answers we will get, then we can define the codomain of our function to be its range. The schematic diagram is given in Figure 1.26.

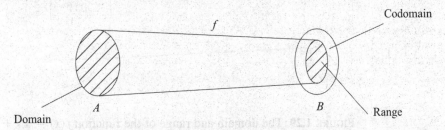

FIGURE 1.26 The domain, codomain, and range of a function f

For example

(i) Let $A = \{1, 2, 3\}, B = \{a, b, c\}$. Also, $f = \{(1, a), (2, b), (3, c)\}$. Then, $f(1) = a$, $f(2) = b$, and $f(3) = c$. Thus, f is a function from A to B (Figure 1.27). Here, the domain of f is A and the range of f is B. The situation is depicted in Figure 1.27, where an arrow from 1 to a means that 1 is related to a. This picture can be named as an *arrow diagram*. For an arrow diagram to be a function, it is to be remembered that there is exactly one arrow from each element in the domain and Figure 1.27 possesses this property.

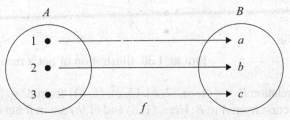

FIGURE 1.27 The arrow diagram of a function f

(ii) Let $A = \{1, 2, 3\}, B = \{a, b, c\}$, and $f = \{(1, a), (2, b), (3, a)\}$. Then f is a function from A to B (Figure 1.28). The domain of f is A and the range of f is $\{a, b\}$.

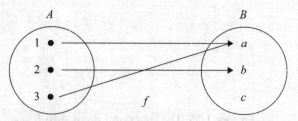

FIGURE 1.28 The domain and range of the function
$f = \{(1, a), (2, b), (3, a)\}$

(iii) Let $f: Z \rightarrow N$, where $f(x) = x^2 + 2$. Then the domain of the function is Z, the codomain is N, and the range is $\{2, 3, 6, 11, 18, \ldots\}$ (Figure 1.29).

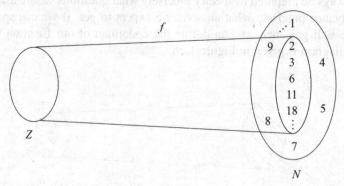

FIGURE 1.29 The domain and range of the function $f(x) = x^2 + 2$

(iv) Let $A = \{1, 2, 3, 4\}, B = \{a, b, c\}$, and $f = \{(1, a), (2, a), (3, b)\}$. Here, f is not a function from A to B, since the domain of f, $\{1, 2, 3\}$, is not equal to A. Also, it may be apparent from the arrow diagram (Figure 1.30) that there exists no arrow from 4. So, f is not a function. However, f will be a function from A' to B if one considers $A' = \{1, 2, 3\}$ and $B = \{a, b, c\}$. Moreover, here, f is the relation R.

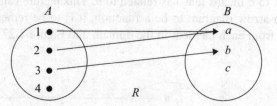

FIGURE 1.30 Illustration of not a function but a relation

(v) The relation $R = \{(1, a), (2, b), (3, c), (1, b)\}$ from $A = \{1, 2, 3\}$ to $B = \{a, b, c\}$ is not a function from A to B. Here, $(1, a)$ and $(1, b)$ are in R but $a \neq b$. Also, it is apparent from the arrow diagram (Figure 1.31) that the given relation is not a function as there are two arrows from 1.

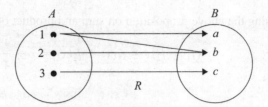

FIGURE 1.31 Illustration of not a function but a relation

(vi) Assume f as the function that assigns the last two bits of a bit string of length 2 or greater to that string, i.e., $f(11010) = 10$. Then the domain of f is the set of all bit strings of length 2 or greater, and both the codomain and range are the set $\{00, 01, 10, 11\}$.

There exist fruitful utility of functions in programming languages; one of its illustrations is given below.

For example The domain and codomain of functions are used in programming languages and are presented below:

The *Java* statement

   Int *floor*(float real){...}

and the *Pascal* statement

   function floor $(x : real)$: *integer*

both demonstrate that the domain of the floor function (discussed in Section 1.31.1) is the set of real numbers and its codomain is the set of integers.

For example Let C be a computer program that accepts an integer as input and produces an integer as output. Let $A = B = Z$. Then C forms a relation f_c which is defined as follows.

It is observed that f_c is a function, since any particular input corresponds to a unique output.

The above illustration can be generalized to a program with any set A of possible inputs and set B of corresponding outputs. In general, therefore, we may think of functions as *input–output* relations.

1.24 ADDITION AND MULTIPLICATION OF FUNCTIONS

Two real-valued functions with the same domain can be added and multiplied.

Let f_1 and f_2 be functions from A to R. Then, $f_1 + f_2$ and $f_1 f_2$ are also functions from A to R and are defined by

$$(f_1 + f_2)x = f_1(x) + f_2(x)$$
$$(f_1 f_2)x = f_1(x) f_2(x)$$

EXAMPLE 1.47 Given that f_1 and f_2 are functions from R to R, in which $f_1(x) = x$ and $f_2(x) = (1/x) - x$. Determine the functions $f_1 + f_2$ and $f_1 f_2$.

Solution Using the above proposition on sum and product of functions,

$$(f_1 + f_2)x = f_1(x) + f_2(x) = x + \frac{1}{x} - x = \frac{1}{x}$$

and

$$(f_1 f_2)x = f_1(x) f_2(x) = x\left(\frac{1}{x} - x\right) = 1 - x^2$$

1.25 CLASSIFICATION OF FUNCTIONS

In this section we discuss several types of functions. These functions possess a number of important applications appearing in later sections of the text.

1.25.1 One-to-One (Injective) Function

A function f is said to be *one-to-one*, or *injective* from A to B, if for each $b \in B$, there exists at most one $a \in A$ with $f(x) = y$.

The condition for a function to be one-to-one is that if $a_1, a_2 \in A$ and $f(a_1) = f(a_2)$, then $a_1 = a_2$.

EXAMPLE 1.48 Investigate the function from $\{a, b, c, d\}$ to $\{1, 2, 3, 4, 5\}$ with $f(a) = 4$, $f(b) = 5, f(c) = 1$, and $f(d) = 3$ for one-to-one.

Solution The function f is one-to-one, since f assigns different values at the four elements of its domain, which is shown in Figure 1.32.

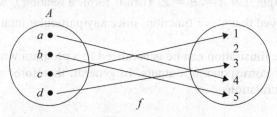

FIGURE 1.32 A one-to-one function

EXAMPLE 1.49 Examine the function $f(x) = x^2$ from the set of integers to the set of integers for one-to-one.

Solution The function $f(x) = x^2$ is not one-to-one, because, for instance, $f(1) = f(-1) = 1$, but $1 \neq -1$. However, the function f is one-to-one if its domain is limited to Z^+.

EXAMPLE 1.50 Test the function $f(x) = x + 1$ for one-to-one.

Solution The function $f(x) = x + 1$ is a one-to-one function, since $x + 1 \neq y + 1 \Rightarrow x \neq y$.

1.25.2 Onto (Surjective) Function

A function from A to B is called *onto*, or *surjective*, if and only if for every element $b \in B$, there is an element $a \in A$ with $f(a) = b$.

EXAMPLE 1.51 Let f be a function from $\{a, b, c, d\}$ to $\{1, 2, 3\}$ defined by $f(a) = 3, f(b) = 2$, $f(c) = 1$, and $f(d) = 3$. Examine the function f for an onto function.

Solution Here, all the three elements of the codomain are images of elements in the domain (Figure 1.33). So, the function f is onto. It may be noted here that if the codomain

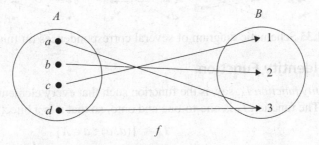

FIGURE 1.33 An onto function

were $\{1, 2, 3, 4\}$, then f would not be onto.

EXAMPLE 1.52 Is the function $f(x) = x + 1$ from the set of integers to the set of integers onto?

Solution The given function is onto, since for every integer y there exists an integer x such that $f(x) = y$. For instance, since $f(x) = y$, then $x + 1 = y$, which is true iff $x = y - 1$.

1.25.3 One-to-One and Onto (Bijective)

The function f is defined as *bijective* if it is both one-to-one and onto.

EXAMPLE 1.53 Let f be the function from $\{a, b, c, d\}$ to $\{1, 2, 3, 4\}$ with $f(a) = 4, f(b) = 2$, $f(c) = 1$, and $f(d) = 3$. Investigate the function f to be bijective.

Solution The given function f is one-to-one and onto. It is one-to-one because the function takes on distinct values. It is also onto since all four elements of codomain are images of elements in the domain. Hence, f is a bijective function.

The above example on bijective function is illustrated in Figure 1.34.

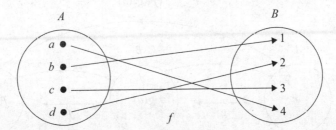

FIGURE 1.34 The bijective function

Figure 1.35 displays four functions, in which the first is one-to-one but not onto, the second is onto but not one-to-one, the third is both one-to-one and onto, and the fourth is neither one-to-one nor onto. The fifth correspondence in the figure is not a function, since it sends an element to two different elements.

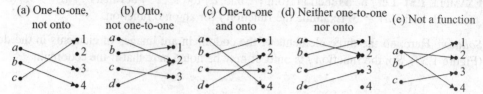

FIGURE 1.35 Schematic diagram of several correspondences on functions

1.25.4 Identity Function

The *identity function* I_A, say, is the function such that every element of A assigns each element to itself. The function I_A is one-to-one and onto, so that it is a bijective function. Symbolically,

$$I_A = \{(a, a) : a \in A\}$$

In other way, the function $f: A \to A$ defined by $f(a) = a$, $\forall a \in A$ is called an *identity function* for A.

1.25.5 Constant Function

Let f be a function from A to B. The function f is said to be a *constant function* if every element of A is assigned to the same element of B. In other words, if the range of f possesses only one element, then f is called a *constant function*.

For example A function $f(x) = 5$, $\forall x \in R$ is a constant function, and R_f (i.e., range of f) = $\{5\}$.

1.26 COMPOSITION OF FUNCTION

Let g be a function from the set A to the set B and let f be a function from the set B to the set C. The *composition* of the functions f and g, denoted by $f \circ g$, is defined by

$$(f \circ g)(a) = f(g(a))$$

Alternatively, $f \circ g$ is the function that assigns to the element a of A and the element is assigned by f to $g(a)$. It may be noted here that the composition $f \circ g$ cannot be defined unless the range of g is a subset of the domain of f. The composition of functions is illustrated in Figure 1.36.

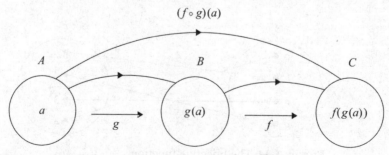

FIGURE 1.36 The composition of functions f and g

Note

(i) The composition of function is not commutative, i.e., $f \circ g \neq g \circ f$, where f and g are functions.
(ii) The composition $g \circ f$ is called the left composition g with f.

EXAMPLE 1.54 Let g be the function from the set $\{a, b, c\}$ to itself such that $g(a) = b$, $g(b) = c$, and $g(c) = a$. Let f be the function from the set $\{a, b, c\}$ to the set $\{1, 2, 3\}$ such that $f(a) = 3$, $f(b) = 2$, and $f(c) = 1$. Determine the composition of f and g and also the composition of g and f.

Solution The composition $f \circ g$ is defined by

$$(f \circ g)(a) = f(g(a)) = f(b) = 2$$

Again,

$$(f \circ g)(b) = f(g(b)) = f(c) = 1$$

Moreover,

$$(f \circ g)(c) = f(g(c)) = f(a) = 3$$

Furthermore, $g \circ f$ is not defined, since the range of f is not a subset of the domain of g.

EXAMPLE 1.55 Let f and g be the functions from the set of integers defined by $f(x) = 2x + 3$ and $g(x) = 3x + 2$. Determine the compositions of f and g, and of g and f.

Solution Here, the compositions $f \circ g$ and $g \circ f$ are defined. Then,

$$(f \circ g)(x) = f(g(x)) = f(3x + 2) = 2(3x + 2) + 3 = 6x + 7$$

and

$$(g \circ f)(x) = g(f(x)) = g(2x + 3) = 3(2x + 3) + 2 = 6x + 11$$

Note From the above example, it is seen that even though $f \circ g$ and $g \circ f$ are defined for the functions f and g, $f \circ g$ and $g \circ f$ are not equal. It implies that the commutative law does not hold for the composition of functions.

EXAMPLE 1.56 Let $f : R \to R$ and $g : R \to R$ be defined by $f(x) = x + 1$ and $g(x) = 2x^2 + 3$. Find $f \circ g$ and $g \circ f$. Is $f \circ g = g \circ f$?

Solution Here,

$$(g \circ f)(x) = g(f(x)) = g(x + 1) = 2(x + 1)^2 + 3$$

and

$$(f \circ g)(x) = f(g(x)) = f(2x^2 + 3) = 2x^2 + 3 + 1 = 2x^2 + 4$$

Also, $g \circ f$ and $f \circ g$ both are defined but $g \circ f \neq f \circ g$.

1.26.1 Associativity of Composition of Functions

Theorem 1.3 Let $f: A \to B$ and $g: B \to C$ be two functions.

I. If both f and g are injective then $g \circ f$ is injective.
II. If both f and g are surjective then $g \circ f$ is surjective.

Proof

I. Let x and y be two distinct elements in A. Since f is injective, $f(x) \neq f(y)$. Also, since g is injective then

$$g(f(x)) \neq g(f(y)) \Rightarrow (g \circ f)(x) \neq (g \circ f)(y)$$

Thus, $g \circ f$ is injective.

II. Let z be any element in C. Then find x in A such that $(g \circ f)(x) = z$.
Since g is surjective, there exists an element y in B such that $g(y) = z$. Again, since f is surjective, there is an element x in A such that $f(x) = y$.
Now, $(g \circ f)(x) = g(f(x)) = g(y) = z \Rightarrow g \circ f$ is surjective. Similarly, it can be shown that if both $f: A \to B$ and $g: B \to C$ are bijective, then $g \circ f$ is bijective.

Corollary 1.1 The converse of Part I of the theorem is not true, i.e., if $g \circ f$ is injective, then it is not necessary that f and g are individually injective. For instance, the functions f, g, and $g \circ f$ are depicted in Figure 1.37.

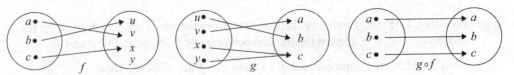

FIGURE 1.37 The functions f, g, and $g \circ f$

From Figure 1.37, we observe that although the composition $g \circ f$ is injective, yet g is not.

EXAMPLE 1.57 If $f: A \to B$, $g: B \to C$ and $h: C \to D$, then show that
$h \circ (g \circ f) = (h \circ g) \circ f$.

Solution Since $f: A \to B$, $g: B \to C$, and $h: C \to D$, then $g \circ f: A \to C$ and $h \circ g: B \to D$.
Hence, $h \circ (g \circ f): A \to D$ and $(h \circ g) \circ f: A \to D$.
Let $x \in A$, $y \in B$, and $z \in C$ such that $f(x) = y$ and $g(y) = z$.
Then,

$$[(h \circ g) \circ f](x) = (h \circ g)[f(x)] = (h \circ g)(y) = h[g(y)] = h(z)$$

and

$$[h \circ (g \circ f)](x) = [h \circ (g \circ f)(x)] = h[(g \circ f)(x)] = h[g(f(x))] = h[g(y)] = h(z)$$

Thus,

$$(h \circ g) \circ f = h \circ (g \circ f), \forall x \in A$$

Theorem 1.4 The composition of any function with the identity function is the function itself.

Proof Assume $x \in A$ and $y \in B$, so that $y = f(x)$. If I_A be the identity function and $I_A: A \to A$, then
$$I_A(x) = x \; \forall x \in A$$
By definition, $f \circ I_A: A \to B$, then
$$f \circ I_A(x) = f[I_A(x)] = f(x) \Rightarrow f \circ I_A(x) = f(x)$$
Similarly, it can be shown that $I_B \circ f(x) = f(x)$ (left to the reader). Thus,
$$f \circ I_A = I_B \circ f = f$$

1.27 INVERSE FUNCTION

Consider a one-to-one correspondence f from the set A to the set B. Since f is an onto function, every element of B is the image of some element in A. Furthermore, as f is also a one-to-one function, every element of B is the image of a unique element of A. Finally, a new function from B to A will be defined from B to A that reverses the correspondence given by f.

Let f be a one-to-one correspondence from the set A to the set B. The *inverse function* of f is defined as the function that assigns to an element $b \in B$, the unique element $a \in A$ such that $f(a) = b$. The inverse function of f is denoted by f^{-1}. Hence, $f^{-1}(b) = a$ whenever $f(a) = b$. The concept of an inverse function is depicted in Figure 1.38.

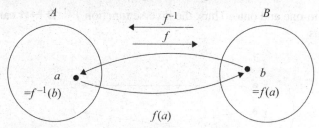

FIGURE 1.38 The inverse function

If a function is not a one-to-one correspondence, one cannot define an inverse function of f. Also, if f is not a one-to-one correspondence, it means that either it may not be one-to-one or it may not be onto. Again, if f is not one-to-one, some element b in the codomain is the image of more than one element in the domain. If f is not onto for some element b in the codomain, no element a in the domain exists for which $f(a) = b$. Subsequently, if f is not a one-to-one correspondence, one cannot assign to each element b in the codomain and a unique element a in the domain such that $f(a) = b$, since for some b there exists either more than one such a or no such a.

Theorem 1.5 Suppose $f: A \to B$ is a bijective function and f^{-1} is its inverse. For each $x \in B$
$$f \circ f^{-1}(x) = x$$
and for each $x \in A$
$$f^{-1} \circ f(x) = x$$
i.e.,
$$f \circ f^{-1} = I_B \quad \text{and} \quad f^{-1} \circ f = I_A$$

Proof Let x be an element of B and let $f^{-1}(x) = z$. From the definition of an inverse function, $f(z) = x$. Thus,

$$f \circ f^{-1}(x) = f(f^{-1}(x)) = f(z) = x$$

Again,

if $x \in A$ and $f(x) = z$, then $f^{-1}(z) = x$

Hence,

$$f^{-1} \circ f(x) = f^{-1}(f(x)) = f^{-1}(z) = x$$

For example Let the function $f: A \rightarrow B$ be defined, as shown in Figure 1.39.

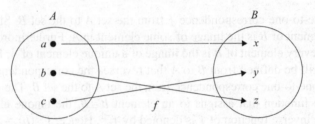

FIGURE 1.39 $f: A \rightarrow B$

Then f is one-to-one and onto. Thus, the inverse function $f^{-1}: B \rightarrow A$ can be shown pictorially (Figure 1.40) as

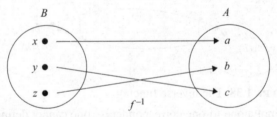

FIGURE 1.40 $f^{-1}: B \rightarrow A$

EXAMPLE 1.58 Let $A = \{a, b, c\}$, $B = \{1, 2, 3\}$, and $f = \{(a, 1), (b, 3), (c, 2)\}$. Determine the inverse.

Solution Here, f is both one-to-one and onto. So,

$$f^{-1} = \{(1, a), (2, c), (3, b)\}$$

is a function from B to A.

EXAMPLE 1.59 Show that the function $f(x) = x^3$ and $g(x) = x^{1/3}$ for all $x \in R$ are inverse of each other.

Solution Here,

$$(f \circ g)(x) = f(x^{1/3}) = (x^{1/3})^3 = x \Rightarrow f = g^{-1}$$

Also,

$$(g \circ f)(x) = g(x^3) = (x^3)^{1/3} = x \Rightarrow g = f^{-1}$$

Thus, $f = g^{-1}$ and $g = f^{-1}$, i.e., the functions f and g are inverses of each other.

EXAMPLE 1.60 Let $f: R \to R$ be defined by $f(x) = 3x - 4$. Find a formula for f^{-1}.

Solution Here, f is one-to-one and onto. So, f has an inverse function f^{-1}.
Let y be the image of x under the function $f: y = f(x) = 3x - 4$. Consequently, x will be the image of y under the inverse function f^{-1}. Solve for x in terms of y from the above equation,

$$x = (y + 4)/3$$

Then,

$$f^{-1}(y) = (y + 4)/3$$

Now, replace y by x to obtain

$$f^{-1}(x) = \frac{y + 4}{3}$$

which is the formula for f^{-1}.

EXAMPLE 1.61 Let $X = \{a, b, c\}$. Define $f: X \to X$ such that $f = \{(a, b), (b, a), (c, c)\}$. Determine
  (i) f^{-1} (ii) f^2 (iii) f^3 (iv) f^4

Solution Given that $f = \{(a, b), (b, a), (c, c)\}$. Here,

$$f(a) = b \quad \text{and} \quad f^{-1}(b) = a$$
$$f(b) = a \quad \text{and} \quad f^{-1}(a) = b$$
$$f(c) = c \quad \text{and} \quad f^{-1}(c) = c$$

(i) $f^{-1} = \{(b, a), (a, b), (c, c)\}$
(ii) $f^2 = f \circ f = \{(a, b), (b, a), (c, c)\} \circ \{(a, b), (b, a), (c, c)\}$
    Now,

$$f^2(a) = (f \circ f)(a) = f(f(a)) = f(b) = a$$
$$f^2(b) = (f \circ f)(b) = f(f(b)) = f(a) = b$$
$$f^2(c) = (f \circ f)(c) = f(f(c)) = f(c) = c$$

Thus,

$$f^2 = \{(a, a), (b, b), (c, c)\}$$

(iii) $f^3 = f \circ f^2$
    So,

$$f^3(a) = (f \circ f^2)(a) = f(f^2(a)) = f(a) = b$$
$$f^3(b) = (f \circ f^2)(b) = f(f^2(b)) = f(b) = a$$
$$f^3(c) = (f \circ f^2)(c) = f(f^2(c)) = f(c) = c$$

Thus,

$$f^3 = \{(a, b), (b, a), (c, c)\}$$

(iv) $f^4 = f \circ f^3$
Then,
$$f^4(a) = (f \circ f^3)(a) = f(f^3(a)) = f(b) = a$$
$$f^4(b) = (f \circ f^3)(b) = f(f^3(b)) = f(a) = b$$
$$f^4(c) = (f \circ f^3)(c) = f(f^3(c)) = f(c) = c$$

Thus,
$$f^4 = \{(a, a), (b, b), (c, c)\}$$

1.27.1 Invertible Function

A function $f: A \to B$ is said to be *invertible* if its inverse f^{-1} is also a function. In other words, one-to-one correspondence is called invertible if one can define the inverse of this function.

EXAMPLE 1.62 Let f be the function from $\{a, b, c\}$ to $\{1, 2, 3\}$ such that $f(a) = 2$, $f(b) = 3$, and $f(c) = 1$. Is f invertible, and if so, find its inverse.

Solution The function f is invertible, since it is a one-to-one correspondence. The required inverse function f^{-1} is then
$$f^{-1}(1) = c, \quad f^{-1}(2) = a, \quad \text{and} \quad f^{-1}(3) = b$$

EXAMPLE 1.63 Let f be the function from the set of integers to the set of integers such that $f(x) = x + 1$. Is f invertible, and if it is, find its inverse.

Solution The function f possesses an inverse, since it is a one-to-one correspondence. Assume that y is the image of x, i.e., $y = x + 1$, which implies that $y - 1$ is the unique element of Z, set of integers, which is sent to y via f. Then, $f^{-1}(y) = y - 1$. Now, replacing y by x, we obtain $f^{-1}(x) = x - 1$, which is the inverse of f, using the usual independent variable x.

1.27.2 Image of a Subset

When f is a function from a set A to a set B, the image of a subset can also be defined.

Let f be a function from the set A to the set B and let S be a subset of A. The *image* of S is the subset of B that consists of the images of the elements of S. Then, the image of S, denoted by $f(S)$, can be expressed as
$$f(S) = \{f(s) : s \in S\}$$

EXAMPLE 1.64 Let $A = \{a, b, c, d, e\}$ and $B = \{1, 2, 3, 4\}$ with $f(a) = 2$, $f(b) = 1$, $f(c) = 4$, $f(d) = 1$, and $f(e) = 1$. Determine the image of the subset $S = \{b, c, d\}$.

Solution The image of the subset $S = \{b, c, d\}$ is given by the set $f(S) = \{1, 4\}$.

1.28 HASH FUNCTION

It is well known that a server in a computer centre at an university supports to keep up records of students. In this respect, a question then arises that how many memory locations are needed to retrieve student records quickly? The solution to this problem is to use a suitably chosen *hashing function*. Actually, records are identified by a *key*. For instance, student records are identified, using the register number/enrolment number of the student as represented by a key. A hashing function, denoted by h, is used to store and retrieve data. The hashing function h assigns memory location $h(k)$ to the record that has k as its key.

The popular hashing function is the function $h(k) = k \bmod n$, where n denotes the number of available memory locations.

Hashing function should be evaluated easily so that the files can be quickly located. This requirement can be met by hashing function. To evaluate $h(k)$, one should only compute the remainder when k is divided by n. Moreover, the hashing function should be onto, so that all memory locations are possible. The function $h(k) = k \bmod n$ also satisfies this property.

For example When $n = 12$, the record of the student with register number 17 is assigned to memory location 5, since $h(17) = 17 \bmod 12 = 5$. Similarly, since $h(5132) = 5132 \bmod 121 = 50$, the record of the student with register number 5132 is assigned to memory location 50.

Since a hashing function is not one-to-one (as there exists more possible keys than memory locations), more than one file may be assigned to a memory location. When this happens, we say that a *collision* occurs. To resolve the collision it is imperative to assign the first free location following the occupied memory location assigned by the hashing function.

For example After making the two earlier assignments, we assign location 51 to the student with register number 2822. To see this, first note that $h(k)$ maps this register number to location 50, because

$$h(2822) = 2822 \bmod 231 = 50$$

But this location is already occupied (by register number 5132 of the student). However, memory location 51, the first location following memory location 50, is free. There are several sophisticated ways to resolve collisions that are more efficient than the method we have described already.

Hashing functions are employed in cryptology as well, in which they are used to produce digital finger prints and other electronic means to verify the authenticity of messages.

1.29 RECURSIVELY DEFINED FUNCTIONS

Sometimes, it is difficult to define a function or a set explicitly. However, it may be easy to define this function or set in terms of itself. This process is called *recursion*. Thus, a function is said to be *recursively defined* if the function refers to itself. We can use recursion to define sequences, functions, and sets.

For example The sequence of powers of 2 is given by $a_n = 2^n$ for $n = 0, 1, 2, \ldots,$. The recursion and its allied are displayed in Section 2.13. Also, the detailed description of the recursive function is presented in Section 2.13.4.

1.30 SOME SPECIAL FUNCTIONS

Two important functions in discrete mathematics, namely, the *floor* and *ceiling functions*, irrespective of some more, will be introduced here. These functions are often used when objects are counted. They play an important role in the analysis of the number of steps used by procedures to solve problems of a particular size. The floor and ceiling functions are useful in a wide variety of applications, including those involving data storage and data transmission.

1.30.1 Floor and Ceiling Functions

The *floor function* or the *floor* of x, denoted by $\lfloor x \rfloor$, is the *greatest integer* less than or equal to x.

The *ceiling function* or the *ceiling* of x, denoted by $\lceil x \rceil$, is the *least integer* greater than or equal to x.

For example

$$\left\lfloor \frac{1}{2} \right\rfloor = 0 \qquad \left\lceil \frac{1}{2} \right\rceil = 1$$

$$\left\lfloor -\frac{1}{2} \right\rfloor = -1 \qquad \left\lceil -\frac{1}{2} \right\rceil = 0$$

$$\lfloor 7.2 \rfloor = 7 \qquad \lceil 8.1 \rceil = 9$$

$$\lfloor -7.6 \rfloor = -8 \qquad \lceil -10.3 \rceil = -10$$

$$\lfloor 7 \rfloor = 7 \qquad \lceil -9 \rceil = -9$$

Actually, the floor of x 'rounds x down' while the ceiling of x 'rounds x up'. Figure 1.41 shows the graphs of the floor and ceiling functions. A closed bracket, i.e., '[ ]' specifies the point to be included in the graph; an open bracket, i.e., '( )' indicates the point to be excluded from the graph.

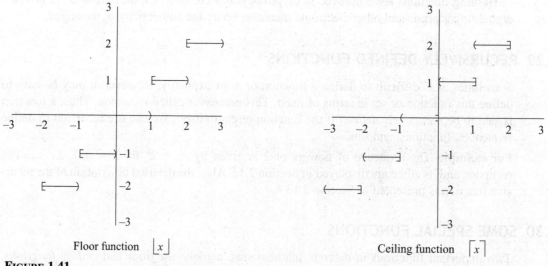

Floor function $\lfloor x \rfloor$ Ceiling function $\lceil x \rceil$

FIGURE 1.41

Mathematical representations of floor and ceiling functions:
1. $\lfloor x \rfloor = n$ iff $n \leq x < n + 1$ (n is an integer)
2. $\lfloor x \rfloor = n$ iff $x - 1 < n \leq x$
3. $\lceil x \rceil = n$ iff $n - 1 < x \leq n$
4. $\lceil x \rceil = n$ iff $x \leq n < x + 1$
5. (a) $\lfloor -x \rfloor = -\lceil x \rceil$ (b) $\lceil -x \rceil = -\lfloor x \rfloor$

The above properties have been framed, using the definitions of floor and ceiling functions.

Another approach for defining the floor function is that to assume $x = n + \varepsilon$, where $n = \lfloor x \rfloor$ is an integer and ε, the fractional part of x, satisfies the inequality $0 \leq \varepsilon < 1$. Similarly, to define the ceiling function, suppose $x = n - \varepsilon$, where $n = \lceil x \rceil$ is an integer and $0 \leq \varepsilon < 1$.

EXAMPLE 1.65 If x is a real number, then show that

$$\lfloor 2x \rfloor = \lfloor x \rfloor + \left\lfloor x + \frac{1}{2} \right\rfloor$$

Solution Let $x = n + \varepsilon$, where n is a positive integer and $0 \leq \varepsilon < 1$. There may be two cases to be tackled.

Case I. $\varepsilon < \frac{1}{2}$ Case II. $\varepsilon \geq \frac{1}{2}$

Case I. $\varepsilon < 1/2$, i.e., $0 \leq \varepsilon < 1/2$

Here, $2x = 2n + 2\varepsilon$. Also, since $0 \leq 2\varepsilon < 1$, $\lfloor 2x \rfloor = 2n$. Similarly,

$$x + \frac{1}{2} = n + \left(\frac{1}{2} + \varepsilon\right)$$

i.e.,

$$\left\lfloor x + \frac{1}{2} \right\rfloor = n, \text{ as } 0 \leq \frac{1}{2} + \varepsilon < 1$$

Thus,

$$\lfloor 2x \rfloor = 2n \text{ and } \lfloor x \rfloor + \left\lfloor x + \frac{1}{2} \right\rfloor = n + n = 2n$$

Case II. $1/2 \leq \epsilon < 1$
Here, $2x = 2n + 2\epsilon = (2n + 1) + (2\epsilon - 1)$, since $0 \leq (2\epsilon - 1) < 1$
Thus, $\lfloor 2x \rfloor = 2n + 1$, because

$$\left\lfloor x + \frac{1}{2} \right\rfloor = \left\lfloor n + \left(\frac{1}{2} + \epsilon\right) \right\rfloor = \left\lfloor n + 1 + \left(\epsilon - \frac{1}{2}\right) \right\rfloor \text{ and } 0 \leq \epsilon - \frac{1}{2} < 1$$

Thus,
$$\left\lfloor x + \frac{1}{2} \right\rfloor = n + 1$$

Consequently,
$$\lfloor 2x \rfloor = 2n + 1 \quad \text{and} \quad \lfloor x \rfloor + \left\lfloor x + \frac{1}{2} \right\rfloor = n + (n + 1) = 2n + 1$$

Hence,
$$\lfloor 2x \rfloor = \lfloor x \rfloor + \left\lfloor x + \frac{1}{2} \right\rfloor$$

EXAMPLE 1.66 Show that for any real number x, $\lfloor x \rfloor + \lfloor -x \rfloor = -1$, if x is not an integer.

Solution Let $\lfloor x \rfloor = n$. Then by definition, if x is not an integer, $n < x < n + 1$.

Multiplying by -1, $(-n - 1) < -x < -n$. Again, by definition, $\lfloor -x \rfloor = -n - 1$. Hence, $\lfloor x \rfloor + \lfloor -x \rfloor = n + (-n - 1) = -1$.

EXAMPLE 1.67 Data stored on a computer disk or transmitted over a data network are usually represented as a string of bytes. Each byte is made up of 8 bits. How many bytes are required to encode 100 bits of data?

Solution To determine the number of bytes needed, we determine the smallest integer that is at least as large as the quotient when 100 is divided by 8, the number of bits in a byte. Consequently, $\lceil 100/8 \rceil = \lceil 12.5 \rceil = 13$ bytes are required.

1.30.2 Integer and Absolute Value Functions

The *integer value* of x (x be any real number), written as INT(x), converts x into an integer by deleting (truncating) the fractional part of the number. Thus,

$$\text{INT}(3.14) = 3, \text{INT}(\sqrt{5}) = 2, \text{INT}(-6.5) = -6, \text{INT}(8) = 8$$

Note INT(x) = $\lfloor x \rfloor$ or INT(x) = $\lceil x \rceil$, depending on x is positive or negative.

The *absolute value* of the real number x, written as ABS(x) or $|x|$, is defined as ABS(0) = 0, and for $x \neq 0$, ABS(x) = x or ABS(x) = $-x$, depending on x is positive or negative. Thus,

$$|-9| = 9, |5| = 5, |-4.44| = 4.44, |-0.032| = 0.032, |5.55| = 5.55$$

Note $|x| = |-x|, x < 0$, and for $x \geq 0$, $|x| = x$.

1.30.3 Remainder Function

Let m be any integer and let M be an integer. Then,

$$m(\text{mod } M)$$

(read m modulo M) will denote the integer remainder when m is divided by M. Specifically, $m(\bmod M)$ is the unique integer q such that
$$m = Mp + q$$
where $0 \leq q < M$. To find the remainder q, divide m (a positive quantity) by M. Thus,
$$12(\bmod 5) = 2,\ 12(\bmod 3) = 0,\ 25(\bmod 7) = 4,\ 35(\bmod 8) = 3$$
When m is negative, divide $|m|$ by M to obtain a remainder q', then
$$m(\bmod M) = M - q',\ q' \neq 0$$
Thus,
$$-26(\bmod 7) = 7 - 5 = 2,\ -371(\bmod 8) = 8 - 3 = 5,\ -39(\bmod 3) = 0$$

1.31 FUNCTIONS OF COMPUTER SCIENCE

Here, we shall restrict ourselves to only those functions whose arguments and values are natural numbers.

1.31.1 Partial and Total Functions

Let us define N^n to be the set of all n-tuples of elements of N and A be the subset of N^n. Then the function $f: A \to N$ is a *partial function*. On the other hand, a function $f: N^n \to N$ is a *total function*.

For example If $g: N \times N \to N$ is defined by $g(x, y) = x - y$, then g is a partial function, since it is defined only for x, y, where $x, y \in N$. If $f: N \times N \to N$ is defined by $f(x, y) = x + y$, then f is a total function.

If we emphasize that a function $f: A \to N$ has domain A, we say that it is *total*. If f is a partial function from A into N and $a \in A - \text{dom}\, f$, then we say that $f(a)$ is *undefined*.

EXAMPLE 1.68 Consider the following program fragment:

```
read (n);
while n > 1 do
  begin
    If n is even then n: = n div 2
    else n: = 2n + 1;
  end;
write (n);
```

What is the implication of the program?

Solution Assume that a positive integer is input. Then the value of n will be halved in the *while* loop, provided it is even. If n becomes odd before it reaches 1, then the second part of the *while* loop is invoked and n remains odd and increases for ever. This implies that if $f: N \to N$ is the function defined by the program, then

$$f(n) = \begin{cases} 0, & \text{if } n = 0 \\ i, & \text{if } n \text{ is a power of } 2 \\ \text{undefined otherwise} \end{cases}$$

Thus, $f: N \to N$ is a partial function with domain $\{0\} \cup \{2^r : r \in N\}$.

1.31.2 Primitive Recursive Function

A function f is called *primitive recursive* if it can be obtained from a set of *initial functions* by a finite number of operations of recursion. However, initial functions over N consist of the following functions:

Zero function Z defined by $Z(x) = 0$

Successor function S defined by $S(x) = x + 1$

Projection function U_i^n defined by $U_i^n(x_1, \ldots, x_n) = x_i$

Note Since $U_1'(x) = x$ for every x in N, U_1' is simply the identity function. So, U_i^n is also termed as a *generalized identity function*.

EXAMPLE 1.69 Show that the function $f(x) = k$, where k is a constant, is primitive recursive.

Solution Let $k = 0$, then $f(x) = 0 = Z(x)$; otherwise,

$$f(x + 1) = k = f(x) = U_2^2(x, f(x))$$

i.e., the given function is primitive recursive.

EXAMPLE 1.70 Show that the function $f(x, y) = x + y$ is primitive recursive. Using it compute $f(2, 3)$.

Solution Since $x + (y + 1) = (x + y) + 1$, so

$$f(x, y + 1) = f(x, y) + 1 = S(f(x, y))$$

Also, $f(x, 0) = x$. Now, we define $f(x, y)$ as

$$f(x, 0) = x = U_1^1(x) \quad \text{and} \quad f(x, y + 1) = S(U_3^3(x, y), f(x, y))$$

Thus, f is primitive recursive.
Now,

$$f(2, 3) = S(f(2, 2)) = S(S(f(2, 1))) = S(S(S(f(2, 0)))) = S(S(S(2))) \text{ [since, } f(2, 0) = 2\text{]}$$
$$= S(S(3))$$
$$= S(4) = 5$$

A popular example of a recursive function, that is not primitive recursive, is the Ackermann function which will be described in the next section.

1.31.3 Ackermann Function

The Ackermann function, $A(x, y)$, is defined by

$A(0, y) = y + 1$, for all non-negative integers y

$A(x + 1, 0) = A(x, 1)$, for all positive integers x

$A(x + 1, y + 1) = A(x, A(x + 1, y))$, for all positive integers x and y

It may be noted that $A(x, y)$ is well defined, i.e., it is computed for finite values of x and y. Also, the function is not primitive, but recursive. Now, we demonstrate the application of the function in computing its value.

EXAMPLE 1.71 Determine $A(1, 1), A(1, 2)$. Finally, compute $A(2, 2)$.

Solution

$$A(1, 1) = A(0 + 1, 0 + 1) = A(0, A(1, 0)) = A(0, A(0, 1)) = A(0, 2) = 3$$
$$A(1, 2) = A(0 + 1, 1 + 1) = A(0, A(1, 1)) = A(0, 3) = 4 = A(1, A(1, 1)) = A(1, 3)$$
$$ = A(0 + 1, 2 + 1) = A(0, A(1, 2)) = A(0, 4) = 5$$
$$A(2, 2) = A(1 + 1, 1 + 1) = A(1, A(2, 1)) = A(1, 5)$$

Now,

$$A(1, 5) = A(0 + 1, 4 + 1) = A(0, A(1, 4)) = 1 + A(1, 4) = 1 + A(0 + 1, 3 + 1)$$
$$= 1 + A(0, A(1, 3)) = 1 + 1 + A(1, 3)$$
$$= 1 + 1 + 1 + A(1, 2) = 1 + 1 + 1 + 4 = 7$$

Thus,

$$A(2, 2) = A(1, 5) = 7$$

1.32 GROWTH OF FUNCTION

Growth of functions is intended to compare relative sizes of functions that is very useful in the analysis of computer algorithms. In particular, it estimates the number of comparisons used by the linear and binary search algorithms to find an element in a sequence of n elements. The time required to solve a problem depends on the number of operations it performs. Also, the time depends on the hardware and software used to run the program that implements the algorithm.

The growth of a function is described by special notation called *asymptotic notation*, such as big-O, big-omega, and theta notations. These establish estimates which are used in the analysis of algorithms.

State-of-art discussion of growth of function is presented in Chapter 8.

1.33 THE INCLUSION–EXCLUSION PRINCIPLE

When the two tasks can be done simultaneously, then both the sum rule and the product rule (will be discussed in detail in Section 2.2) cannot be used. If we add the number of ways to do each task then the ways to do both the tasks are counted twice. Thus, to correct this double counting and to find the number of ways to do one of the tasks, we add the number of ways in which each task can be done and then subtract the number of ways in which both the tasks can be done. This method of counting is called the principle of *inclusion–exclusion*. Sometimes, it is also called the *subtraction* principle.

In *set-theoretic approach* this counting principle can be stated as follows.

Let A and B be two finite sets. To select an element from A there exists $|A|$ ways and $|B|$ ways to select an element from B. The number of ways to select an element from A, or from B, i.e., the number of ways to select an element from their union, is the sum of the number of ways to select an element from A and the number of ways to select an element from B, minus the number of ways to select an element which is both in A and B. Since there are $|A \cup B|$ ways to select an element in either A or B, and $|A \cap B|$ ways to select an element common to both sets, we have

$$|A \cup B| = |A| + |B| - |A \cap B|$$

Considering three sets A, B, and C, the above principle can be formulated as

$$|A \cup B \cup C| = |A| + |B| + |C| - |A \cap B| - |B \cap C| - |C \cap A| + |A \cap B \cap C|$$

In general, if $A_1, A_2, \ldots, A_n$ are n-finite sets, then

$$|A_1 \cup A_2 \cup \cdots \cup A_n| = \sum_{1 \leq i \leq 1} |A_i| - \sum_{1 \leq i \leq j \leq n} |A_i \cap A_j| + \sum_{1 \leq i \leq j \leq k \leq n} |A_i \cap A_j \cap A_k|$$
$$+ \cdots + (-1)^{n-1} |A_1 \cap A_2 \cap \cdots \cap A_n|$$

The following example demonstrates the process of solving a type of counting problem using the above principle.

EXAMPLE 1.72 How many bit strings of length 8 either start with 1 bit or end with 2 bits 00?

Solution Let A be the set containing bit strings of length 8 beginning with 1 bit. We can create a bit string of length 8 that begins with 1 bit, in $2^7 = 128$ ways, i.e., $|A| = 128$. This is possible by the product rule because the first bit can be selected in only one way and each of the other 7 bits can be selected in two ways. Similarly, we can construct a bit string of length 8 ending with 2 bits 00, in $2^6 = 64$ ways, i.e., $|B| = 64$. This is also possible by the product rule, since each of the first 6 bits can be selected in two ways and the last 2 bits can be selected in one way.

Again, the ways to construct a bit string of length 8 starting with 1 bit are the same as the ways to construct a bit string that ends with 2 bits 00. There are $2^5 = 32$ ways to construct such a string, i.e., $|A \cap B| = 32$. This follows by the product rule, because the first bit can be selected in only one way, each of the second through the sixth bits can be chosen in two ways, and the last 2 bits can be selected in one way.

Hence, the number of bit strings of length 8 that begin with 1 bit or end with 2 bits 00, which equals the number of ways to construct a bit string of length 8 that begin with 1 bit or that ends with 2 bits 00, equals

$$|A \cup B| = |A| + |B| - |A \cap B|$$
$$= 128 + 64 - 32 = 160$$

EXAMPLE 1.73 How many positive integers not exceeding 100 are divisible either by 4 or by 6?

Solution The integers which are divisible by 4 are

4, 8, 12, 16, 20, 24, 28, 32, 36, 40, 44, 48, 52, 56, 60, 64, 68, 72, 76, 80, 84, 88, 92, 96, 100

i.e., there are 25 integers not exceeding 100 which are divisible by 4. Also, the integers divisible by 6 are

6, 12, 18, 24, 30, 36, 42, 48, 54, 60, 66, 72, 78, 84, 90, 96

i.e., there exist 16 integers, which are divisible by 6.

Let A be the set possessing the integers which are divisible by 4 and B be the set which contains the integers, divisible by 6, then

$$n(A) = 25, \quad n(B) = 16$$

Thus, the number of integers which are divisible by 4 or 6 are

$$|A \cup B| = |A| + |B| - |A \cap B|$$
$$= 25 + 16 - 8 \qquad \text{[common in both } A \text{ and } B\text{]}$$
$$= 33$$

EXAMPLE 1.74 In a class of discrete mathematics every student possesses a major subject in computer science or mathematics, or both. The number of students having mathematics as a major one (possibly along with mathematics) is 25; the number of students having mathematics as a major one (possibly along with computer science) is 13; and the number of students having both computer science and mathematics as major is 8.

Solution Let A be the set of students in the class majoring in computer science and B be the set of students in the class majoring in mathematics. Then, $A \cap B$ is the set of students in the class who are having both mathematics and computer science as majors. Since every student in the class is majoring in either computer science or mathematics (or both), it follows that number of students in the class is $|A \cup B|$. Thus,

$$|A \cup B| = |A| + |B| - |A \cap B| = 25 + 13 - 8 = 30$$

Hence, there are 30 students in the class. The illustration is shown in Figure 1.42.

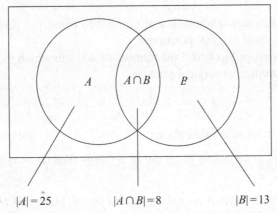

FIGURE 1.42 The set of students in a discrete mathematics class

 EXAMPLE 1.75 A computer company receives 40 applications for a job of programmers. Among them 25 knew JAVA, 28 knew ORACLE, and 7 did not know any of the languages. How many of them knew both the languages?

Solution Let J be the set of programmers who knew JAVA and O be the set of programmers who knew ORACLE. Then $J \cup O$ is the set of programmers who knew JAVA or ORACLE (or both) and $J \cap O$ is the set of programmers who knew both the languages. By the principle of inclusion–exclusion, number of programmers knowing both the languages are

$$|J \cap O| = |J| + |O| - |JUO| = 25 + 28 - 33 \quad (\because |J \cup O| = 40 - 7 = 33)$$
$$= 20$$

1.33.1 Applications of Inclusion–Exclusion Principle

Variety of counting problems can be tackled with the use of inclusion–exclusion principle. This principle can be used in *counting* the number of *onto functions* from one finite set to another. This also counts the permutations of objects that leave no object in its original position, which can be termed as *derangements*.

Counting the Number of Onto Functions

The principle of inclusion–exclusion evaluates number of onto functions from a set with m elements to a set with n elements. This evaluation is based on a prepositional form as given below.

Let m and n be positive integers with $m \geq n$. Then there are

$$n^m - {}^nC_1(n-1)^m + {}^nC_2(n-2)^m - \cdots + (-1)^{n-1}\,{}^nC_{n-1}\,m$$

onto functions from a set with m elements to a set with n elements. The proof is left to the reader.

An onto function from a set with m elements to a set with n elements corresponds to a path of distribution of the m elements in the domain to n indistinguishable boxes so that no box is empty, and then to associate each of the n elements of the codomain to a box. This implies that the number of onto functions from a set with m elements to a set with n elements is the number of ways to distribute m distinguishable objects to n indistinguishable boxes so that no box is empty multiplied by the number of permutation of a set with n elements.

Derangements

The principle of inclusion–exclusion will be used to count the permutation of n objects that leave no objects in their original positions.

If n things are arranged in a row, the number of ways in which they can be deranged so that no one of them occupies its original place is

$$n!\left[1 - \frac{1}{1!} + \frac{1}{2!} - \frac{1}{3!} + \cdots + -1)^n \frac{1}{1!}\right]$$

or, no object goes to its scheduled place.

Note If r things go to wrong place out of n things then $(n-r)$ things go to original place (here $r < n$).

If $D_n =$ number of ways, if all n things go to wrong place and $D_r =$ number of ways, if r things go to wrong place.

If r things go to wrong place out of r, then $(n-r)$ things go to correct places. Then

$$D_n = {}^nC_{n-r}\,D_r$$

If at least p of them are in the wrong places, then

$$D_n = \sum_{r=p}^{n} {}^nC_{n-r}\,D_r$$

where

$$D_r = r!\left[1 - \frac{1}{1!} + \frac{1}{2!} - \frac{1}{3!} + \cdots (-1)^n \frac{1}{r!}\right]$$

EXAMPLE 1.76 A person writes letters to six friends and addresses the corresponding envelopes. In how many ways can the letters be placed in the envelopes so that (i) at least two of them are in the wrong envelopes and (ii) all the letters are in the wrong envelopes.

Solution

(i) The number of ways in which at least two of them are in the wrong envelopes

$$= \sum_{r=2}^{6} {}^nC_{n-r}\,D_r$$

$$= {}^nC_{n-2}D_2 + {}^nC_{n-3}D_3 + {}^nC_{n-4}D_4 + {}^nC_{n-5}D_5 + {}^nC_{n-6}D_6 \text{ [here } n=6]$$

$$= {}^6C_4.2!\left(1 - \frac{1}{1!} + \frac{1}{2!}\right) + {}^6C_3.3!\left(1 - \frac{1}{1!} + \frac{1}{2!} - \frac{1}{3!}\right)$$
$$+ {}^6C_2.4!\left(1 - \frac{1}{1!} + \frac{1}{2!} - \frac{1}{3!} + \frac{1}{4!}\right)$$
$$+ {}^6C_1.5!\left(1 - \frac{1}{1!} + \frac{1}{2!} - \frac{1}{3!} + \frac{1}{4!} - \frac{1}{5!}\right)$$
$$+ {}^6C_0.6!\left(1 - \frac{1}{1!} + \frac{1}{2!} - \frac{1}{3!} + \frac{1}{4!} - \frac{1}{5!} + \frac{1}{6!}\right)$$
$$= 15 + 40 + 135 + 264 + 265 = 719$$

(ii) The number of ways in which all letters be placed in wrong envelopes

$$= 6!\left(1 - \frac{1}{1!} + \frac{1}{2!} - \frac{1}{3!} + \frac{1}{4!} - \frac{1}{5!} + \frac{1}{6!}\right)$$
$$= 720\left(\frac{1}{2} - \frac{1}{6} + \frac{1}{24} - \frac{1}{120} + \frac{1}{720}\right) = 360 - 120 + 30 - 6 + 1 = 265$$

1.34 SEQUENCE AND SUMMATION

Here, we discuss important notions of sequence and summation, precious components of computer science.

1.34.1 Sequence

Sequence comes under the category of special type of function with its own notation. Here, we will study the sequence and also the summation notations.

A *sequence* is a function from the set $N = \{1, 2, 3, \ldots,\}$ of positive integers into a set A. Thus, a sequence is usually expressed as

$$a_1, a_2, a_3, \ldots, \text{ or } \{a_n : n \in N\} \text{ or simply } \{a_n\}$$

where a_n is used to denote the image of the integer n. Occasionally, the domain of a sequence is the set $\{0, 1, 2, \ldots,\}$ of non-negative integers rather than N. In such a case we say n begins with 0 rather than 1.

A finite sequence is a function from the set $M = \{1, 2, \ldots, m\}$ into a set A, and is generally denoted by $a_1, a_2, a_3, \ldots, a_m$. Such a sequence is sometimes called a *list* or an *m-tuple*.

For example

(i) The sequences

$$1, \frac{1}{2}, \frac{1}{3}, \frac{1}{4}, \ldots \text{ and } 1, \frac{1}{2}, \frac{1}{4}, \frac{1}{8}, \ldots$$

can be represented as

$$a_n = \frac{1}{n} \text{ and } b_n = 2^{-n}$$

respectively, where the first term of the first sequence starts with $n = 1$ and the first term of the second sequence begins with $n = 0$.

(ii) The sequence $1, -1, 1, -1$ can be defined by
$$a_n = (-1)^{n+1}, \text{ or equivalently } b_n = (-1)^n$$
where the first sequence begins with $n = 1$ and the second sequence begins with $n = 0$.

(iii) A finite sequence over a finite set A, viewed as a character set or an alphabet, is called a *string* or *word*. This is usually written in the form $a_1, a_2, \ldots, a_m$, in which the subscript m, the number of characters in the string, is called its *length*. The set with zero character as a string is called the *empty string* or *null string*.

The idea of a sequence is important in computer science, where a sequence is sometimes called a *linear array* or *list*.

1.34.2 Summation

Consider a sequence $a_1, a_2, a_3, \ldots$. To add them altogether at a time a symbol 'Σ', read as *sigma* (the Greek letter) will be introduced here. As such, the sums
$$a_1 + a_2 + \cdots + a_n \quad \text{and} \quad a_m + a_{m+1} + \cdots + a_n$$
can be represented, respectively, as
$$\sum_{j=1}^{n} a_j \quad \text{and} \quad \sum_{j=m}^{n} a_j$$

The subscript j is called a *dummy index* or *dummy variable*. The letters i, k, and t are also used as dummy variables.

For example
$$\sum_{i=1}^{n} a_i b_i = a_1 b_1 + a_2 b_2 + \cdots + a_n b_n$$
$$\sum_{j=2}^{5} j^2 = 2^2 + 3^2 + 4^2 + 5^2 = 4 + 9 + 16 + 25 = 54$$
$$\sum_{k=1}^{n} k = 1 + 2 + 3 + \cdots + n = \frac{n(n+1)}{2}$$

SUMMARY

In this chapter, we have introduced a number of concepts which are fundamental in nature and fruitfully applicable in mathematics and computer science. Among these there exist set and relation, in which we have discussed basic operations, namely, formation of unions and intersections, Venn diagrams, complements and power set, and carried on to look at the cardinalities of set, ordered pairs, and cartesian product of sets. Then we discussed an important conception from computer perspective, i.e., computer representation of sets. It is used for storing elements from an arbitrary ordering of the elements of the universal set. It also makes computation of sets in an easy manner. We then continued with relation in which emphasis has been given on binary relation. Through binary relation several types of relational form are described. One of the salient features of the relation lies in the relational model for databases, which has been discussed with some interesting examples. The important notion introduced from a computer science angle is that of a function. Apart from giving a purely set-theoretic definition, we also showed how function can and should be viewed as rules. These rules are utilized to a certain measure to functions of computer science.

EXERCISES

1. Find the subsets and the number of subsets of the set $A = \{a, b, c\}$.
2. Express each of the following sets in roster form:
   (i) $A = \{x : x \text{ is an integer}, -3 < x < 3\}$
   (ii) $B = \{x : x \text{ is a perfect square}, x < 17\}$
   (iii) $C = \{x \in R : x^2 - 4 = 0\}$
   (iv) $D = \{x : x \text{ is appositive integral divisor of } 60\}$
3. Describe the following sets in set-builder form:
   (i) $A = \{2, 4, 6, 8, 10\}$
   (ii) $B = \{3, 5, 7, 9, \ldots, 87, 89\}$
   (iii) $C = \{-5, -4, -3, -2, -1, 0, 1, 2\}$
   (iv) $D = \{1, 4, 9, 16, 25, 36\}$
4. Give some examples of finite and infinite sets.
5. If $A = \{1, 2, 3, 4\}$, $B = \{2, 4, 6, 9\}$, and $C = \{3, 4, 5, 6\}$, determine
   (i) $A \cup B$
   (ii) $B \cap C$
   (iii) $A \cap C$
   (iv) $A \cap (B \cap C)$
   (v) $A \cap (B \cup C)$
6. In a class of 25 students, 12 have taken mathematics and 8 have taken mathematics but not biology. Find the number of students who have taken mathematics and biology, and those who have taken biology but not mathematics.
7. If $A = \{1, 4\}$, $B = \{4, 5\}$, and $C = \{5, 7\}$, determine
   (i) $(A \times B) \cup (A \times C)$
   (ii) $(A \times B) \cap (A \times C)$
8. Find the power set of
   (i) $A = \{1, 2\}$,
   (ii) $B = \{\{a\}, \{b\}\}$,
   (iii) $C = \{(a, b), c\}$,
   (iv) $D = \{1, 3, \{1, 2, 3\}\}$.
9. Let $A = \{2, 3, 4\}$ and $B = \{3, 4, 5, 6, 7\}$. Assume a relation R from A to B such that $(x, y) \in R$ when a divides b (with zero remainder). Determine R, its domain, and range.
10. Let R be the relation on $A = \{1, 2, 3, 4\}$ to $B = \{3, 4, 5, 6\}$ defined as aRb if and only if $a \le b$, $a, b \in A$. Then determine R and its domain and range.
11. Let $A = \{x, y, z\}$, $B = \{X, Y, Z\}$, $C = \{x, y\}$, and $D = \{Y, Z\}$. Let R be a relation from A to B defined by $R = \{(x, X), (x, Y), (y, Z)\}$ and S be a relation from C to D defined by $S = \{(x, Y), (y, Z)\}$. Find R', $R \cup S$, $R \cap S$, and $R - S$.
12. If R be a relation in the set of integers Z defined by $R = \{(x, y) : x \in Z, y \in Z, (x - y) \text{ is divisible by } 6\}$. Then prove that R is an equivalence relation.
13. Consider the relation R on $A = \{4, 5, 6, 7\}$ defined by $R = \{(4,5), (5,5), (5,6), (6, 7), (7, 4)$ and $(7, 7)\}$. Find the symmetric closure of R.
14. Let $R = \{(1,2), (2,3), (3,1)\}$ and $A = \{1,2,3\}$. Find the reflexive, symmetric, and transitive closure of R, using
    (i) composition of relation R
    (ii) composition of matrix relation R
    (iii) graphical representation of R
15. Consider a relation R whose directed graph is shown below. Determine its inverse R^{-1} and complement R'.

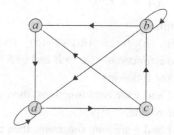

16. Show that the relation $(x, y) R (a, b) \Leftrightarrow x^2 + y^2 = a^2 + b^2$ is an equivalence relation on the plane, and determine the equivalence classes.
17. Let $A = \{1, 2, 3, 4\}$. Investigate the following relations as a function from A to A:
    (i) $f = \{(2, 3), (1, 4), (2, 1), (3, 2), (4, 4)\}$
    (ii) $g = \{(3, 1), (4, 2), (1, 1)\}$
    (iii) $h = \{(2, 1), (3, 4), (1, 4), (2, 1), (4, 4)\}$
18. State whether or not each diagram shown below defines a function from $A = \{a, b, c\}$ into $B = \{x, y, z\}$.

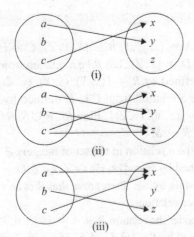

(i)

(ii)

(iii)

19. Is the following function one-to-one?

$$f: N \to N \text{ where } f(n) = \begin{cases} 2n, & \text{if } n \text{ is seven} \\ n, & \text{if } n \text{ is odd} \end{cases}$$

20. Let $A = \{0, -1, 1\}$ and $B = \{0, 1\}$. Let $f: A \to B$, where $f(a) = |a|$. Is f onto?

21. Let the function $f: R \to R$ be defined by

$$f(x) = \begin{cases} 2x + 5, & x > 9 \\ x^2 - |x|, & x \in [-9, 9] \\ x - 4, & x < -9 \end{cases}$$

Determine
(i) $f(3)$ (ii) $f(12)$ (iii) $f(-15)$ (iv) $f^2(5)$

22. Consider functions $f: A \to B$ and $g: B \to C$. Prove the following:
    (i) If f and g are onto functions, then $g \circ f$ is one-to-one.
    (ii) If f and g are onto functions, then $g \circ f$ is an onto function.

23. Let f be a function from $\{a, b, c, d\}$ to $\{1, 2, 3\}$ defined by $f(a) = 3, f(b) = 2, f(c) = 1$ and $f(d) = 3$. Examine the function f for onto.

24. Prove that $f: N \to N$, where $f(x) = x^2$ is injective (one-to-one), but the similar function $g: Z \to Z$, where $g(x) = x^2$ is not injective.

25. Let $f: R \to R$ be defined by $f(x) = 2x - 3$. Also, f is one-to-one and onto; hence, f has an inverse function f^{-1}. Find a formula for f^{-1}.

26. If $f: R^+ \to R^+$ and $g: R^+ \to R^+$ are defined by formulae $f(x) = \sqrt{x}$ and $g(x) = 3x + 1$ $\forall x \in R^+$, find $f \circ g$ and $g \circ f$. Is $f \circ g = g \circ f$?

27. If $f: R \to R$ and $g: R \to R$ be defined by $f(x) = x^3 - 4x$, $g(x) = 1/x^2 + 1$ and $h(x) = x^4$, find the following composition functions:
    (i) $(f \circ g \circ h)(x)$
    (ii) $(g \circ g)(x)$
    (iii) $(h \circ g \circ f)(x)$
    (iv) $(g \circ h)(x)$

28. Let $f: R \to R$ be defined by

$$f(x) = \begin{cases} 2x + 1, & x \leq 0 \\ x^2 + 1, & x > 0 \end{cases}$$

Let $g: R \to R$ be defined by

$$g(x) = \begin{cases} 3x - 7, & x \leq 0 \\ x^3, & x > 0 \end{cases}$$

Then find the composition $g \circ f$.

29. Let A and B be two sets. If $A \to B$ is one-to-one onto, then show that $f^{-1}: B \to A$ is also one-to-one onto.

30. Let $A = \{1, 2, 3\}$, $B = \{a, b\}$, and $C = \{r, s\}$. Also, $f: A \to B$ is defined by $f(1) = a$, $f(a) = 2, f(3) = b$ and $g: B \to C$ is also defined by $g(a) = s, g(b) = r$. Determine $g \circ f: A \to C$.

31. Let $A = [1, 2, 3, 4]$ and $B = \{a, b, c, d\}$, and let $f = \{(1, a), (2, a), (3, d), (4, c)\}$. Show that f is a function but f^{-1} is not.

32. Let $f: R \to R$ be defined by $f(x) = 3x - 4$. Show that f is one-to-one and onto. Also, find a formula for f^{-1}.

33. Show that the mapping $f: R \to R$ be defined by $f(x) = ax + b$, where $a, b, x \in R, a \neq 0$ is invertible. Find its inverse.

2 Combinatorics

LEARNING OBJECTIVES

After reading this chapter you will be able to understand the following:
- The permutations and combinations by which sets are organized to use the data they contain and to interpret them
- Basic concepts of probability consisting of discrete, conditional, random variables, and their components
- About recursion and its use to define sets and sequences
- The use of recursive style for solving difference equation (recurrence relations) and the discretization of differential equations
- About the inclusion–exclusion principle and its use to count the number of elements in a union of sets, and to solve counting problems
- About generating functions and their use to solve recurrence relations

2.1 INTRODUCTION

Combinatorics deals with the study of arrangements of objects. It is an important part of discrete mathematics. Enumeration, the counting of objects to solve a variety of problems, is also a key part of combinatorics. Counting is used to determine the complexity of algorithms, to determine sufficient telephone numbers, or Internet protocol addresses to meet demand. Moreover, counting techniques are useful in computing probabilities of events.

The basic rules of counting, which we will study in this chapter, can solve different types of problems. Another important combinatorial tool is the pigeonhole principle, which we will discuss in this chapter. Here, we will also study the recurrence relations, a tool for the analysis of computer programs. In addition, the generating function, inclusion–exclusion principle, and its applications will also be discussed here.

2.2 BASIC PRINCIPLES OF COUNTING

Counting problems exist throughout mathematics and computer science. Sometimes it is necessary to count the successful outcomes of experiments and all their possible outcomes to determine probabilities of discrete events. It is also needed to count the number of operations used by an algorithm to study its time complexity. Here, we will introduce the basic techniques of counting, the methods of which serve as the foundation for almost all counting techniques.

The basic principles of counting, such as multiplication principle and addition principle are described below.

2.2.1 Multiplication Principle (the Principle of Sequential Counting)

Suppose there is an event E which can occur in m ways and, independent of this event, there is a second event F which can occur in n ways. Then the total number of occurrence of the events E and F in the given order is mn. More generally, suppose an event E_1 can occur in n_1 ways, and, following E_1, a second event E_2 can occur in n_2 ways, and following E_2, a third event E_3 can occur in n_3 ways, and so on. Then the total number of occurrence of the events $E_1, E_2, E_3, \ldots$, in the order indicated is $n_1 n_2 n_3 \cdots$.

The multiplication principle is also named as the *fundamental principle of counting*.

Note For AND → '×' (multiply)

EXAMPLE 2.1 Find the number of four-letter words, with or without meaning, which can be formed out of the letters of the word ROSE, where the repetition of the letters is not allowed.

Solution There are as many words as there are ways of filling in four vacant places by the four letters, keeping in mind that the repetition is not allowed. The first place can be filled in four different ways by any of the four letters R, O, S, and E. Following which, the second place can be filled in by any of the remaining three letters in three different ways, the third place in two different ways, and the fourth place in one way. Thus, the number of ways in which the four places can be filled, by the multiplication principle (or, principle of counting) is $4 \times 3 \times 2 \times 1 = 24$. Hence, the required number of words is 24.

Note If the repetition of the letters was allowed, how many words can be formed? Each of the four vacant places can be filled in succession in four different ways. Hence, the required number of words is $4 \times 4 \times 4 \times 4 = 256$.

EXAMPLE 2.2 Given four flags of different colours, how many different signals can be generated, if a signal requires the use of two flags one below the other?

Solution There will be as many signals as there are ways of filling in two vacant places in succession by the four flags of different colours. The upper vacant place can be filled in four different ways by any of the four flags; following which, the lower vacant place can be filled in three different ways by any of the remaining three different flags. Hence, by multiplication principle, the required number of signals is $4 \times 3 = 12$.

2.2.2 Addition Rule (the Principle of Disjunctive Counting)

Suppose some event E can occur in m ways and a second event F can occur in n ways, and suppose both the events cannot occur simultaneously. Then, E or F can occur in $m + n$ ways. More generally, suppose an event E_1 can occur in n_1 ways, a second event E_2 can occur in n_2 ways, a third event E_3 can occur in n_3 ways, $\ldots$, and suppose no two of the events can occur at the same time. Then, one of the events can occur in $n_1 + n_2 + n_3 + \cdots$ ways.

Note For OR → '+' (addition)

EXAMPLE 2.3 Suppose there are eight male professors and five female professors teaching a history class. In how many ways a student can choose a history professor?

Solution A student can choose a history professor in $8 + 5 = 13$ ways.

EXAMPLE 2.4 In how many ways can we get a sum of 7 or 11 when two distinguishable dice are rolled?

Solution The ordered pairs in which the sum is 7 are (1,6), (2,5), (3,4), (4,5), (5,2), (6,1), which are distinct. There are six ways to obtain the sum 7. Similarly, the ordered pairs in which the sum is 11 are (5,6), (6,5), which are also distinct. The number of ways in which we get a sum 11 with the two dice is 2. Thus, we can get a sum 7 or 11 with two distinguishable dice in $6 + 2 = 8$ ways.

2.3 FACTORIAL NOTATION

If n is a natural number, then the product of all the natural numbers from 1 to n is called 'n-factorial'. It is denoted by the symbol $n!$ or $\lfloor n$.

From the definition,

$$n! = n(n-1)(n-2)\cdots 3.2.1$$

The factorial notation $n!$ can also be defined recursively as follow:

$$0! = 1, (n+1)! = n!(n+1), n \geq 0$$

From the above recursive definition, we get

$$1! = 0!(1) = 1, \quad 2! = 1!(2) = 1.2, \quad 3! = 2!(3) = 1.2.3$$

EXAMPLE 2.5 Find the value of 8!.

Solution For $n \geq 0, (n+1)! = n!(n+1)$

Hence,

$$\begin{aligned} 8! &= (7+1)! = 7!\,8 = 6!\,7.8 = 5!\,6.7.8 \\ &= 4!\,5.6.7.8 = 3!\,4.5.6.7.8 \\ &= 2!\,3.4.5.6.7.8 = 1!\,2.3.4.5.6.7.8 \\ &= 1.2.3.4.5.6.7.8 \\ &= 40{,}320 \end{aligned}$$

EXAMPLE 2.6 Simplify

(i) $\dfrac{n!}{(n-1)!}$

(ii) $\dfrac{(n+1)!}{n!}$

Solution

(i) $\dfrac{n!}{(n-1)!} = \dfrac{n(n-1)!}{(n-1)!} = n$

(ii) $\dfrac{(n+1)!}{n!} = \dfrac{(n+1)n!}{n!} = (n+1)$

EXAMPLE 2.7 Show that $3! + 4! \neq (3+4)!$

Solution

$$3! + 4! = (3 \times 2 \times 1) + (4 \times 3 \times 2 \times 1) = 6 + 24 = 30$$

Also,

$$(3+4)! = 7! = 7.6.5.4.3.2.1 = 5040$$

Thus,

$$3! + 4! \neq (3+4)!$$

2.4 BINOMIAL THEOREM

Theorem 2.1 Let n be a positive integer. Then for all a and b,

$$(a+b)^n = {}^nC_0 a^n + {}^nC_1 a^{n-1} b + {}^nC_2 a^{n-2} b^2 + \cdots + {}^nC_r a^{n-r} b^r + \cdots + {}^nC_n b^n$$

where the symbol ${}^nC_r$, in which r and n are positive integers with $r \leq n$, is defined as

$${}^nC_r = \dfrac{n(n-1)(n-2)\cdots(n-r+1)}{1.2.3\cdots(r-1)r} = \dfrac{n!}{r!(n-r)!}$$

Since

$$n(n-1)(n-2)\cdots(n-r+1) = \dfrac{n(n-1)\cdots(n-r+1)(n-r)(n-r-1)\cdots 3.2.1}{(n-r)(n-r-1)\cdots 3.2.1}$$

$$= \dfrac{n!}{(n-r)!}$$

The proof of the theorem is left to the reader.

Note

(i) The notation

$$\sum_{r=0}^{n} {}^nC_r a^{n-r} b^r$$

stands for $(a+b)^n$. Hence, the theorem can also be expressed as

$$(a+b)^n = \sum_{r=0}^{n} {}^nC_r a^{n-r} b^r$$

Combinatorics 73

(ii) The coefficients ^nC_r occurring in the binomial theorem are known as *binomial coefficients*.
(iii) The number of terms in the expansion of $(a + b)^n$ is $n + 1$, i.e., one more than the index n.
(iv) In the expansion of $(a + b)^n$, the sum of the indices of a and b is $n + 0 = n$ in the first term, $(n - 1) + 1 = n$ in the second term, and so on $0 + n = n$ in the last term. Thus, it can be seen that the sum of the indices a and b is n in every term of the expansion.
(v) We can write $r = n - (n - r)$; hence, we have the following important relation:
$$^nC_{n-r} = {}^nC_r$$
or, in other words, if $a + b = n$, then $^nC_a = {}^nC_b$
(vi) The general term, denoted by T_{r+1}, in the expansion of $(a + b)^n$ is $^nC_r a^{n-r} b^r$, i.e., $(r + 1)$th term is the general term.
(vii) The coefficients of terms equidistant from ends of the expansion are equal.

2.4.1 Pascal's Triangle

Consider the following expansions and write the coefficients of the expansions as follows:

$(a + b)^0 = 1$
$(a + b)^1 = a + b$
$(a + b)^2 = a^2 + 2ab + b^2$
$(a + b)^3 = a^3 + 3a^2b + 3ab^2 + b^3$
$(a + b)^4 = a^4 + 4a^3b + 6a^2b^2 + 6ab^3 + b^4$
$(a + b)^5 = a^5 + 5a^4b + 10a^3b^2 + 10a^2b^3 + 5ab^4 + b^5$
$(a + b)^6 = a^6 + 6a^5b + 15a^4b^2 + 20a^3b^3 + 15a^2b^4 + 6ab^5 + b^6$

```
              1
           1     1
         1    2    1
       1    3    3    1
     1    4    6    4    1
```

It can be seen from the above structures of the expansions that the coefficients follow a definite pattern which can be most easily demonstrated by means of a triangle, known as *Pascal's triangle*.

We can create the lines of the triangle by using the identity

$$^nC_r + {}^nC_{r+1} = {}^{n+1}C_{r+1}$$

2.4.2 Multinomial Theorem

A. *To find the expansion of $(a_1 + a_2 + a_3 + \cdots + a_m)^n$*:
The binomial theorem can be extended to obtain a formula on powers of multinomials $(a_1 + a_2 + a_3 + \cdots + a_m)^n$ as follows:

Theorem 2.2 Let n be a positive integer, then for all $a_1, a_2, a_3, \ldots, a_m$,

$$(a_1 + a_2 + a_3 + \cdots + a_m)^n = \sum \frac{n!}{\alpha_1! \alpha_2! \alpha_3 \cdots \alpha_m!} a_1^{\alpha_1} a_2^{\alpha_2} a_3^{\alpha_3} \cdots a_m^{\alpha_m}$$

where $\alpha_1, \alpha_2, \alpha_3, \ldots, \alpha_m$ are having all positive integral values including zero, consistent with the relation $\alpha_1 + \alpha_2 + \alpha_3 + \cdots + \alpha_m = n$.

Proof The product $(a_1 + a_2 + a_3 + \cdots + a_m)(a_1 + a_2 + a_3 + \cdots + a_m)$ to n factors is the sum of all the products which can be formed by choosing a term out of each of the n factors and multiplying these terms together.

The expansion is therefore the sum of a number of terms of the form $a_1^{\alpha_1} a_2^{\alpha_2} a_3^{\alpha_3} \cdots a_m^{\alpha_m}$, where each index may have any of the values $0, 1, 2, \ldots, n$ subject to the condition

$$\alpha_1 + \alpha_2 + \cdots + \alpha_m = n \qquad (2.1)$$

Choose any positive integral or zero values for $\alpha_1, \alpha_2, \ldots, \alpha_m$ which satisfy this condition. To obtain the coefficient of the term $a_1^{\alpha_1} a_2^{\alpha_2} a_3^{\alpha_3} \cdots a_m^{\alpha_m}$, divide the n factors into groups containing $\alpha_1, \alpha_2, \ldots, \alpha_m$ factors, respectively. Take a_1 out of each factor of the first group, a_2 out of each factor of the second group, and so on. The number of ways in which this can be done is

$$\frac{n!}{\alpha_1! \alpha_2! \cdots \alpha_m!}$$

and this is the coefficient of $a_1^{\alpha_1}, a_2^{\alpha_1} \cdots a_m^{\alpha_m}$ in the expansion.

Hence,

$$(\alpha_1 + \alpha_2 + \cdots + \alpha_m)^n = \sum \frac{n!}{\alpha_1! \alpha_2! \cdots \alpha_3!} a_1^{\alpha_1} a_2^{\alpha_1}, \ldots, a_m^{\alpha_m}$$

where the summation is extended over all m-tuples of non-negative integers $(\alpha_1, \alpha_2, \ldots, \alpha_m)$ such that $\alpha_1 + \alpha_2 + \cdots + \alpha_m = n$.

Note The number of terms in the expansion of $(a_1 + a_2 + \cdots + a_m)^n$ is $^{n+m-1}C_n$.

Corollary 2.1 The binomial theorem is a particular case of multinomial theorem in respect to $m = 2$. Since $\alpha_1 + \alpha_2 = n$, therefore, $\alpha_2 = n - \alpha_1$ and hence

$$(a_1 + a_2)^n = \sum \frac{n!}{\alpha_1!(n - \alpha_1)!} a_1^{\alpha_1} a_2^{n-\alpha_1}$$

$$= \sum \frac{n(n-1)(n-2)\cdots(n-\alpha_1+1)}{\alpha_1!} a_1^{\alpha_1} a_2^{n-\alpha_1}$$

B. *To find the coefficient of x^r in the expansion of $(a_1 + a_2x + a_3x^3 + \cdots + a_m x^{m-1})^n$*: The general term is

$$\frac{n!}{\alpha_1! \alpha_2! \cdots \alpha_m!} a_1^{\alpha_1} (a_2 x)^{\alpha_2} (a_3 x^2)^{\alpha_3} \cdots (a_m x^{m-1})^{\alpha_m}$$

$$= \frac{n!}{\alpha_1! \alpha_2! \cdots \alpha_m!} a_1^{\alpha_1} a_2^{\alpha_2} \cdots a_m^{\alpha_m} x^{\alpha_2 + 2\alpha_3 + \cdots + (m-1)\alpha_m}$$

Hence, we have to take all the terms which are such that

$$\alpha_2 + 2\alpha_3 + \cdots + (m-1)\alpha_m = r \qquad (2.2)$$

and the coefficient of x^r is therefore

$$\sum \frac{n!}{\alpha_1! \alpha_2! \cdots \alpha_m!} a_1^{\alpha_1} a_2^{\alpha_2} \cdots a_m^{\alpha_m}$$

α is being subject to the restrictions (2.1) and (2.2).

Note The coefficient of $a_1^{\alpha_1}, a_2^{\alpha_1}, \ldots, a_m^{\alpha_m}$ in the expansion of $(a_1 + a_2 + \cdots + a_m)^n$ is

$$\frac{n!}{\alpha_1! \alpha_2! \cdots \alpha_m!}$$

EXAMPLE 2.8 Find the number of terms in the expansion of

$$(2x + 3y - 5z)^8$$

Solution Here, $n = 8$ and $m = 3$. Therefore, the number of terms is

$$^{n+m-1}C_n = {}^{10}C_8 = 45$$

EXAMPLE 2.9 Find the coefficient of x^7 in the expansion of $(1 + 3x - 2x^3)^{10}$.

Solution Coefficient of x^7 in the expansion of $(1 + 3x - 2x^3)^{10}$ is

$$\sum \frac{10!}{n_1! n_2! n_3!} (1)^{n_1} (3)^{n_2} (-2)^{n_3}$$

where $n_1 + n_2 + n_3 = 10$ and $n_2 + 3n_3 = 7$; the possible values of $n_1, n_2,$ and n_3 are shown in Table 2.1.

TABLE 2.1

n_1	n_2	n_3
3	7	0
5	4	1
7	1	2

Thus, the coefficient of x^7

$$= \frac{10!}{3! \, 7! \, 0!} (1)^3 (3)^7 (-2)^0 + \frac{10!}{5! \, 4! \, 1!} (1)^5 (3)^4 (-2)^1$$
$$+ \frac{10!}{7! \, 1! \, 2!} (1)^7 (3)^1 (-2)^2$$
$$= 262440 - 204120 + 4320 = 62640$$

2.5 PERMUTATIONS (ARRANGEMENTS OF OBJECTS)

Any arrangement of a set of n objects in a definite order is called a *permutation* of the objects, taken all at a time. Any arrangement of any r of these n objects ($r \leq n$) in a definite order is called an *r-permutation* or a permutation of the n objects taken r at a time.

For example Consider the set of letters a, b, c, and d. Then

(i) $bdca, dcba$, and $acdb$ are permutations of the four letters, taken all at a time
(ii) bad, adb, cbd, and bca are permutations of the four letters taken three at a time
(iii) ad, cb, da, and bd are permutations of the four letters taken two at a time

The number of permutations of n different objects taken r at a time is denoted by ^nP_r.

Theorem 2.3 The number of $r =$ permutations of a set with n distinct elements not responding is

$$^nP_r = n(n-1)(n-2)\cdots(n-r+1) = \frac{n!}{(n-r)!}$$

Proof There will be as many permutations as there are ways of filling in r vacant places by n different objects. The first place can be filled in n ways; following which, the second place can be filled in $(n-1)$ ways, following which the third place can be filled in $(n-2)$ ways, $\ldots$, the rth place can be filled in $\{n(n-1)\}$ ways. Therefore, the number of ways of filling in r vacant places in succession is $n(n-1)(n-2)\cdots\{n-(r-1)\}$ or $n(n-1)(n-2)\cdots(n-r+1)$.

Thus, by the fundamental principle of counting (multiplication rule), there are $n(n-1)(n-2)\cdots(n-r+1), r =$ permutations of the set, i.e.,

$$^nP_r = n(n-1)(n-2)\cdots(n-r+1)$$
$$= \frac{[n(n-1)(n-2)\cdots(n-r+1)][(n-r)(n-r-1)\cdots 2.1]}{[(n-r)(n-r-1)\cdots 2.1]}$$
$$= \frac{n!}{(n-r)!}$$

Hence,

$$^nP_r = \frac{n!}{(n-r)!}$$

Note The number of permutations of n distinct objects taken n at a time is

$$^nP_n = \frac{n!}{(n-n)!} = \frac{n!}{0!} = n!$$

Again, consider three letters a, b, c. Then, there are $3! = 3.2.1 = 6$ permutations of the three letters, such as, abc, acb, bac, bca, cab, and cba.

EXAMPLE 2.10 How many words of three distinct letters can be formed from the letters of the word LAND?

Solution The number of three distinct letters can be formed from the four letters of the word LAND is

$$^4P_3 = \frac{4!}{(4-3)!} = \frac{4!}{1!} = 24$$

2.5.1 Permutations with Repetitions

Theorem 2.4 The number of permutations of n objects, of which n_1, are alike of one kind, n_2 are alike of another kind, n_3 are alike of third kind, ..., n_r are alike of rth kind, is

$$^nP_{n_1,n_2,n_3,\ldots,n_r} = \frac{n!}{n_1!n_2!n_3!\cdots n_r!}$$

where $n = n_1 + n_2 + n_3 + \cdots + n_r$.

Proof Let the required number of permutations be x. If the n_1 like objects are unlike, then for each of these x arrangements, the n_1 like objects can be rearranged among themselves in $n_1!$ ways, without altering the positions of the other objects.

So, the number of permutations will be $xn_1!$. Similarly, if n_2 like objects are unlike, each of these $xn_1!$ permutations will give rise to $n_2!$ permutations. Therefore, the number of permutations will be $xn_1! \, n_2!$. If all the objects are unlike, the number of permutations will be $n_1! \, n_2! \cdots n_r!$. But, if all the objects are unlike, the number of permutations with n objects will be $n!$. Hence,

$$xn_1!n_2!\cdots n_r! 1 = n!$$

$$x = \frac{n!}{n_1!n_2!n_3!\cdots n_r!}$$

i.e.,

$$^nP_{n_1,n_2,n_3,\ldots,n_r} = \frac{n!}{n_1!n_2!n_3!\cdots n_r!}$$

Note The number of permutations of n distinct objects, taken r at a time (when repetitions are allowed) is $(n)^r$.

EXAMPLE 2.11 If there are four black, three green, and five red balls, then in how many ways these colour can be arranged in a row?

Solution Here, the total number of balls $= 4 + 3 + 5 = 12$. Also, the black balls, green balls, and the red balls are all alike. Hence, the balls can be arranged in a row in

$$\frac{12!}{4! \, 3! \, 5!} = 27{,}720 \text{ ways}$$

EXAMPLE 2.12 How many words can be formed using the letter A thrice, the letter B twice, and the letter C once?

Solution Given six objects $A, A, A, B, B,$ and C of which three are alike of the same kind, two are alike of another kind, and one is of its own kind. Hence, the total number of permutations of the requisite number of words formed

$$\frac{6!}{3! \, 2! \, 1!} = 60$$

EXAMPLE 2.13 Find how many arrangements can be made with the letters of the word 'MATHEMATICS'.

Solution There are 11 letters in the word 'MATHEMATICS'. Out of these letters M is repeated twice, A is repeated twice, T is repeated twice, and the rest are all different. So, the required number of arrangements

$$\frac{11!}{2!\,2!\,2!} = 4989600$$

EXAMPLE 2.14 How many four-digit numbers can be formed using the digits 2, 4, 6, and 8 when repetition of digits is allowed.

Solution Here, we have four digits. So,

Number of ways of filling unit's place = 4
Number of ways of filling ten's place = 4
Number of ways of filling hundred's place = 4
Number of ways of filling thousand's place = 4

Thus, the total number of four-digit numbers = $4^4 = 256$.

2.5.2 Circular Permutations

Instead of arranging the objects in a line, if we arrange them in the form of a circle, we call them *circular permutations*. In circular permutations, what really matters is the position of an object relative to the others.

Suppose n persons $(a_1, a_2, ..., a_n)$ are to be arranged around a ring. There are $n!$ ways in which they can be arranged in a row; on the other hand, all the linear arrangements

$$a_1, a_2, a_3, a_4, ..., a_n; \quad a_2, a_3, a_4, ..., a_n, a_1$$
$$a_n, a_1, a_2, ..., a_{n-1}; \quad a_{n-1}, a_n, a_1, a_2, ..., a_{n-2}$$
$$... \quad ... \quad ... \quad ... \quad ...$$

will lead to the same arrangement in a ring. So, each circular arrangement corresponds to linear arrangement. Hence, the total number of circular arrangements of n persons is

$$\frac{n!}{n} = (n-1)!$$

Thus, there are $(n-1)!$ permutations of n distinct objects in a circle.

If we consider the *clockwise* and *anticlockwise* arrangements in the circular permutations, then the following propositions can be possible:
 (i) When distinction is made between the clockwise and the anticlockwise arrangements of n different objects around a circle, then the number of arrangements = $(n-1)$.
 (ii) If no distinction is made between the clockwise and the anticlockwise arrangements of n different objects around a circle, then the number of arrangements = $(1/2)(n-1)!$.

EXAMPLE 2.15 In how many different ways can five men and five women sit around a table, if
(a) there is no restriction?
(b) no two women sit together?

Solution

(a) If there is no restriction then this problem is same as the circular permutation of 10 objects (five men and five women). So, the number of permutations are (10–1)! = 9! = 362880.

(b) If two women are not allowed to sit side by side, that is each woman should occupy a sit between two men.

The number of ways five men can sit around a round table = (5 − 1)! = 4! = 24. Once these five men have already occupied alternate seats, the five women can sit in the five empty seats in 5! = 5 × 4 × 3 × 2 = 120 ways. Thus, the total number of ways = 24 × 120 = 2880.

EXAMPLE 2.16 In how many ways can seven persons sit around a table so that all shall not have the same neighbours in any two arrangements?

Solution Seven persons can sit around a table in 6! ways. But in clockwise and anticlockwise arrangements, each person will have the same neighbour. So, the required number of ways = (1/2) × (6!) = 360.

EXAMPLE 2.17 Find the number of ways in which eight different beads can be arranged to form a necklace.

Solution Fixing the position of one bead, the remaining beads can be arranged in 7! ways. But there is no distinction between the clockwise and anticlockwise arrangements. So, the required number of arrangements = (1/2) × (7!) = 2520.

2.6 COMBINATIONS (SELECTION OF OBJECTS)

Each of the different groups or selections which can be formed by taking some or all of a number of objects, irrespective of their arrangements, is called a *combination*.

Suppose we want to select two out of three persons *A*, *B*, and *C*. We may choose *AB* or *BC* or *AC*. Clearly, *AB* and *BA* represent the same selection or group but they give rise to different arrangements. Clearly, in a groups or selection, the order in which the objects are arranged is immaterial.

For example

(i) The different combinations formed of three letters *a*, *b*, and *c*, taken two at a time, are *ab*, *bc*, and *ca*.
(ii) The only combination that can be formed of three letters *a*, *b*, and *c* taken all at a time, is *abc*.
(iii) Various groups of two out of four persons *A*, *B*, *C*, and *D* are *AB*, *AC*, *AD*, *BC*, *BD*, and *CD*.

Difference between a *permutation* and a *combination*: In a combination, only a group is made and the order in which the objects are arranged is immaterial. On the other hand, in a permutation, not only a group is formed, but also an arrangement in a definite order is considered.

For example

(i) *ab* and *ba* are two different permutations, but each represents the same combination.
(ii) *abc*, *acb*, *bac*, *bca*, *cab*, and *cba* are six different permutations, but each one of them represents the same combination, namely a group of three objects *a*, *b*, and *c*.

Note We use the word 'arrangements' for permutations and 'selections' for combinations.

The number of combinations of n objects, taken r at a time, is denoted by ^nC_r. The symbol ^nC_r is defined only when n and r are integers such that $n \geq r, n > 0$, and $r \geq 0$.

2.6.1 Combinations of n Different Objects

Theorem 2.5 The number of combinations of n distinct objects, taken r at a time, is given by

$$^nC_r = \frac{n!}{r!(n-r)!}$$

Proof Let the number of combinations of n objects, taken r at a time, be x. Then, $^nC_r = x$.

Now, each combination contains r objects, which may be arranged amongst themselves in $r!$ ways. Thus, each combination gives rise to $r!$ permutations. So, x combinations will give rise to $x(r!)$ permutations. Therefore, the number of permutations of n things, taken r at a time, is $x(r!)$.

Consequently, $^nP_r = x(r!) = {^nC_r}(r!)$. Thus,

$$^nC_r = \frac{^nP_r}{r!} = \frac{n!}{r!(n-r)!} \qquad \left(\because {^nP_r} = \frac{n!}{(n-r)!}\right)$$

Note We may write

$$^nC_r = \frac{n(n-1)(n-2)\ldots \text{ to } r \text{ factors}}{r!}$$

In the present topic, we concentrate on counting unordered selection of objects.

For example Consider a query posing like, how many different committees of three students can be formed with four students?

To answer this question, we need only to find the number of subsets with three elements that form the set containing four students. We see that there are four such subjects, one for each of the four students, because choosing four students is the same as choosing one of the four students to leave out of the groups. This means that there are four ways to choose the three students for the committee, where the order in which these students are chosen, is immaterial.

This example illustrates that many counting problems can be solved by finding the number of subsets of a particular size of a set with n elements, where n is a positive integer.

An r combination of elements of a set is an unordered selection of r elements from the set. Thus, an r combination is simply a subset of the set with r elements.

For example It is seen that $^4C_2 = 6$, because the two combinations of $\{a, b, c, d\}$ are the six subsets $\{a, b\}, \{a, c\}, \{a, d\}, \{b, c\}, \{b, d\}$, and $\{c, d\}$.

EXAMPLE 2.18 In how many ways a committee of five members can be selected from six men and five ladies, consisting of three men and two ladies?

Solution Three men out of six and two ladies out of five can be selected in

$$^6C_3 \times {^5C_2} = \left(\frac{6 \times 5 \times 4}{3 \times 2 \times 1}\right) \times \left(\frac{5 \times 4}{2 \times 1}\right) = 200 \text{ ways}$$

EXAMPLE 2.19 How many different committees of five people can be chosen from 20 men and 12 women

(a) if exactly three men must be on each committee?
(b) if at least three women must be on each committee?

Solution

(a) We choose three men from twenty men and then two women from twelve women. So, the number of committees will be

$$^{20}C_3 \times {}^{12}C_2 = 1140 \times 66 = 75240$$

(b) We choose at least three women here, that is, three women, four women, and five women are to be chosen in each case from twelve women. Then, by addition rule, the number of committees will be

$$^{12}C_3 \times {}^{20}C_2 + {}^{12}C_4 \times {}^{20}C_1 + {}^{12}C_5 \times {}^{20}C_0$$
$$= 220 \times 190 + 495 \times 20 + 792 \times 1 = 52492$$

EXAMPLE 2.20 A collection of 10 electric bulbs contains 3 defective ones.
(a) In how many ways can a sample of four bulbs be selected?
(b) In how many ways can a sample of four bulbs be selected which contain two good bulbs and two defective ones?
(c) In how many ways can the sample of four bulbs be selected so that either the sample contains three good ones and one defective ones or one good and three defective ones?

Solution

(a) The four bulbs can be selected out of ten bulbs in

$$^{10}C_4 = \frac{10!}{4!\,6!} = \frac{10 \times 9 \times 8 \times 7}{4 \times 3 \times 2 \times 1} = 210 \text{ ways}$$

(b) Two bulbs can be selected out of seven good bulbs in $^7C_2$ ways and two defective bulbs can be selected out of three defective bulbs in $^3C_2$ ways. Thus, the number of ways in which a sample of four bulbs containing two good bulbs and two defective bulbs can be selected as

$$^7C_2 \times {}^3C_2 = \frac{7!}{2!\,5!} \times \frac{3!}{2!\,1!} = \frac{7 \times 6}{2} \times 3 = 63$$

(c) Three good bulbs can be selected from seven good bulbs in $^7C_3$ ways and one defective bulb can be selected out of three defective ones in $^3C_1$ way.

Similarly, one good bulb can be selected from seven good bulbs in $^7C_1$ ways and three defective ones in $^3C_3$ ways.

So, the number of ways of selecting a sample of four bulbs containing three good ones and one defective or one good and three defective ones are

$$^7C_3 \times {}^3C_1 + {}^7C_1 \times {}^3C_3 = \frac{7!}{3!\,4!} \times \frac{3!}{1!\,2!} + \frac{7!}{1!\,6!} \times \frac{3!}{3!\,0!}$$

$$= \frac{7 \times 6 \times 5}{3 \times 2} \times 3 + 7 = 35 \times 3 + 7 = 112$$

2.6.2 Combinations with Repetitions

When repetition of elements is allowed, the r combinations (with objects of size r) form a set with n elements can be expressed as

$$^{n+r-1}C_r = (n+r-1)!/(r!)(n-1)! = {}^{n+r-1}C_{n-1}$$

We will illustrate the above combinations with the following examples.

EXAMPLE 2.21 Consider a cookie shop in which there are four different kinds of cookies. Find the number of different ways of choosing six cookies (assuming only the type of cookies, and not the individual cookies or the order in which they are chosen).

Solution The number of ways to choose six cookies is the number of six combinations of a set with four elements, which is given by

$$^{4+6-1}C_6 = {}^9C_6 = {}^9C_3 = (9.8.7)/(1.2.3) = 84$$

Thus, there exists 84 different ways to choose the six cookies.

EXAMPLE 2.22 Four boys picked up 30 mangoes. In how many ways can they divide them if all the mangoes be identical?

Solution Clearly, 30 mangoes can be distributed among four boys such that each boy can receive any number of mangoes. Hence, total number of ways = $^{30+4-1}C_{4-1} = {}^{33}C_3 = 5456$.

EXAMPLE 2.23 Assume that a valid computer password consists of seven characters, the first of which is a letter selected from the set $\{A, B, C, D, E, F, G\}$ and the remaining characters are letters chosen from the English alphabet or a digit. Find the number of possible passwords.

Solution A password can be constructed by the following sequences:

Step 1 Select a starting letter from the given set.
Step 2 Select a sequence of letter and digits with repetitions.

Step 1 can be performed in $^7C_1$ or seven ways. Since there are 26 letters and 10 digits that can be selected for each of the remaining six characters, and since repetitions are allowed, sequence 2 can be performed in 36^6 or $2,176,782,336$ ways. By the multiplication rule, there are 7×2176782336 or $15,237,476,352$ different passwords.

The combinations with repetition can also be used to find the number of solutions of certain linear equations in which variables are integers subject to constraints.

EXAMPLE 2.24 How many solutions are there in $x + y + z + u = 29$ subject to the constraints $x \geq 1, y \geq 2, z \geq 3$, and $u \geq 0$?

Solution Since

$$x + y + z + u = 29 \qquad (2.3)$$

where x, y, z, and u are integers such that $x \geq 1, y \geq 2, z \geq 3$, and $u \geq 0$, i.e., $x - 1 \geq 0$, $y - 2 \geq 0, z - 3 \geq 0$, and $u \geq 0$.

Assume $x_1 = x - 1, x_2 = y - 2, x_3 = z - 3$. Then,

$$x = x_1 + 1, y = x_2 + 2, z = x_3 + 3 \quad \text{and} \quad x_1 \geq 0, x_2 \geq 0, x_3 \geq 0, u \geq 0$$

From Eq. (2.3),

$$x_1 + 1 + x_2 + 2 + x_3 + 3 + u = 29$$
$$\Rightarrow \qquad x_1 + x_2 + x_3 + u = 23$$

Hence, the number of solutions $= {}^{23+4-1}C_{4-1} = {}^{26}C_3 = (26 \cdot 25 \cdot 24)/(1 \cdot 2 \cdot 3) = 2600$.

EXAMPLE 2.25 How many integral solutions are there to the system of equations

$$x_1 + x_2 + x_3 + x_4 + x_5 = 20 \quad \text{and} \quad x_1 + x_2 = 15$$

where $x_k \geq 0, k = 1, 2, 3, 4, 5$.

Solution We have

$$x_1 + x_2 + x_3 + x_4 + x_5 = 20 \qquad (2.4)$$

and

$$x_1 + x_2 = 15 \qquad (2.5)$$

Then from Eqs (2.4) and (2.5),

$$x_3 + x_4 + x_5 = 5 \qquad (2.6)$$

$$x_1 + x_2 = 15 \qquad (2.7)$$

and given $x_1 \geq 0, x_2 \geq 0, x_3 \geq 0, x_4 \geq 0, x_5 \geq 0$. Then, the number of solutions in Eq. (2.6):

$$ {}^{5+3-1}C_{3-1} = {}^{7}C_2 = 21$$

and the number of solutions in Eq. (2.7):

$$ {}^{15+2-1}C_{2-1} = {}^{16}C_1 = 16$$

Hence, the total number of solutions of the given system of equations $= 21 \times 16 = 336$.

2.7 DISCRETE PROBABILITY

The theory of probability has its origin in the games of chance such as gambling and since then it has developed so much that we find its applications in almost all fields of knowledge. In particular, it is useful in solving problems related to mortality and insurance. Even in our daily life, there exists a number of phenomena where we cannot make prediction with certainty or complete reliability.

For example

(i) Probably, it may rain today.
(ii) Possibly, you will catch the train.
(iii) There is a good chance that Sushil will get the job next month.

All the above sentences with words like 'probably', 'possibly', and 'good chance' are expressions indicating a degree of uncertainty.

The theory of probability has become so rich and developed that it is being used in different branches of engineering, such as information theory, automatic control, and ship behaviour in navigation.

Actually, before defining probability, we will discuss the basic concepts related to probability.

2.7.1 Terminology (Basic Concepts)

Random Experiment

It is an experiment which, if conducted repeatedly, under identical conditions does not give essentially the same result.

For example Tossing of a coin, throwing a die, drawing a card from a pack of playing cards. In all these cases, there are a number of possible results which can occur but there is an uncertainty as to which one of then will actually occur.

Outcome

The result of a random experiment is called an *outcome*.

Sample Space

The set of all possible outcomes of a random experiment is known as *sample space*.

For example

(i) The sample space in a throw of a die is the set $\{1, 2, 3, 4, 5, 6\}$.
(ii) When a coin and a die are tossed together, there are 12 *sample points* (every element of the sample space is called a *sample point*) in the sample space described by the set.

$\{(T,1), (T,2), (T,3), (T,4), (T,5), (T,6), (H,1), (H,2), (H,3), (H,4), (H,5), (H,6),\}$

Trial and Event

The performance of a random experiment is called a *trial* and the *outcome*, an event.

For example The throwing of a dice is a trial and turning up of 1, 2, 3, 4, 5, or 6 is an event.

Simple (or Elementary) Events and Compound Events

An event consisting of only one sample point of a sample space is called a *simple event* or an *elementary event*. An event consisting of more than one sample point is called *compound event*.

Note The empty set or the null set is sometimes called *impossible event* or the *null event*.

For example If a die is rolled once, then the event of getting '4' is a simple event whereas the event of getting 'an even number' is a compound event since it consists of three simple events such as 2, 4, and 6.

Mutually Exclusive Events

Events are said to be mutually exclusive if the occurrence of any one of them rules out the occurrence of all other, i.e., two mutually exclusive events cannot occur simultaneously in the same trial.

For example In tossing of a coin, there exists only one possibility, either head or tail. Both head and tail cannot occur simultaneously.

Equally Likely Events

The events are said to be equally likely if there is no reason to accept any one in preference to any other.

For example In tossing of a coin, appearance of a head or a tail is equally likely events.

Exhaustive Events

All possible outcomes of an event are known as *exhaustive events*.

For example In throwing of a single die the exhaustive events are six, such as 1, 2, 3, 4, 5, and 6. If two dice are thrown, the exhaustive events will be $6 \times 6 = 36$.

Independent Events

Two or more events are said to be independent if the occurrence or non-occurrence of any one does not depend on the occurrence or non-occurrence of any other event.

For example The outcome of first toss of a coin does not influence the outcome of second toss of a coin.

Dependent Events

If the occurrence of one event influences the occurrence of the other event, the events are said to be dependent events.

For example If a card is drawn from a pack of shuffled cards, and not replaced before drawing the second card, then the second draw is dependent on the first one.

2.8 FINITE PROBABILITY SPACES

Let $S = \{a_1, a_2, \ldots, a_n\}$, be a finite sample space. A finite probability space is obtained by assigning to each point a_i, in S a real number p_i, satisfying the following properties:

(i) Each p_i is non-negative, i.e., $p_i \geq 0$.
(ii) The sum of the p_i is 1, i.e., $p_1 + p_2 + \cdots + p_n = 1$.

The probability of an event A, written as P(A), is then defined to be sum of the probabilities of the points in A.

 EXAMPLE 2.26 Three coins are tossed and the number of heads are observed. Find the probability that
(i) at least one head appears
(ii) all heads or all tails appear

Solution Here, the sample space is $S = \{0, 1, 2, 3\}$. Then the probability space obtained from the elements of S is

$$P(0) = \frac{1}{8}, \quad P(1) = \frac{3}{8}, \quad P(2) = \frac{3}{8}, \quad P(3) = \frac{1}{8}$$

It is observed, here, that each probability is non-negative, and the sum of the probabilities is 1. Let A be the event that at least one head appears, and let B be the event that all heads or tails appear, i.e., let

$$A = \{1, 2, 3\} \quad \text{and} \quad B = \{0, 3\}$$

Then, by definition,

$$P(A) = P(1) + P(2) + P(3) = \frac{3}{8} + \frac{3}{8} + \frac{1}{8} = \frac{7}{8}$$

and

$$P(B) = P(0) + P(3) = \frac{1}{8} + \frac{1}{8} = \frac{1}{4}$$

2.9 PROBABILITY OF AN EVENT

If there exists n exhaustive, mutually exclusive, and equally likely outcomes of an experiment and m of them are favourable to an event A, then the probability of the event A is defined by the ratio m/n. Thus, the probability of occurrence of an event A, denoted by $P(A)$, is defined as

$$P(A) = \frac{\text{Number of outcomes favourable to } A}{\text{Total number of possible outcomes}} = \frac{n(A)}{n(S)}$$

2.9.1 Axioms of Probability

The probability function P defined on the class of all events in a finite probability space satisfies the following axioms:
A. The probability of an impossible event is zero, i.e., $P(\varphi) = 0$.
B. The probability of a certain event is 1, i.e., $P(S) = 1$.
C. The probability of any event A is a non-negative real number never greater than 1, i.e., $0 \leq P(A) \leq 1$.

2.9.2 Odds in Favour and Odds Against an Event

If a of the outcomes are favourable to an event A and b of the outcomes are against it as a result of an experiment, then we say that odds are a to b in favour of A, or odds are b to a against A.

2.9.3 Addition Principle

If A and B are events of a sample space S, then

$$P(A \cup B) = P(A) + P(B) - P(A \cap B)$$

Similarly, for any three events A, B, and C,

$$P(A \cup B \cup C) = P(A) + P(B) + P(C) - P(A \cap B) - P(A \cap C) - P(C \cap A) + P(A \cap B \cap C)$$

If the events A and B are mutually exclusive, then $A \cap B = \phi$. So, $P(A \cap B) = 0$; therefore, for the probability of occurrence of either event A or event B is given by

$$P(A \cup B) = P(A) + P(B)$$

If A and B are mutually exclusive and exhaustive events, then $A \cup B = S$, i.e.,

$$P(A \cup B) = P(S) = 1$$

The set of all outcomes not in A is called the *complement* of A in S and is denoted by $\bar{A}$, or (not A), or A^c. Since the complementary events A and $\bar{A}$ are mutually exclusive and exhaustive, it follows that

$$A \cup \bar{A} = S \Rightarrow P(A \cup \bar{A}) = P(S) = 1 \Rightarrow P(A) + P(\bar{A}) = 1$$
$$\Rightarrow P(A) = 1 - P(\bar{A})$$

This is generally referred to as the *complementation* rule.

2.10 CONDITIONAL PROBABILITY

Let E be an event in a sample space S with $P(E) > 0$. The probability that an event A occurs, once E has already occurred, or, specifically, the conditional probability of A given E, denoted by $P(A/E)$, is defined as

$$P(A/E) = \frac{P(A \cap E)}{P(E)}$$

where $P(A/E)$ measures the relative probability of A with respect to the given event E.

Now, let $n(A)$ denote the number of elements in the event A and $n(E)$ denote the number of elements in the event E. Then,

$$P(A \cap E) = \frac{n(A \cap E)}{n(S)}, P(E) = \frac{n(E)}{n(S)}$$

So,

$$P(A/E) = \frac{P(A \cap E)}{P(E)} = \frac{n(A \cap E)}{n(E)}$$

EXAMPLE 2.27 Find the probability of the occurrence of a number greater than 2 in a throw of a die, if it is known that only even numbers can occur.

Solution Let A and B be the events of occurrence of even number and even number greater than 2, respectively. So,

$$A = \{2, 4, 6\} \quad \text{and} \quad B = \{4, 6\}$$

Then

$$A \cap B = \{4, 6\}$$

Simultaneously,

$$n(A \cap B) = 2, n(S) = n(A) = 3$$

Thus,

$$P(B/A) = \frac{n(A \cap B)}{n(A)} = \frac{2}{3}$$

EXAMPLE 2.28 Five persons a, b, c, d, and e are contesting an election. Three persons are to be elected. If one of them, say, d, has been elected uncontested, find the probability that c would be elected.

Solution Let A be the event of electing three persons, one of then being d. This event may occur in $^4C_2 = 6$ ways (since one person d is already elected), i.e., $A = \{abd, acd, aed, bcd, bed, ced\}$. So, $n(A) = 6$. Again, let C be the event of c, being one of the elected persons. Then C has $^3C_1$, i.e., three elements, which are $C = \{acd, bed, ced\}$. So, $n(C) = 3$ and $A \cap C = \{acd, bed, ced\}$, i.e., $n(A \cap C) = 3$. Thus, the required probability is

$$P(C/A) = \frac{n(A \cap C)}{n(A)} = \frac{3}{6} = \frac{1}{2}$$

2.10.1 Multiplication Rule

Suppose A and B are events in a sample space S with $P(A) > 0$. By definition of conditional probability,

$$P(B/A) = \frac{P(A \cap B)}{P(A)}$$

Multiplying both sides by $P(A)$ we get the useful result as

$$P(A \cap B) = P(A) \cdot P(B/A)$$

which yields a formula for the probability that both events A and B occur. It can be extended to m number of events such as $A_1, A_2, \ldots, A_m$, i.e.,

$$P(A_1 \cap A_2 \cap \cdots \cap A_m) = P(A_1) \cdot P(A_2/A_1) \cdots (P(A_m/A_1 \cap A_2 \cdots \cap A_{m-1}))$$

Note Events A and B are *independent* if $P(A \cap B) = P(A)P(B)$, otherwise they are dependent.

EXAMPLE 2.29 Show that the events A, the occurrence of head on the first coin B, and the occurrence of tail on the second coin in the toss of two coins are independent.

Solution Here, sample space $S = \{HH, TH, HT, TT\}$, $n(S) = 4$, where (HH) implies that both the coins exhibit heads, etc. Also,

$$A = \{HH, HT\}, \quad n(A) = 2$$
$$B = \{HT, TT\}, \quad n(B) = 2$$
$$A \cap B = \{HT\}, \quad n(A \cap B) = 1$$

Then,

$$P(A \cap B) = \frac{n(A \cap B)}{n(S)} = \frac{1}{4}$$

and

$$P(A) = \frac{n(A)}{n(S)} = \frac{2}{4} = \frac{1}{2}, \quad P(B) = \frac{n(A)}{n(S)} = \frac{2}{4} = \frac{1}{2}$$

So,

$$P(A \cap B) = P(A) \cdot P(B)$$

Thus, A and B are independent.

EXAMPLE 2.30 A can hit a target four times in five shots; B, three times in four shots; C, two in three shots. They fire one each. Find the probability that at least two shots hit the target.

Solution Given that for A, $n(S) = 5$, $n(A) = 4$.
So,

$$P(A) = \frac{n(A)}{n(S)} = \frac{4}{5}$$

Similarly,

$$P(B) = \frac{3}{4}, \quad P(C) = \frac{2}{3}$$

Thus,

$$P(\bar{A}) = 1 - P(A) = 1 - \frac{4}{5} = \frac{1}{5}, \quad P(\bar{B}) = 1 - P(B) = 1 - \frac{3}{4} = \frac{1}{4}$$

and

$$P(\bar{C}) = 1 - P(C) = 1 - \frac{2}{3} = \frac{1}{3}$$

Now, we will investigate the probable cases.

Case I If all the three shots hit the target then the probability of this event $A \cap B \cap C$ is given by

$P(A \cap B \cap C) = P(A)P(B)P(C)$ (since A, B, and C are independent)

$$= \frac{4}{5} \times \frac{3}{4} \times \frac{2}{3} = \frac{2}{5}$$

Case II If the shots of A and B hit the target and that of C does not, then the probability of this event $A \cap B \cap \overline{C}$ is given by

$P(A \cap B \cap \overline{C}) = P(A)P(B)P(\overline{C})$ (since A, B, and $\overline{C}$ are independent)

$$= \frac{4}{5} \times \frac{3}{4} \times \frac{1}{5} = \frac{1}{5}$$

Case III If the shots of A and C hit the target and that of B does not, then the probability of this event $A \cap \overline{B} \cap C$ is given by

$P(A \cap \overline{B} \cap C) = P(A)P(\overline{B})P(C)$ (since A, $\overline{B}$, C are independent)

$$= \frac{4}{5} \times \frac{1}{4} \times \frac{2}{3} = \frac{2}{15}$$

Case IV If the shots of B and C hit the target and that of A does not, then the probability of this event $\overline{A} \cap B \cap C$ is given by

$P(\overline{A} \cap B \cap C) = P(\overline{A})P(B)P(C)$ (since $\overline{A}$, B, and C are independent)

$$= \frac{1}{5} \times \frac{3}{4} \times \frac{2}{3} = \frac{1}{10}$$

All the above events are mutually exclusive. So, the required probability is

$$\frac{2}{5} + \frac{1}{5} + \frac{2}{15} + \frac{1}{10} = \frac{25}{30} = \frac{5}{6}$$

EXAMPLE 2.31 An array of length 10 is searched for a keyword. On an average, how many steps will it take to find the key?

Solution Assuming that the key is equally likely to be in any position of the array, the expected value of the trial is

$$1 \times \frac{1}{10} + 2 \times \frac{1}{10} + \cdots + 10 \times \frac{1}{10} \quad \text{or} \quad \frac{55}{10}, \text{ i.e., } 5.5$$

Thus, on an average, we can expect to find a keyword in 5.5 steps.

2.11 INDEPENDENT REPEATED TRIALS AND BINOMIAL DISTRIBUTION

We have already discussed the probability spaces associated with an experiment repeated a finite number of times, such as the tossing of a coin three times. This concept of repetition is expressed as follows.

Let S be a finite probability space. By the space of *n independent repeated trials*, we mean the probability space S_n consisting of ordered n-tuple of elements of S, with the probability of an n-tuple defined to be the product of the probabilities of its components,

$$P(S_1, S_2, \ldots, S_n) = P(S_1)P(S_2) \cdots P(S_n)$$

EXAMPLE 2.32 A man throws two dice. Find the probability of not getting 1 (ace) on both the dice.

Solution Let A be the event of getting 1 (ace) in one throw from one die. Then $A = \{1\}$, $n(A) = 1$ and $S = \{1, 2, 3, 4, 5, 6\}$, $n(S) = 6$.

So,
$$P(A) = \frac{n(A)}{n(S)} = \frac{1}{6}$$

Similarly, the probability of getting 1 (ace) in one throw from the second die B,
$$P(B) = \frac{1}{6}$$

So, the probability of getting 1 (ace) on both the dice in one throw is
$$P(A)P(B) = \frac{1}{6} \times \frac{1}{6} = \frac{1}{36}$$

Hence, the probability of not getting 1 (ace) on both the dice is
$$1 - \frac{1}{36} = \frac{35}{36}$$

2.11.1 Repeated Trials with Two Outcomes and Bernoulli Trials

Let us consider an experiment with only two outcomes. Independent repeated trials of such an experiment are called *Bernoulli trials*, named after the Swiss mathematician Jacob Bernoulli (1654–1705). The term *independent trial* means that the outcome of any trial does not depend on the previous outcomes such as tossing a coin. We will name one of the outcomes as success and the other outcome as failure.

Let p denote the probability of success in a Bernoulli trial, and so $q = 1 - p$ is the probability of failure. A *binomial experiment* consists of a fixed number of Bernoulli trials. The notation $B(n, p)$ will be used to denote a binomial experiment with n trials and probability p of success.

Frequently, we are interested in the number of successes in a binomial experiment and not in the order in which they occur.

Thus, the probability of exactly k success in a binomial experiment $B(n, p)$ is given by $P(k) = P(k \text{ successes}) = {}^nC_k p^k q^{n-k}$. The probability of one or more successes is $1 - q^n$.

 The function $P(k) = {}^nC_k p^k q^{n-k}$, $k = 0, 1, 2, \ldots, n$, for a binomial experiment $B(n, p)$, is called the *binomial distribution* since it corresponds to the successive terms of the binomial expansion:
$$(q + p)^n = q^n + {}^nC_1 q^{n-1}p + {}^nC_2 q^{n-2}p^2 + \cdots + p^n$$

The use of the term *distribution* will be explained later in this chapter.

EXAMPLE 2.33 If a coin is tossed n times, find the probability that not more than r heads will appear.

Solution Let A be the event of appearing head in one toss, then
$$P(A) = \frac{n(A)}{n(S)} = \frac{1}{2}, \text{ i.e., } p = \frac{1}{2}$$

Similarly, the probability of appearing tail in one toss, $q = 1/2$. According to the given question, not more than r heads appearing implies that the number of heads may be $0, 1, 2, \ldots, r$.

Thus, the probability of appearing 0 head (E_0) is

$$P(E_0) = {^nC_0} p^0 q^{n-0} = {^nC_0} \left(\frac{1}{2}\right)^0 \left(\frac{1}{2}\right)^n = {^nC_0} \left(\frac{1}{2}\right)^n$$

The probability of appearing 1 head (E_1),

$$P(E_1) = {^nC_1} p^1 q^{n-1} = {^nC_1} \left(\frac{1}{2}\right) \left(\frac{1}{2}\right)^{n-1} = {^nC_1} \left(\frac{1}{2}\right)^{n-1}$$

The probability of appearing r heads (E_r),

$$P(E_r) = {^nC_r} p^r q^{n-r} = {^nC_r} \left(\frac{1}{2}\right)^r \left(\frac{1}{2}\right)^{n-r} = {^nC_r} \left(\frac{1}{2}\right)^n$$

All these events are mutually exclusive, so the required probability

$$= P(E_0) + P(E_1) + \cdots + P(E_r)$$

$$= {^nC_0} \left(\frac{1}{2}\right)^n + {^nC_1} \left(\frac{1}{2}\right)^n + \cdots + {^nC_r} \left(\frac{1}{2}\right)^n$$

$$= ({^nC_0} + {^nC_1} + \cdots + {^nC_r}) \left(\frac{1}{2}\right)^n$$

2.12 RANDOM VARIABLES

A *random variable* is a real-valued function defined over the sample space S of an experiment, i.e., a variable whose value is a number determined by the sample point (outcome of the experiment) of a sample space S is called a *random variable*. A random variable is denoted by $X, Y, Z, \ldots$, etc.

Let R_X denote the set of numbers assigned by a random variable X, and simultaneously R_X is referred to as the range space.

Note X is a function from S to the real numbers x and R_X is the range of X.

For example
(i) Let X be a random variable which is the number of heads obtained in two independent tosses of an unbiased one, i.e., $S = \{HH, HT, TH, TT\}$. Then, $X(HH) = 2$, $X(HT) = 1$, $X(TH) = 1$, and $X(TT) = 0$. Therefore, X can take values 0, 1, and 2.
(ii) A box contains 12 items of which 3 are defective. A sample of three items is selected from the box. The sample space S consists of ${^{12}C_3} = 220$ different samples of size 3. Let X denote the number of defective items in the sample, then X is a random variable with range space, $R_X = \{0, 1, 2, 3\}$.

2.12.1 Probability Distribution of a Random Variable

Let a discrete random variable X take all the possible values $x_1, x_2, \ldots, x_n$, which is also the range space $R_X(= \{x_1, x_2, \ldots, x_n\})$ of X defined on a finite sample space S. Then X with probabilities $p_1, p_2, \ldots, p_n$ satisfies the following conditions:

(i) $p_i = P(X = x_i) \geq 0, i = 1, 2, 3, \ldots, n$, i.e., p_i are all non-negative.

(ii) $\sum_i p_i = p_1 + p_2 + \cdots + p_n = 1$.

The function $p_i = P(X = x_i)$, or $P(x)$ is called the *probability function* of the random variable X. The set of all ordered pairs $(x_1, p_1), \ldots, (x_i, p_i)$, or $(x, P(x))$, usually given in Table 2.2

TABLE 2.2

x_1	x_2	$\cdots$	x_i
p_1	p_2	$\cdots$	p_i

is called the *probability distribution* of the random variable X (Table 2.2).

For example The probability distribution of a random variable X, the number of heads, in two tosses of a coin can be described as follows:

Here, X can take the values 0, 1, or 2. Now,

$$P(X = 0) = P(\text{no heads}) = \frac{1}{4}, \quad P(X = 1) = P(\text{one head}) = \frac{1}{2}$$

$$P(X = 2) = P(\text{two heads}) = \frac{1}{4}$$

Thus, the probability distribution of X is shown in Table 2.3.

TABLE 2.3

X	0	1	2
$P(X)$	$\frac{1}{4}$	$\frac{1}{2}$	$\frac{1}{4}$

EXAMPLE 2.34 Two numbers from 1 to 3 are chosen at random with repetition allowed. Let X denote the sum of the numbers. Find the probability distribution of X.

Solution There are nine ordered pairs summing up the sample space. Here, X assumes the values 2, 3, 4, 5, and 6 with the following probabilities:

$$P(2) = P(1, 1) = \frac{1}{9}, P(3) = P\{(1,2), (2, 1)\} = \frac{2}{9}$$

$$P(4) = P\{(1,3), (2,2), (3,1)\} = \frac{3}{9}$$

$$P(5) = P\{(2,3), (3,2)\} = \frac{2}{9}, P(6) = P(3,3) = \frac{1}{9}$$

Hence, the probability distribution is shown in Table 2.4.

TABLE 2.4

x_i	2	3	4	5	6
$P(x_i)$	$\dfrac{1}{9}$	$\dfrac{2}{9}$	$\dfrac{3}{9}$	$\dfrac{2}{9}$	$\dfrac{1}{9}$

EXAMPLE 2.35 Five defective bulbs are accidentally mixed with twenty good ones. It is not possible to just look at a bulb and tell whether or not it is defective. Find the probability distribution of the number of defective bulbs, if four bulbs are drawn at random from this lot.

Solution Let X denote the number of defective bulbs drawn from four bulbs at random. Here X can take the values 0, 1, 2, 3, or 4. Given,

$$\text{Number of defective bulbs} = 5$$
$$\text{Number of good bulbs} = 20$$
$$\text{Total number of bulbs} = 25$$

Now, $P(X = 0) = P(\text{no defective}) = P(\text{all four good ones})$

$$= \frac{{}^{20}C_4}{{}^{25}C_4} = \frac{20 \times 19 \times 18 \times 17}{25 \times 24 \times 23 \times 22}$$

$$= \frac{969}{2530}$$

$$P(X = 1) = P(\text{one defective and three good ones}) = \frac{{}^{5}C_1 \times {}^{20}C_3}{{}^{25}C_4} = \frac{1140}{2530}$$

$$P(X = 2) = P(\text{two defective and two good ones}) = \frac{{}^{5}C_2 \times {}^{20}C_2}{{}^{25}C_4} = \frac{380}{2530}$$

$$P(X = 3) = P(\text{three defective and one good ones}) = \frac{{}^{5}C_3 \times {}^{20}C_1}{{}^{25}C_4} = \frac{40}{2530}$$

$$P(X = 4) = P(\text{all four defective}) = \frac{{}^{5}C_4}{{}^{25}C_4} = \frac{1}{2530}$$

Thus, the probability distribution of the random variable X is

X :	0	1	2	3	4
$P(X)$:	$\dfrac{969}{2530}$	$\dfrac{1140}{2530}$	$\dfrac{380}{2530}$	$\dfrac{40}{2530}$	$\dfrac{1}{2530}$

2.12.2 Expectation of a Random Variable

There exists two important parameters (or measurements) associated with a random variable X, the *mean* of X, denoted by μ, and the *standard deviation* of X, denoted by σ. The mean μ is also called the *expectation* of X, written as $E(x)$. Actually, the mean μ measures the *central tendency* of X, and the standard deviation σ measures the *dispersion* of X.

Let X be a random variable on a probability space $S = \{a_1, a_2, \ldots, a_m\}$. The mean or expectation of X is defined by

$$\mu = E(X) = X(a_1)P(a_1) + X(a_2)P(a_2) + \cdots + X(a_m)P(a_m)$$

$$= \sum_i X(a_i)P(a_i)$$

In particular, if X possesses the distribution like Table 2.5,

TABLE 2.5

x_1	x_2	$\cdots$	x_n
p_1	p_2	$\cdots$	p_n

then the expectation of X is

$$\mu = E(X) = x_1 p_1 + x_2 p_2 + \cdots + x_n p_n = \sum_i x_i p_i$$

EXAMPLE 2.36 If X is the number of tails in three tosses of a coin, find the expectation of X.

Solution The number of tails which can occur with their respective probabilities are shown in Table 2.6.

TABLE 2.6

x_i	0	1	2	3
P_i	$\frac{1}{8}$	$\frac{3}{8}$	$\frac{3}{8}$	$\frac{1}{8}$

Then the expectation or expected number of tails is

$$\mu = E(X) = 0\left(\frac{1}{8}\right) + 1\left(\frac{3}{8}\right) + 2\left(\frac{3}{8}\right) + 3\left(\frac{1}{8}\right) = \frac{12}{8} = \frac{3}{2}$$

2.12.3 Variance and Standard Deviation of a Random Variable

Let X be a random variable with mean μ and probability distribution as shown in Table 2.7.

TABLE 2.7

x_i	x_2	x_3	$\cdots$	x_n
p_1	p_2	p_3	$\cdots$	p_n

The variance, Var(X), and standard deviation σ of X are defined by

$$\text{Var}(X) = (x_1 - \mu)^2 p_1 + (x_2 - \mu)^2 p_2 + \cdots + (x_n - \mu)^2 p_n$$

$$= \sum_i (x_i - \mu)^2 p_i = E(x - \mu)^2$$

and

$$\sigma = \sqrt{\text{Var}(x)}$$

Comparatively, the following formula is convenient for computing Var(x):

$$\text{Var}(X) = x_1^2 p_1 + x_2^2 p_2 + \cdots + x_n^2 p_n - \mu^2 = E(x^2) - \mu^2$$

EXAMPLE 2.37 If X is the number of tails in three tosses of a coin, find the variance of X.

Solution Let X denote the number of times tails occurs when a coin is tossed three times. The distribution of X appears in Example 2.36, where its mean, $\mu = 3/2$ is computed. The variance of X is computed as follows.

$$\text{Var}(X) = \left(0 - \frac{3}{2}\right)^2 \times \frac{1}{8} + \left(1 - \frac{3}{2}\right)^2 \times \frac{3}{8} + \left(2 - \frac{3}{2}\right)^2 \times \frac{3}{8} + \left(3 - \frac{3}{2}\right)^2 \times \frac{1}{8} = 0.75$$

Alternatively,

$$\text{Var}(X) = 0^2 \times \frac{1}{8} + 1^2 \times \frac{3}{8} + 2^2 \times \frac{3}{8} + 3^2 \times \frac{1}{8} - \left(\frac{3}{2}\right)^2 = 0.75$$

Thus, the standard deviation is $\sigma = \sqrt{0.75} = 0.866$.

2.12.4 Binomial Distribution

The binomial distribution holds under the following conditions:
(i) Trials are repeated under identical conditions for a fixed number of times.
(ii) There are only two mutually exclusive outcomes, such as, success or failure for each trial.
(iii) The probability of success in each trial remains constant and does not change from trial to trial.
(iv) The trials are independent, i.e., the probability of an event in any trial is not affected by the results of any other trial.

The number X of k successes is a random variable with distribution appearing in Table 2.8.

TABLE 2.8

k	0	1	2	$\cdots$	n
$P(k)$	q^n	${}^nC_1 q^{n-1} p$	${}^nC_2 q^{n-2} p^2$	$\cdots$	p^n

Consider the binomial distribution $B(n, p)$. Then
(i) Expected value $E(X) = \mu = np$
(ii) Variance $\text{Var}(X) = \sigma^2 = npq$
(iii) Standard deviation $\sigma = \sqrt{npq}$

2.13 RECURSION

Let n be a natural number. We define $n!$ as

$$n! = n(n-1)!, \quad \text{for } n \geq 1$$

which is most commonly used for its precise notation. It is easy to define it in terms of itself, i.e., recursively.

Recursion is an elegant and powerful method for problem-solving technique. This is used extensively in both discrete mathematics and computer science. We can use recursion to define sequences, functions, sets, algorithms, and many more.

2.13.1 Recursively Defined Sequences

A sequence is a discrete structure which is used to represent ordered list of elements. A sequence can be defined as a function whose domain is some infinite set of integers (often N) and whose range is a set of real numbers. The notation $\{a_n\}$ will be considered here to denote the sequence, where a_n is its term.

Consider the following norms for generating a sequence:
(a) Initialize with 1
(b) Given a term of sequence, create the next term by adding 2 to it

Following the above rules, we can generate terms of sequence as

$$1, 3, 5, 7, 9, 11, \ldots$$

If we denote the sequence as

$$a_1, a_2, a_3, a_4, \ldots$$

Then we can express the above norms, in general (particularly for the above sequences), as

$$a_n = a_{n-1} + 2, \quad \text{for } n \geq 2 \quad \text{and} \quad a_1 = 1$$

which is an example of recurrence relation and $a_1 = 1$ is an initial condition.

2.13.2 Recursive Definition

Let us consider the following sequence

$$2, 2^2, 2^4, 2^8, 2^{14}, \ldots$$

which has three characteristic steps as follows:

Step I $2 \in$ sequence.
Step II If $n \in$ sequence then $2^n \in$ sequence.
Step III Every term (or number) of the sequence is generated by the applications of properties (a) and (b) only. This type of definition is known as a *recursive* definition. Thus, the above properties can be used to define a sequence or a set or a function recursively.

Recursive definition consists of the following three steps:

Basis step (Step I): Specify the primitive element(s) (at least one).

Recursive step (Step II): Rule(s) for generating new element from existing elements.

Terminal step (Step III): Conforms that the first two steps, basis and recursive clauses, are the only ways to obtain the elements.

EXAMPLE 2.38 Show the recursive definitions of the following sequence:
(a) 2, 4, 6, 8, 10, ...
(b) 0, 1, 00, 01, 10, 11, 000, 001, 010, 011, ...

Solution
(a) The recursive definition for sequence 2, 4, 6, 8, 10, ... is as follows:
  Step I $2 \in$ sequence.
  Step II If $n \in$ sequence then the $n + 2 \in$ sequence.

(b) Let the sequence for given binary numbers be B. Then it can be defined recursively as follows:
  Step I $0 \in B$ and $1 \in B$.
  Step II If $b \in B$ then $b\,0 \in$ and $b\,1 \in B$ (*concatenation*).
  It may be noted here that the terminal step (Step III) is omitted for convenience.

2.13.3 Recursively Defined Sets

Recursive definition of sets possesses two parts, one a basis step and the other a recursive one (omitting the terminal step). In the basis step, primitive elements (at least one) are specified. In the recursive step, rules for generating new elements in the set from those already known to be in the set are provided.

For example Consider a set of multiples of 3 which is denoted by a set A, say. Then the recursive definition to define A is

Step I $3 \in A$.
Step II If $x \in A$ then $x + 3 \in A$.

We can enumerate elements of A as $\{3, 6, 9, 12, 15, \ldots\}$.

EXAMPLE 2.39 Let B be the set defined recursively as follows:

Step I $2 \in B$.
Step II If $n \in B$ then $n^2 \in B$.
Describe the set by listing method.

Solution From the basis step, $2 \in B$.

From the recursive step, $4 \in B$, considering $n = 2$.
Again, let $n = 4$, then $16 \in B$. Repeating the process, we get

$$B = \{2, 4, 16, 256, 65536, \ldots\}$$

EXAMPLE 2.40 Define the following sets recursively:

(i) $\{0, 00, 10, 100, 110, 0000, 1010, \ldots\}$
(ii) $\{b, bb, bbb, bbbb, \ldots\}$

Solution

(i) Let $A = \{0, 00, 10, 100, 110, 0000, 1010, \ldots\}$
  (a) $0 \in A$
  (b) if $x \in A$ then $1x \in A$ and $xx \in A$
(ii) Let $B\{b, bb, bbb, bbbb, \ldots\}$
  (a) $b \in B$
  (b) if $x \in B$ then $bx \in B$

2.13.4 Recursively Defined Functions

We have already mentioned the recursive function in Section 1.29. In recursive definition of sets, we have used recursive step to generate new elements from the given elements. The same notion will be implemented here to describe in detail the recursive function.

I. *Basis step* A sum of initial values of the function $f(x), f(x + 1), \ldots, f(x + k - 1)$ are prescribed. An equation that specifies such initial values is called an *initial condition*.

II. *Recursive step* A formula to find $f(n)$ from the k preceding functional values $f(n - 1)$, $f(n - 2), \ldots, f(n - k)$ is created. Such a formula is called *recursive formula or recurrence relation*. It may be noted here that to define a sequence recursively, a recursive formula must be accompanied by the information about the beginning of the sequence. Hence, the recursive function definition has one or more initial conditions and a recurrence relation.

For example

(i) Consider the most popular factorial notation. The factorial notation f is defined by $f(n) = n!$, where $f(0) = 1!$.

Here, f is recursively defined as
(a) $f(0) = 1$ (initial value)
(b) $f(n) = n.f(n-1)$, $n \geq 1$ (recurrence relation)

(ii) *Fibonacci numbers*: The number $1, 1, 2, 5, 3, 8, \ldots$ are Fibonacci numbers. They possess a Fibonacci number, except the first two and is the sum of two immediately preceding Fibonacci numbers. This yields the recursive definition of nth Fibonacci number as
(a) $f_1 = f_2 = 1$ (initial condition)
(b) $f_n = f_{n-1} + f_{n-2}$ (recurrence relation).

EXAMPLE 2.41 Let f be defined recursively by

$$f(n+1) = 2f(n) + 3, \quad f(0) = 3$$

Determine $f(1), f(2), f(3),$ and $f(4)$.

Solution From the recursive definition, it follows that

$$f(1) = 2f(0) + 3 = 2.3 + 3 = 9$$
$$f(2) = 2f(1) + 3 = 2.9 + 3 = 21$$
$$f(3) = 2f(2) + 3 = 2.21 + 3 = 45$$
$$f(4) = 2f(3) + 3 = 2.45 + 3 = 93$$

2.14 RECURRENCE RELATION

Permutations and combinations are one of the most fundamental tools for counting the elements of finite sets. However, these tools are inadequate to handle problems of computer science. So, an alternative like, *recurrence relation* is inevitable to define the terms of a sequence.

In recursive definition, the recursive step supports to find new terms from the existing (preceding) terms. This is defined in terms of a formula called *recurrence relation*. This is often called *recurrence equation* or *difference equation*.

Definition A recurrence relation for the sequence $\{a_n\}$ is a formula that expresses a_n in terms of one or more preceding terms of the sequence, namely, $a_0, a_1, \ldots, a_{n-1}$, for $n \geq n_0$. Here n_0, a non-negative integer is used to define the initial condition. A sequence is called a *solution of a recurrence relation* if its terms satisfy the recurrence relation.

For example

(i) The recurrence relation $a_n = a_{n-1} + 2$ with initial condition $a_1 = 1$ recursively defines the sequence $1, 3, 5, 7, 9, 11, \ldots$.
(ii) The recurrence relation ${}^nC_r = {}^{n-1}C_r + {}^{n-1}C_{r-1}$, for all n defines binomial coefficient with initial conditions ${}^nC_0 = 1$ and ${}^nC_n = 1$.

2.14.1 Order and Degree of Recurrence Relation

The order of a recurrence relation is defined to be the difference between the highest and the lowest subscripts of the dependent variable (a_r or y_k) appearing in the relation.

For example

(i) The equation $a_r - 3a_{r-1} + 2a_{r-2} = 0$ is a recurrence relation of order 2

$$(\text{i.e., } r - (r-2) = 2)$$

(ii) The equation $2y_r + 3y_{r-1} = 0$ is a first-order recurrence relation.

The degree of a recurrence relation is defined to be the highest power of a_r occurring in the relation.

For example The recurrence relation $a_r^4 + 2a_{r-1}^2 + 18a_{r-1}^3 + a_{r-3} = 0$ has the degree 4, as the highest power of a_r is 4.

2.14.2 Linear Homogeneous and Non-Homogeneous Recurrence Relations

A recurrence relation is called *linear* recurrence relation if its degree is one. In general, a recurrence relation of the form

$$c_0 a_r + c_1 a_{r-1} + \cdots + c_k a_{r-k} = f(r)$$

where $c_0, c_1, \ldots, c_k$ are constants (real numbers), $c_k \neq 0$ is called *linear* recurrence relation of order k with constant coefficients.

The above recurrence relation is said to be *linear homogeneous relation* of order k if $f(r) = 0$; otherwise, it is called a *non-homogeneous relation* provided $f(r) \neq 0$.

For example The recurrence relation

$$a_{r+2} + 6a_{r+1} + 18a_r = 0$$

is a linear homogeneous equation of order 2. Also, the recurrence relation

$$a_{r+2} - 6a_{r+1} - a_r = 3r + 2^r$$

is a linear non-homogeneous equation of order 2.

2.14.3 Solution of Linear Recurrence Relation with Constant Coefficients

The complete solution of a linear recurrence relation (or, equation) with constant coefficients consists of two parts, like
(i) The homogeneous solution in absence of $f(r)$, i.e., when $f(r) = 0$.
(ii) The particular solution in presence of $f(r)$, i.e., when $f(r) \neq 0$.

Let

$$a^{(h)} = \left(a_0^{(h)}, a_1^{(h)}, \ldots, a_r^{(h)}, \ldots\right)$$

be the homogeneous solution of the equation

$$c_0 a_r + c_1 a_{r-1} + \cdots + c_k a_{r-k} = f(r)$$

and
$$a^{(p)} = \left(a_0^{(p)}, a_1^{(p)}, \ldots, a_r^{(p)}, \ldots\right)$$
denote the particular solution, then
$$c_0 a_r^{(h)} + c_1 a_{r-1}^{(h)} + \cdots + c_k a_{r-k}^{(h)} = 0 \quad (2.8)$$
and
$$c_0 a_r^{(p)} + c_1 a_{r-1}^{(p)} + \cdots + c_k a_{r-k}^{(p)} = f(r) \quad (2.9)$$

Adding Eqs (2.8) and (2.9),
$$c_0[a_r^{(h)} + a_r^{(p)}] + c_1[a_{r-1}^{(h)} + a_{r-1}^{(p)}] + \cdots + c_k[a_{r-k}^{(h)} + a_{r-k}^{(p)}] = f(r)$$

which shows that $a_r = a_r^{(h)} + a_r^{(p)}$ is the complete solution of the recurrence relation.

2.14.4 Homogeneous Solution

Consider
$$c_0 a_r + c_1 a_{r-1} + \cdots + c_k a_{r-k} = 0 \quad (2.10)$$
as a linear homogeneous recurrence relation of order k with constant coefficients.

Let the solution of the above equation be of the form $a_r = A\alpha_1^r$. Substituting this into Eq. (2.10), we get
$$c_0(A\alpha_1^r) + c_1(A\alpha_1^{r-1}) + c_2(A\alpha_1^{r-2}) + \cdots + c_k(A\alpha_1^{r-k}) = 0$$
$$\Rightarrow A\alpha_1^{r-k}[c_0\alpha_1^k + c_1\alpha_1^{k-1} + \cdots + c_k] = 0$$
$$\Rightarrow c_0\alpha_1^k + c_1\alpha_1^{k-1} + \cdots + c_k = 0$$

which is called the *characteristic equation* of the recurrence relation and α_1 is called the *characteristic root*. The solution $A\alpha_1^r$ of the recurrence equation (2.10) is a homogeneous solution.

Note A characteristic equation of the kth degree possesses k characteristic roots.

The homogeneous solution can be written in different forms depending on the nature of the characteristic roots of the characteristic equation. In case the roots are *real* and *distinct*, i.e., the roots are $\alpha_1, \alpha_2, \ldots, \alpha_n$, then the homogeneous solution of the recurrence relation (2.10) is given by
$$a_r^{(h)} = A_1\alpha_1^r + A_2\alpha_2^r + \cdots + A_k\alpha_k^r$$

If the roots are *repeated*, then let α_1 be a root of multiplicity m and the corresponding homogeneous solution is given by
$$a_r^{(h)} = \left(A_1 r^{m-1} + A_2 r^{m-2} + \cdots + A_{m-1} r + A_m\right)\alpha_1^r$$

where $A_1, A_2, \ldots, A_m$ are constants to be determined by the initial conditions.

Thus, every recurrence equation has a homogeneous part. On the other hand, if the given recurrence equation is non-homogeneous, then its homogeneous part can be obtained by assigning right-hand side to zero.

To find the characteristic equation, we use the following procedures:
(i) Find the order of the recurrence equation.
(ii) Consider any variable and substitute it. For example, replace a_r, a_{r-1}, a_{r-2} by x^3, x^2, x, respectively, in homogeneous part of recurrence equation of order 3. Therefore, the characteristic equation of the recurrence equation $a_r = 3a_{r-1} + 2a_{r-2}$ is given by $x^2 - 3x - 2 = 0$, because it is of order 2. Also, the characteristic equation of the recurrence equation $a_n - 4a_{n-1} + 4a_{n-2} = 2^n + n$ is given by $x^2 - 4x + 4 = 0$.

EXAMPLE 2.42 Solve the recurrence relation $a_r - 6a_{r-1} + 8a_{r-2} = 0$.

Solution The characteristic equation is

$$\alpha^2 - 6\alpha + 8 = 0 \text{ or } x^2 - 6x + 8 = 0$$
$$\Rightarrow (x-2)(x-4) = 0$$
$$\Rightarrow x = 2 \text{ or } 4$$

Hence, the homogeneous solution of the equation is

$$a_r = A_1 2^r + A_2 4^r$$

EXAMPLE 2.43 Solve the recurrence relation $a_r - 4a_{r-1} + 4a_{r-2} = 0$.

Solution The characteristic equation is given by

$$x^2 - 4x + 4 = 0$$
$$\Rightarrow (x-2)^2 = 0$$
$$\Rightarrow x = 2 \text{ or } 2$$

Hence, the homogeneous solution of the equation is

$$a_r = (A_1 + A_2 r) 2^r$$

EXAMPLE 2.44 Solve $a_r - 5a_{r-1} + 6a_{r-2} = 0$, where $a_0 = 2$ and $a_1 = 5$.

Solution The characteristic equation of the given recurrence relation is

$$x^2 - 5x + 6 = 0 \Rightarrow (x-2)(x-3) = 0 \Rightarrow x = 2 \text{ or } 3$$

Since the roots are real and distinct, the homogenous solution of recurrence relation is

$$a_r = A_1 2^r + A_2 3^r$$

We have

$$a_0 = 2, \text{ i.e., } A_1 + A_2 = 2 \qquad (2.11)$$

and

$$a_1 = 5, \text{ i.e., } A_1 (2)^1 + A_2 (3)^1 = 5$$
$$\Rightarrow 2A_1 + 3A_2 = 5 \qquad (2.12)$$

Solving Eqs (2.11) and (2.12), we get $A_1 = 1$ and $A_2 = 1$.
Thus, the required solution is $a_r = 2^r + 3^r$.

EXAMPLE 2.45 Solve the difference equation $a_r + a_{r-1} + a_{r-2} = 0$.

Solution The characteristic equation is $x^2 + x + 1 = 0$.
The roots are imaginary, i.e.,
$$x = \frac{-1 + i\sqrt{3}}{2} \quad \text{or} \quad \frac{-1 - i\sqrt{3}}{2}$$

Thus, the homogeneous solution of the equation is
$$a_r = A_1 \left[\frac{-1 + i\sqrt{3}}{2}\right]^r + A_2 \left[\frac{-1 - i\sqrt{3}}{2}\right]^r$$

EXAMPLE 2.46 Consider the Fibonacci sequence $\{1, 1, 2, 3, 5, \ldots,\}$. The recurrence relation for the Fibonacci sequence of numbers is $a_r = a_{r-1} + a_{r-2}, r \geq 2$ with initial conditions $a_0 = 1, a_1 = 1$. Determine the solution.

Solution The roots of the characteristic equation $x^2 - x - 1 = 0$ are
$$x_1 = \frac{1 + \sqrt{5}}{2} \quad \text{and} \quad x_2 = \frac{1 - \sqrt{5}}{2}$$

So, the homogeneous solution of the given recurrence relation is
$$a_r = A_1 \left(\frac{1 + \sqrt{5}}{2}\right)^r + A_2 \left(\frac{1 - \sqrt{5}}{2}\right)^r \tag{2.13}$$

where A_1 and A_2 are the constants whose values can be determined, using the initial conditions $a_0 = 1$ and $a_1 = 1$.
Put $r = 0$ in Eq. (2.13)
$$A_1 + A_2 = 1 \tag{2.14}$$
Put $r = 1$ in Eq. (2.13)
$$A_1 \left(\frac{1 + \sqrt{5}}{2}\right) + A_2 \left(\frac{1 + \sqrt{5}}{2}\right) = 1 \tag{2.15}$$

Solving Eqs (2.14) and (2.15) for A_1 and A_2, we get
$$A_1 = \frac{1}{\sqrt{5}} \left(\frac{1 + \sqrt{5}}{2}\right) \quad \text{and} \quad A_2 = \frac{1}{\sqrt{5}} \left(\frac{1 - \sqrt{5}}{2}\right)$$

Thus, the explicit formula for Fibonacci sequence is
$$A_1 = \frac{1}{\sqrt{5}} \left[\left(\frac{1 + \sqrt{5}}{2}\right)^{r+1\neq} - \left(\frac{1 - \sqrt{5}}{2}\right)^{r+1}\right], \quad r = 0, 1, 2, \ldots$$

2.14.5 Particular Solution

Homogeneous Equations

When the difference equation is of homogeneous linear type, then the particular solution of the equation can be determined by putting the values of the initial conditions in the homogeneous solution.

EXAMPLE 2.47 Solve the difference equation $a_r - 4a_{r-1} + 4a_{r-2} = 0$ and find the particular solution, given that $a_0 = 1$ and $a_1 = 6$.

Solution The characteristic equation is

$$x^2 - 4x + 4 = 0 \implies (x-2)^2 = 0 \implies x = 2, 2$$

Therefore, the homogeneous solution is

$$a_r^{(h)} = (A_1 + A_2 r) \cdot 2^r \qquad (2.16)$$

Putting $r = 0$ and $r = 1$ in Eq. (2.16), we get

$$a_0 = (A_1 + 0) \cdot 2^0 = 1 \implies A_1 = 1$$
$$a_1 = (A_1 + A_2) \cdot 2^1 = 6 \implies A_1 + A_2 = 3, \text{ i.e., } A_2 = 2$$

Hence, the particular solution is

$$a_r^{(p)} = (1 + 2r) \cdot 2^r$$

EXAMPLE 2.48 Solve the difference equation $a_r - 7a_{r-1} + 10a_{r-2} = 0$ satisfying the conditions $a_0 = 0$ and $a_1 = 6$.

Solution The characteristic equation is

$$x^2 - 7x + 10 = 0 \implies (x-2)(x-5) = 0 \implies x = 2 \text{ or } 5$$

Therefore, the homogeneous solution is

$$a_r^{(h)} = A_1 \cdot 2^r + A_2 \cdot 5^r \qquad (2.17)$$

Putting $r = 0$ and $r = 1$ in Eq. (2.17), we get

$$a_0 = A_1 + A_2 = 0 \implies A_1 + A_2 = 0 \implies A_1 = -A_2$$
$$a_1 = A_1 \cdot 2 + A_2 \cdot 5 = 6 \implies 2A_1 + 5A_2 = 6 \implies A_2 = 2, A_1 = -2$$

Hence, the particular solution is $a_r^{(p)} = -2 \cdot 2^r + 2 \cdot 5^r$.

Non-Homogeneous Equations

To find the particular solution of non-homogeneous linear difference equation for which there exists couple of methods as follows:

  I. Undetermined coefficients method
 II. E and Δ operator method

I. *Undetermined coefficients method* The procedure for obtaining the particular solution of a recurrence relation depends on the form of $f(r)$, i.e., the right-hand side of the given relation. We consider a final solution according to the form of $f(r)$, containing a number of unknown constant coefficients, which have to be determined. Consequently, we use the following rules.

Rule 1 When $f(r)$ is of the form of $f(r) = C$, a constant, and 1 is not a root of characteristic equation, then the particular solution to be assumed as P, a constant.

Rule 2 When $f(r)$ is of the form of β^r, in which β is not a characteristic root, then the particular solution to be assumed as $P\beta^r$.

Rule 3 When $f(r)$ is a polynomial of degree t in r, i.e.,

$$f_1 r^t + f_2 r^{t-1} + \cdots + f_t r + f_{t+1}$$

then the particular solution to be assumed as

$$P_1 r^t + P_2 r^{t-1} + \cdots + P_t r + P_{t+1}$$

Rule 4 When $f(r)$ is of the form of $(F_1 r^t + F_2 r^{t-1} + \cdots + F_t r + F_{t+1})\beta^r$, where β is not a characteristic root of the recurrence relation, then the particular solution to be assumed as

$$(P_1 r^t + P_2 r^{t-1} + \cdots + P_t r + P_{t+1})\beta^r$$

Rule 5 When $f(r)$ is of the form of $(F_1 r^t + F_2 r^{t-1} + \cdots + F_t r + F_{t+1})\beta^r$, where β is a characteristic root of multiplicity $m-1$, then the particular solution to be assumed as

$$r^{m-1}(P_1 r^t + P_2 r^{t-1} + \cdots + P_t r + P_{t+1})\beta^r$$

The general form of particular solution to be assumed for the special forms of $f(r)$ can also be represented in Table 2.9.

TABLE 2.9 Particular soulution in tabular form

$f(r)$	Assumed solution
C	P
β^r	$P\beta^r$
$f_1 r^t + f_2 r^{t-1} + \cdots + f_t r + f_{t+1}$	$P_1 r^t + P_2 r^{t-1} + \cdots + P_t r + P_{t+1}$
$(F_1 r^t + F_2 r^{t-1} + \cdots + F_t r + F_{t+1})\beta^r$	$(P_1 r^t + P_2 r^{t-1} + \cdots + P_t r + P_{t+1})\beta^r$
$(F_1 r^t + F_2 r^{t-1} + \cdots + F_t r + F_{t+1})\beta^r$	$r^{m-1}(P_1 r^t + P_2 r^{t-1} + \cdots + P_t r + P_{t+1})\beta^r$

Here, we will present some numericals illustrating the evaluation of particular solutions.

EXAMPLE 2.49 Find the particular solution of the recurrence relation $a_r - 5a_{r-1} + 6a_{r-2} = 1$.

Solution Here, the right-hand side is a constant quantity. So, the general form of the particular solution is assumed as $a_r^{(p)} = P$, a constant.

Putting the value in the given recurrence relation, we get

$$P - 5P + 6P = 1 \implies 2P = 1 \implies P = 1/2$$

Thus, the particular solution of the given relation is

$$a_r^{(p)} = \frac{1}{2}$$

EXAMPLE 2.50 Determine the particular solution of the recurrence relation

$$a_r + 5a_{r-1} + 6a_{r-2} = 3r^2$$

Solution Since the order of the recurrence relation is 2, so we can assume the general form of the particular solution as

$$a_r^{(p)} = P_1 r^2 + P_2 r + P_3$$

Substituting this solution into the given relation, we get

$$(P_1 r^2 + P_2 r + P_3) + 5\lfloor P_1(r-1)^2 + P_2(r-1) + P_3 \rfloor$$
$$+ 6\lfloor P_1(r-2)^2 + P_2(r-2) + P_3 \rfloor = 3r^2$$

$$\implies 12 P_1 r^2 + (12 P_2 - 34 P_1) r + (29 P_1 - 17 P_2 + 12 P_3) = 3r^2$$

Comparing the coefficients of r^2, r, and constants, we obtain the following equations:

$$12 P_1 = 3, \quad 12 P_2 - 34 P_1 = 0, \quad 29 P_1 - 17 P_2 + 12 P_3 = 0$$

Solving these equations, we get

$$P_1 = \frac{1}{4}, \quad P_2 = \frac{17}{24}, \quad P_3 = \frac{115}{288}$$

Thus, the required particular solution is

$$a_r^{(p)} = \frac{1}{4} r^2 + \frac{17}{24} r + \frac{115}{288}$$

II. E and Δ operator method By E and Δ operator method, we will find the solution of

$$C_0 y_{n+r} + C_1 y_{n+1} + \cdots + C_n y_n = f(n) \qquad (2.18)$$

Equation (2.18) can also be written as

$$C_0 E^r y_n + C_1 E_{r-1} y_n + \cdots + C_n y_n = f(n)$$
$$\implies (C_0 E^r + C_1 E^{r-1} + \cdots + C_n) y_n = f(n)$$
$$\implies P(E) y_n = f(n)$$

where $P(E) = C_0 E^r + C_1 E^{r-1} + \cdots + C_n$

Thus,

$$y_n = \frac{f(n)}{P(E)} \qquad (2.19)$$

To find the particular solution of Eq. (2.19) for different forms of $f(n)$, we have the following rules.

Rule 1 When $f(n)$ is constant, A, say, we know that the operation of E on the constant is equal to the constant itself, i.e., $EA = A$.

Therefore,

$$P(E)A = (C_0 E^r + C_1 E^{r-1} + \cdots + C_n)A$$
$$= (C_0 + C_1 + \cdots + C_n)A$$
$$= P(1)A$$

$$\frac{A}{P(E)} = \frac{1}{P(1)} A, \quad P(E) \neq 0$$

Using Eq. (2.19), the particular solution of Eq. (2.18) is

$$y_n = \frac{A}{P(1)}, \quad P(1) \neq 0$$

in which $P(1)$ is obtained by putting $E = 1$ in $P(E)$.

Rule 2 When $f(n)$ is of the form Az^n, we have

$$P(E)(Az^n) = \left[C_0 E^r + C_1 E^{r-1} + \cdots + C_n\right](Az^n)$$
$$= A\left[C_0 z^{r+n} + C_1 z^{r+n-1} + \cdots + C_n z^n\right]$$
$$= A\left[C_0 z^r + C_1 z^{r-1} + \cdots + C_n\right] z^n$$
$$= A P(z) z^n$$

$$\Rightarrow \quad \frac{Az^n}{P(E)} = \frac{Az^n}{P(z)}, \text{ provided } P(z) \neq 0$$

Thus,

$$y_n = \frac{Az^n}{P(z)}, \quad P(z) \neq 0$$

Let $A = 1$, then

$$y_n = \frac{z^n}{P(z)}$$

When $P(z) = 0$

(i) $(E - z)y_n = Az^n$

For this, the particular solution becomes

$$y_n^{(p)} = A \frac{1}{E - z} z^n = Anz^{n-1}$$

(ii) $(E - z)^2 y_n = A z^n$
The particular solution is then
$$y_n^{(p)} = A \frac{1}{(E - z)^2} z^n = A \frac{n(n - 1)}{2!} z^{n-2}$$

(iii) $(E - z)^3 y_n = A z^n$
The particular solution then becomes
$$y_n^{(p)} = A \frac{1}{(E - z)^3} z^n = A \frac{n(n - 1)(n - 2)}{3!} z^{n-3} \text{ and so on}$$

Rule 3 When $f(n)$ be a polynomial of degree m in n.

We know that $E = 1 + \Delta$. So, $P(E) = P(1 + \Delta)$ or
$$\frac{1}{P(E)} = \frac{1}{P(1 + \Delta)} = b_0 + b_1 \Delta + \cdots + b_m \Delta^m + \cdots$$
expanding in ascending powers of Δ up to Δ^m.

$$\Rightarrow \frac{1}{P(E)} f(n) = (b_0 + b_1 \Delta + \cdots + b_m \Delta^m) f(n)$$
$$= b_0 f(n) + b_1 \Delta f(n) + \cdots + b_m \Delta^m f(n)$$

remaining terms vanish because $f(n)$ is a polynomial of degree m.

Thus, the particular solution, in this case, will be
$$y_n^{(p)} = b_0 f(n) + b_1 \Delta f(n) + \cdots + b_m \Delta^m f(n)$$

Rule 4 When $f(n)$ is of the form $f(n) z^n$, where $f(n)$ is a polynomial of degree m, we have,
$$E^r[f^n f(n)] = f^{r+n} f(n + r) = f^r f^n E^r f(n) = z^n (zE)^r f(n)$$
Similarly,
$$\frac{1}{P(E)} [z^n f(n)] = z^n \frac{1}{P(zE)} [f(n)] = z^n \frac{1}{P[z(1 + \Delta)]} [f(n)]$$
$$(\because E = 1 + \Delta)$$

$$\Rightarrow \frac{1}{P(E)} [z^n f(n)] = z^n [P(z + z\Delta)]^{-1} f(n)$$

EXAMPLE 2.51 Solve the recurrence relation $a_r - 3 a_{r-1} + 2 a_{r-2} = 2^r$.

Solution The given equation can be rewritten as $(E^2 - 3E + 2) a_r = 2^r$. Therefore, $P(E) = E^2 - 3E + 2 = (E - 2)(E - 1)$.

The characteristic equation of the given recurrence relation is $P(E) = 0$,
$$\Rightarrow \qquad (E - 2)(E - 1) = 0 \quad \Rightarrow E = 1 \text{ or } 2$$

So, the homogeneous solution is

$$a_r^{(h)} = A_1 1^n + A_2 2^n$$

The particular solution is

$$\text{P.I.} = a_r^{(p)} = \frac{1}{P(E)} 2^r = \frac{1}{(E-2)(E-1)} 2^r$$

$$= r \frac{1}{(E-1)} 2^{r-1} = r \frac{1}{(2-1)} 2^{r-1} = r 2^{r-1}$$

Hence, the complete solution is

$$a_r = a_r^{(h)} + a_r^{(p)} = A_1 1^n + A_2 2^n + r 2^{r-1}$$

EXAMPLE 2.52 Solve the recurrence relation $a_{r+2} - 7a_{r+1} + 10a_r = 12e^{3r} + 4^r$.

Solution The given recurrence relation can be rewritten as $a_r = 12e^{3r} + 4^r (E^2 - 7E + 10)$. The characteristic equation of the given recurrence relation is

$$m^2 - 7m + 10 = 0 \implies (m-5)(m-2) = 0$$

$$\implies m = 2 \text{ or } 5$$

So, the homogeneous solution is $A_1 2^r + A_2 5^r$.

The particular solution is

$$\text{P.I.} = \frac{1}{E^2 - 7E + 10} (12 e^{3r} + 4^r)$$

$$= 12 \frac{1}{E^2 - 7E + 10} e^{3r} + \frac{1}{E^2 - 7E + 10} 4^r$$

$$= 12 \frac{e^{3r}}{(e^3)^2 - 7(e^3) + 10} + \frac{4^r}{(4)^2 - 7(4) + (10)}$$

$$= \frac{12 e^{3r}}{e^6 - 7e^3 + 10} - \frac{4^r}{2}$$

Thus, the complete solution of the given recurrence relation is

$$= A_1 2^r + A_2 5^r + \frac{12 e^{3r}}{e^6 - 7e^3 + 10} - \frac{4^r}{2}$$

2.15 GENERATING FUNCTIONS

Generating functions are a powerful and a useful tool for representing sequences efficiently by coding the terms of a sequence as coefficients of powers of a variable x in a power series. These functions can also be used to solve counting problems such as the one of them is the number of ways to select or distribute objects of different kinds. Generating functions can be utilized fruitfully to solve recurrence relations by transforming the terms of a sequence into an equation. This equation can then be solved to find a closed form for generating function. With the help of this closed form, the coefficients of the power series for the generating function can be determined, solving the original recurrence relation.

If $a_0, a_1, a_2, \ldots, a_n$ is an infinite sequence of numbers, then the *generating function* $G(x)$, for the infinite sequence $(a_0, a_1, a_2, \ldots, a_n, \ldots,)$ is the infinite series (or power series), which is represented as

$$G(x) = a_0 + a_1 x + a_2 x^2 + \cdots + a_n x^n + \cdots = \sum_{n=0}^{\infty} a_n x^n$$

Note The generating function for $\{a_n\}$, stated above, is at times called the *ordinary generating function* of $\{a_n\}$ to distinguish it from other types of generating functions for this sequence, since mathematicians often use other kinds of generating functions.

For any sequence $\{a_n\}$, we write $G(x)$ to denote the generating function of $\{a_n\}$. Given a sequence, we can obtain its generating function and its inverse.

For example The generating function of $a_n = \alpha^n, n \geq 0$ is

$$\alpha^0 + \alpha^1 x + \alpha^2 x^2 + \alpha^3 x^3 + \cdots$$

This infinite series can be written in a closed form as $1/(1 - \alpha x)$ which is rather a compact way to represent the sequence $\{a_n\}$ or $(1, \alpha, \alpha^2, \ldots,)$.

Therefore, a generating function in which the coefficients of x^n are sequential terms of the sequence $\{a_n\}$ is called the *binomial generating function* of the sequence $\{a_n\}$.

For example The binomial generating function for the sequence $\{a_n\} = \{1, 2, 3, \ldots, r, \ldots,\}$ can be evaluated as follows.

Let $B(x)$ be the binomial generating function of the given sequence. Then,

$$B(x) = \sum_{n=0}^{\infty} a_n x^n = \sum_{n=0}^{\infty} r x^n \quad \text{(since } a_n = r\text{)}$$

$$= 1 + 2x + 3x^2 + \cdots \quad (2.20)$$

Multiplying Eq. (2.20) by x and then subtracting the resulting one from Eq. (2.20), we get

$$(1 - x)B(x) = 1 + x + x^2 + \cdots + \cdots = \frac{1}{1 - x}$$

i.e.,

$$B(x) = \frac{1}{(1 - x)^2}$$

which is the required binomial generating function.

Let $E(x)$ be a generating function for a sequence $\{a_n\}$ as follows:

$$E(x) = a_0 + a_1 x + a_2 \frac{x^2}{2!} + a_3 \frac{x^3}{3!} + \cdots = \sum_{n=0}^{\infty} a_n \frac{x^n}{n!}$$

which is an exponential series.

Thus, the generating function $E(x)$, defined above, for a sequence $\{a_n\}$ is called the *exponential generating function*.

For example The exponential generating function for the sequence $\{a_n\} = \{1, 3, 3^2, \ldots, 3^n, \ldots,\}$ can be determined as follows.

Here, the general term is $a_n = 3^n$. Let $E(x)$ be the exponential generating function for the given sequence. Then,

$$E(x) = \sum_{n=0}^{\infty} a_n \frac{x^n}{n!} = \sum_{n=0}^{\infty} 3^n \frac{x^n}{n!} = 1 + 3x + \frac{(3x)^2}{2!} + \frac{(3x)^3}{3!} + \cdots$$
$$= e^{3x}$$

which is the exponential generating function for the given sequence.

Some more sequences and their generating functions are presented below:

$$(0, 0, 0, \ldots,) \leftrightarrow 0.x^0 + 0.x^1 + 0.x^2 + \cdots = 0$$
$$(1, 0, 0, 0, \ldots,) \leftrightarrow 1.x^0 + 0.x^1 + 0.x^2 + \cdots = 1$$
$$(2, 3, 1, 0, \ldots,) \leftrightarrow 2.x^0 + 3.x^1 + 1.x^2 + 0.x^3 + \cdots = 2 + 3x + x^2$$

EXAMPLE 2.53 Find the generating function for the sequence $(1, 1, 1, 1, 1, 1)$.

Solution By definition, the generating function of the sequence $(1, 1, 1, 1, 1, 1)$ is

$$G(x) = 1 + x + x^2 + x^3 + x^4 + x^5 = \frac{x^6 - 1}{x - 1}$$

EXAMPLE 2.54 Let $u_k = {}^nC_k, k = 0, 1, 2, \ldots, n$; n being a positive integer. Find the generating function for the sequence $(u_0, u_1, u_2, \ldots, u_n)$.

Solution The generating function for the sequence $(a_0, a_1, a_2, \ldots, a_n)$ is

$$G(x) = {}^nC_0 + {}^nC_1 x + {}^nC_2 x^2 + \cdots + {}^nC_n x^n$$
$$= (1 + x)^n \quad \text{(by binomial theorem)}$$

Note The above example is perhaps the best example of a generating function.

EXAMPLE 2.55 Find the generating function for the infinite sequence $(1, \alpha, \alpha^2, \alpha^3, \ldots,)$, where α is constant.

Solution The generating function for the given sequence is

$$G(x) = 1 + \alpha x + \alpha^2 x^2 + \alpha^3 x^3 + \cdots \quad (2.21)$$

To find a closed-form expression of $G(x)$:

From Eq. (2.21),
$$G(x) - 1 = \alpha x + \alpha^2 x^2 + \alpha^3 x^3 + \cdots$$
$$\Rightarrow \frac{G(x) - 1}{\alpha x} = 1 + \alpha x + \alpha^2 x^2 + \cdots = G(x) \quad \text{[by Eq. (2.21)]}$$
$$\Rightarrow G(x) - 1 = \alpha x\, G(x) \quad \Rightarrow G(x) - \alpha x\, G(x) = 1$$
$$\Rightarrow G(x) = \frac{1}{1 - \alpha x}$$

Thus, the required generating function is $1/(1 - \alpha x)$.

EXAMPLE 2.56 Find the generating function (in closed form) of the Fibonacci sequence $\{a_n\}$ defined by
$$a_n = a_{n-1} + a_{n-2}; \quad a_0 = 0, a_1 = 1$$

Solution The generating function for the given sequence is
$$G(x) = a_0 + a_1 x + a_2 x^2 + a_3 x^3 + \cdots = \sum_{n=0}^{\infty} a_n x^n \qquad (2.22)$$

Consider,
$$a_n = a_{n-1} + a_{n-2}, \quad n \geq 0$$

Multiply both sides by x^n, and sum over all $n \geq 2$,
$$\sum_{n=2}^{\infty} a_n x^n = \sum_{n=2}^{\infty} a_{n-1} x^n + \sum_{n=2}^{\infty} a_{n-2} x^n$$

Consider the first sum
$$\sum_{n=2}^{\infty} a_n x^n = a_2 x^2 + a_3 x^3 + \cdots$$
$$= G(x) - a_0 - a_1 x \quad \text{[from Eq. (2.22)]}$$

Similarly,
$$\sum_{n=2}^{\infty} a_{n-1} x^n = a_1 x^2 + a_2 x^3 + \cdots$$
$$= x(a_1 x + a_2 x^2 + \cdots)$$
$$= x[G(x) - a_0] \quad \text{[from Eq. (2.22)]}$$

and
$$\sum_{n=2}^{\infty} a_{n-2} x^n = a_0 x^2 + a_1 x^3 + a_2 x^4 + \cdots$$
$$= x^2(a_0 + a_1 x + a_2 x^2 + \cdots)$$
$$= x^2 G(x)$$

Substituting these three expressions in Eq. (2.22), we get
$$G(x) - a_0 - a_1 x = x[G(x) - a_0] + x^2 G(x)$$

Since $a_0 = 0$ and $a_1 = 1$, so,

$$G(x) - x = xG(x) + x^2 G(x), \Rightarrow G(x)[1 - x - x^2] = x \Rightarrow G(x) = \frac{x}{1 - x - x^2}$$

Thus, the required generating function is

$$G(x) = \frac{x}{1 - x - x^2}$$

We now illustrate the solutions of different recurrence relations using the generating function method.

EXAMPLE 2.57 Solve the recurrence relation $a_r - 3a_{r-1} + 2a_{r-2} = 0, r \geq 2$ by the generating function method with the initial conditions $a_0 = 2$ and $a_1 = 3$.

Solution Let $G(x)$ be the generating function of the sequence $\{a_n\}$, i.e.,

$$G(x) = \sum_{r=0}^{\infty} a_r x^r$$

Multiply the given recurrence relation by x^r,

$$a_r x^r - 3a_{r-1} x^r + 2a_{r-2} x^r = 0$$

Sum over all $r \geq 2$,

$$\sum_{r=2}^{\infty} a_r x^r - 3 \sum_{r=2}^{\infty} a_{r-1} x^r + 2 \sum_{r=2}^{\infty} a_{r-2} x^r = 0$$

$\Rightarrow \quad [G(x) - a_0 - a_1 x] - 3x[G(x) - a_0] + 2x^2 G(x) = 0$

$\Rightarrow \quad (2x^2 - 3x + 1)G(x) - a_0 - a_1 x + 3a_0 x = 0$

Now, using the given initial conditions, i.e., $a_0 = 2, a_1 = 3$ we get,

$$(2x^2 - 3x + 1) G(x) - 2 - 3x + 6x = 0$$

$\Rightarrow \quad G(x) = \dfrac{2 - 3x}{2x^2 - 3x + 1} = \dfrac{1}{1 - x} + \dfrac{1}{1 - 2x}$

Thus, $a_r = 1 + 2^r$.

EXAMPLE 2.58 Using generating function method, solve the recurrence relation $y_{n+2} - 4y_{n+1} + 3y_n = 0$ with the initial conditions, $y_0 = 2$ and $y_1 = 4$.

Solution Let $A(x) = \sum_{n=0}^{\infty} y_n x^n$ be the generating function of the sequence $\{y_n\}, n \geq 0$.

The given recurrence relation is

$$y_{n+2} - 4y_{n+1} + 3y_n = 0 \tag{2.23}$$

Multiply Eq. (2.23) by x^n and summing over all $n \geq 2$, we get

$$\sum_{n=2}^{\infty} y_{n+2} x^n - 4 \sum_{n=2}^{\infty} y_{n+1} x^n + \sum_{n=2}^{\infty} y_n x^n = 0$$

$$\Rightarrow \frac{1}{x^2}[A(x) - y_0 - y_1 x] - 4 \cdot \frac{1}{x}[A(x) - y_0] + 3A(x) = 0$$

Now, using the conditions, i.e., $y_0 = 2$ and $y_1 = 4$, we get

$$\frac{1}{x^2}[A(x) - 2 - 4x] - \frac{4}{x}[A(x) - 2] + 3A(x) = 0$$

$$\Rightarrow A(x)[1 - 4x + 3x^2] = 2 - 4x$$

$$\Rightarrow A(x) = \frac{2 - 4x}{1 - 4x + 3x^2} = \frac{2 - 4x}{(1-x)(1-3x)}$$

$$\Rightarrow A(x) = \frac{1}{1-x} + \frac{1}{1-3x} = \sum_{n=0}^{\alpha} x^n + \sum_{n=0}^{\alpha} 3^n x^n$$

$$= \sum_{n=0}^{\infty} (1 + 3^n) x^n$$

Comparing this with the assumed generating function, we get

$$y_n = 1 + 3^n$$

EXAMPLE 2.59 Solve the recurrence relation $a_{r+2} - 5a_{r+1} + 6a_r = 2$ by the method of generating functions satisfying the initial conditions $a_0 = 1$ and $a_1 = 3$.

Solution Let $G(x) = \sum_{n=0}^{\infty} a_n x^n$ be the generating function.

Given the recurrence relation as

$$a_{r+2} - 5a_{r+1} + 6a_r = 2 \tag{2.24}$$

Multiply Eq. (2.24) by x^r and summing up from $r = 0$ to ∞, we get

$$\sum_{r=0}^{\infty} a_{r+2} x^r - 5 \sum_{r=0}^{\infty} a_{r+1} x^r + 6 \sum_{r=0}^{\infty} a_r x^r = 2 \sum_{r=0}^{\infty} x^r$$

$$\Rightarrow (a_2 + a_3 x + a_4 x^2 + \cdots) - 5(a_1 + a_2 x + a_3 x^2 + \cdots)$$
$$+ 6(a_0 + a_1 x + a_2 x^2 + \cdots) = 2(1 + x + x^2 + \cdots)$$
$$[\because G(x) = a_0 + a_1 x + a_2 x^2 + \cdots]$$

$$\Rightarrow \frac{G(x) - a_0 - a_1 x}{x^2} - 5\left(\frac{G(x) - a_0}{x}\right) + G(x) = \frac{2}{1-x}$$

Put $a_0 = 1$ and $a_1 = 3$ in the above equation and after little simplification, we get

$$G(x) = \frac{5x^2 - 4x + 1}{(1 - x)(1 - 2x)(1 - 3x)}$$

By partial fractions, we get

$$G(x) = \frac{1}{(1 - x)} + \frac{-1}{(1 - 2x)} + \frac{1}{(1 - 3x)}$$

Thus, by applying inverse transformation, the solution is $a_r = 1 - 2^r + 3^r$.

EXAMPLE 2.60 Solve the recurrence relation $a_{r+2} - 2a_{r+1} + a_r = 2^r$ by the method of generating functions with the initial conditions $a_0 = 2$ and $a_1 = 1$.

Solution By considering generating functions of the given recurrence relation on both the sides, we have

$$\frac{G(x) - a_0 - a_1 r}{x^2} - 2\left(\frac{G(x) - a_0}{x}\right) + G(x) = \frac{1}{1 - 2x}$$

Now, put $a_0 = 2$ and $a_1 = 1$ in the above equation and after simplification, we get

$$G(x) - 2 - x - 2xG(x) - 2x + x^2 G(x) = \frac{x^2}{1 - 2x}$$

$$\Rightarrow \quad (1 - x)^2 G(x) = 2 + 3x + \frac{x^2}{1 - 2x}$$

$$\Rightarrow \quad G(x) = \frac{2}{(1 - x)^2} + \frac{3x}{(1 - x)^2} + \frac{x^2}{(1 - 2x)(1 - x)^2}$$

By partial fraction, we get

$$\frac{x^2}{(1 - 2x)(1 - x)^2} = \frac{1}{(1 - 2x)} - \frac{1}{(1 - x)^2}$$

Hence,

$$G(x) = \frac{1}{(1 - x)^2} + \frac{3x}{(1 - x)^2} + \frac{1}{(1 - 2x)}$$

Thus, $a_r = (r + 1) + 3r + 2^r$, i.e., $a_r = 1 + 4r + 2^r$.

EXAMPLE 2.61 Assume that there exists a valid codeword of an n-digit number in decimal notation having an even number of 0s. Let a_r denote the member of valid codeword of length n. If the sequence $\{a_r\}$ satisfies the recurrence relation $a_r = 8a_{r-1} + 10^{r-1}$, with the initial condition $a_1 = 9$, then find an explicit formula for a_r using the generating function.

Solution To make our job simpler with the generating functions, we extend the given sequence by setting $a_0 = 1$; and using the recurrence relation, we have $a_1 = 8a_0 + 10^0 = 8 + 1 = 9$, which is consistent with our original initial condition.

Let the generating function for the sequence $\{a_r\}$ be

$$G(x) = \sum_{r=0}^{\infty} a_r x^r \qquad (2.25)$$

Multiply both sides of the recurrence relation $a_r = 8a_{r-1} + 10^{r-1}$ by x^r and sum for all $r \geq 1$, we get

$$\sum_{r=1}^{\infty} a_r x^r = 8 \sum_{r=1}^{\infty} a_{r-1} x^r + \sum_{r=1}^{\infty} 10^{r-1} x^r \qquad (2.26)$$

$$\Rightarrow \quad [G(x) - a_0] = 8xG(x) + x \sum_{r=1}^{\infty} (10x)^{r-1}$$

Since $a_0 = 1$ and $a_1 = 9$, it reduces to

$$G(x) - 1 = 8xG(x) + x(1 - 10x)^{-1}$$

$$\Rightarrow \quad G(x) = \frac{1 - 9x}{(1 - 8x)(1 - 10x)} \qquad (2.27)$$

By partial fraction, we get

$$\frac{1 - 9x}{(1 - 8x)(1 - 10x)} = \frac{A}{(1 - 8x)} + \frac{B}{(1 - 10x)}$$

$$\Rightarrow \quad 1 - 9x = A(1 - 10x) + B(1 - 8x)$$

$$\Rightarrow \quad A + B = 1 \text{ and } 10A + 8B = 9$$

Solving for A and B, we get

$$G(x) = \frac{1}{2}\left[\frac{1}{1 - 8x} + \frac{1}{1 - 10x}\right] = \frac{1}{2}\left[\sum_{r=0}^{\infty} 8^r x^r + \sum_{r=0}^{\infty} 10^r x^r\right]$$

$$\Rightarrow \quad \sum_{r=0}^{\infty} a_r x^r = \sum_{r=0}^{\infty} \frac{1}{2}(8^r + 10^r) x^r$$

Comparing, we get

$$a_r = \frac{1}{2}(8^r + 10^r)$$

Generating functions are also useful in studying the number of different types of an integer n. A *partition* of a positive integer is a way to write this integer as the sum of positive integers where repetition is allowed and the order of the integers in the sum does not matter.

For example The partitions of 5 (with no restrictions) are $1 + 1 + 1 + 1 + 1, 1 + 1 + 1 + 2, 1 + 1 + 3, 1 + 2 + 2, 1 + 4, 2 + 3,$ and 5.

2.16 COUNTING (COMBINATORIAL) METHOD

Generating functions are useful in solving variety of counting problems. They are generally used to count the number of combinations of various types. One of the varieties of this counting is to count the solutions of the form

$$e_1 + e_2 + e_3 + \cdots + e_n = k$$

where k is a constant and each e_i is a non-negative integer that may be subjected to a specified constraint. After all, generating function will be used to tackle this type of counting problems.

EXAMPLE 2.62 Find the number of non-negative integral solutions to $e_1 + e_2 + e_3 + \cdots + e_n = r$ where $0 \leq e_i \leq 1$.

Solution Let $G_i(x) = 1 + x$ for each $i = 1, 2, \ldots, r$. Thus, the generating function is $G_1(x)G_2(x) \cdots G_n(x) = (1 + x)^n$ and the number of solutions is ^nC_r.

EXAMPLE 2.63 Find the number of solutions of $e_1 + e_2 + e_3 + e_4 + e_5 = 21$ where $e_1, e_2, e_3, e_4,$ and e_5, are non-negative integers with $0 \leq e_1 \leq 3$, $0 \leq e_2 \leq 3$, $2 \leq e_3 \leq 6$, $2 \leq e_4 \leq 6$, e_5 is odd, and $1 \leq e_5 \leq 9$.

Solution Let

$$G_1(x) = 1 + x + x^2 + x^3$$
$$G_2(x) = 1 + x + x^2 + x^3$$
$$G_3(x) = x^2 + x^3 + x^4 + x^5 + x^6$$
$$G_4(x) = x^2 + x^3 + x^4 + x^5 + x^6$$
$$G_5(x) = x + x^3 + x^5 + x^7 + x^9$$

Thus, the generating function is $G_1(x)G_2(x)G_3(x)G_4(x)G_5(x) = (1 + x + x^2 + x^3)^2 (x^2 + x^3 + x^4 + x^5 + x^6)^2 (x + x^3 + x^5 + x^7 + x^9)$. The number of solutions with the given constraints is the coefficient of x^{21} in the expansion of $(1 + x + x^2 + x^3)^2 (x^2 + x^3 + x^4 + x^5 + x^6)^2 (x + x^3 + x^5 + x^7 + x^9)$, i.e., we will obtain a term equal to x^{21} by choosing terms from the first two sums x^{e_1} and x^{e_2}, terms from the second two sums x^{e_3} and x^{e_4}, term from the last sum x^{e_5}, so that the exponents, $e_1, e_2, e_3, e_4,$ and e_5 satisfy the equation $e_1 + e_2 + e_3 + e_4 + e_5 = 21$ and the given constraints.

It is observed that the coefficients of x^{21} in the product is 4. Hence, the number of solutions is 4.

EXAMPLE 2.64 Find a generating function for a_r, the number of ways to select r balls from a pile of three green, three white, three blue, and three red balls.

Solution Here, the generating function, will be a multiplication of four factors corresponding to each colour green, white, blue, and red.

Since there are three balls of each colour, each factor will be $(1 + x + x^2 + x^3)$. Hence, the required generating function is $(1 + x + x^2 + x^3)^4$.

EXAMPLE 2.65 Find the generating function for a_r, the number of ways to select r objects from n objects with unlimited repetitions, and also find a_r.

Solution Since each object can be selected unlimitedly, the generating function is then

$$G(x) = (1 + x + x^2 + \cdots)^n = \{(1-x)^{-1}\}^n = (1-x)^{-n}$$

To find a_r, we shall find the coefficient of x^r in $G(x)$, i.e., coefficient of

$$x^r = (-1)^r \frac{(-n)(-n-1)\cdots(-n+1-r)}{r!}$$

$$= \frac{(n+r-1)!}{r!(n-r)!} = {}^{n+r-1}C_r$$

2.17 PIGEONHOLE PRINCIPLE

The *pigeonhole principle* states that if there are more pigeons than pigeonholes, then there must be at least one pigeonhole with at least two pigeons in it. This principle is applicable to other objects besides pigeons and pigeonholes. Thus, it may be stated that if $(n + 1)$ or more objects are placed into n boxes then there exists at least one box containing two or more objects.

The pigeonhole principle (also known as *Dirichlet drawer* principle or *shoe box* principle) is at times useful in counting principles.

In the *set-theoretic approach*, the pigeonhole principle can be expressed as follows.

Let X and Y be any two finite sets such that $|X| < |Y|$. Then, a function $f:X \to Y$ cannot be one-to-one, i.e., there exists at least two elements x_1, x_2 in X so that $f(x_1) = f(x_2)$.

For example
(i) Suppose the department of mathematics contains 13 professors. Then 2 of the professors (pigeons) were born in the same month (pigeonholes) out of 12 months.
(ii) Suppose a laundry bag contains many red, white, and blue socks. Then one needs to grab only four socks (pigeons) to be certain of getting a pair with the same colour (pigeonholes).

Corollary 2.2 If n pigeons are assigned to m pigeonholes, then at least one pigeonhole contains two or more pigeons $(m < n)$.

Proof Let m pigeonholes be labelled with the numbers 1 to m, beginning with the pigeon 1, each pigeon is assigned with respect to the pigeonholes with the same number. Since $m < n$, i.e., the number of pigeonholes is less than the number of pigeons, $n - m$ pigeons are left without assigning a pigeonhole. Thus, at least one pigeonhole will be assigned to a second pigeon.

2.17.1 Generalized Pigeonhole Principle

If a pigeonhole is occupied by $kn + 1$ or more pigeons, where k is a positive integer, then at least one pigeonhole is occupied by $k + 1$ or, more pigeons.

EXAMPLE 2.66 Find the minimum number of students in a class to be sure that three of them are born in the same month.

Solution Here, $n = 12$ months are the pigeonholes and $k + 1 = 3$ or $k = 2$. Hence, among any $kn + 1 = 25$ students (pigeons), three of them are born in the same month.

SUMMARY

Combinatorial views exist in the heart of computer science and modern discrete mathematics. The simple query of how many objects are there having a certain property can transfer into a deep problem about algorithm analysis or the structural symmetry of some digital hardware. After introducing the elementary aspects of the theory, the addition and multiplication rules, we consider counting problems, which are expressed in terms of ordered or unordered arrangements of the objects of a set. These arrangements are called *permutations and combinations*, and are used in many counting problems. We have also visualized probability from the point of view that in assessing the likelihood of an event one often needs to count how many ways it can occur. We discussed an important process called *recursion* by which one can express objects in terms of itself. We also treat the powerful pigeonhole principle. In this chapter, a number of techniques have been outlined which are required in analysing algorithm complexity, in numerical applications, and in solving difference equations (recurrence relations).

EXERCISES

1. In how many different ways one can answer all the questions of a true–false test consisting of four questions?
2. Find the number, n, of license plates that can be made from each plate containing a sequence of four letters followed by three different digits 0–9 (not necessarily distinct).
3. How many four-digit numbers can be formed using the digits 1, 3, 4, 6, 7, and 8? Also, find the numbers if no digit is repeated.
4. In how many ways can we choose a prime number or an even number between 10 and 20 (excluding both numbers)?
5. Find the value of m after the execution of the following code:
   $m = 0$
   for $i = 1$ to n_1
      $m = m + 1$
   for $j = 1$ to n_2
      $m = m + 1$
   for $k = 1$ to n_3
      $m = m + 1$
6. Find the value of $n!/r!(n-r)!$, when $n = 6$ and $r = 4$.
7. Find the value of x, if $(1/4!) + (1/5!) = (x/6!)$.
8. There are ten persons called on an interview. Each one is capable of to be selected for the job. How many permutations are there to select four from the ten persons?
9. How many permutations can be made out of the letter of the word 'COMPUTER'? How many of these
   (i) begin with C? (ii) end with R?
   (iii) begin with C and end with R?
   (iv) C and R occupy the end places?
10. How many different variable names can be formed by using the letters
    $a, a, a, b, b, b, b, c, c, c$?
11. Show that the total number of permutations of n different things taken not more than r at a time, when each thing may be repeated any number of times is $n(n^r - 1)/(n - 1)$.
12. There are ten candidates for an examination, out of which four are appearing in mathematics and remaining six are appearing in different subjects. In how many ways can they be seated in a row so that no two mathematics' candidates are together?
13. A tea party is arranged for sixteen people along two sides of a large table with eight chairs on each side. Four men will have to sit on one particular side and two on the other side. In how many ways can they be seated?
14. There are n points in a plane, out of these points no three are in the same straight line except p points which are collinear. How many (i) straight lines and (ii) triangles can be formed by joining them?

15. A man has seven relatives, four of them are ladies and three gentlemen; his wife has also seven relatives, three of them are ladies and four gentlemen. In how many ways can they invite a dinner party of three ladies and three gentlemen so that there are three of them man's relatives and three of the wife's relatives?

16. Let there be n different things which are arranged around a circle. Then, in how many ways can three objects be selected when no two of the selected objects are consecutive?

17. A five-digit number is formed by the digits 1, 2, 3, 4, 5 without repetition. Find the probability that the number formed is divisible by 4.

18. Find the probability of getting a sum of 5 or 7 in a toss of two dice.

19. A coin is tossed n times. What is the probability that the head will present itself an odd number of times?

20. Let A and B be two independent events. The probability that both A and B happen is 1/2 and the probability that neither A nor B happen is 1/2. Then, find the probability of occurrence of A and B.

21. In a class, 30% students fail in English, 20% students fail in Hindi, and 10% both in English and Hindi. A student is chosen at random, then what is the probability that he will fail in English, if he has failed in Hindi?

22. A man is known to speak the truth 3 out of 4 times. He throws a die and reports that it is a 6. Find the probability that it is actually a 6.

23. In a test, an examinee either guesses or copies or knows the answer to a multiple-choice question with four choices. The probability that he makes a guess is 1/3 and the probability that he copies the answer is 1/6. The probability that his answer is correct, given that he copied it, is 1/8. Find the probability that he knew the answer to the question, given that he correctly answered it.

24. A man takes a step forward with probability 0.4 and backward with probability 0.6. Find the probability that at the end of eleven steps he is one step away from the starting point.

25. Solve the recurrence relation $a_r - 6a_{r-1} + 9a_{r-2} = 0$.

26. Solve the recurrence relation $a_r - 7a_{r-1} + 12a_{r-2} = 0$.

27. Solve the recurrence relation $a_r - 3a_{r-1} + 2a_{r-2} = 0$.

28. Solve the recurrence relation $2a_r - 5a_{r-1} + 2a_{r-2} = 0$; find the particular solution such that $a_0 = 0$ and $a_1 = 1$.

29. Solve the recurrence relation $a_r - 7a_{r-1} + 10a_{r-2} = 0$. Find the particular solution such that $a_0 = 0$ and $a_1 = 6$.

30. Find the particular solution of the difference equation $a_{r+2} - 2a_{r-1} + a_r = 3r + 5$.

31. Find the particular solution of the difference equation $a_r - 4a_{r-1} + 4a_{r-2} = (r+1)2^r$.

32. Show that the particular solution of the recurrence relation $a_r + 5a_{r-1} + 6a_{r-2} = 3r^2 - 2r + 1$ is $a_r = (1/4)r^2 + (13/24)r + 71/288$.

33. Solve the recurrence relation $a_{r+2} - a_{r+1} - 2a_r = r^2 2^r$.

34. Solve the recurrence relation $a_{r+1} - a_r = r^2$.

35. Solve the recurrence relation $a_{r+2} - 2a_{r+1} + a_r = r^2$.

36. Solve the recurrence relation $a_r - 2a_{r-1} + a_{r-2} = 2^r, r \geq 2$, by the method of generating function satisfying the boundary conditions $a_0 = 2, a_1 = 1$.

37. Solve the recurrence relation $a_r - 5a_{r-1} + 6a_{r-2} = 2, r \geq 2$ by the method of generating function with the conditions $a_0 = 1$ and $a_1 = 1$.

38. Using generating function, solve the recurrence relation $a_r - 7a_{r-1} + 10a_{r-2} = 0$, with initial conditions $a_0 = 3$ and $a_1 = 3$.

39. Given a linear homogeneous recurrence relation $a_{r+2} - 6a_{r+1} + 9a_r = f(r)$. Determine the particular solution of the above relation for $f(r) = 3^r$, $f(r) = r.3^r$, and $f(r) = (r^2 + 1).3^r$, respectively.

40. Consider 32 people who save paper or bottles (or both) for recycling. Among 32 people, 30 save paper and 14 save bottles. Find the number of people who (i) save both, (ii) save only paper, and (iii) save only bottles.

3 Mathematical Logic

> **LEARNING OBJECTIVES**
>
> After reading this chapter you will be able to understand the following:
> - The rules of logic that help to understand and reason with statements
> - The circumstances under which the formulae will be regarded as being 'true', which is the semantics of the theory
> - The compound propositions that play in response to logical equivalences
> - The propositional formula that can be expressed in certain normal forms, namely disjunctive and conjunctive involving connectives
> - The use of both direct and indirect arguments to derive new results from those already known to be true
> - The predicate logic that can be used to express the meaning of statements in mathematics and computer science by the ways in reasoning and exploring relationships between objects
> - The necessity of quantification to create a proposition from a propositional function
> - The rules of inference that are used to produce valid arguments in propositional logic
> - Several methods of proof that are implemented to show the validity of statements

3.1 INTRODUCTION

Logic is the basis of all mathematical and automated reasoning. In fact, logical reasoning is the essence of mathematics. The *Oxford English Dictionary* defines the word 'logical' as 'correctly reasoned'. It has practical applications in the design of computing machines, in the specification of systems, in artificial intelligence, in computer programming, and in the other areas of computer science.

Mathematical logic is a subarena of mathematics with close interactions with computer science. It includes the study of the expressive power of formal systems and the deductive power of formal proof systems. To understand mathematics, we must think about a correct mathematical argument, i.e., a proof for a theorem. Students of computer science often find it imperative as how important proofs are in computer science. In fact, theorems along with their proofs play essential roles when we verify that computer programs produce the correct output for all possible input values, when we show that algorithm always produces the correct result, and when we create artificial intelligence.

For hardware design, it is a propositional logic which, at present, is most applicable and it enables us to handle the desired behaviour of gates and circuits. We discuss the fundamental ideas in Sections 3.2–3.5, and describe the method of truth tables for testing the validity of a formula. In Sections 3.6–3.8, we present the views of semantics (meaning of a propositional statement), namely a formula which is provable in the system if and only if it is always true (a tautology); or, more generally, that it is formally derivable from a set of assumptions if and only if it is true under every assignment of truth values making all the assumptions true. We have then carried on to look on logical equivalence and logical implication in respective Sections 3.9–3.10 of which the former one shows the equivalence of two compound propositions by using truth tables. In Section 3.11, we give a standard form for propositional formulae called *normal forms* which are accepted as a better option to construct truth tables. We introduce arguments in Section 3.12, which hold only compound propositions to be valid. Then we have investigated a collection of rules of inference in propositional logic in Section 3.13, which are among the most important components in producing valid arguments. Next, we set to concentrate on an important way to create a proposition from a propositional function, that is quantifications in Section 3.16. It expresses the extent to which a predicate is true over a range of elements. Finally, we introduce the notion of proof, and describe methods and constructive proofs in Section 3.17.

3.2 STATEMENT (PROPOSITIONS)

A statement (or a proposition) is a declarative sentence (i.e., a sentence that declares a fact) which is either true or false but not both; and which is also sufficiently objective, meaningful, and precise. The truth or falsity of a statement is called its *truth value*. The truth values 'True' and 'False' of a statement are denoted by T and F, respectively. Also, the value of a statement if true is denoted by 1 and false if expressed by 0.

For example Consider the following sentences:
   (i) Kolkata is in India
  (ii) $4 + 2 = 6$
 (iii) $5 < 7$
 (iv) Bangalore is in West Bengal
  (v) $x + 2 = 5$
 (vi) Where are you going?
(vii) Roses are red
(viii) Go to bed

The sentences (i), (ii), (iii), (iv), and (vii) are statements; among them (iv) is false and others are true. The item (v) is not a proposition (or a statement), since it is neither true nor false. It can be a proposition if we assign a value to the variable x. The item (vi) is a question, not a declarative sentence, hence it is not a statement. Finally, (viii) is not a statement, but a command only.

Statements are usually denoted by the letters $p, q, r, \ldots$. If p denotes the statement 'Bangalore is in West Bengal', then instead of saying that the above statement is false, we can simply represent the value of p as F.

3.3 LAWS OF FORMAL LOGIC

Here, we will state two famous laws of formal logic.
1. *Law of contradiction* For every proposition p it is not the same notion that p is both true and false.
2. *Law of intermediate exclusion* If p is a statement (proposition), then either p is true or false, and there is no possibility of intermediate exclusion.

3.4 BASIC SET OF LOGICAL OPERATORS/OPERATIONS

In this section, we will discuss three basic logical operators/operations namely, conjunction ($\wedge$), disjunction ($\vee$), and negation ($\sim$) which correspond to the English words like 'and', 'or', and 'not', respectively.

3.4.1 Conjunction (AND, $p \wedge q$)

If any two propositions can be combined by the word 'and', then we can create a new proposition called *compound proposition*. This proposition is actually called the *conjunction* of the original propositions. Symbolically, $p \wedge q$ is read as 'p and q' and represents the conjunction of p and q. Since $p \wedge q$ is a proposition, it has a truth value, and this truth value depends only on the truth values of p and q. Thus, if p and q are true, then $p \wedge q$ is true, otherwise $p \wedge q$ is false.

The truth value of $p \wedge q$ is shown in Table 3.1. Here, the first row says that if p is true and q is true, then $p \wedge q$ is also true. The second row implies that if p is true and q is false, then $p \wedge q$ is false, and so on. It may be noted that there are four rows corresponding to the four possible combinations of T and F for the two propositions p and q.

TABLE 3.1 $p \wedge q$

p	q	$p \wedge q$
T	T	T
T	F	F
F	T	F
F	F	F

For example Consider the following four statements:

(i) p: Kolkata is in West Bengal and q: $4 + 4 = 8$
(ii) p: Kolkata is in West Bengal and q: $4 + 4 = 9$
(iii) p: Kolkata is in Orissa and q: $4 + 4 = 8$
(iv) p: Kolkata is in Orissa and q: $4 + 4 = 9$

Here, only the first statement (i) is true. Each of the other statements is false, since at least one of its substatements is false.

EXAMPLE 3.1 Find the conjunction of the propositions p and q when p is the proposition 'Today is Saturday' and q is the proposition 'It is raining heavily today'.

Solution The conjunction of these propositions, $p \wedge q$, is the proposition 'Today is Saturday and it is raining heavily today'. This proposition is true on rainy Saturday and is false on any day, but not on Saturday, and on Saturday when it does not rain.

3.4.2 Disjunction (OR, $p \vee q$)

If any two propositions can be combined by the word 'or' to form a compound proposition, then the proposition is called the *disjunction* of the original propositions. Symbolically, $p \vee q$ is read as 'p or q' and denotes the disjunction of p and q. The truth value of $p \vee q$ depends only on the truth values of p and q. Thus, if p and q are false, then $p \vee q$ is false, otherwise $p \vee q$ is true.

The truth value of $p \vee q$ is represented in Table 3.2. It may be observed from the table that $p \vee q$ is false in the fourth row when both p and q are false.

TABLE 3.2 $p \vee q$

p	q	$p \vee q$
T	T	T
T	F	T
F	T	T
F	F	F

For example Consider the following four statements:

(i) p: Kolkata is in West Bengal and q: $4 + 4 = 8$
(ii) p: Kolkata is in West Bengal and q: $4 + 4 = 9$
(iii) p: Kolkata is in Orissa and q: $4 + 4 = 8$
(iv) p: Kolkata is in Orissa and q: $4 + 4 = 9$

Here, only the last statement (iv) is false. Each of the other statements is true, since at least one of its substatements is true.

EXAMPLE 3.2 Find the disjunction of the propositions p and q where p is the proposition 'Today is Saturday' and q is the proposition 'It is raining heavily today'.

Solution The disjunction of p and q, $p \vee q$, is the proposition 'Today is Saturday or it is raining heavily today'.

The English word 'or' can be used in two different notions, namely, as an *inclusive* ('and/or') or *exclusive* ('either/or').

For example Consider the following two statements:

(i) p: Ashish will go to Kolkata or to Bangalore.
(ii) q: There is something wrong with the fan or switch.

In the statement (i), the disjunction of the statement p is used in exclusive sense (p or q but not both); i.e., one or other possibility exists but not both.

In the statement (ii), the disjunctive is implemented in an inclusive sense (p or q or both). Here, at least one of the two possibilities will occur. However, both could have occurred. In the inclusive sense, we shall always use 'or' unless it is stated.

EXAMPLE 3.3 Let p be 'Rekha speaks Bengali' and let q be 'Rekha speaks Oriya'. Give a simple verbal sentence which describes each of the following:
 (i) $p \vee q$ (ii) $p \wedge q$

Solution

(i) Rekha speaks Bengali or Oriya.
(ii) Rekha speaks Bengali and Oriya.

Example 3.4 Assign a truth value to each of the following statements:
(i) $6 + 4 = 10 \vee 0 > 2$
(ii) $5 \times 4 = 21 \vee 9 + 7 = 17$

Solution

(i) True, since one of its components, i.e., 6 + 4 = 10 is true.

(ii) False, since both of its components are false.

3.4.3 Negation (NOT, ~p)

Given any proposition p, a new proposition, called the *negation* of p, can be formed by stating 'It is not the case that . . .' or 'It is false that . . .' for p, or, if possible by inserting in p the word 'not'. Symbolically, $\sim p$ is read as 'not p', and denotes the negation of p. Thus, if p is true, then $\sim p$ is false; and if p is false, then $\sim p$ is true. The truth value of $\sim p$ is represented in Table 3.3.

TABLE 3.3 $\sim p$

p	$\sim p$
T	F
F	T

For example Consider the following six statements:
 (i) Kolkata is in India
 (ii) It is not the case that Kolkata is in India
 (iii) Kolkata is not in India
 (iv) 4 + 4 = 9
 (v) It is not the case that 4 + 4 = 9
 (vi) 4 + 4 ≠ 9

Here, (ii) and (iii) are each the negation of (i); and (v) and (vi) are each the negation of (iv). Since (i) is true, (ii) and (iii) are false; and since (iv) is false, (v) and (vi) are true.

EXAMPLE 3.5 Find the negation of the following propositions:
 (i) Today is Saturday.
 (ii) It is a rainy day.
 (iii) If it snows, Mona does not drive the car.

Solution

(i) Today is not Saturday.
(ii) It is not a rainy day.
(iii) It snows and Mona drives the car.

EXAMPLE 3.6 Let p: Priya is tall and q: Priya is beautiful. Write the following statements in symbolic form.
 (i) Priya is tall and beautiful.
 (ii) Priya is tall but beautiful.
 (iii) It is false that Priya is short or beautiful.
 (iv) Priya is tall or Priya is short and beautiful.

Solution

(i) $p \wedge q$ (ii) $p \wedge \sim q$ (iii) $\sim(\sim p \vee q)$ (iv) $p \vee (\sim p \wedge q)$

3.5 PROPOSITIONS AND TRUTH TABLES

Let $P(p, q)$ denote an expression constructed from logical variables $p, q, \ldots$, which take on the value TRUE (T) or FALSE (F), and which operate on the logical connectives $\wedge, \vee, \sim$ (and others discussed subsequently). Such an expression is called a *proposition*.

One of the important characteristics of a proposition $P(p, q)$ is that its truth value depends upon the truth value of its variables, i.e., we can determine the truth value of a proposition once we know the truth values of each of its variables. We will construct this relationship through a *truth table*.

For example Consider the proposition $\sim(p \vee q)$. Table 3.4 shows how the truth table $\sim(p \vee q)$ is constructed. First of all, columns are to be marked by p, q, $p \vee q$, and $\sim(p \wedge q)$. Then fill up the p and q columns with the logically possible combinations of T's and F's and subsequently the $(p \vee q)$ and $\sim(p \vee q)$ columns with the appropriate truth values. Hence, the truth table, $\sim(p \vee q)$, is shown in Table 3.5.

TABLE 3.4 Construction of $\sim(p \vee q)$

p	q	$p \vee q$	$\sim(p \vee q)$
T	T	T	F
T	F	T	F
F	T	T	F
F	F	F	T

TABLE 3.5 $\sim(p \vee q)$

p	q	$\sim(p \vee q)$
T	T	F
T	F	F
F	T	F
F	F	T

3.5.1 Connectives

Statements can be connected by words like 'not', 'and', etc. These words are known as *logical connectives*. The statements which do not contain any of the connectives are called *atomic statements* or *simple statements* (or, *primitive statement*).

The common connectives used are negation ($\sim$), and ($\wedge$), or ($\vee$), if ... then ($\rightarrow$ or $\Rightarrow$), if and only if ($\leftrightarrow$ or $\Leftrightarrow$), equivalence ($\equiv$). We will use these connections along with symbols to combine various simple statements.

EXAMPLE 3.7 Write the following statements in symbolic form:
  (i) If Avinash is not in a good mood or he is not busy, then he will go to Kharagpur.
  (ii) If Sayantan knows object-oriented programming and oracle, then he will get a job.

Solution
  (i) Let p: Avinash is in a good mood, q: Avinash is busy and r: Avinash will go to Kharagpur, then $(\sim p \vee \sim q) \rightarrow r$.
  (ii) Let p: Sayantan know object-oriented programming, q: Sayantan know oracle, r: Sayantan will get a job, then $(p \wedge q) \rightarrow r$.

EXAMPLE 3.8 Let p: Babu is rich, q: Babu is happy.
Give a simple verbal sentence which describes each of the following statements:
  (i) $p \vee q$ (ii) $p \wedge q$ (iii) $q \rightarrow p$ (iv) $p \vee \sim q$
  (v) $q \leftrightarrow \sim q$ (vi) $\sim p \rightarrow q$ (vii) $\sim \sim p$ (viii) $(\sim p \wedge q) \rightarrow p$

Solution The meaning of the symbols $\wedge$, $\vee$, $\sim$, $\rightarrow$, and $\leftrightarrow$ are 'and', 'or', 'it is false' 'if ... then', and 'if and only if', respectively. Then the above statements are expressed as

(i) Babu is rich or Babu is happy.
(ii) Babu is rich and Babu is happy.
(iii) Babu is happy then Babu is rich.
(iv) Babu is rich or Babu is not happy.
(v) Babu is happy if and only if Babu is not rich.
(vi) Babu is not rich then Babu is happy.
(vii) It is not true that Babu is not rich.
(viii) If Babu is not rich and happy then Babu is rich.

3.5.2 Compound Propositions

Many statements (or propositions) are composite that is composed of subpropositions by means of logical operators or connections. Such statements are referred to *compound* (or *composite*) statements. The fundamental property of a compound statement is that its truth value is completely determined by the truth values of its subpropositions together with the way in which they are combined to construct the compound statement.

For example

(i) 'Sohan is intelligent or studies every night' is a compound proposition with subpropositions 'Sohan is intelligent' and 'Sohan studies every night'.
(ii) 'The sun is shining and the sky is blue' is a compound proposition with subpropositions 'The sun is shining' and 'the sky is blue'.

EXAMPLE 3.9 Construct a truth table for each of the following compound propositions:
  (i) $(p \wedge q) \vee (p \wedge r)$ (ii) $\sim(p \vee q) \vee (\sim p \wedge \sim q)$

Solution Tables 3.6 and 3.7

(i) **TABLE 3.6** $(p \wedge q) \vee (p \wedge r)$

p	q	r	$p \wedge q$	$p \wedge r$	$(p \wedge q) \vee (p \wedge r)$
F	F	F	F	F	F
F	F	T	F	F	F
F	T	F	F	F	F
F	T	T	F	F	F
T	F	F	F	F	F
T	F	T	F	T	T
T	T	F	T	F	T
T	T	T	T	T	T

(ii) **TABLE 3.7** $\sim(p \vee q) \vee (\sim p \wedge \sim q)$

p	q	$\sim p$	$\sim q$	$p \vee q$	$\sim(p \vee q)$	$(\sim p \wedge \sim q)$	$\sim(p \vee q) \vee (\sim p \wedge \sim q)$
F	F	T	T	F	T	T	T
F	T	T	F	T	F	F	F
T	F	F	T	T	F	F	F
T	T	F	F	T	F	F	F

EXAMPLE 3.10 Find the truth table of $p \wedge (q \vee r)$.

Solution Table 3.8

TABLE 3.8 $p \wedge (q \vee r)$

p	q	r	$q \vee r$	$p \wedge (q \vee r)$
T	T	T	T	T
T	T	F	T	T
T	F	T	T	T
T	F	F	F	F
F	T	T	T	F
F	T	F	T	F
F	F	T	T	F
F	F	F	F	F

3.5.3 Conditional Statement

If p and q are any two statements, then the statement $p \rightarrow q$ which is read as 'If p then q' is called a *conditional statement*. The symbol $\rightarrow$ is used to denote the connective 'If ... then'.

The conditional $p \rightarrow q$ can also be read:

(i) p only if q (ii) p implies q (iii) p is sufficient for q (iv) q if p

The statement p is called the *antecedent* and the statement q is called the *consequent* (or *conclusion*). If p is true and q is false, then the conditional $p \rightarrow q$ is false, otherwise $p \rightarrow q$ is true. The truth values of $p \rightarrow q$ are given in Table 3.9.

TABLE 3.9 $p \rightarrow q$

p	q	$p \rightarrow q$
T	T	T
T	F	F
F	T	T
F	F	T

For example

(i) Let p: Arpita is a graduate and q: Arpita is a lawyer then, $p \rightarrow q$: If Arpita is a graduate, then she is a lawyer.
(ii) Let p: The function is differentiable and q: The function is continuous. The statement has to be interpreted as 'p is sufficient for q' or in other words, $p \rightarrow q$.

3.5.4 Converse, Contrapositive, and Inverse

We can arrange some new conditional statements using a conditional statement $p \rightarrow q$. In particular, there exist three interrelated conditional statements that occur frequently carrying special names as follows:

The *converse* of $p \rightarrow q$ is the proposition $q \rightarrow p$.
The *contrapositive* of $p \rightarrow q$ is the proposition $\sim q \rightarrow \sim p$.
The *inverse* of $p \rightarrow q$ is the proposition $\sim p \rightarrow \sim q$.

It may be seen that among the three conditional statements formed from $p \rightarrow q$, only the contrapositive always possesses the same truth value.

The contrapositive, $\sim q \rightarrow \sim p$, of a conditional statement $p \rightarrow q$ always has the same truth value as $p \rightarrow q$. It may be noted that the contrapositive is false only when $\sim p$ is false and $\sim q$ is true, i.e., only when p is true and q is false. Again, neither the converse, $q \rightarrow p$, nor the inverse, $\sim p \rightarrow \sim q$, has the same truth value as $p \rightarrow q$ for all possible truth values of p and q. It is also seen that when p is true and q is false, the original conditional statement is false, but the converse and the inverse both are true.

The usage of conditional statements is illustrated in the following example.

EXAMPLE 3.11 Determine the contrapositive, the converse, and the inverse of the conditional statement 'The home team wins whenever it is raining'.

Solution The given statement can be modified as
 'If it is raining, then the home team wins'.
Since 'q whenever p' is one of the ways to express the conditional statement $p \rightarrow q$. Consequently, the contrapositive of this conditional statement is
 'If the home team does not win, then it is not raining'.
The converse is
 'If the home team wins, then it is raining'.
The inverse is
 'If it is not raining, then the home team does not win'.
Here, only the contrapositive is equivalent to the original statement.

3.5.5 Biconditional Statement

A statement of the form 'p if and only if q' is called a *biconditional statement*. It is denoted by $p \leftrightarrow q$

A biconditional statement contains the connective 'if and only if' and has two conditions. If p and q have the same truth value, then $p \leftrightarrow q$ is true. The truth values of $p \leftrightarrow q$ are given in Table 3.10.

TABLE 3.10 $p \leftrightarrow q$

p	q	$p \leftrightarrow q$
T	T	T
T	F	F
F	T	F
F	F	T

For example
 (i) $3 + 3 = 6$ if and only if $4 + 3 = 7$.
 (ii) An integer is even if and only if it is divisible by 2.
 (iii) Two lines are parallel if and only if they have the same slope.

EXAMPLE 3.12 Is it true that $\sin(\theta) = \cos(\theta)$ if and only if $\theta = 45°$?

Solution If $\theta = 45°$, then $\sin(\theta) = 1/\sqrt{2}$ and $\cos(\theta) = 1/\sqrt{2}$, so $\sin(\theta) = \cos(\theta)$. If we think of 'sin' and 'cos' in terms of the angles of a triangle, we might conclude that $\theta = 45°$ is the only solution to the equation $\sin(\theta) = \cos(\theta)$, and the proposition of this example would indeed be true. However, the mathematical functions cos ( ) and sin ( ) are defined for all real numbers θ, and the equation $\sin(\theta) = \cos(\theta)$ is satisfied if θ is expressed as $\theta = (45 + 180n)°$ for any integer n, so the proposition is false.

EXAMPLE 3.13 Show that $p \Rightarrow q$ is the same as $\sim q \Rightarrow \sim p$.

Solution Table 3.11

TABLE 3.11 Similarity of $\sim q \Rightarrow \sim p$ and $p \Rightarrow q$

p	q	$\sim p$	$\sim q$	$\sim q \Rightarrow \sim p$	$p \Rightarrow q$
T	T	F	F	T	T
T	F	F	T	F	F
F	T	T	F	T	T
F	F	T	T	T	T

3.6 ALGEBRA OF PROPOSITIONS

Propositions that satisfy various laws are listed in Table 3.12. They are similar to algebraic laws and useful to simplify expression. In fact, the connective $\vee$ is often treated like $+$, and the connective $\wedge$ is often dealt like $\cdot$.

TABLE 3.12 Laws of the algebra of propositions

1. (a) $p \vee p \equiv p$	**Idempotent laws**	(b) $p \wedge p \equiv p$
2. (a) $(p \vee q) \vee r \equiv p \vee (q \vee r)$	**Associative laws**	(b) $(p \wedge q) \wedge r \equiv p \wedge (q \wedge r)$
3. (a) $p \vee q \equiv q \vee p$	**Commutative laws**	(b) $p \wedge q \equiv q \wedge p$
4. (a) $p \vee (q \wedge r) \equiv (p \vee q) \wedge (p \vee r)$	**Distributive laws**	(b) $p \wedge (q \vee r) \equiv (p \wedge q) \vee (p \wedge r)$
5. (a) $p \vee T \equiv T$	**Identity laws**	(b) $p \wedge T \equiv p$
6. (a) $p \vee F \equiv p$		(b) $p \wedge F \equiv F$
7. (a) $p \vee \sim p \equiv T$	**Complement laws**	(b) $p \wedge \sim p \equiv F$
8. (a) $\sim T \equiv F$		(b) $\sim F \equiv T$
9. $\sim \sim p \equiv p$	**Involution law**	
10. (a) $\sim (p \vee q) \equiv \sim p \wedge \sim q$	**De Morgan's laws**	(b) $\sim (p \wedge q) \equiv \sim p \vee \sim q$

3.7 PROPOSITIONAL FUNCTIONS

Let A be a given set. A *propositional function* (or, an *open sentence* or *condition*) defined on A is an expression $P(x)$ which has the property that $P(a)$ is true or false for each $a \in A$. Then, $P(x)$ becomes a statement (with a truth value) whenever any element $a \in A$ is substituted for the variable x. The set A is called the *domain* of $P(x)$, and the set T_P of all elements of A for which $P(a)$ is true, is called the *truth set* of $P(x)$. In other words,

$$T_P = \{x : x \in A, P(x) \text{ is true}\} \quad \text{or} \quad T_P = \{x : P(x)\}$$

When A is some set of numbers, then $P(x)$ takes the form of an equation or inequality by involving the variable x.

EXAMPLE 3.14 Let $P(x)$ denote the sentence '$x + 4 > 3$'. Investigate $P(x)$ as a propositional function on each of the following sets:

(i) N, the set of natural numbers
(ii) $M = \{-1, -2, -3, \ldots\}$
(iii) $C = \{\text{the set of complex numbers}\}$

Solution

(i) Yes, $P(x)$ is a propositional function on N.
(ii) Although $P(x)$ is false for every element in M, yet $P(x)$ is still a propositional function on M.
(iii) No, since $i + 4 > 3$ (where $i = \sqrt{-1}$ is an imaginary quantity) does not have any meaning. In other words, inequalities are not defined for complex numbers.

EXAMPLE 3.15 Find the truth set of each of the following propositional function $P(x)$ defined on the set N of positive integers:

(i) $P(x): x + 3 < 7$ (ii) $P(x): x + 5 > 8$ (iii) $P(x): x + 4 < 1$

Solution

(i) The truth set of $P(x)$ is

$$P(x) = \{x : x \in N, x + 3 < 7\} = \{1, 2, 3\}$$

consisting of integers less than 4.

(ii) The truth set of $P(x)$ is

$$P(x) = \{x : x \in N, x + 5 > 8\} = \{4, 5, 6, \dots\}$$

consisting of all integers greater than 3.

(iii) The truth set of $P(x)$ is

$$P(x) = \{x : x \in N, x + 4 < 1\} = \phi$$

the null set. In other words, $P(x)$ is not true for any positive integer in N.

3.8 TAUTOLOGIES AND CONTRADICTIONS

In some proposition $P(p, q)$, if the last column of their truth tables contain only T, i.e., the propositions are true for any truth values of their variables, then such propositions are called *tautologies*. On the other hand, a proposition $P(p, q, \dots)$ is called a *contradiction* if it contains only F in the last column of its truth table, i.e., if it is false for any truth values of its variables. Furthermore, a proposition that is neither a tautology nor a contradiction is called a *contingency*. So, the proposition 'p or not p', i.e., $p \vee \sim p$, is a tautology, and the proposition 'p and not p', i.e., $p \wedge \sim p$ is a contradiction. This can be verified by their truth tables, as shown in Tables 3.13 and 3.14. Here, the truth tables possess only two rows since each proposition has only one variable p.

TABLE 3.13 $p \vee \sim p$

p	$\sim p$	$(p \vee \sim p)$
T	F	T
F	T	T

TABLE 3.14 $p \wedge \sim p$

p	$\sim p$	$(p \wedge \sim p)$
T	F	F
F	T	F

Note The negation of a tautology is a contradiction, since it is always false. Again, the negation of a contradiction is a tautology, since it is always true.

EXAMPLE 3.16 Verify that the proposition $p \vee \sim(p \wedge q)$ is a tautology.

Solution

TABLE 3.15 Illustration of a tautology

p	q	$p \wedge q$	$\sim(p \wedge q)$	$p \vee \sim(p \wedge q)$
F	F	F	T	T
F	T	F	T	T
T	F	F	T	T
T	T	T	F	T

Since, in Table 3.15, the truth value of $p \vee \sim(p \wedge q)$ is true for all values of p and q and, the proposition is a tautology.

3.8.1 System Specifications (Consistency)

Translating sentences in natural language (such as English) into logical expressions is an essential part of specifying both hardware and software systems. System and software engineers consider natural language and produce precise and unambiguous specifications that can be used as the basis for system development. The following example illustrates how compound propositions can be used in this process.

EXAMPLE 3.17 Translate the specification 'The automated reply cannot be sent when the file system is full' using logical connectives.

Solution Let p: The automated reply can be sent and q: The file system is full. Then, $\sim p$: It is not the case that the automated reply can be sent. Again, $\sim p$: The automated reply cannot be sent. Consequently, the specification can be represented by the conditional statement $q \rightarrow \sim p$.

EXAMPLE 3.18 Investigate the following system specifications as consistent one:

'The diagnostic message is stored in the buffer or it is retransmitted'
'The diagnostic message is not stored in the buffer'
'If the diagnostic message is stored in the buffer, then it is retransmitted'

Solution Let p: The diagnostic message is stored in the buffer and q: The diagnostic message is retransmitted. The specifications can then be written as $p \vee q$, $\sim p$, and $p \rightarrow q$. All the three specifications will be true if p would have been false, so that $\sim p$ is true. To make $p \vee q$ true, p must be made false and q is true. Finally, because $p \rightarrow q$ is true when p is false and q is true, we conclude that the given specifications are consistent.

3.8.2 Principle of Substitution

We have already discussed the construction of tautology. However, there exists another procedure of creating tautology through the principle of substitution (or replacement process), which we will describe now.

The *principle of substitution* illustrates a substitution (or replacement) process of a formula F by another formula G, i.e. if F can be obtained from G by substituting (or replacing) formulae for some variables of G following the condition that the same formula is replaced for the same variables each time, if needed. From logical point of view, we demonstrate the above definition as follows.

Assume a formula $F: P \rightarrow (q \rightarrow r)$. If we now replace $(q \rightarrow r)$ by an equivalent formula $\sim q \vee r$ in F, we get another formula $G: p \pm (\sim q \vee r)$. Then, we can verify that formulae F and G are equivalent to each other. This process of obtaining G from F is known as the *substitution principle*.

Note Actually, to construct a substitution instance of a formula, substitutions are made for the simple proposition (without connectives) and not for the compound proposition. For instance, $p \rightarrow q$ is not a substitution instance of $p \rightarrow \sim r$, because it must be replaced by r but not by $\sim r$.

EXAMPLE 3.19 Show that the proposition $(p \wedge \sim q) \vee \sim (p \wedge q)$ is a tautology.

Solution The given proposition can be expressed in the form $p \wedge \sim p$, where $p = p \wedge \sim q$. Since $p \wedge \sim p$ is a tautology, then by the principle of substitution, $(p \wedge \sim q) \vee \sim (p \wedge q)$ is also a tautology.

EXAMPLE 3.20 Prove that $p \to (q \to r) \Leftrightarrow p \to (\sim q \vee r) \Leftrightarrow (p \wedge q) \to r$.

Solution It is known that $q \to r \Leftrightarrow \sim q \vee r$. Now, replacing $q \to r$ by $\sim q \vee r$, we get $p \to (\sim q \vee r)$, which is equivalent to $\sim p \vee (\sim q \vee r)$. Then,

$$\sim p \vee (\sim q \vee r) \Leftrightarrow (\sim p \vee \sim q) \vee r \Leftrightarrow \sim (p \wedge q) \vee r \Leftrightarrow (p \wedge q) \to r$$

by associativity of $\vee$, De Morgan's law etc.

Note It is possible to substitute more than one variable by other variables provided all the substitutions are assumed to be occurred simultaneously.

EXAMPLE 3.21 Show that $(p \to q) \wedge (r \to q) \Leftrightarrow (p \vee r) \to q$.

Solution It is known that $(p \to q) \wedge (r \to q) \Leftrightarrow (\sim p \vee q) \wedge (\sim r \vee q)$. Then, replacing $p \to q$ and $r \to q$ by $(\sim p \vee q)$ and $(\sim r \vee q)$, respectively, we get

$$(p \to q) \wedge (r \to q) \Leftrightarrow (\sim p \wedge \sim r) \vee q \quad [\because (A_1 \vee A_2) \wedge (A_3 \vee A_2) \Leftrightarrow (A_1 \vee A_3) \vee A_2]$$
$$\Leftrightarrow \sim (p \vee r) \vee q \quad [\text{replacing } (\sim p \wedge \sim r) \text{ by } \sim (p \vee r)]$$
$$\Leftrightarrow (p \vee r) \to q \quad [\because \sim p \vee q \Leftrightarrow (p \to q)]$$

3.9 LOGICAL EQUIVALENCE

Two propositions $P(p, q, \ldots)$ and $Q(p, q, \ldots)$ are said to be *logically equivalent*, or simply *equivalent* or *equal*, denoted by $P(p, q, \ldots) \equiv Q(p, q, \ldots)$, if they have the identical truth tables. The notion can also be defined as the propositions $P(p, q, \ldots)$ and $Q(p, q, \ldots)$ are logically equivalent if $P \leftrightarrow Q$ is a tautology. The equivalence of P and Q is also denoted by $P \Leftrightarrow Q$.

For example

(i) Consider the truth values of $\sim(p \wedge q)$ and $(\sim p) \vee (\sim q)$ as shown in Tables 3.16 and 3.17. Here, it is seen that both the truth values are the same, i.e., both the propositions are false in the first row and true in the other three rows. Accordingly, we can write $\sim(p \wedge q) \equiv (\sim p) \vee (\sim q)$. In other words, the propositions are logically equivalent.

(ii) Consider the statement 'It is not the case that Kolkata is in West Bengal and $4 + 4 = 9$'. This statement can be written in the form $\sim(p \wedge q)$, where p: Kolkata is in West Bengal and q: $4 + 4 = 9$. However, $\sim(p \wedge q) \equiv (\sim p) \vee (\sim q)$. Thus, the statement 'Kolkata is not in West Bengal, or $4 + 4$ is not equal to 9' carries the same meaning as the given statement, i.e., the propositions are logically equivalent, as shown in Tables 3.16 and 3.17.

TABLE 3.16 ~(p∧q)

p	q	p∧q	~(p∧q)
T	T	T	F
T	F	F	T
F	T	F	T
F	F	F	T

TABLE 3.17 (~p)∨(~q)

p	q	~p	~q	(~p)∨(~q)
T	T	F	F	F
T	F	F	T	T
F	T	T	F	T
F	F	T	T	T

3.9.1 De Morgan's Laws

We generally use a truth table to show the equivalency of two compound propositions. In particular, the compound propositions p and q are equivalent if and only if the columns in the truth tables agree. There are two laws of logical equivalences, known as *De Morgan's laws*. The laws are

I. $\sim(p \vee q) \equiv (\sim p) \wedge (\sim q)$
II. $\sim(p \wedge q) \equiv (\sim p) \vee (\sim q)$

Actually, the first law computes the negation of a disjunction by taking the conjunction of the negations of the compound propositions. This has been depicted in Table 3.18. Similarly, in the second one, the negation of a conjunction is formed by taking the disjunction of the negation of the component propositions. The usage of De Morgan's laws is shown in the following example.

EXAMPLE 3.22 By using De Morgan's laws, express the negation of

'Rina has a cellphone and she has a laptop computer'

and

'Papu will go to the concert or Babu will go to the concert'

Solution Let p: Rina has a cellphone and q: Rina has a laptop computer. Then, 'Rina has a cellphone and she has a laptop computer' can be represented by $p \wedge q$. By De Morgan's laws, $\sim(p \wedge q) \equiv (\sim p) \vee (\sim q)$. Consequently, the negation of the original statement can be expressed as 'Rina does not have a cellphone and she does not have a laptop computer'.

Let r: Papu will go to the concert and s: Babu will go to the concert. Then, 'Papu will go to the concert or Babu will go to the concert' can be represented by $r \vee s$. By De Morgan's laws, $\sim(r \vee s) \equiv (\sim r) \wedge (\sim s)$. Consequently, 'Papu will not go to the concert or Babu will not go to the concert'.

EXAMPLE 3.23 By using truth table, show that $\sim(p \vee q)$ is equivalent to $(\sim p) \wedge (\sim q)$.

Solution

TABLE 3.18 Equivalence of $\sim(p \vee q)$ and $(\sim p) \wedge (\sim q)$

p	q	p∨q	~(p∨q)	~p	~q	(~p)∧(~q)
T	T	T	F	F	F	F
T	F	T	F	F	T	F
F	T	T	F	T	F	F
F	F	F	T	T	T	T

It may be noticed from Table 3.18 that the truth tables in the two columns headed by $\sim(p \vee q)$ and $(\sim p) \wedge (\sim q)$ are identical.

EXAMPLE 3.24 By using Table 3.19, show that $p \wedge (q \vee r)$ is equivalent to $(p \wedge q) \vee (p \wedge r)$.

Solution

TABLE 3.19 Equivalence of $p \wedge (q \vee r)$ and $(p \wedge q) \vee (p \wedge r)$

p	q	r	$q \vee r$	$p \wedge (q \vee r)$	$p \wedge q$	$p \wedge r$	$(p \wedge q) \vee (p \wedge r)$
T	T	T	T	T	T	T	T
T	T	F	T	T	T	F	T
T	F	T	T	T	F	T	T
T	F	F	F	F	F	F	F
F	T	T	T	F	F	F	F
F	T	F	T	F	F	F	F
F	F	T	T	F	F	F	F
F	F	F	F	F	F	F	F

Since the two columns, headed by $p \wedge (q \vee r)$ and $(p \wedge q) \vee (p \wedge r)$, of the truth Table 3.19 agree, the two compound propositions are identical. This shows that $p \wedge (q \vee r)$ is equivalent to $(p \wedge q) \vee (p \wedge r)$.

EXAMPLE 3.25 Show that $p \Leftrightarrow q$ and $(p \Rightarrow q) \wedge (q \Rightarrow p)$ are equivalent.

Solution Table 3.20

TABLE 3.20 Equivalence of $p \Leftrightarrow q$ and $(p \Rightarrow q) \wedge (q \Rightarrow p)$

p	q	$p \Leftrightarrow q$	$p \Rightarrow q$	$q \Rightarrow p$	$(p \Rightarrow q) \wedge (q \Rightarrow p)$
F	F	T	T	T	T
F	T	F	T	F	F
T	F	F	F	T	F
T	T	T	T	T	T

EXAMPLE 3.26 Among the two restaurants next to each other, one has a sign that says 'Good food is not cheap' and the other has a sign that 'Cheap food is not good'. Investigate the signs regarding their equivalence.

Solution Let p: Food is good and q: Food is cheap. The first sign says $p \rightarrow \sim q$ and the second one says $q \rightarrow \sim p$.

TABLE 3.21 Equivalence of $p \rightarrow \sim q$ and $q \rightarrow \sim p$

p	q	$\sim p$	$\sim q$	$p \rightarrow \sim q$	$q \rightarrow \sim p$
F	F	T	T	T	T
F	T	T	F	T	T
T	F	F	T	T	T
T	T	F	F	F	F

From the truth table shown in Table 3.21, it is observed that both the signs are equivalent.

3.10 LOGICAL IMPLICATION

A proposition $P(p, q, \ldots)$ is said to *logically imply* a proposition $Q(p, q, \ldots)$, written as $P(p, q, \ldots) \Rightarrow Q(p, q, \ldots)$. If $Q(p, q, \ldots)$ is true whenever $P(p, q, \ldots)$ is true. The notion can also be defined as: A proposition P is said to logically imply a proposition Q if $P \Rightarrow Q$ is a tautology.

For example Consider Table 3.22. It is observed that p is true in rows 1 and 2, and consequently, $p \vee q$ is also true in these cases. Thus, $p \Rightarrow p \vee q$. Now, if $Q(p, q, \ldots)$ is true whenever $P(p, q, \ldots)$ is true, then the argument $P(p, q, \ldots) \vdash Q(p, q, \ldots)$ is valid and converse. Moreover, the argument $P \vdash Q$ is valid if and only if the conditional statement $P \Rightarrow Q$ is always true, i.e., a tautology.

TABLE 3.22 $p \Rightarrow p \vee q$

p	q	$p \vee q$
T	T	T
T	F	T
F	T	T
F	F	F

EXAMPLE 3.27 Show that $p \leftrightarrow q$ logically implies $p \Rightarrow q$.

Solution First of all the tables of $p \leftrightarrow q$ and $p \Rightarrow q$ are constructed as shown in Table 3.23. From the table, it is observed that $p \leftrightarrow q$ is true in rows 1 and 4 and $p \Rightarrow q$ is also true in these cases.

TABLE 3.23 $(p \leftrightarrow q) \rightarrow (p \Rightarrow q)$

p	q	$p \leftrightarrow q$	$p \Rightarrow q$
F	F	T	T
F	T	F	T
T	F	F	F
T	T	T	T

EXAMPLE 3.28 Show that $p \wedge q$ logically implies $p \leftrightarrow q$.

Solution Consider the truth tables of $p \wedge q$ and $p \leftrightarrow q$ as shown in Table 3.24. From the table, it is seen that $p \wedge q$ is true only in row 1 and simultaneously the proposition $p \leftrightarrow q$ is also true in this case. Thus, $p \wedge q$ logically implies $p \leftrightarrow q$.

TABLE 3.24 $(p \wedge q) \rightarrow (p \leftrightarrow q)$

p	q	$p \wedge q$	$p \leftrightarrow q$
T	T	T	T
T	F	F	T
F	T	F	T
F	F	F	T

3.11 NORMAL FORMS

One can examine the equivalency of logical expressions P and Q using the truth table. However, the process is tedious to tackle, when the number of variables increases. Convenient option lies in the transformation of the expressions P and Q to standard forms of expressions P^* and Q^*, and then a comparison of P^* and Q^* shows whether $P \equiv Q$. These standard forms are known as *normal forms* or *canonical forms*.

For simplicity we use the words *product* in place of conjunction and *sum* in place of disjunction. Some of the basic normal forms are as follows:

I. Disjunctive normal form (dnf)
II. Conjunctive normal form (cnf)

3.11.1 Disjunctive Normal Form

In a logical expression, a product of the variables and their negation is called an *elementary product*. For instance, $P \wedge \sim R$, $Q \wedge P \wedge \sim R$, etc. are elementary products. Also, the sum of the variables and their negations is called an *elementary sum*. For instance, $P \vee \sim Q \sim P \vee \sim Q \vee \sim R$ are elementary sums.

The elementary sums or products satisfy the following properties:

(i) An elementary sum is identically *true* if and only if it contains at least one pair of factors in which one is the negation of the other. (A part of the elementary sum of product which is itself an elementary sum of product is called a *factor* of the original sum or product.)

(ii) An elementary product is identically *false* if and only if it contains at least one pair of factors in which one is negation of the other.

A logical expression is called a *disjunctive normal form*, abbreviated as *dnf*, if it is a sum of elementary products.

For example The forms $p \vee (q \wedge r)$ and $p \vee (\sim q \wedge r)$ are in dnf, but $p \wedge (q \vee r)$ is not in dnf. For example, consider the 'exclusive OR' $P \oplus Q$, as defined in Table 3.25. That is, $P \oplus Q$ is true iff one of P or Q is true but not both. Thus, $P \oplus Q$ is equivalent to the form $(\sim P \wedge Q) \vee (P \wedge \sim Q)$ which is its dnf.

TABLE 3.25 Exclusive OR/$P \oplus Q$

P	Q	$P \oplus Q$
F	F	F
F	T	T
T	F	T
T	T	F

In general, we obtain the dnf for an n-variable propositional form $f(P_1, P_2, \ldots, P_n)$ from its truth table as follows.

For each row in which $f(P_1, P_2, \ldots, P_n)$ assumes the value T, we form the conjunction $P_1 \wedge P_2 \wedge \cdots \wedge \sim P_k \cdots \wedge \ldots \sim P_n$, where we take P_k if there is a T in the kth position in the row and $\sim P_k$ if there is an F there. This conjunction is called a *minterm*. Then we form the disjunction of the minterms as

$$(P_1 \wedge \sim P_2 \wedge \cdots P_n) \vee (\sim P_1 \wedge P_2 \wedge \cdots \sim P_n) \vee \cdots \vee (P_1 \wedge P_2 \wedge \cdots P_n)$$

Thus, the disjunction of the minterms yields the dnf.

EXAMPLE 3.29 Find the dnf for the propositional form $f(P, Q, R)$ defined as (see Table 3.26).

TABLE 3.26 dnf of $f(P, Q, R)$

P	Q	R	$f(P, Q, R)$
F	F	F	T
F	F	T	F
F	T	F	T
F	T	T	F
T	F	F	T
T	F	T	F
T	T	F	F
T	T	T	T

Solution The dnf is expressed as $(\sim P \wedge \sim Q \wedge \sim R) \vee (\sim P \wedge Q \wedge \sim R) \vee (P \wedge \sim Q \wedge \sim R) \vee (P \wedge Q \wedge R)$, which is the required dnf.

EXAMPLE 3.30 Find the dnf of $\sim(P \vee Q) \Leftrightarrow (P \wedge Q)$.

Solution

$$\sim(P \vee Q) \Leftrightarrow (P \wedge Q) = (\sim(P \vee Q) \wedge (P \wedge Q)) \vee ((P \vee Q) \wedge \sim(P \wedge Q))$$

[Since $R \Leftrightarrow S = (R \wedge S) \vee (\sim R \wedge \sim S)$]

$$= (\sim P \wedge \sim Q \wedge P \wedge Q) \vee ((P \vee Q) \wedge (\sim P \vee \sim Q))$$
$$= (\sim P \wedge \sim Q \wedge P \wedge Q) \vee (P \vee Q \wedge \sim P) \vee (P \vee Q \wedge \sim Q)$$
$$= (\sim P \wedge \sim Q \wedge P \wedge Q) \vee (P \wedge \sim P) \vee (Q \wedge \sim P) \vee (P \wedge \sim Q) \vee (Q \wedge \sim Q)$$

EXAMPLE 3.31 Determine the dnf of $p \Rightarrow ((p \Rightarrow q) \wedge \sim(\sim q \vee \sim p))$.

Solution

$$p \Rightarrow ((p \Rightarrow q) \wedge \sim(\sim q \vee \sim p))$$
$$\equiv \sim p \vee ((p \Rightarrow q) \wedge \sim(\sim q \vee \sim p)) \equiv \sim p \vee ((\sim p \vee q) \wedge \sim(\sim q \vee \sim p))$$
$$\equiv \sim p \vee ((\sim p \vee q) \wedge (q \wedge p)) \equiv \sim p \vee ((\sim p \wedge (q \wedge p)) \vee (q \wedge (q \wedge p)))$$
$$\equiv \sim p \vee ((\sim p \wedge q) \wedge p) \vee ((q \wedge q) \wedge p) \equiv \sim p \vee (F) \vee (p \wedge q)$$
$$\equiv \sim p \vee (p \wedge q)$$

which is the required dnf.

3.11.2 Conjunctive Normal Form

If a form is a product of elementary sums then that form is called a *conjunctive normal form*. It is abbreviated as *cnf*.

For example The forms $p \wedge r$ and $\sim p \wedge (q \vee r)$ are in cnf.

EXAMPLE 3.32 Find the cnf of the following:
  (i) $p \wedge (p \Rightarrow q)$
  (ii) $(q \vee (p \wedge r)) \wedge \sim((p \vee r) \wedge q)$

Solution
  (i) $p \wedge (p \Rightarrow q) \equiv p \wedge (\sim p \vee q)$ (which is in the cnf)
  (ii) $(q \vee (p \wedge r)) \wedge \sim((p \vee r) \wedge q) \equiv (q \vee (p \wedge r)) \wedge (\sim(p \vee r) \vee \sim q)$
  $$\equiv q \vee (p \wedge r) \vee (\sim p \wedge \sim r) \vee \sim q$$
  $$\equiv (q \vee p) \wedge (q \vee r) \wedge (\sim p \vee \sim q) \wedge (\sim r \vee \sim q)$$

(which is the required cnf)

3.12 ARGUMENTS

An *argument* is a positive declaration (or an assertion) that a given set of propositions $P_1, P_2, \ldots, P_n$, called *premises*, yields another proposition Q, called the *conclusion*. Such an argument can be denoted by either of its *tautological form*, i.e., $P_1, P_2, \ldots, P_n \vdash Q$ (here, '$\vdash$' means then) or,

$$P_1 \wedge P_2 \wedge \cdots \wedge P_n \Rightarrow Q$$

Thus, an argument
$$P_1, P_2, \ldots, P_n \vdash Q$$
is said to be *valid* if Q is true whenever all the premises $P_1, P_2, \ldots, P_n$ are true; or, in other way, the propositions $P_1, P_2, \ldots, P_n$ together with another proposition Q will be a *valid argument* if
$$P_1 \wedge P_2 \wedge \cdots \wedge P_n \Rightarrow Q$$
is a tautology. An argument which is not valid is called a *fallacy*.

For example

(i) The argument represented as $p, p \Rightarrow q \vdash q$ is a valid one. Here, the rule is called the *law of detachment*. The proof of this law can be readily available from Table 3.27. In particular, both p and $p \Rightarrow q$ are true in row 1, and in this case q is also true.

(ii) The following argument is a fallacy: $p \Rightarrow q \vdash q$ because $p \Rightarrow q$ and q, both are true in row 3 in Table 3.27. However, in this case p is false.

Thus, the argument $P_1, P_2, \ldots, P_n \vdash Q$ is valid if and only if the proposition $P_1 \wedge P_2 \wedge \cdots \wedge P_n \Rightarrow Q$ is a tautology.

TABLE 3.27 The law of detachment

p	q	$p \Rightarrow q$
T	T	T
T	F	F
F	T	T
F	F	T

A *fundamental principle of logical reasoning* states that 'if p implies q and q implies r then p implies r'; i.e., the following argument $p \Rightarrow q$, $q \Rightarrow r \vdash p \Rightarrow r$ is valid. This rule is known as *law of syllogism*. The rule will be verified in the next example.

EXAMPLE 3.33 Verify the law of syllogism by a truth table, i.e., show that the proposition $(p \rightarrow q) \wedge (q \rightarrow r) \rightarrow (p \rightarrow r)$ is a tautology.

Solution The truth table of the law is given below. It can be seen from Table 3.28 that the premises $p \Rightarrow q$ and q are true in rows numbered 1, 5, 7, 8. Also, the conclusion $p \Rightarrow r$ is true in the said rows. So, the argument is valid. It may be observed from the table that, since there exists three variables p, q, and r, the truth table requires $2^3 = 8$ rows.

TABLE 3.28 Verification of law of syllogism

p	q	r	[$(p \rightarrow q)$	$\wedge$	$(q \rightarrow r)$]	$\rightarrow$	$(p \rightarrow r)$
T	T	T	T	T	T	T	T	T	T	T	T	T
T	T	F	T	T	T	F	T	F	F	T	T	F
T	F	T	T	F	F	F	F	T	T	T	T	T
T	F	F	T	F	F	F	F	T	F	T	T	F
F	T	T	F	T	T	T	T	T	T	F	T	T
F	T	F	F	T	T	F	T	F	F	F	T	F
F	F	T	F	T	F	T	T	T	T	F	T	T
F	F	F	F	T	F	T	F	T	F	F	T	F
		Step	1	2	1	3	1	2	1	4	1	2	1

Now, we use the above theory to arguments involving statements. We make stress on the validity of an argument which neither depends upon the truth values nor upon the content of the statements appearing in the argument, but upon the particular form of the argument. This is illustrated in the following example.

EXAMPLE 3.34 Consider the following argument:
p: If a person is illiterate, he is unhappy
q: If a person is unhappy, he dies young
... ...
r: Illiterate persons die young
Investigate the validity of the argument.

Solution Here, the statements p and q denote the premises and the statement r represents the conclusion of the argument. The given argument is of the form
$$p \Rightarrow q, \quad q \Rightarrow r \vdash p \Rightarrow r$$
Hence, by Example 3.33 the argument (law of syllogism) is valid.

EXAMPLE 3.35 Show that the following argument is valid:
$$p \to \sim q, r \to q, r \vdash \sim p$$

Solution First of all, the truth tables of the premises and the conclusion are constructed, which is shown in Table 3.29. It is observed from the table that $p \to \sim q, r \to q$, and r are true simultaneously in the fifth row, in which $\sim p$ is also true. Thus, the argument is valid.

TABLE 3.29 Validation of the argument $p \to \sim q, r \to q, r \vdash \sim p$

p	q	r	$p \to \sim q$	$r \to q$	$\sim q$
T	T	T	F	T	F
T	T	F	F	T	F
T	F	T	T	F	F
T	F	F	T	T	F
F	T	T	T	T	T
F	T	F	T	T	T
F	F	T	T	F	T
F	F	F	T	T	T

EXAMPLE 3.36 Show that the following argument is a fallacy:
$$p \to q, \sim p \vdash \sim q$$

Solution The truth table of $[(p \to q) \land \sim p] \to \sim q$ is first constructed (Table 3.30). Since the proposition $[(p \to q) \land \sim p] \to \sim q$ is not a tautology, the argument is a fallacy. In the same fashion, the argument is a fallacy since in the third line of the truth table $p \to q$ and $\sim p$ are true, but $\sim q$ is false.

TABLE 3.30 Illustration of the fallacy of an argument $p \to q, \sim p \vdash \sim q$

p	q	$p \to q$	$\sim p$	$(p \to q) \land \sim p$	$\sim q$	$[(p \to q) \land \sim p] \to \sim q$
T	T	T	F	F	F	T
T	F	F	F	F	T	T
F	T	T	T	T	F	F
F	F	T	T	T	T	T

3.13 RULES OF INFERENCE

It is observed in logical reasoning that a certain number of propositions are assumed to be true, and based on that assumption some other propositions are derived or inferred. The propositions that are assumed to be true are called *premises* or *hypothesis* and the proposition derived by using the rules of inference is called a *valid argument*; in other words, an argument is valid if and only if it is impossible for all the premises to be true and the conclusion to be false. *Rules of inference* can be used to derive new statements from the existing statements. These are the basic tools for constructing valid arguments and establishing the truth of sentences.

We will present here some of the important rules of inference, as shown in Table 3.31.

TABLE 3.31 Several rules of inference

Rules of inference	Tautological form	Name
$\dfrac{p}{\therefore p \vee q}$	$p \rightarrow (p \vee q)$	Addition
$\dfrac{p \wedge q}{\therefore p}$	$(p \wedge q) \rightarrow p$	Simplification
$\dfrac{\begin{array}{c}p\\q\end{array}}{\therefore p \wedge q}$	$((p) \wedge (q)) \rightarrow (p \wedge q)$	Conjunction
$\dfrac{\begin{array}{c}p\\p \rightarrow q\end{array}}{\therefore q}$	$[(p \rightarrow q) \wedge p] \rightarrow q$	Modus Ponens
$\dfrac{\begin{array}{c}p \rightarrow q\\\sim q\end{array}}{\therefore \sim p}$	$[(p \rightarrow q) \wedge \sim q] \rightarrow \sim p$	Modus Tollens
$\dfrac{\begin{array}{c}p \rightarrow q\\q \rightarrow r\end{array}}{\therefore p \rightarrow r}$	$[(p \rightarrow q) \wedge (q \rightarrow r)] \rightarrow (p \rightarrow r)$	Hypothetical Syllogism
$\dfrac{\begin{array}{c}p \vee q\\\sim p\end{array}}{\therefore q}$	$[(p \vee q) \wedge \sim p] \rightarrow q$	Disjunction Syllogism

Now, we will discuss some of the important rules of inference in propositional logic, considering those statements, which are logically correct arguments.

3.13.1 Law of Detachment (or Modus Pones)

Here, the tautology is $(p \wedge (p \rightarrow q)) \rightarrow q$, which is the basis of the rule of inference called the *law of detachment*, or *modus pones* (modus pones is a *Latin word* for *mode that affirms*). This tautology is presented in the following valid argument form

$$\dfrac{\begin{array}{c}p\\p \rightarrow q\end{array}}{\therefore q}$$

The assertions above the horizontal line are called *premises* (or *hypothesis*). The assertion below the line is called the *conclusion* (the symbol '∴' denotes 'therefore'). In particular, modus pones expresses that if a conditional statement and the hypothesis of this conditional statement are both true, then the conclusion must also be true.

For example Assume that the conditional statement, 'If Tanu gets a first class', then 'Tanu will get a job' and its hypothesis, 'Tanu gets a first class', are true. Then from modus pones, it follows that the conclusion of the conditional statement, 'Tanu will get a job', is true.

Let p: Tanu gets a first class and q: Tanu will get a job. Then the premises are p and $p \to q$, the conclusion is q. The inferential form is thus

$$\begin{array}{c} p \\ p \to q \\ \hline \therefore q \end{array}$$

It may also be noted that if one or more of the valid argument's premises is false, then the valid argument can lead to an incorrect solution.

EXAMPLE 3.37 Determine whether the argument given here is valid and also determine whether its conclusion must be true because of the validity of the argument,

'If $\sqrt{3} > \dfrac{7}{4}$, then $(\sqrt{3})^2 > \left(\dfrac{7}{4}\right)^2$. Consequently, $3 > \dfrac{49}{16}$'.

Solution Let

$$p : \sqrt{3} > \frac{7}{4}$$

and

$$q : 3 > \left(\frac{7}{4}\right)^2$$

The premises of the argument are p and $p \to q$, and q is its conclusion. This argument is valid because it is constructed by using modes pones, a valid argument form. However, one of its premises $\sqrt{3} > 7/4$ is false. Consequently, we cannot conclude that the conclusion is true. Furthermore, the conclusion of the argument is false, because $3 < 49/16$.

3.13.2 Law of Contraposition (Modus Tollens)

Here, the tautology is $(p \to q) \land \sim q \to \sim p$, which is based on the *law of contraposition* or *modus tollens* (modus tollens is a *Latin word* meaning *mode that denies*). The tautology is expressed in the following form:

$$\begin{array}{c} p \to q \\ \sim q \\ \hline \therefore \sim p \end{array}$$

The validity of modus tollens can be shown to follow from the modus pones together with the fact that a conditional statement is logically equivalent to its contrapositive.

For example 'If Kiran gets a first class, then Kiran will get a job'.

Let p be the proposition 'Kiran gets a first class' and q the proposition 'Kiran will get a job'. Then, the inferential form of the argument is

$$p \to q$$
$$\sim q$$
$$\therefore \sim p$$

Hence, the given argument is valid.

3.13.3 Disjunctive Syllogism

Here, the tautology is $(p \lor q) \land \sim p \to q$ which is formed from the *law of disjunctive syllogism*. The rule of inference is presented in the form

$$p \lor q$$
$$\sim p$$
$$\therefore q$$

This argument tells us that there exists two possibilities; if one of them is ruled out, then the other can also be ruled out. Hence, the argument is valid.

For example Consider the proposition 'Either it is raining or it is windy'. The proposition expresses that if it is not raining, therefore it is windy.

Let p: It is raining and q: It is windy. Then the inferential form is

$$p \lor q$$
$$\sim p$$
$$\therefore q$$

Hence, the argument is valid.

3.13.4 Hypothetical Syllogism

If the two implications $p \to q$ and $q \to r$ are true, then the implication $p \to r$ is also true. Symbolically, it can be expressed as

$$p \to q$$
$$q \to r$$
$$\therefore p \to r$$

Hence, the argument is valid. This argument is known as *hypothetical syllogism*.

For example Consider the proposition 'If Hari studies sincerely, Hari will obtain a first class. If Hari obtains a first class, Hari will get a good job'.

Let p: Hari studies sincerely
q: Hari will obtain a first class
r: Hari will get a good job

Thus, the inferential form is

$$p \to q$$
$$q \to r$$
$$\therefore p \to r$$

Hence, the argument is valid.

EXAMPLE 3.38 Investigate the rule of inference for the basis of the following argument: 'It is below freezing now. So, it is either below freezing or raining now'.

Solution Let p: It is below freezing now and q: It is raining now. Then, this argument is of the form

$$p$$
$$\therefore p \vee q$$

This is an argument that uses the addition rule.

EXAMPLE 3.39 Investigate the rule of inference for the basis of the following argument: 'It is below freezing and raining now. So, it is below freezing now'.

Solution Let p: It is below freezing now and q: It is raining now. Then, this argument is of the form

$$p \wedge q$$
$$\therefore p$$

This argument uses the simplification rule.

EXAMPLE 3.40 Investigate the validity of the following argument:

$$p \to q$$
$$r \to \sim q$$
$$\therefore p \to \sim r$$

Solution Given that $r \to \sim q$. By contrapositive law, we can write this as $q \to \sim r$ and $p \to q$, and $q \to \sim r$ gives $p \to \sim r$ by the law of hypothetical syllogism.

Thus, we have derived $p \to \sim r$ from the given premises. Hence, the given argument is valid.

EXAMPLE 3.41 Represent and investigate the validity of the following argument:

'Either Puneet is not guilty or Pankaj is telling the truth'

'Pankaj is not telling the truth'

... ...

∴ 'Puneet is not guilty'

Solution Let p: Puneet is not guilty and q: Pankaj is telling the truth. Then, the argument can be written as

$$p \lor q$$
$$\sim q$$
$$\overline{\qquad\qquad}$$
$$\therefore p$$

Thus, by disjunctive syllogism, the argument is valid.

EXAMPLE 3.42 Show the validity of the following argument 'If Gora gets the job and works hard, then he will be promoted. If Gora gets promotion, then he will be happy. He will not be happy. Therefore, either he will not get the job or he will not work hard'.

Solution Let p: Gora gets the job, q: Gora works hard, r: Gora gets promotion, and s: Gora will be happy.

The given argument can be written in symbolic form as

$$(p \land q) \to r$$
$$r \to s$$
$$\sim s$$

$(p \land q) \to r$ and $r \to s$ gives $(p \land q) \to s$ (by hypothetical syllogism)

$(p \land q) \to s$ and $\sim s$ gives $\sim(p \land q)$ (by modus tollens)

So, $\sim p \lor \sim q$, i.e., either he will not get the job or he will not work hard. Hence, the argument is valid.

3.14 WELL-FORMED FORMULAE

Statement formulae consist of one or more simple statements and some connectives too. If p and q are any statements, then $p \lor q$, $(p \land q) \lor (\sim p)$, and $(\sim p) \land q$ are statement formulae derived from the statement variables p and q. These variables are called the *components of the statement formulae*. A statement formula is called a *string* consisting of variables, parentheses, and connective symbols.

A statement formula is called a *well-formed formulae* (wff), if it can be generated by the following norms:

Norm I: A statement variable p alone is a wff.
Norm II: If p is a wff, then $\sim p$ is also a wff.
Norm III: If p and q are wff, then $(p \land q)$, $(p \lor q)$, $(p \to q)$, and $(p \leftrightarrow q)$ are also wff.
Norm IV: A string of symbol is a wff if and only if it is formed by finitely many applications of the Norms I–III.

According to the above recursive definition of a wff, the formulae $\sim(p \lor q)$, $(\sim p \land q)$ and $(p \to (p \lor q))$ are also wff.

EXAMPLE 3.43 Investigate the following as wff:
(i) ~(P ∧ Q)
(ii) ~P ∧ Q
(iii) (P ∧ Q) ⇒ Q
(iv) (P ⇒ Q) ⇒ (∧Q)

Solution
(i) ~(P ∧ Q) is a wff.
(ii) ~P ∧ Q is not a wff because if P and Q are wff, then with connective ∧, (~P ∧ Q) or ~(P ∧ Q) is wff.
(iii) (P ∧ Q) ⇒ Q is not wff because parenthesis is missing.
(iv) (P ⇒ Q) ⇒ (∧Q) is not wff, since ∧Q is not a wff.

3.15 PREDICATE CALCULUS

Consider the two statements 'Dr Roy is an effective teacher' and 'Dr Chakraborty is an effective teacher'. In respect to propositions, there exists no relation between them, but they have some common part. Instead of thinking of two statements, we can think of a single statement like 'x is an effective teacher' because both Dr Roy and Dr Chakraborty possess the same characteristic namely, effective teacher. By replacing x with any other name we can form many properties. The common feature expressed by 'an effective teacher' is called *predicate*. A part of a declarative sentence describing the properties of an object or relation among objects can be referred as predicate.

In this section, we will discuss the ways by which propositions can be created from statements.

Consider the following statements involving variables, '$x > 5$', '$x = y + 5$', and '$x + y = z$'. These statements are neither true nor false, when the values of the variables are not specified.

Let us first consider the statement '$x > 5$'.

The statement 'x is greater than 5' has two parts. The first part, the variable x, is the subject of the statement. The second part, the predicate, 'is greater than 5', refers to a property that the subject of the statement can have. We denote the statement 'x is greater than 5' by $P(x)$, where P denotes the predicate 'is greater than 5' and x is the variable. The statement $P(x)$ is also said to be the value of the *propositional function* P at x. Once a value has been assigned to the variable x, the statement $P(x)$ becomes a proposition and has a truth value. Thus, we can define a predicate $P(x)$ as an expression having the quality that on an assignment values to the variable x, from an appropriate domain, a statement results. This can be illustrated by the following examples.

EXAMPLE 3.44 Assume $P(x)$ as the statement '$x > 5$'. Find the truth values of $P(6)$ and $P(4)$.

Solution The statement $P(6)$ is obtained by setting $x = 6$ in the statement '$x > 5$'. Hence, $P(6)$, which is the statement '$6 > 5$', is true. However, $P(4)$, which is the statement '$4 > 5$', is false.

We also have statements that involve more than one variable. For instance, consider the statement '$x = y + 3$'. We denote this statement by $Q(x, y)$, where x and y are variables and

Q is the predicate. When the values are assigned to the variables x and y, the statement $Q(x, y)$ has a truth value.

EXAMPLE 3.45 Let $Q(x, y)$ denote the statement '$x = y + 5$'. Find the truth values of $Q(2, 3)$ and $Q(5, 0)$.

Solution To find $Q(2, 3)$, set $x = 2$ and $y = 3$ in the statement $Q(x, y)$. Thus, $Q(2, 3)$ is the statement '$2 = 3 + 5$', which is false. Also, the statement $Q(5, 0)$ is the proposition '$5 = 0 + 5$', which is true.

In the above fashion, we can extend to n-variables, i.e., a statement involving n-variables, $x_1, x_2, \ldots, x_n$ can be represented by

$$P(x_1, x_2, \ldots, x_n)$$

A statement of the form $P(x_1, x_2, \ldots, x_n)$ is the value of the propositional function P at the n-tuple $(x_1, x_2, \ldots, x_n)$ and P is also called an n-ary predicate.

The application of propositional function to computer programs is illustrated in the following example.

For example Consider the statement
If $x > 0$ then $x := x + 5$

To encounter the statement into a program, the value of the variable x at that point in the execution of the program is to be inserted into $P(x)$, which is '$x > 0$'. If $P(x)$ is true for this value of x, the assignment statement $x := x + 5$ is executed, so the value of x is increased by 5. If $P(x)$ is false for this value of x, the assignment statement is not executed, so the value of x is unchanged.

Predicates are used for the verification of computer programs which produces the desired output by the given valid input.

EXAMPLE 3.46 Let $A = \{1, 2, 3, 4, 5\}$. Determine the truth value of the following statements:
(i) $x + 3 < 10$, $\forall x \in A$
(ii) $x + 3 \leq 7$, $\forall x \in A$
(iii) $x + 3 = 7$, $\forall x \in A$
(iv) $x + 3 > 10$, $\forall x \in A$

Solution
(i) True; for every number in A satisfies $x + 3 < 10$.
(ii) False; if $x_0 = 5$, then $x_0 + 3 \neq 7$, i.e., 5 is not a solution to the given condition.
(iii) True; if $x = 4$, then $4 + 3 = 7$.
(iv) False; for every number in A, $x + 3 > 10$ is not satisfied.

Example 3.47 Let $P(x)$ denote the sentence '$x + 2 > 5$'.
Investigate $P(x)$, the propositional function, on each of the following sets:
(i) $M = \{-1-, 2, -3, \ldots\}$
(ii) Z^+, the set of positive integers
(iii) C, the set of complex number

Solution

(i) Here, $P(x)$ is false for every element in $P(x)$, still $P(x)$ is a propositional function on M.
(ii) $P(x)$ is the propositional function.
(iii) Let $3 + 2i$, $i = \sqrt{-1}$ is an imaginary number, be a complex number. It may be noted that $3 + 2i > 7$ does not have any meaning. In other words, inequalities are not defined for complex numbers.

3.16 QUANTIFIER

If in a propositional function the variables are prescribed, the resulting statement yields a proposition with a certain truth table. However, there is another way, called *quantification*, by which we can create a proposition from a propositional function. Quantification expresses a measure with which a predicate is true over a range of elements. In English literature, the words some, few, many, all and none are used in quantification. Here, we will discuss two types of quantification: universal which says that a predicate is true for every element under consideration, and existential quantification, which informs that there is one or more element under consideration for which the predicate is true. The area of logic that deals with predicates and quantifiers is called the *predicate calculus*.

3.16.1 Universal Quantifier

It has been observed that several mathematical statements affirm that a property is true for all values of a variable in a particular domain, called the *domain of discourse* (or the *universe of discourse*), and often just referred to as the *domain*. The universal quantification of $P(x)$ for a particular domain is the proposition that asserts $P(x)$ is true for all values of x in this domain. The meaning of the universal quantification of $P(x)$ varies when we vary the domain. Without domain, the universal quantifications are undefined.

Consider the expression

$$(\forall x \in A)P(x) \quad \text{or} \quad \forall x P(x) \tag{3.1}$$

which reads 'for every x in A, $P(x)$ is a true statement', or simply, 'for all x, $P(x)$'. The symbol $\forall$ which reads 'for all' or 'for every' is called the *universal quantifier*. The statement (3.1) is equivalent to the statement

$$T_P = \{x : x \in A, P(x)\} = A \tag{3.2}$$

i.e., the truth set of $P(x)$ is the entire set A.

The expression $P(x)$ by itself is an open sentence or condition and therefore has no truth value. However, it is preceded by the quantifier, does have a truth value which follows from the equivalence of Eqs (3.1) and (3.2). Specifically,

Q_1: If $\{x : x \in A, P(x)\} = A$

then $\forall x\, P(x)$ is true; otherwise $\forall x\, P(x)$ is false.

Note An element for which $P(x)$ is false is called a *counter example* of $\forall x\, P(x)$.

For example
(i) The proposition $(\forall x \in N)(x + 4 > 3)$ is true since
$$\{x : x + 4 > 3\} = \{1, 2, 3, \ldots\} = N$$
(ii) The proposition $P(x): x + 5 < 9$, then for all $x \geq 0$, is a false statement, since $P(5)$ is not true.

EXAMPLE 3.48 Let $P(x)$ be the statement '$2x + 1 > 2x$'. What is the truth value of the quantification $\forall x\, P(x)$, where the domain consists of all real numbers?

Solution The quantification $\forall x\, P(x)$ is true because $P(x)$ is true for all real numbers x.

EXAMPLE 3.49 Let $Q(x)$ be the statement '$x < 3$'. What is the truth value of the quantification $\forall x\, Q(x)$, where the domain consists of all real numbers?

Solution The statement $Q(x)$ is not true for every real number x because, for instance, $Q(4)$ is false, i.e., $x = 4$ is a counter example for the statement $\forall x\, Q(x)$. Thus, $\forall x\, Q(x)$ is false.

Example 3.50 Consider the statement $\forall x\, N(x)$, where $N(x)$ is 'Computer is connected to the network'. What is the implication of the statement?

Solution The statement $\forall x\, N(x)$ means that for every computer x on campus, that computer x is connected to the network (see Table 3.32). This statement can be expressed in English as 'every computer on campus is connected to the network'.

3.16.2 Existential Quantifier

It is observed from the mathematical statement that there is an element with a certain property. Such statement can be expressed by prefixing $P(x)$ with the proposition 'there exists an element x'. The proposition 'there exists an x' is called an *existential quantifier*. The existential quantification of $P(x)$ is the proposition 'there exists a value of x' for which $P(x)$ is valid.

Consider the expression
$$(\exists x \in A)P(x) \quad \text{or} \quad \exists x, P(x) \tag{3.3}$$

which reads 'there exists an x in A such that $P(x)$ is a true statement' or, simply, 'for some x, $P(x)$'. The symbol $\exists$ which reads 'there exists' or 'for some' or 'for at least one' is called the *existential quantifier*. The statement (3.3) is equivalent to the statement

$$T_P = \{x : x \in A, P(x)\} \neq \phi \tag{3.4}$$

i.e., the truth set of $P(x)$ is not empty. Accordingly, $\exists x\, P(x)$, i.e., $P(x)$ is preceded by the quantifier $\exists$, does have a truth value. Specifically,

$$Q_2: \text{If}\{x: P(x)\} \neq \phi$$

then $\exists x\, P(x)$ is true; otherwise $\exists x\, P(x)$ is false.

For example
(i) The proposition $(\exists x \in N)(x + 4 < 7)$ is true, since $\{x: x + 4 < 7\} = \{1, 2\} = \phi$.
(ii) The proposition $(\exists x \in Z)(-1 < x < 1)$ is true, since $\{x: -1 < x < 1\} = \{-1, 0, 1\} = \phi$.

EXAMPLE 3.51 Let $P(x)$ denote the statement '$x > 3$'. Find the truth value of the quantification $\exists x \, P(x)$ where the domain consists of all real numbers.

Solution Because '$x > 3$' is sometimes true; for instance, when $x = 4$, the existential quantification of $P(x)$, which is $\exists x \, P(x)$ is also true.

EXAMPLE 3.52 Find the truth value of $\exists x \, P(x)$, where $P(x)$ is the statement '$x^2 > 8$' and the universe of discourse consists of the positive integers not exceeding 3.

Solution Because of the given domain $\{1, 2, 3\}$, the proposition $\exists x \, P(x)$ is the same as the disjunction
$$P(1) \vee P(2) \vee P(3)$$
Thus, $P(3)$, which is the statement '$3^2 > 8$', is true; it follows that $\exists x \, P(x)$ is also true.

TABLE 3.32 Quantifier

Statement	True whenever	False whenever
$\forall x \, P(x)$	$P(x)$ is true for every x	There is an x for which $P(x)$ is false
$\exists x \, P(x)$	There is an x for which $P(x)$ is true	$P(x)$ is false for every x

3.16.3 Bound Variables

If a quantifier is applied on the variable x, then this occurrence of the variable is called *bound*. Also, an occurrence of a variable that is not bound by a quantifier or set is said to be *free*. The variables occurring in a propositional function must be bound or set equal to a particular value to convert it into a proposition. It can be made possible by the combination of universal quantifiers, existential quantifiers, and value assignments.

For example Let x and y be the variable in the statement $\exists x \, (x + y = 1)$. Actually, the variable x is bound by the existential quantification $\exists x$. However, the variable y is free, since it is not bound by a quantifier and no value is assigned to this variable. This shows that, in the statement $\exists x \, (x + y = 1)$, x is bound, but y is free.

3.17 INTRODUCTION TO PROOFS

In this section, we discuss the notion of a proof and describe methods for constructing proofs. A proof is a valid argument that establishes the truth of a mathematical statement.

The techniques of proof, we will discuss in this chapter, are important not only because they are used to prove mathematical theorems, but also for their many applications to computer science. These applications are used in verifying that computer programs are correct, establishing that operating systems are secure, making inferences in artificial intelligence, showing that system specifications are consistent, and so on. To understand the technique and its use in proofs applications are essential both in mathematics and in computer science.

Mathematical Logic

3.17.1 Brief Status of Terminology

A *theorem* is a statement that can be expected to show that it is true. The theorems which are less important are sometimes called *propositions*. An *axiom* (or *postulates*) is a statement which is assumed to be true, the *premises*, if any, of the theorem. *Rules of inference* are used to draw *conclusions*. The final step of a proof is usually just the conclusion of the theorem. *Lemma* (plural *lemmas* or *lemmata*) is useful in the proof of other results. A *corollary* is a theorem that can be established directly from a theorem that has already been proved. A *conjecture* is a statement that may be proposed to be a true statement based on the partial evidence.

3.17.2 Methods of Proof

There exist several types of *proofs* of mathematical theorems. We shall introduce here some of the common type of proofs.

3.17.3 Direct Proof

In a direct proof, we assume that p is true and axioms, definitions, and previously proven theorems, together with rules of inference, are used to show that q must also be true. Direct proofs consist of straightforward sequence of steps leading from hypothesis to the conclusion.

Here, we will discuss examples of several direct proofs. To do the same, let us first give a definition.

The integer n is even if there exist another integer m such that $n = 2m$, and n is odd if there exist another integer m such that $n = 2m + 1$.

Note An integer cannot be both even and odd at a time.

EXAMPLE 3.53 Present a direct proof of 'If n is an even integer then n^2 is an even integer'.

Solution Let p: n is an even integer and q: n^2 is an even integer. Consider the hypothesis p: if n is an even integer, then by definition of an even integer, $n = 2m$ for some integer m. Thus, $n^2 = (2m)^2 = 4m^2$, which is divisible by 2. Hence, n^2 is an even integer. Thus, $p \to q$.

EXAMPLE 3.54 Give a direct proof of 'If m and n are odd integers, then mn is an odd integer'.

Solution To present a proof of the given conditional statement, we assume that the hypothesis of this statement is true, i.e., m and n are odd integers. By the definition of an odd integer, $m = 2k_1 + 1$ and $n = 2k_2 + 1$ for some integers k_1 and k_2. Then,

$$mn = (2k_1 + 1)(2k_2 + 1) = 2(2k_1 k_2 + k_1 + k_2) + 1$$

which is odd. Hence, the proof.

3.17.4 Consistency

An extremely important notion in mathematical logic is the *consistency*, which is defined as follows.

A collection of statements is *consistent* if the statements can all be true simultaneously.

A set of formulae $H_1, H_2, H_3, \ldots, H_n$ is said to be consistent if their conjunction $H_1 \wedge H_2 \wedge \cdots \wedge H_n$ has the truth value T for some assignment of the truth values to the atomic variables appearing in $H_1, H_2, \ldots, H_n$. On the other hand, a set of formulae

$H_1, H_2, \ldots, H_n$ is *inconsistent* if their conjunction $H_1 \wedge H_2 \wedge \cdots \wedge H_n$ implies a contradiction, i.e., $H_1 \wedge H_2 \wedge \cdots \wedge H_n \Rightarrow S \wedge \sim S$ (a contradiction), where S is any formula.

We use the conception of inconsistency in methods of proof, namely proof by contradiction and indirect proof with few examples.

3.17.5 Contraposition

Proofs of statement/theorems that are not direct, i.e., that do not start with the hypothesis and end with the conclusion, are called *indirect proofs*.

A versatile and a powerful type of indirect proof is known as *proof by contraposition*. This proof is based on the fact that the conditional statement $p \to q$ is equivalent to its contrapositive $\sim q \to \sim p$ is true. To prove by contraposition of $p \to q$, we first consider $\sim q$ as a hypothesis, and using axioms, definitions, and previously proven theorems, together the rules of inference, we show that $\sim p$ must follow.

EXAMPLE 3.55 If k^2 is an even integer, then k is an even integer.

Solution Let p: k^2 is an even integer and q: k is an even integer. Assume first that $\sim q$ is true, then k is not an even integer. So, k must be odd, i.e., $k = 2m + 1$, for some integer m. Then,

$$k^2 = (2m + 1)^2 = 4m^2 + 4m + 1 = 2(2m^2 + 2m) + 1$$

i.e., $k^2 = 2n + 1$, where $n = 2m^2 + 2m$. So, k^2 is odd and $\sim q \to \sim p$. Hence, by contraposition, k is an even integer.

3.17.6 Contradiction (reductio ad absurdum)

In this method of proof, we assume the negation of what we are trying to prove and get a logical contradiction. So, our assumption must have been false and what we were originally required to prove must be true. To prove $p \to q$ is true, we construct the procedure as follows:

(i) Assume $p \wedge (\sim q)$ is true.
(ii) Using the assumption, search some conclusion that is false.
(iii) The contradiction obtained in step (ii) leads us to the conclusion that $p \wedge (\sim q)$ is false which powers that $p \to q$ is true.

EXAMPLE 3.56 Show that if $x^2 - 4 = 0$, then $x \neq 0$.

Solution Assume that $x = 0$, then $0^2 - 4 = -4 \neq 0$, which contradicts $x^2 - 4 = 0$. Hence, the assumption that $x = 0$ is false and we have proved that $x \neq 0$.

EXAMPLE 3.57 Prove that $\sqrt{2}$ is not a rational number.

Solution Suppose that $\sqrt{2}$ is a rational number. Then, we can find integers a and b such that

$$\sqrt{2} = \frac{a}{b}$$

and we can assume that any common factors in a and b have been cancelled. We now square both sides of the equation to get

$$2 = \left(\frac{a}{b}\right)^2$$

Thus, $2b^2 = a^2$. Hence, a^2 is a multiple of 2, and is therefore even. So, a must be even, since the square of any odd number is also odd. Thus, 2 divides a and we can write $a = 2m$ for some integer m. Hence,

$$2b^2 = (2m)^2 \quad \text{or,} \quad 2b^2 = 4m^2 \quad \text{or,} \quad b^2 = 2m^2$$

Thus, b^2 is even. Hence, b is even. But now a and b have a common factor of 2, which is a contradiction to the statement that a and b have no common factors.

Hence, our initial assumption that $\sqrt{2}$ is a rational number is false. Thus, $\sqrt{2}$ is an *irrational* number.

3.17.7 Mathematical Induction

Statement: A sentence which can be judged to be true or false is called a *statement*. We generally denote a statement holding for $n \in N$ by $p(n)$

For example

(i) $p(n) : 2^n$ is divisible by 2 for all $n \in N$.
   It is clearly a true statement.
(ii) $p(n) : (10n + 3)$ is prime.

   Clearly, $p(3) = (10.3 + 3) = 33$, which is not prime. So, the given statement does not hold for all natural numbers.

In general, mathematical induction can be used to prove statements that assert that $p(n)$ is true for all positive integers n, where $p(n)$ is a propositional function.

To proof a mathematical statement (in the form of a formula) by mathematical induction, it requires two steps, a *basis step*, where we show that $p(1)$ is true, and an *inductive step*, where we show that for all positive integers k, if $p(k)$ is true, then $p(k + 1)$ is true.

Note In a proof by mathematical induction, it is only to be shown, if it is assumed that $p(k)$ is true, then $p(k + 1)$ is also true.

The *principle of mathematical induction* can be expressed as follows. Let $p(n)$ be a statement which is defined for the positive integers $n = 1, 2, 3, \ldots$. Then $p(n)$ is true for all positive integers n provided that

I. $p(1)$ is true.
II. $p(k + 1)$ is true whenever $p(k)$ is true.

Therefore, three steps are required to prove mathematical statement, using the principle of mathematical induction:

Step 1 (inductive basis step): Verify that $p(1)$ is true.
Step 2 (inductive hypothesis step): Assume that $p(k + 1)$ is true for an arbitrary value of k.
Step 3 (inductive step): Verify that $p(k + 1)$ is true on the basis of inductive hypothesis.

Note (replacement of basis step) The principle of mathematical induction chooses first $n = 1$ and then proves that $p(n)$ is true for $n \geq 1$. On the other hand, we can choose an integer different from 1, say, $n = p$ and prove that for $n = k + 1$ assuming that the statement is true for $n = k$ ($k \geq p$).

The *summation* formula by mathematical induction can be proved as follows. Mathematical induction is well suited for proving the validity of summation formulae. However, the disadvantage of this method lies in the fact that we cannot use it to derive a summation formula. That means, you must have the formula with you before attempting to prove it by mathematical induction.

EXAMPLE 3.58 Show that if n is a positive integer, then

$$1 + 2 + 3 + \cdots + n = \frac{n(n+1)}{2}$$

Solution Let $p(n)$ be the proposition or the statement that the sum of the first n positive integers is

$$\frac{n(n+1)}{1}$$

We will then opt for two steps to prove that $p(n)$ is true for $n = 1, 2, 3, \ldots$. First of all, we will show that $p(1)$ is true and, secondly, that the conditional statement $p(k)$ implies that $p(k+1)$ is true for $k = 1, 2, 3, \ldots$.

Inductive basis step $p(1)$ is true because

$$1 = \frac{1(1+1)}{2}$$

Inductive hypothesis step Assume that $p(k)$ holds for an arbitrary positive integer k, i.e.,

$$1 + 2 + \cdots + k = \frac{k(k+1)}{2}$$

Inductive step Under the assumption of inductive hypothesis step, it will be shown that $p(k+1)$ is true, namely, that

$$1 + 2 + \cdots + k + (k+1) = \frac{(k+1)[(k+1)+1]}{2} = \frac{(k+1)(k+2)}{2}$$

is also true.

Adding $(k+1)$ to both sides of the equation in $p(k)$, we obtain

$$1 + 2 + \cdots + k + (k+1) = \frac{k(k+1)}{2} + (k+1)$$

$$= \frac{k(k+1) + 2(k+1)}{2} = \frac{(k+1)(k+2)}{2}$$

which shows that $p(k+1)$ is true under the assumption that $p(k)$ is true. Thus, by the use of mathematical induction, it is shown that $p(k)$ is true for all positive integer n.

EXAMPLE 3.59 Show by mathematical induction

$$\frac{1}{1.2} + \frac{1}{2.3} + \cdots + \frac{1}{n(n+1)} = \frac{n}{n+1}$$

Solution Let $p(n)$ be the given statement.

Inductive basis step For

$$n = 1, \frac{1}{1.2} = \frac{1}{1+1} = \frac{1}{2}$$

i.e., $p(1)$ is true.

Inductive hypothesis step Assume that $p(k)$ is true, i.e.,

$$\frac{1}{1.2} + \frac{1}{1.3} + \cdots + \frac{1}{k(k+1)} = \frac{k}{k+1}$$

is true.

Adding

$$\frac{1}{(k+1)(k+2)}$$

on both sides of $p(k)$, we obtain

$$\frac{1}{1.2} + \frac{1}{2.3} + \cdots + \frac{1}{k(k+1)} + \frac{1}{(k+1)(k+2)}$$

$$= \frac{k}{k+1} + \frac{1}{(k+1)(k+2)} = \frac{k(k+2)+1}{(k+1)(k+2)}$$

$$= \frac{k^2 + 2k + 1}{(k+1)(k+2)} = \frac{(k+1)(k+1)}{(k+1)(k+2)}$$

$$= \frac{k+1}{k+2}$$

Thus, $p(k + 1)$ is true whenever $p(k)$ is true. Hence, by the principle of mathematical induction $p(n)$ is true for all positive integers n.

EXAMPLE 3.60 Prove by induction that the expression for the number of diagonals in a polygon of n sides is

$$\frac{n(n-3)}{2}$$

Solution Given that there are n sides in a polygon. Among them consider two sides for the diagonals, possibly by $^nC_2$ ways.

Thus, the total number of diagonals is $^nC_2 - n$, i.e.,

$$\frac{n(n-1)}{2} - n = \frac{n(n-3)}{2}$$

EXAMPLE 3.61 (extension of one of De Morgan's laws) Let $A_1, A_2, \ldots, A_n$ be any n sets. Show by mathematical induction that

$$\overline{\left(\bigcup_{i=1}^{n} A_i\right)} = \bigcap_{i=1}^{n} \overline{A_i}$$

Solution Let $p(n)$ be the given statement.

Inductive basis step For $n = 1$, $p(1)$ is the statement $\overline{A_1} = \overline{A_1}$, which is true.

Inductive hypothesis step Assume that $p(k)$ is true, i.e.,

$$\overline{\left(\bigcup_{i=1}^{k} A_i\right)} = \bigcap_{i=1}^{k} \overline{A_i}$$

is true.

Inductive step Here, we will show that the statement $p(n)$ is true for $n = k + 1$. Then

$$\overline{\left(\bigcup_{i=1}^{n+1} A_i\right)} = \overline{A_1 \cup A_2 \cup \cdots \cup A_k \cup A_{k+1}}$$

$$= \overline{(A_1 \cup A_2 \cup \cdots \cup A_k) \cup A_{k+1}} \quad \text{(associative property of } \cup\text{)}$$

$$= \overline{(A_1 \cup A_2 \cup \cdots \cup A_k)} \cap \overline{A_{k+1}} \quad \text{(by De Morgans's law for two sets)}$$

$$= \left(\bigcap_{i=1}^{k} \overline{A_i}\right) \cap \overline{A_{k+1}} \quad \text{[using } p(k)\text{]}$$

$$= \left(\bigcap_{i=1}^{k+1} \overline{A_i}\right)$$

Thus, by the principle of mathematical induction $p(n)$ is true for all $n \geq 1$.

EXAMPLE 3.62 Show that the statement $2 + 4 + \cdots + 2n = (n + 2)(n - 1)$, for all $n \geq 1$, satisfies the inductive step but has no basis.

Solution Let $p(n)$ be the proposition that $2 + 4 + \cdots + 2n = (n + 2)(n - 1)$.

(i) $p(1)$ implies that $2 = (1 + 2)(1 - 1)$, which is not true.

(ii) If $p(n)$ were true, then $2 + 4 + \cdots + 2n = (n + 2)(n - 1)$ would be true, and by adding $2(n + 1)$ to both sides we would get

$$2 + 4 + \cdots + 2n = (n + 2)(n - 1) + 2(n + 1)$$
$$= (n + 3)n = [(n + 1) + 2][(n + 1) - 1]$$

Hence, $p(n + 1)$ would also be true.

Thus, the inductive step is satisfied, but the basis for the induction fails and the result is false.

EXAMPLE 3.63 Let $N = n^2 + n + 41$. Show that there are some values of n for which N is a prime number, and others for which it is not. It follows that there is no inductive step which would show that $n^2 + n + 41$ is a prime number for all possible n.

Solution Let $p(n)$ imply that $n^2 + n + 41$ is a prime number.

(i) $p(1)$ implies that $1^2 + 1 + 41 = 43$ is a prime. This is true.

(ii) Now, we have to find a value of n for which $n^2 + n + 41$ is not a prime number!

In fact, $p(1), p(2), \ldots, p(39)$ are all true, but $p(40)$ and $p(41)$ are false. For instance, if $n = 39$ then $N = 39^2 + 39 + 41 = 1601$, which is prime. However, if $n = 40$, then

$N = 40^2 + 40 + 41 = 1681$, and $1681 = 41^2$. Hence, N is not prime when $n = 40$, i.e., the truth of $p(39)$ does not imply the truth of $p(40)$ and no inductive step is possible.

Thus, the basis for the induction holds but the inductive step fails and the result is false.

3.17.8 Proof by Cases

Whenever if it is not possible to take into account all the cases of a proof at a time, a proof by cases should come into the mind. This occurs when there is no way to begin a proof; however, allied information may encourage by moving the proof forward.

To prove $p \rightarrow q$ by cases, we take p to be in the form $p_1 \vee p_2 \vee \cdots \vee p_n$ by proving separately each of the following $p_1 \rightarrow q, p_2 \rightarrow q, \ldots, p_n \rightarrow q$, we can establish $(p_1 \vee p_2 \vee \cdots \vee p_n) \rightarrow q$.

EXAMPLE 3.64 Prove that for every positive integer n, $n^3 + n$ is even.

Solution

Case I Suppose n is even. Then, $n = 2m$ for some positive integer m. Now,
$$n^3 + n = 8m^3 + 2m = 2(4m^3 + m)$$
which is even.

Case II Suppose n is odd. Then $n = 2m + 1$ for some positive integer m. Now,
$$n^3 + n = (2m + 1)^3 + 2m + 1 = (8m^3 + 12m^2 + 6m + 1) + 2m + 1$$
$$= 2(4m^3 + 6m^2 + 4m + 1)$$
which is even. Hence, the sum $n^3 + n$ is even.

EXAMPLE 3.65 Show that if n is an integer then $n^2 \geq n$.

Solution

Case I When $n = 0$, then $0^2 = 0$. We see that $0^2 \geq 0$. It follows that $n^2 \geq n$ is true.

Case II When $n \geq 1$, then we multiply both sides of the inequality $n \geq 1$ by the positive integer n and we obtain $n \cdot n \geq n \cdot 1$. So, $n^2 \geq n$ for $n \geq 1$.

Case III Assume $n \leq -1$. However, here $n^2 \geq 0$. It follows that $n^2 \geq n$. Thus, since the inequality $n^2 \geq n$ holds in all the three cases, we can conclude that if n is an integer, then $n^2 \geq n$.

SUMMARY

In this chapter, we have concentrated on the basic ideas of propositional and predicate logic. We discussed basic logical operators/operations which provide rules of logic. These rules give precise meaning to mathematical statements and make distinction between valid and invalid mathematical arguments. We have also covered the semantics of propositional logic via truth tables and applications to the analysis of logical arguments. We have visualized an important type of method for logical equivalences in which a statement can be replaced by another statement with the same truth value. In continuation, De Morgan's laws have also been discussed. Propositional logic, discussed earlier, cannot adequately express the meaning of statements in mathematics and natural language. So, we have introduced a powerful type of logic, called *predicate logic*, in which the quantifiers ∀ (for all) and ∃ (there exists) may be applied to variables intended to range over some domain of interpretation. This logic permits us to reason and explore relationships between objects. Here, we have discussed the background necessary to understand the statement of predicate logic. Then we studied quantifiers which create a proposition from a propositional function. Later in the chapter, we have studied proofs and their methods in mathematics, which are valid arguments. These establish the truth of mathematical statements.

EXERCISES

1. Classify the following statements as propositions or non-propositions:
   (i) The population of India goes up to 100 million in year 2000
   (ii) $x + y = 10$
   (iii) Go to Kolkata
   (iv) The Intel Pentium-III is a 64-bit computer
2. Consider the following statements:
   p: He is rich and q: He is generous. Write the new propositions using conjunction (∧), disjunction (∨), and negation (~).
3. Let p: It is warm day and q: The temperature is 37°C. Write in simple sentences the meaning of the following:
   (i) ~ p
   (ii) ~ ($p ∨ q$)
   (iii) ~ ($p ∧ q$)
   (iv) ~ (~ p)
   (v) $p ∨ q$
   (vi) $p ∧ q$
   (vii) ~ $p ∧$ ~ q
   (viii) ~ (~ $p ∨$ ~ q)
4. Consider the following statements:
   p: Ramen is coward.
   q: Ramen is lazy.
   r: Ramen is rich.
   Write the following compound statements in the symbolic form:
   (i) Ramen is either coward or poor.
   (ii) Ramen is neither coward nor lazy
   (iii) It is false that Ramen is coward but not lazy.
   (iv) Ramen is coward or lazy but not rich.
   (v) It is false that Ramen is coward or lazy but rich.
   (vi) It is not true that Ramen is not rich.
   (vii) Ramen is rich or else Ramen is both coward and lazy.
5. Translate the following statement into symbolic form:
   'If the utility cost goes up or the request for additional funding is desired, then a new computer will be purchased if and only if we can show that the current computing facilities are indeed not adequate'.
6. Construct the truth tables of the following:
   (i) $p ∧ \sim p$
   (ii) $p ∨ \sim p$
   (iii) $\sim(\sim p)$
7. Construct the truth tables of the following:
   (i) $\sim(p ∨ q)$
   (ii) $\sim(p ∨ \sim q)$
   (iii) $(p ∧ q) ∨ (p ∧ q)$
   (iv) $(p ∨ q) ∨ \sim q$
8. Find the negation of propositions p and q:
   p: All people are intelligent.
   q: No student is graduate.
9. If p: It is cold and q: It is raining, write simple verbal sentence which describes each of the following statements:
   (i) $\sim p$
   (ii) $p ∧ q$
   (iii) $p ∨ q$
   (iv) $p ∨ \sim q$
10. Express the following statement in symbolic form
    'If p implies q and q implies r then p implies r'

11. Show that $p \vee p \leftrightarrow p$ is a tautology.
12. Verify that proposition $p \vee \sim(p \wedge q)$ is a tautology.
13. From the following formulae, find the tautology, contingency, and contradiction:
    (i) $(p \wedge \sim q) \vee (\sim p \wedge q)$
    (ii) $\sim(p \vee q) \vee (\sim p \wedge \sim q)$
14. Investigate the following as a tautology, contingency, and contradiction:
    (i) $p \to (p \to q)$
    (ii) $p \to (q \to p)$
    (iii) $p \wedge \sim p$
15. Let p: It is raining, q: It is Monday, and r: If it is raining then it is Monday or it is raining. Express the proposition r in terms of p and q and show that it is tautology.
16. Show that the following argument is invalid: 'If I buy stocks, I will lose money, therefore I will lose money, I buy stocks'.
17. If p and q are false propositions, verify $(p \vee q) \wedge (\sim p \vee \sim q)$ as true or false.
18. Let the truth values of p and q be 'T', and that of r and s be 'F'. Find the truth values of the following:
    (i) $p \vee (q \wedge r)$
    (ii) $p \to (r \wedge s)$
19. Write the contrapositive, converse, and inverse of the following:
    'Indian team win whenever match is played in Kolkata, home town of Sourav Ganguly'.
20. Using the laws of propositions prove the following:
    (i) $\sim(p \vee q) \vee (\sim p \wedge q) \equiv \sim p$
    (ii) $(p \to q) \wedge (r \to q) \equiv (p \vee r) \to q$
21. Determine the truth value of the following statements, taking the set of real numbers as an universal set. Also, negate each of the statement.
    (i) $\forall x, |x| = x$
    (ii) $\exists x, x^2 = x$
    (iii) $\forall x, x + 1 > x$
    (iv) $\exists x, x + 1 = x$
22. Find the dnf of the following forms:
    (i) $(p \to q) \wedge (\sim p \wedge q)$
    (ii) $\sim(p \to (q \wedge r))$
23. Determine the cnf of the following forms:
    (i) $\sim(p \to r) \wedge (p \leftrightarrow q)$
    (ii) $(p \wedge q) \vee (\sim p \wedge q \wedge r)$
24. Prove that the argument $p \to (q \vee r), (s \wedge t \to q, (q \vee r) \to (s \wedge t) \vdash p \to q$ is valid without using truth tables.
25. Prove that the argument $p \vee (q \to p), \sim p \wedge r \vdash \sim q$ is valid without using truth tables.
26. Investigate the validity of the following argument: 'If Shankha is selected in IAS examination, then he will not be able to go to London. Since Shankha is going to London, he will not be selected in IAS examination'.
27. Let $K(x) : x$ is a two-wheeler, $L(x) : x$ is a scooter, $M(x) : x$ is manufactured by Bajaj. Express the following using quantifiers:
    (i) Every two-wheeler is a scooter.
    (ii) There is a two-wheeler that is not manufactured by Bajaj.
    (iii) There is a two-wheeler manufactured by Bajaj that is not a scooter.
    (iv) Every two-wheeler that is a scooter is manufactured by Bajaj.
28. Show that s is a valid conclusion from the premises

    $p \to q$
    $p \to r$
    $\sim(q \wedge r)$
    $s \vee p$

29. Investigate the validity of the following argument:

    $p \to r$
    $\sim p \to q$
    $q \to s$
    $\overline{\therefore \sim r \to s}$

30. Using the rules of inference, show that the following argument is valid:

    $(p \vee q) \to s$
    $s \to r$
    $\sim (r \vee q)$
    $\overline{\therefore \sim p}$

31. Show that $n^2 > 2n + 1$, $\forall n \geq 3$.
32. Prove that $2^n > 2n$, $\forall n \geq 5$.
33. Prove each of the following statements, using the principle of mathematical induction:
    (i) $6^n - 5n + 4$ is divisible by 5 for all $n \geq 1$.
    (ii) $1 + a + a^2 + a^3 + a^4 + \cdots + a^n = (a^{n+1} - 1)/(a - 1)$ for all $n \geq 1$.
34. Using the principle of mathematical induction, prove that

$$\forall n \in N, \quad \frac{n^5}{5} + \frac{n^3}{3} + \frac{7n}{15}$$

is a natural number.

35. Show, by the principle of mathematical induction, that any positive integer n, greater than or equal to 2, is either prime or product of primes.

4 Algebraic Structures

LEARNING OBJECTIVES

After reading this chapter you will be able to understand the following:
- Implementation of binary operations to define algebraic system
- Conception of group and its utility in variety of its characteristics, namely cyclic group, permutation group, symmetric group, etc.
- Semigroup and its fruitfulness in various aspects such as isomorphism and homomorphism
- Important components of computer science, such as group codes, coding binary information, decoding, and error correction
- Development of algebraic structures by two binary operations

4.1 INTRODUCTION

The study of algebraic structures is important in computer science, not only because of the variety of applications of computing techniques that involve massive algebra, namely operational research, numerical methods, and relational databases, but also because of the way a computer operates, or is formally described by means of programming languages, requires an algebraic perspective. In this chapter, we will discuss several kinds of algebraic systems which are of significance in their own right, and which are also needed in later parts of this book.

Modern algebra, at present, is much more structured. In view of the central role which algebra plays in current computing theory and practice, it is important for us to describe what some of these structures are, and how they fit into different aspects of computer science. The important types of structure examined here are groups, semigroups, cyclic groups, and permutation groups. It is worth mentioning here that groups naturally arise in the guise of permutation groups when studying combinatorics and elsewhere, and semigroups arise particularly in the study of formal languages and automata. This chapter also covers normal subgroups, cosets, group codes, parity and generator matrices, and decoding and error correction.

4.2 BINARY OPERATIONS

Let A be a non-empty set. An operation on A is a function $f: A \times A \to A$ (denotes operation through a function from $A \times A \to A$) is called a *binary operation* on A.

A *unary* operation is a function $g: A \times A$ (function from A into A). Also, a function $h: A \times A \times A \to A$ is called *ternary operation* on A. More generally, an N-ary operation is a function from $A \times A \times \cdots \times A$ (n factors) into A.

Thus, a binary operation on A is a function that assigns each ordered pairs of elements of A to an element of A. The symbols '+, •, *, ○', etc. are used to denote binary operations on a set. The operation + will be a binary operation on A iff

$$a + b \in A \quad \forall a, b \in A \quad \text{and} \quad a + b \text{ is unique}$$

Similarly, the operation * will be a binary operation on A iff

$$a * b \in A \quad \forall a, b \in A \quad \text{and} \quad a * b \text{ is unique}$$

For example

(i) The operations such as addition (+) and multiplication (•) are binary operations on Z. However, subtraction (−) and division (/) are not binary operations on Z, since the difference and the quotient of positive integers need not be positive integers, for instance $3 - 4$ and $-4/3$ are not positive integers.

(ii) Let A and B denote, respectively, the set of even and odd positive integers. Then A is closed under addition and multiplication, since the sum and product of any even numbers are even. On the other hand, B is closed under multiplication but not addition, since $3 + 5 = 8$ is even.

(iii) Let $A = \{-2, 0, 2\}$. Then the addition (+) is not a binary operation on A, since $(-2) + (-2) = -4$ is not an element in A. On the other hand, multiplication (•) is a binary operation on A.

A binary operation on a set A is sometimes called a *composition* in A. This composition under the binary operation can be described by a table, known as the *composite table* or a *composition table* for a finite set.

Let $A = \{a_1, a_2, \ldots, a_n\}$ be the finite set. Then we define the binary operation on A by a table which can be constructed as follows.

The elements of A are arranged horizontally in a row called the *initial row* or *0-row*. These are again arranged vertically in a column, called the initial column or 0-column. The (i, j)th position in Table 4.1 is determined by the intersection of the ith row and the jth column. The table, finally, constructed is presented below.

TABLE 4.1 Composition table on a set $A = \{a_1, a_2, \ldots, a_n\}$ under the binary operation '*'

*	a_1	a_2	...	a_j	...	a_n
a_1	$a_1 * a_1$			⋮		⋮
a_2		$a_2 * a_2$		⋮		⋮
⋮				⋮		⋮
a_i	...	...	...	$a_i * a_j$		⋮
⋮						
a_n	...	...	...	...	...	$a_n * a_n$

For example Consider the set $A = \{1, 2, 3\}$. Define a binary operation '$*$' on the set A by
$$a * b = 2a + 3b$$
The operation $*$ can be represented in Table 4.2 on the set A.

TABLE 4.2 Binary operation '$*$' on a set $A = \{1, 2, 3\}$

*	1	2	3
1	5	8	11
2	7	10	13
3	9	12	15

TABLE 4.3 Binary operation '$*$' on a set $A = \{a, b\}$

*	a	b
a		
b		

To determine the number of binary operations on a set, let us consider a set $A = \{a, b\}$. The binary operation '$*$' on A can be described in Table 4.3.

Since each blank space can be filled in with the element a or b, we conclude that the table can be constructed completely in $2 \times 2 \times 2 \times 2 = 2^4 = 16$ ways. Thus, there exists 16 binary operations on A.

4.2.1 Properties of Binary Operations

This section presents a number of important properties of binary operations.

I. *Closure property* An operation '$*$' on a set A is *closed* if $a * b \in A$, where a and b are elements of A.

For example The operation of addition on the set of integers is a closed operation.

EXAMPLE 4.1 Consider the set $A = \{-1, 0, 1\}$. Investigate the set A as closure under (i) addition and (ii) multiplication.

Solution

(i) The sum of the elements -1 and -1 is $(-1) + (-1) = -2$ and the elements $1, 1$ is $1 + 1 = 2$. Both -2 and 2 are not the elements of A. Hence, the set A is not closed under addition.

(ii) The multiplication of every two elements of the set are
$$(-1)*0 = 0, (-1)*1 = -1, (-1)*(-1) = 1, 0*(-1) = 0, 0*1 = 0, 0*0$$
$$= 0, 1*(-1) = -1, 1*0 = 0, 1*1 = 0$$

Since each element of the multiplication belongs to A, hence A is closed under multiplication.

II. *Commutative law* A binary operation '$*$' on the elements of the set is *commutative* or satisfies *commutative law* if for any two elements a and $b \in A$
$$a * b = b * a$$

EXAMPLE 4.2 Let the operation '$*$' be defined on the set Z^+ of non-negative integers as
$$a * b = a + b + 2 \quad \text{for} \quad a, b \in Z^+$$
Show that the operation '$*$' is commutative.

Solution The operation '∗' is commutative, since

$$a*b = a + b + 2 = b + a + 2 = b*a \quad \text{(since addition of integers is commutative)}$$

EXAMPLE 4.3 Let the binary operation '∗' be defined on the set of integers Z as $a*b = a|b|$. Show that the operation '∗' is not commutative.

Solution Here, for example, assume $a = 2, b = -3$. Then,

$$a*b = 2*(-3) = 2|-3| = 2 \times 3 = 6 \quad \text{and} \quad b*a = (-3)*2 = (-3)|2|$$
$$= (-3) \times 2 = -6$$

Thus, $a*b \neq b*a$. Hence, the operation '∗' is not commutative.

III. *Associative law* An operation '∗' on a set A is said to be *associative* or to satisfy the *associative law* if, for any elements a, b, c in A, we have

$$(a*b)*c = a*(b*c)$$

EXAMPLE 4.4 Consider the operation '∗', which is defined on the set Z^+ of non-negative integers by

$$a*b = a + b + 2 \quad \text{for} \quad a, b \in Z^+$$

Show that the operation '∗' is associative.

Solution Assume $a, b, c \in Z^+$. Then,

$$(a*b)*c = (a + b + 2)*c = (a + b + 2) + c + 2 = a + b + c + 4 \quad \text{and}$$
$$a*(b*c) = a*(b + c + 2) = a + (b + c + 2) + 2 = a + b + c + 4$$

Thus, $(a*b)*c = a*(b*c)$, i.e., the operation '∗' is associative.

EXAMPLE 4.5 Assume that the binary operation '∗' on Z^+, the set of non-negative integers, is defined by

$$a*b = a^2 + b \quad \forall a, b \in Z^+$$

Investigate the operation '∗' as associative.

Solution Let us assume that the elements $a, b, c \in Z^+$. Then,

$$(a*b)*c = (a^2 + b)*c = (a^2 + b)^2 + c = a^4 + b^2 + 2a^2b + c \quad \text{and}$$
$$a*(b*c) = a*(b^2 + c) = a^2 + (b^2 + c) = a^2 + b^2 + c$$

i.e.,

$$(a*b)*c \neq a*(b*c)$$

⇒ the operation '∗' is not associative.

EXAMPLE 4.6 Let the binary operation '∗' be defined on the set of integers Z by

$$a*b = a|b|$$

Show that the operation is associative.

Solution To check the associative property, let us consider $a, b, c \in Z$. Then,

$$(a*b)*c = (a*b)|c| = a|b||c| \quad \text{and} \quad a*(b*c) = a*(b|c|) = a|b||c| = a|b||c|$$

Thus, the given operation is associative.

IV. *Identity element* An element e in a set A is called an *identity element* with respect to the binary operation '$*$' if, for any element a in A,

$$a*e = e*a = a$$

If $a*e = a$, then e is called the *right identity element* for the operation '$*$'. On the other hand, if $e*a = a$, then e is called the *left identity element* for the operation '$*$'.

Again, suppose e is the left identity and f is a right identity for the operation '$*$' on a set A. Then, $e = f$.

Note In particular, an identity element is unique.

V. *Inverse element* Suppose an operation '$*$' on a set possesses an identity element e. If corresponding to each element $a \in A$, there exists an element $b \in A$ such that

$$a*b = b*a = e$$

then b is called the *inverse* of a and is usually denoted by a^{-1}.

For example
(a) Consider the set of integers.
   (i) The integer 0 is an identity element with respect to an additive operation $+$ on Z, since

$$0 + a = a + 0 \quad \forall a \in Z$$

In other way, $-a \in Z$ is the inverse element of a, since

$$a + (-a) = (-a) + a = 0 \quad \forall a \in Z$$

   (ii) The integer 1 is an identity element with respect to the multiplication operation $\bullet$ on Z, since

$$1 \bullet a = a \bullet 1 = a \quad \forall a \in Z$$

However, inverse of an element a does not exist in Z.

(b) Consider the set Q of rational numbers. The element 0 is the identity 1 with respect to the addition; and $-a$ and a are additive inverses, since

$$(-a) + a = a + (-a) = 0$$

On the other hand, 1 is the identity element with respect to multiplication; and $-a$ and $-1/a$ are multiplicative inverses, since

$$(-a)\left(-\frac{1}{a}\right) = \left(-\frac{1}{a}\right)(-a) = 1$$

Note

(i) The element 0 has no multiplicative inverse.
(ii) The inverse of the identity element is the identity element itself.

EXAMPLE 4.7 Let the binary operation '$*$' be defined on the set $A = \{a, b, c\}$ in Table 4.4.

TABLE 4.4 Binary operation '$*$' on a set $A = \{a, b, c\}$

$*$	a	b	c
a	b	c	b
b	a	b	c
c	c	a	b

Show that (i) the operation '$*$' is not commutative and (ii) the operation '$*$' is associative.

Solution (i) From the table, it is observed that

$$a * b = c \text{ and } b * a = a \Rightarrow a * b \neq b * a$$

Thus, the operation '$*$' is not commutative.

(ii) From the table, it is observed that

$$a * (b * c) = a * (c) = b$$

Again,

$$(a * b) * c = (c) * c = b$$
$$\Rightarrow a * (b * c) = (a * b) * c$$

Hence, the operation '$*$' is associative.

EXAMPLE 4.8 Construct a composition table, with respect to multiplication, from the set $A = \{1, \omega, \omega^2\}$, where ω is the cube root of unity. Prove that the same multiplication satisfies the closure property, commutative property, associative property, and also show that 1 is the inverse element. Moreover, find the multiplicative inverse of each element.

Solution Since ω is the cube root of unity, then $\omega^3 = 1$. The composition Table 4.5 for multiplication • can be constructed as follows:

From Table 4.5 we can conclude that

(i) *Closure property* Since all the entries in the table are in A, so closure property is satisfied.
(ii) *Commutative property* Since rows coincide with columns, respectively, so multiplication • is commutative on A.
(iii) *Associative property* Since multiplication is associative on complex numbers and A is a set of complex numbers, so multiplication is associative on A.
(iv) *Identity element* Since row headed by 1 is same as the initial row, so 1 is the identity element.
(v) *Inverse* Clearly,

$$1^{-1} = 1, \ \omega^{-1} = \omega^2, \ (\omega^2)^{-1} = \omega$$

TABLE 4.5 Composition table of the set $A = \{1, \omega, \omega^2\}$ with respect to the operator •

•	1	ω	ω^2
1	1	ω	ω^2
ω	ω	ω^2	1
ω^2	ω^2	1	ω

EXAMPLE 4.9 Let the binary operation '$*$' be defined on the set $A = \{a, b, c, d\}$ by the composition Table 4.6.

TABLE 4.6 Binary operation '$*$' on a set $A = \{a, b, c, d\}$

$*$	a	b	c	d
a	a	c	b	d
b	d	a	b	c
c	c	d	a	a
d	d	b	a	c

Compute (i) $d*a$ and $a*d$, (ii) $b*c$ and $c*b$, (iii) $d*(b*c)$, (iv) $[(d*a)*b]*a$.
Further, (v) investigate whether the operation '$*$' is either commutative or associative?

Solution

(i) $d*a = a, a*d = d$
(ii) $b*c = b, c*b = d$
(iii) $d*(b*c) = d*b = b$
(iv) $[(d*a)*b]*a = [a*b]*a = c*a = c$
(v) Neither commutative nor associative. Since

$$a*b = c \quad \text{and} \quad b*a = d, \text{ i.e., } a*b \neq b*a$$

Also,

$$a*(b*c) = a*b = c \quad \text{and} \quad (a*b)*c = c*c = a$$

i.e.,

$$a*(b*c) \neq (a*b)*c$$

VI. *Cancellation law* A binary operation '$*$' on a set A is said to satisfy the *left cancellation law* if

$$a*b = a*c \Rightarrow b = c$$

and the same is said to satisfy the *right cancellation law* if

$$b*a = c*a \Rightarrow b = c$$

Addition, subtraction, and multiplication of integers in Z do satisfy both the left and right cancellation laws. On the other hand, matrix multiplication does not satisfy the cancellation laws.

For example Assume

$$A = \begin{bmatrix} 1 & 1 \\ 0 & 0 \end{bmatrix}, \quad B = \begin{bmatrix} 1 & 1 \\ 0 & 1 \end{bmatrix}, \quad C = \begin{bmatrix} 0 & -3 \\ 1 & 5 \end{bmatrix}, \quad D = \begin{bmatrix} 1 & 2 \\ 0 & 0 \end{bmatrix}$$

It can be shown that $AB = AC = D$, however $B \neq C$.

4.3 GROUPS

In this section, we investigate mathematical structure, consisting of a set accompanied with a binary operation, that has a lot of important applications. Among the structures, *group* is the key part of it that acts in every area in which symmetry occurs. Versatility of groups can lay the basis of mathematics, physics, and chemistry, as well as in less visible areas such as sociology. Recent applications of group theory can be viewed in the fields of particle physics and in the solutions of puzzles like Rubik's cube. In this book, we shall demonstrate an important application of group theory to binary codes in Section 4.12.1.

Let $(G, *)$ be an algebraic system with '$*$' as a binary operation. Then, $(G, *)$ is called a group if the following axioms hold:

G1 *Closure law* If $a \in G, b \in G$, then $a * b \in G$ $\forall a, b \in G$.
G2 *Associative law* If $a, b, c \in G$, then $a * (b * c) = (a * b) * c$ $\forall a, b, c \in G$.
G3 *Identity element* There exists an element $e \in G$, such that $a * e = e * a = a$ $\forall a \in G$.
G4 *Inverse element* For each $a \in G$, there exists an element $a^{-1} \in G$ (the inverse of a) such that $a * a^{-1} = a^{-1} * a = e$.

A group with additive binary operation is known as *additive group* and that with multiplicative binary operation is known as *multiplicative group*.

4.3.1 Abelian Group

A group G is said to be *abelian* (or commutative) if the commutative law holds, i.e.,

G5 *Commutativity* If $a, b \in G$ then $a * b = b * a$ $\forall a, b \in G$.

Abelian group is also called *commutative group*.

For example
(i) The set Z of integers is an abelian group under addition. The identity element is 0 and $-a$ is the additive inverse of a in Z.
(ii) The non-zero rational numbers, $Q - \{0\}$, form an abelian group under multiplication. The number 1 is the identity element and q/p is the multiplicative inverse of the rational number p/q.

An algebraic system $(G, *)$ is said to be a *groupoid*, if it satisfies the closure law only. An algebraic system $(G, *)$ is called a *monoid*, if it satisfies the closure, associative, and identity laws.

The number of elements in a group G, denoted by $|G|$ or, $O(G)$ is called the *order* of G, and G is called a *finite group* if its order is finite. If A and B are subsets of G, then we write

$$AB = \{ab : a \in A, b \in B\} \quad \text{or} \quad A + B = \{a + b : a \in A, b \in B\}$$

Note A group is a special type of monoid.

From the conception of monoids, it is known that if a group G possesses a finite number of elements, then its binary operation can be presented in a table, known as a *multiplication table*, with operational symbol $\bullet$.

Let us now find the multiplication tables of groups of orders 1, 2, 3.

Let G be a group of order 1. Then, $G = \{e\}$ and we have $e \bullet e = e$. Assume $G = \{e, a\}$ as a group of order 2. Then there exists a multiplication table [Table 4.7(a)] in which one needs to fill up the blank. The blank can be filled in by the element e or by a. Since there should not be repetition in any row or column, we must write e in the blank. The multiplication Table 4.7(b) satisfies the associativity property and the other properties of a group, so it is the multiplication table of a group of order 2.

TABLE 4.7 Multiplication tables of group of orders 2, 3

(a)			(b)			(c)				(d)			
•	e	b	•	e	a	•	e	a	b	•	e	a	b
e	e	a	e	e	a	e	e	a	b	e	e	a	b
a	a		a	a	e	a	a			a	a	b	e
						b	b			b	b	e	a

Finally, if $G = \{e, a, b\}$ is a group of order 3, then there is a multiplication table [Table 4.7(c)] in which we have to fill up four blanks. Here, we can complete Table 4.7(d) by a little calculative way. It can also be shown that the above table satisfies the associative property and the other properties of a group. Thus, it is the multiplication table of a group of order 3.

EXAMPLE 4.10 Show that $(Z, +)$ forms a group.

Solution

Closure property Let $a, b \in Z$, then $a + b \in Z$ $\forall a, b$.
Associative property Let $a, b, c \in Z$. Then,

$a * (b * c) = a + (b * c) = a + b + c$ and $(a*b)*c = (a*b)*c = (a*b) + c$
$\qquad = a + b + c$
$\Rightarrow a*(b*c) = (a*b)*c$

Thus, the operation $+$ is associative in Z.
Identity property The element $0 \in Z$ is the identity, since

$$0 + a = a + 0 = a \quad \forall a \in Z$$

Inverse property Let $a \in Z$, then there exists $-a \in Z$, such that

$$a + (-a) = (-a) + a = 0, \quad \forall a \in Z$$

Hence, $(Z, +)$ is a group.

Corollary 4.1 Since $a + b = b + a$ $\forall a, b \in Z$, so $(Z, +)$ is an abelian group.

Note $(Q, +)$ and $(R, +)$ are abelian groups.

EXAMPLE 4.11 Investigate the set of all non-zero rational numbers as a group with respect to multiplication, i.e., $(Q - \{0\}, \bullet)$ as a group.

Solution

Closure Let $a, b \in Q - \{0\}$. Then, $a \bullet b \in Q - \{0\}$ $\forall a, b$
Associative The operation $\bullet$ is associative in $Q - \{0\}$ (left to the reader)

Identity The element $1 \in Q - \{0\}$ is the identity, since
$$a \bullet 1 = 1 \bullet a = a \quad \forall a \in Q - \{0\}$$

Inverse Let $a \in Q - \{0\}$. Then,
$$\frac{1}{a} \in Q - \{0\}$$

such that
$$a \bullet \frac{1}{a} = \frac{1}{a} \bullet a = 1 \quad \forall a \in Q - \{0\}$$

Hence, $(Q - \{0\}, \bullet)$ is a group.

As an extension, $a \bullet b = b \bullet a \quad \forall a, b \in Q - \{0\}$. Thus, $(Q - \{0\}, \bullet)$ is an abelian group.

Note

(i) $(Q, \bullet)$ is a monoid but not a group, since $0 \in Q$ does not have an inverse in Q.
(ii) $(R - \{0\}, \bullet)$ is an abelian group.

EXAMPLE 4.12 Show that the cube roots of unity, namely $\{1, \omega, \omega^2\}$ form an abelian group under multiplication of complex numbers.

Solution Let $G = \{1, \omega, \omega^2\}$ where $1, \omega, \omega^2$ are cube roots of unity, i.e., $\omega^3 = 1$ and $1 + \omega + \omega^2 = 0$. The composition Table 4.8 by Cauchy's name is furnished as follows.

TABLE 4.8 Composition table of the set $G = \{1, \omega, \omega^2\}$ by Cauchy's name

$\bullet$	1	ω	ω^2
1	1	ω	ω^2
ω	ω	ω^2	1
ω^2	ω^2	1	ω

From Table 4.8 it is observed that the multiplication $\bullet$ is closed. It is known that the multiplication of complex numbers is associative and commutative (left to the reader).

The element $1 \in G$ is the identity. Moreover,
$$1^{-1} = 1, \quad \omega^{-1} = \omega^2, \quad (\omega^2)^{-1} = \omega$$

So, the inverse exists in G for every element in G. Hence, $(G, \bullet)$ is an abelian group.

EXAMPLE 4.13 Let G be the set of all non-zero real numbers and let
$$a * b = \frac{1}{2} ab$$

Prove that $(G, *)$ is an abelian group.

Solution Let us first verify that '$*$' is a binary operation. If $a, b \in G$, then $ab/2$ is a non-zero real number and hence belongs to G.

Next, to show the associativity.

Associative Here,
$$(a*b)*c = \left(\frac{ab}{2}\right)*c = \frac{(ab)c}{4} \quad \text{and} \quad a*(b*c) = a*\left(\frac{bc}{2}\right) = \frac{a(bc)}{4}$$

Thus, $(a*b)*c = a*(b*c)$, and consequently, the operation '$*$' is associative.

Identity If $a \in G$, then
$$a*2 = \frac{(a)(2)}{2} = a = \frac{(2)(a)}{2} = 2*a$$

$\Rightarrow$ The number 2 is the identity in G.

Inverse If $a \in G$, then $a' = 4/a$ is an inverse of a, since
$$a*a' = a*\frac{4}{a} = \frac{a(4/a)}{2} = 2 = \frac{(4/a)a}{2} = \frac{4}{a}*a = a'*a$$

Since $a*b = b*a$ (here $b \equiv a'$) $\forall a, b \in G$, we conclude that G is an abelian group.

EXAMPLE 4.14 Assume an algebraic system $([0, 1], +)$ in which the operation $+$ is defined in Table 4.9. Show that the system is a group.

TABLE 4.9 Binary operation $+$ on an algebraic system $([0, 1], +)$

+	0	1
0	0	1
1	1	0

Solution The given alegebraic system $([0, 1], +)$ is a group, since here 0 is an identity element and every element is its own inverse.

4.3.2 Properties of Groups

We will now discuss here some important properties of groups through statements and proofs.

Theorem 4.1 Let $(G, *)$ be a group. Then

(i) the identity element of G is unique
(ii) for any $a \in G$, the inverse of a is unique

Proof

(i) Let e_1 and e_2 be the two identity elements of G. Then
$$\begin{aligned} e_1 &= e_2 * e_1 \quad \text{(since } e_2 \text{ is the identity)} \\ &= e_2 \quad \text{(since } e_1 \text{ is the identity)} \end{aligned}$$
$$\Rightarrow e_1 = e_2$$

Thus, the identity element of G is unique.

(ii) Let a' and a'' be inverses of a. Then
$$a'*(a*a'') = a'*e \quad \text{(since } a'' \text{ is the inverse of } a\text{)}$$
$$= a' \quad (e \text{ is the identity element})$$
and
$$(a'*a)*a'' = e*a'' \quad \text{(since } a' \text{ is the inverse of } a\text{)}$$
$$= a'' \quad (e \text{ is the identity element})$$
Hence, by associativity,
$$a' = a''$$
Thus, the inverse of a is unique.

Note The inverse of a can be denoted by a^{-1}. Thus, in a group G, $a*a^{-1} = a^{-1}*a = e$.

EXAMPLE 4.15 Investigate the algebraic system $(Z, +)$ as a group, where Z is the set of all integers and $+$ is an ordinary operation of integers.

Solution
(i) *Closure law* The operation $+$ is a closed one, since addition of any two integers yields the element of set Z.
(ii) *Associative law* The operation $+$ is associative, since $(4 + 5) + 7 = 4 + (5 + 7) = 16$.
(iii) *Identity law* The operation $+$ has the identity element 0.
(iv) *Inverse law* For every integer $n \in Z$, there exists an inverse $-n$ such that
$$n + (-n) = 0$$
an identity element of Z.

Thus, we can conclude that $(Z, +)$ is a group.

Theorem 4.2 (Cancellation law)

If $(G, *)$ is a group and if a, b, c are elements of G, then

(i) $a*b = a*c \Rightarrow b = c$ (left cancellation law)
(ii) $b*a = c*a \Rightarrow b = c$ (right cancellation law)

Proof

(i) Assume that $a*b = a*c$. Multiplying both the sides by a^{-1} on the left, we obtain
$$a^{-1}*(a*b) = a^{-1}*(a*c)$$
$$\Rightarrow (a^{-1}*a)*b = (a^{-1}*a)*c \quad \text{(by associativity)}$$
$$\Rightarrow e*b = e*c \quad \text{(by the definition of inverse)}$$
$$\Rightarrow b = c \quad (e \text{ is the identity element})$$
$$\Rightarrow a*b = a*c$$
$$\Rightarrow b = c \quad \text{(left cancellation law)}$$

(ii) The proof is similar to that of part (i).

Theorem 4.3 If G is a group and if a and b be any elements of G, then

(i) the equation $a*x = b$ has a unique solution $x = a^{-1}*b$ in G
(ii) the equation $y*a = b$ has a unique solution $y = b*a^{-1}$ in G

Proof

(i) The element $x = a^{-1}*b$ is a solution of the equation $a*x = b$ since

$$a*(a^{-1}*b) = (a*a^{-1})*b \quad \text{(by associativity)}$$
$$= e*b \quad \text{(by the definition of inverse)}$$
$$= b \quad \text{(by the definition of identity)}$$

For uniqueness of the solution, now suppose that x_1 and x_2 are two solutions of the equation $a*x = b$, then

$$a*x_1 = b = a*x_2$$
$$\Rightarrow \quad x_1 = x_2 \quad \text{(by the left cancellation law)}$$

Thus, $x = a^{-1}*b$ is the unique solution of $a*x = b$.

(ii) The proof is analogous to that of part (i).

Theorem 4.4 Let G be a group and let $a, b \in G$. Then,

(i) $(a^{-1})^{-1} = a$
(ii) $(a*b)^{-1} = b^{-1}*a^{-1}$

Proof

(i) To show that a is the inverse of a^{-1}:

$$a^{-1}*a = a*a^{-1} = e$$

Also, since the inverse of an element is unique, it may be concluded that

$$(a^{-1})^{-1} = a$$

(ii) Since $a, b \in G$ and $a^{-1}, b^{-1} \in G$, then

$$(a*b)*(b^{-1}*a^{-1}) = a*[b*(b^{-1}*a^{-1})] \quad \text{(by associativity)}$$
$$= a*[(b*b^{-1})*a^{-1}] \quad \text{(by associativity)}$$
$$= a*(e*a^{-1}) \quad \text{(since } b*b^{-1} = e)$$
$$= a*a^{-1} = e.$$

Similarly,

$$(b^{-1}*a^{-1})*(a*b) = e$$

Thus,

$$(a*b)^{-1} = b^{-1}*a^{-1}$$

Corollary 4.2 If $a_1, a_2, \ldots, a_n \in G$, then

$$(a_1*a_2*\cdots*a_n)^{-1} = a_n^{-1}*a_{n-1}^{-1}*\cdots*a_2^{-1}*a_1^{-1}$$

This is known as the *inverse reversal law*.

EXAMPLE 4.16 Show that if $a^2 = a$, then $a = e$, a being an element of a group.

Solution Let a be an element of a group G such that $a^2 = a$. Now, to prove that $a = e$,

$$a^2 = a \Rightarrow a.a = a \Rightarrow (aa)a^{-1} = aa^{-1} \Rightarrow a.(aa^{-1}) = e \Rightarrow ae = e \Rightarrow a = e$$

Theorem 4.5 If every element of a group $(G, *)$ is its own inverse, then G is abelian. Or, if G be a group with identity e and if $a^2 = e$ for all $a \in G$, then G is abelian.

Proof Given $a^2 = e \ \forall a \in G$, i.e., $a*a = e$. Pre-multiplying both the sides by a^{-1}, we get $a^{-1}*(a*a) = a^{-1}*e$

$\Rightarrow \quad (a^{-1}*a)*a = a^{-1}$ (by associativity and identity)

$\Rightarrow \quad e*a = a^{-1}$ (by inverse)

$\Rightarrow \quad a = a^{-1}$ [by identity, part (i)]

i.e., every element is inverse of itself. Now, for $a, b \in G$

$$a*b = a^{-1}*b^{-1} \quad \text{[by part (i)]}$$
$$= (b*a)^{-1}$$
$$= b*a \quad \text{[since } a, b \in G \text{ and by part (i)]}$$

Thus, G is an abelian group.

Note The converse of the theorem need not be true. For instance, $(Z, +)$ is an abelian group. There exists no element, except 0, in Z which is its own inverse.

EXAMPLE 4.17 For any group G, prove that G is abelian if $(ab)^2 = a^2 b^2$ for $a, b \in G$.

Solution Let G be an abelian group, then $ab = ba$ Now,

$$(ab)^2 = (ab)(ab) = a[b(ab)] \quad \text{(by associative law)}$$
$$= a[(ba)b] = a[(ab)b] \quad \text{(since } G \text{ is abelian)}$$
$$= a[a(bb)] = (aa)(bb)$$
$$= a^2 b^2$$

4.3.3 Products and Quotients of Groups

Here, we shall discuss the formation of new groups from the existing groups using the notions of product and quotient.

Theorem 4.6 If $(G_1, *_1)$ and $(G_2, *_2)$ are groups, then $G = G_1 \times G_2$, i.e., $(G, *)$ is a group with binary operation '$*$' defined by

$$(a_1, b_1) * (a_2, b_2) = (a_1 *_1 a_2, b_1 *_2 b_2)$$

Proof To prove that $(G, *)$ is a group, we have to show that $G = G_1 \times G_2$ possesses associative property, an identity, and an inverse of every element.

Associativity Let $a, b, c \in G_1 \times G_2$, then

$$a*(b*c) = (a_1, a_2) * ((b_1, b_2) * (c_1, c_2))$$
$$= (a_1, a_2) * (b_1 *_1 c_1, b_2 *_2 c_2)$$
$$= (a_1 *_1 (b_1 *_1 c_1), a_2 *_2 (b_2 *_2 c_2))$$
$$= ((a_1 *_1 b_1) *_1 c_1), ((a_2 *_2 b_2) *_2 c_2)$$
$$= (a_1 *_1 b_1, a_2 *_2 b_2) * (c_1, c_2)$$
$$= ((a_1, a_2) * (b_1, b_2)) * (c_1, c_2)$$
$$= (a*b)*c$$

Identity Let e_1 and e_2 be identities of G_1 and G_2, respectively. Then, the identity of $G_1 \times G_2$ is $e = (e_1, e_2)$. Assume some $a \in G_1 \times G_2$. Then,

$$a * e = (a_1, a_2) * (e_1, e_2) = (a_1 *_1 e_1, a_2 *_2 e_2) = (a_1, a_2) = a$$

Similarly, it can be shown that $e * a = a$.

Inverse To determine the inverse of an element, we proceed as follows:

$$a^{-1} = (a_1, a_2)^{-1} = (a_1^{-1}, a_2^{-1})$$

Now, to verify that this is the exact inverse, we will compute $a * a^{-1}$ and $a^{-1} * a$.
So,

$$a * a^{-1} = (a_1, a_2) * (a_1^{-1}, a_2^{-1}) = (a_1 *_1 a_1^{-1}, a_2 *_2 a_2^{-1}) = (e_1, e_2) = e$$

Similarly, we have $a^{-1} * a = e$. Thus, $(G_1 \times G_2, *)$ is a group.

In general, if $G_1, G_2, \ldots, G_n$ are groups, then $G = G_1 \times G_2 \times \cdots \times G_n$ is also a group.

For example Consider $B = [0, 1]$ as the group defined in Example 4.14 in which the operation $+$ is illustrated in Table 4.9. Then $B^n = B \times B \times \cdots \times B$ (n factors) is a group with operation $\oplus$ defined by

$$(a_1, a_2, \ldots, a_n) \oplus (b_1, b_2, \ldots, b_n) = (a_1 + b_1, a_2 + b_2, \ldots, a_n + b_n)$$

The identity of B^n is $(0, 0, \ldots, 0)$ and every element is its own inverse. However, the binary operation is very dissimilar from $\wedge$ and $\vee$.

EXAMPLE 4.18 If G is a group such that $(ab)^n = a^n b^n$ for three consecutive integers, then show that $ab = ba$.

Solution We have

$$(ab)^n = a^n b^n$$

Then,

$$(ab)^{n+1} = a^{n+1} b^{n+1}$$

Also,

$$(ab)^{n+2} = a^{n+2} b^{n+2}$$

Now,

$$(a^n b^n)(ab) = (ab)^{n+1} = a^{n+1} b^{n+1} \Rightarrow b^n a = a b^n$$

by cancellation. Similarly,

$$b^{n+1} a = a b^{n+1}$$

Again,

$$b^{n+1} a = b(b^n a) = b(a b^n)$$

i.e., $ab^{n+1} = bab^n$. This yields $ab = ba$.

Groups possess important application in coding theory. Group theory is also used in finite machines in compiler design for the recognition of syntactically correct language structures.

4.4 SEMIGROUPS

In this section, we discuss an important mathematical structure which consists of a set along with a binary operation. This structure has several productive applications.

Let S be a non-empty set together with a binary operation '$*$' defined on S. Then, the algebraic structure $(S, *)$ is called a *semigroup* if the following conditions are satisfied.

S1 *Closure law* The binary operation '$*$' is a closed operation, i.e., $a * b \in S$

S2 *Associative law* The binary operation '$*$' is an associative operation, i.e.,
$$a * (b * c) = (a * b) * c \; \forall \, a, b, c \in S$$

If the operation '$*$' has an identity element, then S is called a *monoid*.

For example

(i) $(N, +)$ is a semigroup.

(ii) $(Z, -)$ is not a semigroup. Since it is not associative in Z.
Assume $2, 5, 6 \in Z$. Then,
$$(2 - 5) - 6 = -3 - 6 = -9 \text{ and } 2 - (5 - 6) = 2 - (-1) = 3$$
i.e., $(2 - 5) - 6 \neq 2 - (5 - 6)$.

(iii) Let $(L, \leq)$ be a lattice. Define a binary operation on L by $a * b = a \vee b$. Then L is a semigroup.

The following theorem demonstrates the necessary and sufficient conditions for a semigroup to be a group.

Theorem 4.7 A semigroup $(S, *)$ is a group if and only if

(i) there exists $e \in S$ such that $e * a = a, \; \forall a \in S$
(ii) there exists $b \in S$ such that $b * a = e, \; \forall a \in S$

Proof Assume that $(S, *)$ is a semigroup that satisfies the above two propositions. Then, for $b \in S$, there exists $c \in S$ such that $c * b = e$, by the second part. Next,

$$a = e * a = (c * b) * a = c * (b * a) = c * e$$

and

$$a * b = (c * e) * b = c * (e * b) = c * b = e$$

Hence,

$$a * b = e = b * a$$

Also,

$$a * e = a * (b * a) = (a * b) * a = e * a = a$$

Thus, $a * e = a = e * a$, which implies that e is the identity element of S. Again, since
$$a * b = e = b * a$$
so, $b = a^{-1}$. Hence, $(S, *)$ is a group.

The proof of the converse is left to the reader (proof can be made possible from the definition of a group).

An equivalence relation R on the semigroup $(S, *)$ is called a *congruence relation* if aRa' and bRb' imply $(a * b)R(a' * b')$.

We shall now discuss the *quotient structures* determined by a congruence relation on a group.

Theorem 4.8 Let R be a congruence relation on the group $(G, *)$. Then the semigroup $(G/R, \Theta)$ is a group in which the operation Θ is defined on G/R by

$$[a] \, \Theta \, [b] = [a*b]$$

Proof Since a group is a monoid, then we can say that G/R is also a monoid. Here, it is necessary to show that each element of G/R contains an inverse. Let

$$[a] \in G/R$$

Then,

$$[a^{-1}] \in G/R$$

and

$$[a]\Theta[a^{-1}] = [a*a^{-1}] = [e]$$

Therefore,

$$[a]^{-1} = [a^{-1}]$$

Hence, $(G/R, \Theta)$ is a group.

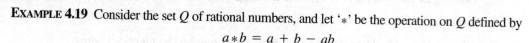

EXAMPLE 4.19 Consider the set Q of rational numbers, and let '$*$' be the operation on Q defined by
$$a*b = a + b - ab$$
Find $3*4, 2*(-5)$ and $7*1/2$. Also, show that $(Q, *)$ is a semigroup. Is it commutative?

Solution Using the definition of the operation '$*$', $3*4 = 3 + 4 - 3 \cdot 4 = 3 + 4 - 12 = -5$

$$2*(-5) = 2 + (-5) - 2 \cdot (-5) = 2 - 5 + 10 = 7$$

$$7*\frac{1}{2} = 7 + \frac{1}{2} - 7\left(\frac{1}{2}\right) = 4$$

For the next part, we have,

$$(a*b)*c = (a + b - ab)*c = (a + b - ab) + c - (a + b - ab)c$$
$$= a + b - ab + c - ac - bc + abc$$
$$= a + b + c - ab - ac - bc + abc$$
$$a*(b*c) = a*(b + c - bc) = a + (b + c - bc) - a(b + c - bc)$$
$$= a + b + c - bc - ab - ac + abc$$

Hence, '$*$' is associative and $(Q, *)$ is a semigroup. Also,

$$a*b = a + b - ab = b + a - ba = b*a$$

Let $A = \{a_1, a_2, \ldots, a_n\}$ be a non-empty set. Let A' be a set of all finite sequences of elements of A, i.e., the set A' consists of all words that can be created from the alphabet A. Let u and v be elements of A'. It may be observed that there exists a binary operation $\bullet'$, known as catenation, on A. If $u = a_1a_2 \cdots a_n$ and $v = b_1b_2 \cdots b_k$, then $u\bullet'v = a_1a_2 \cdots a_nb_1b_2 \cdots b_k$. It can also be seen that if $u, v,$ and w are any elements of A^*, then

$$u\bullet'(v\bullet'w) = (u\bullet'v)\bullet'w$$

so that the operation •′ is an associative binary operation and $(A', \bullet')$ is a semigroup. The semigroup $(A', \bullet')$ is called the *free semigroup* generated by A.

Let $(S, *)$ be a semigroup and let T be a subset of S. If T is closed under the operation '$*$' (i.e., $a*b \in T$ whenever a and b are elements of T), then $(T, *)$ is called a *subsemigroup* of $(S, *)$. Similarly, let $(S, *)$ be a monoid with identity e, and let T be a non-empty subset of S. If T is closed under the operation '$*$' and $e \in T$, then $(T, *)$ is called a *submonoid* of $(S, *)$.

It may be observed that the associative property holds in any subset of a semigroup so that a subsemigroup $(T, *)$ of a semigroup is itself a semigroup. Similarly, a submonoid of a monoid is itself a monoid.

For example

(i) If T is the set of all even integers, then $(T, \times)$ is a subsemigroup of the monoid $(Z, \times)$, where $\times$ is ordinary multiplication, but it is not a submonoid since the identity of Z, the number 1, does not belong to T.

(ii) Let $Z^+ = \{0, 1, 2, 3, \ldots\}$; $(Z^+, +)$ is a monoid. Let $Z' = \{0, 2, 4, 6, \ldots\}$. Then, $Z' \subset Z^+$ is nonempty and $(Z', +)$ itself is a monoid. So, $(Z', +)$ is a submonoid of $(Z^+, +)$.

EXAMPLE 4.20 Consider the algebraic system $(\{0, 1\}, *)$, where '$*$' is a multiplication operation. Investigate the system $(\{0, 1\}, *)$ as a semigroup.

Solution

Closure property The operation '$*$' is a closed one on the given set, since

$$0*0 = 0; \quad 0*1 = 0; \quad 1*0 = 0; \quad 1*1 = 1$$

Associative property The operation '$*$' is associative, since we have

$$(a*b)*c = a*(b*c) \quad \forall a, b, c$$

Since the algebraic system is closed and associative, so it is a semigroup.

EXAMPLE 4.21 Let $(S, *)$ be a semigroup. Show that for a, b, c in S, if $a*c = c*a$ and $b*c = c*b$, then

$$(a*b)*c = c*(a*b)$$

Solution Consider the left-hand side, i.e., $(a*b)*c$. Then,

$$\begin{aligned}
(a*b)*c &= a*(b*c) \quad &(\text{'}*\text{' is associative}) \\
&= a*(c*b) \quad &(\text{since } b*c = c*b) \\
&= (a*c)*b \quad &(\text{'}*\text{' is associative}) \\
&= c*(a*b) \quad &(\text{since } a*c = c*a) \\
& &(\text{'}*\text{' is associative})
\end{aligned}$$

Thus,

$$(a*b)*c = c*(a*b)$$

4.4.1 Isomorphism and Homomorphism

An isomorphism between two posets will be discussed in Section 6.4 as a one-to-one correspondence that preserved order relations, i.e., the salient feature of posets. We now define an isomorphism between two semigroups as a one-to-one correspondence that preserves the binary operations. In general, an isomorphism between two mathematical structures of the same type should prevail the distinguishing features of the structures.

Let $(S, *)$ and $(T, *')$ be two semigroups. A function $f: S \to T$ is called an *isomorphism* from $(S, *)$ to $(T, *')$ if it is a one-to-one correspondence from S to T, and if

$$f(a*b) = f(a) *' f(b)$$

for all a and b in S.

If f is an isomorphism from $(S, *)$ to $(T, *')$, then, since f is a one-to-one correspondence, it follows from Sections 1.25 and 6.4.1 that f^{-1} exists and is a one-to-one correspondence from T to S. We will now show that f^{-1} is an isomorphism from $(T, *')$ to $(S, *)$. Let a' and b' be any elements of T.

Since f is onto, we can find elements a and b in S such that $f(a) = a'$ and $f(b) = b'$. Then, $a = f^{-1}(a')$ and $b = f^{-1}(b')$. Now,

$$f^{-1}(a' *' b') = f^{-1}(f(a) *' f(b)) = f^{-1}(f(a*b)) = (f^{-1} \circ f)(a*b)$$
$$= a*b = f^{-1}(a') * f^{-1}(b')$$

Hence, f^{-1} is an isomorphism. Thus, it can be inferred that the semigroups $(S, *)$ and $(T, *')$ are isomorphic and can be represented symbolically as $S \cong T$

To show that two semigroups $(S, *)$ and $(T, *')$ are isomorphic, let us follow the procedure given below:

Step I Define a function $f: S \to T$ with Dom $(f) = S$.
Step II Show that f is one-to-one.
Step III Show that f is onto.
Step IV Show that $f(a*b) = f(a) *' f(b)$.

EXAMPLE 4.22 Consider T as a set of all even integers. Show that the semigroups $(Z, +)$ and $(T, +)$ are isomorphic.

Solution

Step I Define the function $f: Z \to T$ by $f(a) = 2a$.
Step II Show that f is one-to-one.
 Assume that $f(a_1) = f(a_2)$. Then, $2a_1 = 2a_2$, i.e., $a_1 = a_2$. Thus, f is one-to-one.
Step III Show that f is onto.
 Assume that b is any even integer. Then, $a = b/2 \in Z$ and $f(a) = f(b/2) = 2(b/2) = b$. Thus, f is onto.
Step IV Here, $f(a + b) = 2(a + b) = 2a + 2b = f(a) + f(b)$.

Hence, $(Z,+)$ and $(T,+)$ are isomorphic semigroups.

Like the lattice isomorphisms, the two semigroups $(S, *)$ and $(T, *')$, when they are isomorphic, differ only in the character of their elements. If S and T are finites semigroups, their respective binary operations can be represented by tables. Moreover, S and T will be isomorphic if one can rearrange and relabel the elements of S so that its table is identical with that of T.

EXAMPLE 4.23 Let $S = \{a, b, c\}$ and $T = \{x, y, z\}$. Verify that the following tables (Table 4.10) yield semigroup structures for S and T, which are isomorphic.

TABLE 4.10 Isomorphism between semigroup structures for S and T

*	a	b	c	*	x	y	z
a	a	b	c	x	z	x	y
b	b	c	a	y	x	y	z
c	c	a	b	z	y	z	x

Solution Replacing the elements in S by their images and rearranging the table such that
$$f(a) = y \quad f(b) = x \quad f(c) = z$$
one obtains exactly the table for T. Thus, S and T are isomorphic.

EXAMPLE 4.24 Let $(G_1, *)$ and $(G_2, *)$ be the two algebraic systems described in Table 4.11. Investigate that the two systems are isomorphic.

TABLE 4.11 Isomorphism between algebraic systems $(G_1, *)$ and $(G_2, \bullet)$

*	a	b	c	$\bullet$	1	ω	ω^2
a	a	b	c	1	1	ω	ω^2
b	b	c	a	ω	ω	ω^2	1
c	c	a	b	ω^2	ω^2	1	ω

Solution The two algebraic systems $(G_1, *)$ and $(G_2, \bullet)$ are isomorphic and $(G_2, \bullet)$ is an isomorphic image of G_1, such that
$$f(a) = 1 \quad f(b) = \omega \quad f(c) = \omega^2$$

Theorem 4.9 Let $(S, *)$ and $(T, *')$ be monoids with identities e and e', respectively. Let $f : S \to T$ be an isomorphism. Then, $f(e) = e'$.

Proof Let b be any element of T. Since f is onto, there exists an element a in S such that $f(a) = b$. Then,
$$a = a * e, b = f(a) = f(a * e) = f(a) *' f(e) = b *' f(e)$$
Similarly, since $a = e * a$ and $b = f(e) * b$, so, for any $b \in T$,
$$b = b *' f(e) = f(e) *' b$$
which implies that $f(e)$ is an identity for T. Thus, since the identity is unique, it follows that
$$f(e) = e'$$

If $(S, *)$ and $(T, *')$ are semigroups in which S possesses an identity and T does not, then from the above theorem, it can be concluded that $(S, *)$ and $(T, *')$ cannot be isomorphic.

For example Let T be the set of all even integers and let $\times$ be the ordinary multiplication. Then the semigroups $(Z, \times)$ and $(T, \times)$ are not isomorphic, since Z has an identity and T does not.

Let $(S, *)$ and $(T, *')$ be two semigroups. A mapping $f: S \to T$ is called a *homomorphism* from $(S, *)$ to $(T, *')$ if

$$f(a*b) = f(a)*'f(b)$$

for all a and b in S. Also, if f is onto, then T is a *homomorphic image* of S.

For example Let $G_1 = R$ and $G_2 = R - \{0\}$. Let $(G_1, \cdot)$ and $(G_2, \cdot)$ be two semigroups with respect to multiplication $\cdot$. Consider a mapping $f: G_1 \to G_2$, defined by $f(a) = a^n \; \forall a \in G_1$, where $n \in Z$. The mapping is a homomorphism one, for

$$f(ab) = (ab)^n = a^n \cdot b^n = f(a) \cdot f(b)$$

EXAMPLE 4.25 Let $N' = \{2, 3, 4\}$. Consider $(N, +)$ and $(N', +)$ as two semigroups. Define $f: N \to N'$ by $f(a) = a + 1, \; \forall a, b \in N$. Show that f is not a homomorphism.

Solution From the given mapping, $f(a + b) = a + b + 1$. Also,

$$f(a) + f(b) = a + 1 + b + 1 = a + b + 2$$

So,

$$f(a + b) \neq f(a) + f(b)$$

Thus, f is not a homomorphism.

The following theorems describe the strong algebraic resemblance between $(S, *)$ and $(T, *')$.

Theorem 4.10 Let f be a homomorphism from a semigroup $(S, *)$ to a semigroup $(T, *')$. If S' is a subsemigroup of $(S, *)$, then,

$$f(S') = \{t \in T : t = f(s) \text{ for some } s \in S'\}$$

The image of S' under f is a subsemigroup of $(T, *')$.

Proof If t_1 and t_2 are any elements of $f(S')$, then there exists s_1 and s_2 in S' such that $t_1 = f(s_1)$ and $t_2 = f(s_2)$.
Now,

$$t_1 *' t_2 = f(s_1) *' f(s_2) = f(s_1 * s_2) = f(s_3)$$

where $s_3 = s_1 * s_2 \in S'$. Hence, $t_1 *' t_2 \in f(S')$.

Thus, $f(S')$ is closed under the operation $*'$. Since the associative property holds in T, it holds in $f(S')$; so, $f(S')$ is a subsemigroup of $(T, *')$.

Theorem 4.11 If f is a homomorphism from a commutative semigroup $(S, *)$ onto a semigroup $(T, *')$, then $(T, *')$ is also commutative.

Proof Let t_1 and t_2 be any elements of T. Then there exists s_1 and s_2 in S such that $t_1 = f(s_1)$ and $t_2 = f(s_2)$.
Thus,

$$t_1 *' t_2 = f(s_1) *' f(s_2) = f(s_1 * s_2) = f(s_2 * s_1) = f(s_2) *' f(s_1) = t_2 *' t_1$$

Hence, $(T, *')$ is also commutative.

4.4.2 Products and Quotients of Semigroups

In this section, we shall form new semigroups from existing semigroups, that can be enunciated by the following theorem.

Theorem 4.12 If $(S, *_1)$ and $(T, *_2)$ are semigroups, then $(S \times T, *)$ is a semigroup with direct product $S \times T$, where $*$ is defined by

$$(a, b) * (a', b') = (a *_1 a', b *_2 b'), \quad \text{or, simply} \quad (a, b)(a', b') = (aa', bb')$$

where $a, a' \in S$ and $b, b' \in T$.

Proof The semigroup $S \times T$ is closed under the operation '$*$'.
Let $a, b, c \in S \times T$. Then,

$$\begin{aligned} a*(b*c) &= (a, a') * ((b, b') * (c, c')) = (a, a') * (b *_1 c, b' *_2 c') \\ &= (a *_1 (b *_1 c), a' *_2 (b' *_2 c')) = ((a *_1 b) *_1 c, (a' *_2 b') *_2 c') \\ &= (a *_1 b, a' *_2 b') * (c, c') = ((a, a') * (b, b')) * (c, c') \\ &= (a * b) * c \quad \text{(associative)} \end{aligned}$$

Since the operation '$*$' is closed and associative. Hence, $S \times T$ is a semigroup.

An equivalence relation R on the semigroup $(S, *)$ is called a *congruence relation* if aRa' and bRb' imply $(a*b)R(a'*b')$.

EXAMPLE 4.26 Let $(Z, +)$ be a semigroup and R be an equivalence relation on Z defined by aRb if and only if $a \equiv b \pmod{2}$. Recall that if $a \equiv b \pmod{2}$, then we have $a - b = kn$, k being an integer. Show that the relation is a congruence relation.

Solution If $a \equiv b \pmod 2$ and $c \equiv d \pmod 2$, then 2 divides $a - b$ and 2 divides $c - d$, so

$$a - b = 2m \quad \text{and} \quad c - d = 2n$$

where m and n are in Z. Adding both we obtain

$$(a - b) + (c - d) = 2m + 2n$$
$$\Rightarrow \quad (a + c) - (b + d) = 2(m + n)$$
$$\Rightarrow \quad a + c \equiv b + d \pmod 2$$

Hence, the relation is a congruence relation.

EXAMPLE 4.27 Consider the semigroup $(Z, +)$, where $+$ is an ordinary addition. Let $f(x) = x^2 - 2x - 3$. Also, let R be a relation on Z defined by aRb iff $f(a) = f(b)$. Investigate on R as a congruence relation.

Solution It can be shown that the relation R is an equivalence relation on Z. However, R is not a congruence relation because we have

$$2R0, \text{ since } f(2) = f(0) = -3 \quad \text{and} \quad -2R4, \text{ since } f(-2) = f(4) = 5$$

also,

$$(2 + (-2))R(0 + 4)$$

i.e., $0R4$, since $f(0) = -3$ and $f(4) = 5$.

It is known that $[a]$ is the equivalence class represented by a, which is an integer. Let $[a] = R(a)$ be the equivalence class containing a and S/R denote the set of all equivalence class. The applicability of the notation $[a]$ is illustrated in the following theorem.

Note The operation Θ is a kind of *quotient binary relation* on S/R which is created from the original binary relation '$*$' on S by the congruence relation.

Theorem 4.13 Let R be a congruence relation on the semigroup $(S, *)$. Consider the relation Θ from $S/R \times S/R$ to S/R in which the ordered pair $([a], [b])$ is, for a and b in S, related to $[a * b]$ such that
  (i) Θ is a function from $S/R \times S/R$ to S/R, and $[a]\Theta[b] = [a*b]$
  (ii) $(S/R, \Theta)$ is a semigroup

Proof Assume that $([a], [b]) = ([a'], [b'])$. Then aRa' and bRb', so that we have $a*bRa'*b'$, since R is a congruence relation. Thus, $[a*b] = [a'*b']$, i.e., Θ is a function. This implies that Θ is a binary operation on S/R.

Now, to verify that Θ is an associative operation, we proceed as follows:
$$[a]\Theta([b]\Theta[c]) = [a]\Theta[b*c] = [a*(b*c)] = [(a*b)*c] = [a*b]\Theta[c] = ([a]\Theta[b])\Theta[c]$$
Hence, S/R is a semigroup.

The quotient S/R is the *quotient semigroup* or *factor semigroup*.

Corollary 4.3 Let R be a congruence relation on the monoid $(S, *)$. If we define the operation Θ in S/R by
$$[a]\Theta[b] = [a*b]$$
then $(S/R, \Theta)$ is a monoid.

Semigroup is very much applicable to formal languages and automata theory.

4.5 SUBGROUP

Let $(G, *)$ be a group and H be a non-empty subset of G. If $(H, *)$ is itself a group, then $(H, *)$ is called a *subgroup* of $(G, *)$. In other words, if H be a subset of a group G such that

  (i) the identity e of G belongs to H
  (ii) if a and b belong to H, then $a*b \in H$
  (iii) if $a \in H$, then $a^{-1} \in H$

where H is called a *subgroup* of G. Parts (i) and (ii) imply that H is a submonoid of G. Again, if G is a group and H is a subgroup of G, then H is also a group with respect to the operation in G, since the associative property in G also holds in H.

For example Let $G = \{1, -1, i, -i\}$ and $H = \{1, -1\}$. Here, G and H are groups with respect to the binary operation, such as multiplication $\cdot$. H is a subset of G, and so, $(H, \cdot)$ is a subgroup of $(G, \cdot)$.

 EXAMPLE 4.28 Let $(Z, +)$ be a group. Investigate on the following subsets of G
  (i) the set G_1 of all odd integers
  (ii) the set G_2 of all positive integers
     as subgroups of G.

Solution
  (i) The set G_1 does not satisfy the closure law, since addition of two integers is always even. Thus, G_1 is not a subgroup of G.
  (ii) *Closure law* The set G_2 is closed under the operation $+$, since addition of two even integers is even.

Associative law The operation + is associative, since $(a + b) + c = a + (b + c), \forall a, b, c \in G_2$.

Identity law The element 0 is the identity element. Hence, $0 \in G_2$.

Inverse law The inverse of every element $a \in G_2$ is $-a \notin G_2$. Hence, the inverse of every element does not exist.

Since the structure $(G_2, +)$ does not satisfy all the conditions of a subgroup, so $(G_2, +)$ is not a subgroup of $(Z, +)$.

It is known that every set is a subset of itself. So, if G is a group, then G is itself a subgroup of G. Also, if e is the identity element of G, then G and $H = \{e\}$ are subgroups of G. These two subgroups $(G, *)$ and $(H, *)$ are called *trivial* subgroups of G.

The following theorem asserts the authenticity of the subgroup of a group.

Theorem 4.14 A non-empty subset H of a group G is a subgroup of G if and only if

(i) $a \in H, b \in H \Rightarrow a * b \in H$
(ii) $a \in H \Rightarrow a^{-1} \in H$
    where a^{-1} is the inverse of a in G.

Proof

Necessary condition Let H be a subgroup of G. Then H must be closed with respect to the operation '*', i.e., $a \in H, b \in H \Rightarrow a * b \in H$. Let $a \in H$ and let a^{-1} be the inverse of a in G. Then a^{-1} is also the inverse of a in H. Since H is itself a group, each element of H must possess inverse. Therefore, $a \in H \Rightarrow a^{-1} \in H$.

Sufficient condition It may be seen here that the binary operation '*' in G is also a binary operation in H. Hence, H is closed under the operation '*'.

As the elements of H are also the elements of G and the elements of G satisfy the associative law for the binary operation, therefore the elements of H will also satisfy the associative law.

Now, $a \in H \Rightarrow a^{-1} \in H$. Again, from part (i), we have

$$a \in H, a^{-1} \in H \Rightarrow aa^{-1} = e \in H$$

which shows the existence of identity element in H.

Thus, all the conditions are satisfied. Therefore, H is a subgroup of G.

Theorem 4.15 Let $(G, *)$ be a group and $H(\neq \phi) \subseteq G$. Then, $(H, *)$ is a subgroup of $(G, *)$ if and only if $a, b \in H \Rightarrow a * b^{-1} \in H$.

Proof If $(H, *)$ is a subgroup and $a, b \in H$, then $b^{-1} \in H$ and so $a * b^{-1} \in H$, by closure law. Conversely, suppose H is a non-empty subset of G which contains the element $a * b^{-1} \in H$, whenever $a, b \in H$. Also, $a \in H \Rightarrow a * a^{-1} = e \in H$, by hypothesis. Again,

$$e, a \in H \Rightarrow e * a^{-1} = a^{-1} \in H$$

Finally,

$$a, b \in H \Rightarrow a * b^{-1} \in H \Rightarrow a * (b^{-1})^{-1} = a * b \in H$$

As a subset of G, the set H 'inherits' the associative law. Thus, all the group axioms are satisfied, and $(H, *)$ is, therefore, a subgroup of $(G, *)$.

Theorem 4.16 If $(H_1, *)$ and $(H_2, *)$ are both subgroups of the group $(G, *)$, then $(H_1 \cap H_2, *)$ is also a subgroup.

Proof The set $H_1 \cap H_2 \neq \phi$, since $e \in H_1 \cap H_2$. Suppose that $a, b \in H_1 \cap H_2$, then $a, b \in H_1$ and $a, b \in H_2$, since $(H_1, *)$ and $(H_2, *)$ are subgroups. Also,

$$a, b \in H_1 \Rightarrow a * b^{-1} \in H_1 \quad \text{and} \quad a, b \in H_2 \Rightarrow a * b^{-1} \in H_2$$

i.e.,

$$a * b^{-1} \in H_1 \cap H_2$$

Thus, $H_1 \cap H_2$ is a subgroup of $(G, *)$.

The intersection of subgroups is illustrated in the following example.

For example Let $Z = \{0, \pm 1, \pm 2, \pm 3, \ldots\}$ and let $H_1 = \{0, \pm 2, \pm 4, \pm 6, \ldots\}$ and $H_2 = \{0, \pm 3, \pm 6, \pm 9, \ldots\}$. Then, $H_1 \cap H_2 = \{0, \pm 6, \pm 12, \ldots\}$ Here, $(H_1, +)$ and $(H_2, +)$ are subgroups of $(Z, +)$.

On the other hand, $(H_1 \cup H_2, +)$ is not a subgroup of $(Z, +)$. Because $H_1 \cup H_2 = \{0, \pm 2, \pm 3, \pm 4, \ldots\}$ and $2, 3 \in H_1 \cup H_2$; but $2 + 3 = 5 \notin H_1 \cup H_2$. Thus, $H_1 \cup H_2$ is not closed under $+$.

EXAMPLE 4.29 Let H be a subset of a group G. Show that H is a subgroup of G if H has the following three properties:
(i) the identity element e belongs to H
(ii) H is closed under the operation of G, i.e., if $a, b \in H$ then $ab \in H$
(iii) H is closed under inverses, i.e., if $a \in H$ then $a^{-1} \in H$

Solution H is non-empty and has an identity element by property (i). The operation is well-defined in H by property (ii). Inverses exist in H by property (iii). Lastly, the associative law holds in H, since it holds in G. Thus, H is a subgroup of G.

4.6 CYCLIC GROUP

A group is called *cyclic* if for some $a \in G$, there exists element $x \in G$, of the form a^n where n is some integer. The element a, here, is called a *generator* of G. A cyclic group G, generated by a, can be denoted by $G = \langle a \rangle$. If $(G, *)$ is a finite cyclic group generated by the element a, then the elements of G can be expressed in the form

$$G = \{a, a^2, a^3, \ldots, a^n\}$$

Again, if $(G, *)$ is an infinite cyclic group generated by a, then we can write

$$G = \{\ldots, a^{-3}, a^{-2}, a^{-1}, a^0, a^1, a^2, a^3, \ldots\}$$

There may exist more than one generator in a cyclic group.

For example The multiplicative group $G = \{1, -1, i, -i\}$ is cyclic. We can re-write it as $G = \{i, i^2, i^3, i^4\}$. Thus, G is a cyclic group and i is a generator. The group G can also be modified as $G = \{-i, (-i)^2, (-i)^3, (-i)^4\}$. Thus, $-i$ is also a generator of G. Hence, there may exist more than one generator in a cyclic group. Here, the generators can be represented as

$$G = \langle i \rangle \quad G = \langle -i \rangle$$

Note If $(G, +)$ is a cyclic group, then each element of G can be expressed in the form na, where n is an integer.

For example The set of integers with respect to $+$, i.e., $(Z, +)$ is a cyclic group. Here, $1^0 = 1$, $1^1 = 1$, $1^2 = 1 + 1 = 2$, $1^3 = 1 + 1 + 1$, and so on.

Similarly, 1^{-1} = inverse of $1 = -1$, $1^{-2} = (1^2)^{-1} = -2$, $1^{-3} = (3)^{-1} = -3$ and so on.

If $(G, *)$ is a finite group of order n, then

$$a^i * a^j = \begin{cases} a^{i+j}, & \text{if } i + j < n \\ a^0, & \text{if } i + j = 0 \\ a^{i+j}, & \text{if } i + j > n \end{cases}$$

where a is the generator of G.

We will now discuss some important properties of cyclic group through the following theorems.

Theorem 4.17 Every cyclic group is an abelian.

Proof Let G be a cyclic group and let a be a generator of G so that

$$G = \langle a \rangle = \{a^n : n \in Z\}$$

Let g_1 and g_2 be any two elements of G. Then, there exists some integers p and q such that $g_1 = a^p$ and $g_2 = a^q$. Now,

$$g_1 g_2 = a^p \cdot a^q = a^{p+q} = a^{q+p} = g_2 g_1$$

$\Rightarrow$ G is an abelian group.

Theorem 4.18 If a is a generator of a cyclic group G, then a^{-1} is also a generator of G.

Proof Let $G = \langle a \rangle$ be a cyclic group generated by a. Let a^p be any element of G, where p is some integer. We can write $a^p = (a^{-1})^{-p}$, in which $-p$ is also some integer. So, each element of G, here, is generated by a^{-1}. Thus, a^{-1} is also a generator of G.

Theorem 4.19 If a cyclic group G is generated by an element a of the order n, then a^m is a generator of G if and only if the greatest common divisor (g.c.d) of m and n is 1, i.e., if and only if m and n are relative primes.

Proof Assume that m is relatively prime to n. Consider the cyclic subgroup $H = \{a^m\}$ of G, which is generated by a^m. Since each integral power of a^m will also be an integral power of a, so $H \subseteq G$.

Since m is relatively prime to n, there exists two integers p and q such that $pm + qn = 1$. So,

$$a^{pm+qn} = 1 \Rightarrow a^{pm} \cdot a^{qn} = 1 \Rightarrow (a^m)^p = a, \text{ since } a^{qn} = (a^n)^q = e^q = e$$

Therefore, each integral power of a will also be some integral power of a^m, i.e., $G \subseteq H$. Hence, $H = G$ and a^m is a generator of G. Conversely, suppose a^m is a generator of G. Let the g.c.d of m and n be d, and $d \neq 1$, i.e., $d > 1$. Then, m/d and n/d must be integers. Now,

$$(a^m)^{n/d} = (a^n)^{m/d} = e^{m/d} = e$$

i.e., n/d is a positive integer less than n itself. Thus, the order of a^m, i.e., $O(a^m) < n$. Therefore,

a^m cannot be a generator of G, because the order of a^m is not equal to the order of G. Hence, d must be equal to 1. Thus, m is prime to n.

EXAMPLE 4.30 The set of integers $(Z, +)$ is a cyclic group of which the generator is 1.

Solution We have $1^0 = 1$, $1^1 = 1$, $1^2 = 1 + 1 = 2$, $1^3 = 1 + 1 + 1 = 3$, and so on. Similarly, 1^{-1} (= inverse of 1) $= -1$, $1^{-2} = (1^2)^{-1} = -2$, $1^{-3} = (3)^{-1} = -3$ and so on.

Thus, each element of G can be expressed as some integral power of 1. Similarly, we can show that -1 is also a generator.

EXAMPLE 4.31 The multiplicative group $1, \omega, \omega^2$ is a cyclic group.

Solution We have

$$\omega^0 = 1, \omega^1 = \omega, \omega^2 = \omega^2, \omega^3 = 1 \quad \text{and} \quad (\omega^2)^0 = 1, (\omega^2)^1 = \omega^2, (\omega^2)^2 = \omega^4 = \omega^3 \cdot \omega = \omega$$

Thus, each element of G can be expressed as some integral powers of ω and ω^2. Hence, the group is a cyclic group with generators ω and ω^2.

EXAMPLE 4.32 How many generators are there in the cyclic group G of order 8?

Solution Let G be a generator of G. Then $O(a) = 8$ and we can express G as

$$G = \{a, a^2, a^3, a^4, a^5, a^6, a^7, a^8\}$$

Here, 7 is prime to 8, i.e., a^7 is a generator of G. Also, 5 is prime to 8, i.e., a^5 is a generator of G. Finally, 3 is prime to 8, i.e., a^3 is a generator of G. Thus, there exists only four generators of G, such as a, a^3, a^5, a^7.

4.7 PERMUTATION GROUPS

Before the discussion of permutation group we first of all define permutation.

A *permutation* is a one-to-one mapping of a non-empty set P, say, onto itself. When a set P is finite (confine here with finite sets only) possessing n elements, then we speak of a permutation of n symbols.

A group $(G, *)$ is called a *permutation group* on a non-empty set P if the elements of G are permutations of P and the operation '$*$' is the composition of two functions.

If ϕ is a permutation on n symbols, then ϕ can be determined completely by its values $\phi(1), \phi(2), \ldots, \phi(n)$ and is expressed in the form of

$$\phi = \begin{pmatrix} 1 & 2 & 3 & \ldots & \ldots & n \\ \phi(1) & \phi(2) & \phi(3) & \ldots & \ldots & \phi(n) \end{pmatrix}$$

Here, the images $\phi(1), \phi(2), \ldots, \phi(n)$ are the elements of P arranged in some order. The order of symbols in the first row of a permutation is immaterial but columns should not be affected.

For example Let

$$\phi = \begin{pmatrix} 1 & 2 & 3 & 4 \\ 2 & 1 & 4 & 3 \end{pmatrix}$$

denote the permutation on the four symbols {1, 2, 3, 4} which maps 1 on 2, 2 on 1, 3 on 4, and 4 on 3. This permutation actually corresponds to the symmetry of the square which is nothing but a reflection along the vertical bisector.

The number of elements of a finite set is the degree of the permutation. Moreover, every permutation of a finite set of n symbols may be written in $n!$ ways.

For example Let
$$S = \{1, 2, 3\}$$
and
$$\phi = \begin{pmatrix} 1 & 2 & 3 \\ 2 & 3 & 1 \end{pmatrix}$$

Then, we can write
$$\phi = \begin{pmatrix} 1 & 2 & 3 \\ 2 & 3 & 1 \end{pmatrix} = \begin{pmatrix} 2 & 3 & 1 \\ 3 & 1 & 2 \end{pmatrix} = \begin{pmatrix} 3 & 1 & 2 \\ 1 & 2 & 3 \end{pmatrix} = \begin{pmatrix} 1 & 3 & 2 \\ 2 & 1 & 3 \end{pmatrix} = \begin{pmatrix} 2 & 1 & 3 \\ 3 & 2 & 1 \end{pmatrix} = \begin{pmatrix} 3 & 2 & 1 \\ 1 & 3 & 2 \end{pmatrix}$$

Hence, there are $3! = 6$ ways of pattern of expressing ϕ.

4.7.1 Equality of Permutations

Let f and g be two permutations defined on a non-empty set P. Then, $f = g$ if and only if $f(x) = g(x)\ \forall x \in P$.

For example Let $S = \{1, 2, 3, 4\}$ and let
$$f = \begin{pmatrix} 1 & 2 & 3 & 4 \\ 3 & 1 & 2 & 4 \end{pmatrix} \qquad g = \begin{pmatrix} 4 & 1 & 3 & 2 \\ 4 & 3 & 2 & 1 \end{pmatrix}$$

be two permutations of degree 4. Here, we see that
$$f(1) = 3 = g(1) \qquad f(3) = 2 = g(3)$$
$$f(2) = 1 = g(2) \qquad f(4) = 4 = g(4)$$

Thus, $f(x) = g(x)\ \forall x \in \{1, 2, 3, 4\}$, which implies that $f = g$.

4.7.2 Permutation Identity

An *identity permutation* on S, denoted by I, is defined as
$$I(a) = a\ \forall a \in S$$

For example

(i) Let $S = \{a_1, a_2, \ldots, a_n\}$. Then,
$$I = \begin{pmatrix} a_1 & a_2 & \cdots & a_n \\ a_1 & a_2 & \cdots & a_n \end{pmatrix}$$
is the identity permutation on S.

(ii) Let $S = \{1, 2, 3, 4\}$. Then

$$f = \begin{pmatrix} 1 & 2 & 3 & 4 \\ 1 & 2 & 3 & 4 \end{pmatrix}$$

is the identity permutation on S.

4.7.3 Composition of Permutations (or, Product of Permutations)

Let f and g be two arbitrary permutations of like degree, given by,

$$f = \begin{pmatrix} a_1 & a_2 & a_3 & \cdots & \cdots & a_n \\ b_1 & b_2 & b_3 & \cdots & \cdots & b_n \end{pmatrix} \quad g = \begin{pmatrix} b_1 & b_2 & b_3 & \cdots & \cdots & b_n \\ c_1 & c_2 & c_3 & \cdots & \cdots & c_n \end{pmatrix}$$

on a non-empty set A. Then the *composition* (or, *product*) of f and g is defined as

$$f \circ g = \begin{pmatrix} a_1 & a_2 & a_3 & \cdots & \cdots & a_n \\ b_1 & b_2 & b_3 & \cdots & \cdots & b_n \end{pmatrix} \circ \begin{pmatrix} b_1 & b_2 & b_3 & \cdots & \cdots & b_n \\ c_1 & c_2 & c_3 & \cdots & \cdots & c_n \end{pmatrix}$$

$$= \begin{pmatrix} a_1 & a_2 & a_3 & \cdots & \cdots & a_n \\ c_1 & c_2 & c_3 & \cdots & \cdots & c_n \end{pmatrix}$$

Here, f replaces a_1 by b_1 and then g replaces b_1 by c_1 so that $f \circ g$ replaces a_1 by c_1. Similarly, $f \circ g$ replaces a_2 by c_2, a_3 by $c_3, \ldots, a_n$ by c_n.

The composition $f \circ g$ is also a permutation on P. It is observed that the permutation g is formed in such a way that the second row of f should coincide with the first row of g. This feature is most important for the determination of $f \circ g$. Also, if one is interested to find $g \circ f$, then f should be expressed in such a manner that the second row of g must coincide with the first row of f.

EXAMPLE 4.33 Find the composition (or, product) of the following two permutations and show that it is not commutative

$$f = \begin{pmatrix} 1 & 2 & 3 & 4 \\ 2 & 1 & 4 & 3 \end{pmatrix} \quad g = \begin{pmatrix} 1 & 2 & 3 & 4 \\ 3 & 2 & 1 & 4 \end{pmatrix}$$

Solution We compute $f \circ g$ and $g \circ f$ as follows:

$$f \circ g = \begin{pmatrix} 1 & 2 & 3 & 4 \\ 2 & 1 & 4 & 3 \end{pmatrix} \circ \begin{pmatrix} 1 & 2 & 3 & 4 \\ 3 & 2 & 1 & 4 \end{pmatrix} = \begin{pmatrix} 1 & 2 & 3 & 4 \\ 2 & 1 & 4 & 3 \end{pmatrix} \circ \begin{pmatrix} 2 & 1 & 4 & 3 \\ 2 & 3 & 4 & 1 \end{pmatrix} = \begin{pmatrix} 1 & 2 & 3 & 4 \\ 2 & 3 & 4 & 1 \end{pmatrix}$$

and

$$g \circ f = \begin{pmatrix} 1 & 2 & 3 & 4 \\ 3 & 2 & 1 & 4 \end{pmatrix} \circ \begin{pmatrix} 1 & 2 & 3 & 4 \\ 2 & 1 & 4 & 3 \end{pmatrix} = \begin{pmatrix} 1 & 2 & 3 & 4 \\ 3 & 2 & 1 & 4 \end{pmatrix} \circ \begin{pmatrix} 3 & 2 & 1 & 4 \\ 4 & 1 & 2 & 3 \end{pmatrix} = \begin{pmatrix} 1 & 2 & 3 & 4 \\ 4 & 1 & 2 & 3 \end{pmatrix}$$

Thus, $f \circ g \neq g \circ f$, i.e., the composition (or, product) of two permutations is not commutative. But it can be shown that the permutation (group) with respect to multiplication is associative.

For example Let

$$P_1 = \begin{pmatrix} 1 & 2 & 3 \\ 1 & 2 & 3 \end{pmatrix} \quad P_2 = \begin{pmatrix} 1 & 2 & 3 \\ 2 & 3 & 1 \end{pmatrix} \quad P_3 = \begin{pmatrix} 1 & 2 & 3 \\ 3 & 1 & 2 \end{pmatrix}$$

Then,

$$\begin{aligned} P_1 \circ (P_2 \circ P_3) &= \begin{pmatrix} 1 & 2 & 3 \\ 1 & 2 & 3 \end{pmatrix} \circ \left[\begin{pmatrix} 1 & 2 & 3 \\ 2 & 3 & 1 \end{pmatrix} \circ \begin{pmatrix} 1 & 2 & 3 \\ 3 & 1 & 2 \end{pmatrix} \right] \\ &= \begin{pmatrix} 1 & 2 & 3 \\ 1 & 2 & 3 \end{pmatrix} \circ \left[\begin{pmatrix} 1 & 2 & 3 \\ 2 & 3 & 1 \end{pmatrix} \circ \begin{pmatrix} 2 & 3 & 1 \\ 1 & 2 & 3 \end{pmatrix} \right] \\ &= \begin{pmatrix} 1 & 2 & 3 \\ 1 & 2 & 3 \end{pmatrix} \circ \begin{pmatrix} 1 & 2 & 3 \\ 1 & 2 & 3 \end{pmatrix} = \begin{pmatrix} 1 & 2 & 3 \\ 1 & 2 & 3 \end{pmatrix} \end{aligned}$$

$$\begin{aligned} (P_1 \circ P_2) \circ P_3 &= \left[\begin{pmatrix} 1 & 2 & 3 \\ 1 & 2 & 3 \end{pmatrix} \circ \begin{pmatrix} 1 & 2 & 3 \\ 2 & 3 & 1 \end{pmatrix} \right] \circ \begin{pmatrix} 1 & 2 & 3 \\ 3 & 1 & 2 \end{pmatrix} \\ &= \begin{pmatrix} 1 & 2 & 3 \\ 2 & 3 & 1 \end{pmatrix} \circ \begin{pmatrix} 1 & 2 & 3 \\ 3 & 1 & 2 \end{pmatrix} \\ &= \begin{pmatrix} 1 & 2 & 3 \\ 1 & 2 & 3 \end{pmatrix} \end{aligned}$$

Hence, $P_1 \circ (P_2 \circ P_3) = (P_1 \circ P_2) \circ P_3$.

4.7.4 Inverse Permutation

Since a permutation is one-to-one, onto map, and hence it is invertible, i.e., every permutatation f on a set $P = \{a_1, a_2, \ldots, a_n\}$ possesses a unique inverse permutation, denoted by f^{-1}. Thus, if

$$f = \begin{pmatrix} a_1 & a_2 & \cdots & a_n \\ b_1 & b_2 & \cdots & b_n \end{pmatrix}$$

then

$$f^{-1} = \begin{pmatrix} b_1 & b_2 & \cdots & b_n \\ a_1 & a_2 & \cdots & a_n \end{pmatrix}$$

EXAMPLE 4.34 Find the inverse of the permutation

$$\begin{pmatrix} 1 & 2 & 3 & 4 & 5 \\ 2 & 3 & 1 & 5 & 4 \end{pmatrix}$$

Solution Let the inverse of the given permutation be

$$f = \begin{pmatrix} 1 & 2 & 3 & 4 & 5 \\ x & y & z & u & v \end{pmatrix}$$

Then,

$$\begin{pmatrix} 1 & 2 & 3 & 4 & 5 \\ 2 & 3 & 1 & 5 & 4 \end{pmatrix} \begin{pmatrix} 1 & 2 & 3 & 4 & 5 \\ x & y & z & u & v \end{pmatrix} = \begin{pmatrix} 1 & 2 & 3 & 4 & 5 \\ 1 & 2 & 3 & 4 & 5 \end{pmatrix}$$

$$\Rightarrow \begin{pmatrix} 1 & 2 & 3 & 4 & 5 \\ y & z & x & v & u \end{pmatrix} = \begin{pmatrix} 1 & 2 & 3 & 4 & 5 \\ 1 & 2 & 3 & 4 & 5 \end{pmatrix}$$

$$\Rightarrow x = 3, y = 1, z = 2, u = 5, v = 4$$

Hence, the required inverse is

$$\begin{pmatrix} 1 & 2 & 3 & 4 & 5 \\ 3 & 1 & 2 & 5 & 4 \end{pmatrix}$$

4.7.5 Cyclic Permutations

Let $t_1, t_2, \ldots, t_r$ be r distinct elements of the set $P = \{t_1, t_2, \ldots, t_n\}$. Then the permutation $p : P \to P$ is defined by

$$p(t_1) = t_2, p(t_2) = t_3, \ldots, p(t_{r-1}) = t_r, p(t_r) = t_1$$

is called a *cyclic permutation* of length r, or simply a cycle of length r. It is generally denoted by

$$(t_1, t_2, \ldots, t_r)$$

If the elements $t_1, t_2, \ldots, t_r$ are arranged on a circle, as shown in Figure 4.1, then a cycle p of length r rotates these elements in a clockwise direction so that t_1 moves to t_2, t_2 to $t_3, \ldots, t_{r-1}$ to t_r, and t_r to t_1, which is shown in Figure 4.1.

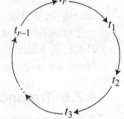

FIGURE 4.1 A cycle of length r in a clockwise direction

For example Let $P = \{1, 2, 3\}$, and the possible permutations are

$$\begin{pmatrix} 1 & 2 & 3 \\ 1 & 2 & 3 \end{pmatrix}, \begin{pmatrix} 1 & 2 & 3 \\ 1 & 3 & 2 \end{pmatrix}, \begin{pmatrix} 1 & 2 & 3 \\ 3 & 2 & 1 \end{pmatrix}, \begin{pmatrix} 1 & 2 & 3 \\ 2 & 1 & 3 \end{pmatrix}, \begin{pmatrix} 1 & 2 & 3 \\ 2 & 3 & 1 \end{pmatrix}, \begin{pmatrix} 1 & 2 & 3 \\ 3 & 1 & 2 \end{pmatrix}$$

Thus, a permutation which replaces n objects cyclically is called a cyclic permutation of degree n which means that each element in the bracket is replaced by the element following it and the last element by the first. It is observed that $P = (1 \ 2 \ 3) = (2 \ 3 \ 1) = (3 \ 1 \ 2)$. Hence, a circular permutation may be denoted by more than one rowed symbols.

The number of elements permuted by a cycle is called its *length*. A cycle of length 1 means that the image of an element is the element itself and represents *identity permutation*. Cycles of length 1 are generally omitted.

Disjoint cycles are those which have no common elements. Every permutation of a finite set can be expressed as a cycle or as a product of disjoint cycles.

For example

(i) The permutation

$$p = \begin{pmatrix} 1 & 2 & 3 & 4 & 5 & 6 \\ 2 & 1 & 4 & 6 & 5 & 3 \end{pmatrix}$$

is written as $(1, 2) \ (3, 4, 6) \ (5)$.

The cycle (1, 2) has length 2, the cycle (3, 4, 6) has length 3, and the cycle (5) has length 1. None of the cycles have a symbol common and hence they are disjoint cycles.

(ii) Let $P = \{1, 2, 3, 4, 5, 6\}$. Then the cycles (1, 2, 5) and (3, 4, 6) are disjoint, whereas the cycles (1, 2, 5) and (2, 4, 6) are not.

If $p_1 = (a_1, a_2, \ldots, a_m)$ and $p_2 = (b_1, b_2, \ldots, b_n)$ then we can show that $p_1 \circ p_2 = p_2 \circ p_1$. We shall now state a fundamental theorem whose proof is supported by the following illustration.

Theorem 4.20 *A permutation of a finite set, neither identity nor a cycle, can be expressed as a product of disjoint cycles of length ≥ 2.*

Proof The proof can be performed by using the permutation

$$p = \begin{pmatrix} 1 & 2 & 3 & 4 & 5 \\ 2 & 1 & 4 & 6 & 5 \end{pmatrix}$$

of the set $P = \{1, 2, 3, 4, 5, 6\}$, which is to be expressed as a product of disjoint cycles of length ≥ 2. The process is as follows.

Let us first start with 1 and we find that $p(1) = 2, p(2) = 1$. Then, we select the third element of P, i.e., 3, which is not appeared in the previous cycle, and we have $p(3) = 4, p(4) = 6, p(6) = 3$. Now, we write p as a product of disjoint cycles as $p = (1, 2) \circ (3, 4, 6)$.

4.7.6 Transposition

A cyclic permutation (a, b) which interchanges the symbols keeping all other intact is called a *transposition*. In other way, transposition is a cycle of length 2 of the form (a, b), i.e., it acts as a mapping which maps each object onto itself excepting two, each of which is mapped on the other.

For example The cyclic permutation (1, 2) is a transposition.

It may be observed that if $p = (a_i, a_j)$ is a transposition of P, then $p \circ p = 1_p$, the identity permutation of P.

Every permutation can be resolved as a product of finite number of transpositions but the decomposition is not unique. In fact,

$$(t_1, t_2, \ldots, t_r) = (t_1, t_r) \circ (t_1, t_{r-1}) \circ \cdots \circ (t_1, t_3) \circ (t_1, t_2)$$

For example $(1, 2, 3, 4, 5) = (1, 5) \circ (1, 4) \circ (1, 3) \circ (1, 2)$

Theorem 4.21 *Every permutation of n symbols can be expressed as a product of disjoint cycles.*

Proof Let f be an element of the set P. Consider the cycle

$$(1, f(1), f^2(1), \ldots)$$

Since $O(f)$ is finite, $f^k = I$ for some k, i.e., $f^k(1) = 1$ for some k. Then the cycle

$$(1, f(1), f^2(1), \ldots, f^{k-1}(1))$$

and the permutation f will have the same effect on the symbols

$$1, f(1), f^2(1), \ldots, f^{k-1}(1)$$

Again, choose a symbol i, such that $f(i) \neq i$ and consider the cycle

$$(i, f(i), f^2(i), \ldots, f^{m-1}(i))$$

where m is the least positive integer such that $f^m(i) = i$. If

$$1, f(1), f^2(1), \ldots, f^{k-1}(1), i, f(i), f^2(i), \ldots, f^{m-1}(i)$$

do not exhaust all the symbols, then the procedure must terminate as there are only a finite number of symbols. Thus,

$$f = (1, f(1), f^2(1), \ldots, f^{k-1}, i, f(i), f^2(i), \ldots, f^{m-1}(i))$$

is a product of disjoint cycles.

Corollary 4.4 Every permutation of a finite set having at least two elements can be expressed as a product of transpositions, which may not be disjoint.

EXAMPLE 4.35 Consider the set $P = \{1, 2, 3, 4, 5, 6, 7, 8\}$. Express the following permutation p, of the set P, as a product of transpositions

$$p = \begin{pmatrix} 1 & 2 & 3 & 4 & 5 & 6 & 7 & 8 \\ 3 & 4 & 6 & 5 & 2 & 1 & 8 & 7 \end{pmatrix}$$

Solution We have $p = (7,8) \circ (2,4,5) \circ (1,3,6)$. Since we can write $(1,3,6) = (1,6) \circ (1,3)$ and $(2,4,5) = (2,5) \circ (2,4)$ therefore, $p = (7,8) \circ (2,5) \circ (2,4) \circ (1,6) \circ (1,3)$. Thus, it is observed that every cycle can be expressed as a product of transpositions. However, this can also be possible in different ways.

For example

$$(1, 2, 3) = (1, 3) \circ (1, 2) = (2, 1) \circ (2, 3) = (1, 3) \circ (3, 1) \circ (1, 3) \circ (1, 2) \circ (3, 2) \circ (2, 3)$$

Hence, it follows that every permutation on a set of two or more elements can be written as a product of transpositions in various ways.

4.7.7 Even and Odd Permutations

A permutation is called an *even permutation* if it can be expressed as a product of an even number of transpositions, otherwise it is said to be an *odd permutation*.

For example The permutation

$$\begin{pmatrix} 1 & 2 & 3 & 4 & 5 & 6 \\ 5 & 6 & 2 & 4 & 1 & 3 \end{pmatrix}$$

is odd; while the permutation

$$\begin{pmatrix} 1 & 2 & 3 & 4 & 5 & 6 \\ 6 & 3 & 4 & 5 & 2 & 1 \end{pmatrix}$$

is even; for

$$\begin{pmatrix} 1 & 2 & 3 & 4 & 5 & 6 \\ 5 & 6 & 2 & 4 & 1 & 3 \end{pmatrix} = (1\ 5)(2\ 6\ 3) = (1\ 5)(2\ 6)(2\ 3)$$

is the product of an odd number of transpositions and

$$\begin{pmatrix} 1 & 2 & 3 & 4 & 5 & 6 \\ 6 & 3 & 4 & 5 & 2 & 1 \end{pmatrix} = (1\ 6)(2\ 3\ 4\ 5) = (1\ 6)(2\ 3)(2\ 4)(2\ 5)$$

is the product of an even number of transpositions.

From the definition of even and odd permutations, it follows that
(i) the product of two even permutations is even
(ii) the product of two odd permutations is even
(iii) the product of an even and an odd permutation is odd

EXAMPLE 4.36 Is the permutation

$$p = \begin{pmatrix} 1 & 2 & 3 & 4 & 5 & 6 & 7 \\ 2 & 4 & 5 & 7 & 6 & 3 & 1 \end{pmatrix}$$

even or odd?

Solution We first write p as a product of disjoint cycles, obtaining

$$p = (3, 5, 6) \circ (1, 2, 4, 7)$$

Next, we write each of the cycles as a product of transpositions:

$$(1, 2, 4, 7) = (1, 7) \circ (1, 4) \circ (1, 2)$$
$$(3, 5, 6) = (3, 6) \circ (3, 5)$$

Then,

$$p = (3, 6) \circ (3, 5) \circ (1, 7) \circ (1, 4) \circ (1, 2)$$

Since p is a product of an odd number of transpositions, it is an odd permutation.

Note The set of all even permutations of degree n forms a finite group of order $n!/2$ with respect to the composition of permutation and is called *alternating group*. It is denoted by A_n.

For example The alternating group on four symbols, i.e., $(A_4, \circ)$ is a group and $O(A_4) = 4!/2 = 12$.

4.8 SYMMETRIC GROUP

The set G of all permutations on P forms a group with respect to composition of functions as operation. This group is called the *symmetric group* of degree n and is denoted by P_n.

By considering the symmetries of regular polygons, we can form certain other permutation groups known as *dihedral groups*.

To construct dihedral group, let us first consider an equilateral triangle whose vertices are denoted by 1, 2, and 3 with O as centroid (Figure 4.2). The symmetries of this triangle are described as follows.

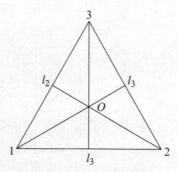

FIGURE 4.2 Given triangle

First of all, there is a counterclockwise rotation f_2 of the triangle about O through $120°$. Then, f_2 takes the form as the permutation (Figure 4.3)

$$f_2 = \begin{pmatrix} 1 & 2 & 3 \\ 2 & 3 & 1 \end{pmatrix}$$

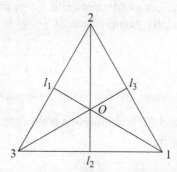

FIGURE 4.3 Triangle formed by f_2

Next, a counterclockwise rotation f_3 about O through $240°$ gives the permutation as

$$f_3 = \begin{pmatrix} 1 & 2 & 3 \\ 3 & 1 & 2 \end{pmatrix}$$

Lastly, there is a counterclockwise rotation f_1 about O through $360°$ yields the following permutation

$$f_1 = \begin{pmatrix} 1 & 2 & 3 \\ 1 & 2 & 3 \end{pmatrix}$$

These three symmetries are called *rotational symmetries*.

We may also obtain three additional symmetries of the triangle (known as *axial symmetries*), $g_1, g_2,$ and $g_3,$ by reflecting about the lines $l_1, l_2,$ and $l_3,$ respectively (Figure 4.4). We may denote these reflections as the following permutations:

$$g_1 = \begin{pmatrix} 1 & 2 & 3 \\ 1 & 3 & 2 \end{pmatrix} \quad g_2 = \begin{pmatrix} 1 & 2 & 3 \\ 3 & 2 & 1 \end{pmatrix} \quad g_3 = \begin{pmatrix} 1 & 2 & 3 \\ 2 & 1 & 3 \end{pmatrix}$$

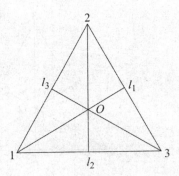

FIGURE 4.4 Triangle formed by g_2

Thus, $G = P_3 = \{f_1, f_2, f_3, g_1, g_2, g_3\}$ is a group with respect to the composition of functions $\circ$ and $(G, \circ)$ is called the *dihedral group*.

With the operation of the composition of function $\circ$, we shall now form a multiplication table shown in Table 4.12. Entries in the table can be made either of the two ways: geometrically or algebraically. Let us first compute $f_2 \circ g_2$ geometrically, and we proceed as in Figures 4.2–4.4. Algebraically, computation of $f_2 \circ g_2$ is being done as follows.

$$f_2 \circ g_2 = \begin{pmatrix} 1 & 2 & 3 \\ 2 & 3 & 1 \end{pmatrix} \circ \begin{pmatrix} 1 & 2 & 3 \\ 3 & 2 & 1 \end{pmatrix} = \begin{pmatrix} 1 & 2 & 3 \\ 1 & 3 & 2 \end{pmatrix} = g_1$$

Similarly, $g_3 \circ g_3 = f_1$. Thus, the following composition table is formed.

TABLE 4.12 Multiplication table using composition of function $\circ$

$\circ$	f_1	f_2	f_3	g_1	g_2	g_3
f_1	f_1	f_2	f_3	g_1	g_2	g_3
f_2	f_2	f_3	f_1	g_3	g_1	g_2
f_3	f_3	f_1	f_2	g_2	g_3	g_1
g_1	g_1	g_2	g_3	f_1	f_2	f_3
g_2	g_2	g_3	g_1	f_3	f_1	f_2
g_3	g_3	g_1	g_2	f_2	f_3	f_1

4.9 COSETS

Let $(G, *)$ be a group and H be any subgroup of G. Also, let a be any element of G, then the set

$$a * H = \{a * h : h \in H\}$$

is called the *left coset* of H in G generated by a. Similarly, the set

$$H * a = \{h * a : h \in H\}$$

is called the *right coset* of H in G generated by a. If e is the identity element of G, then $H * e = H = e * H$. So, H itself is a right as well as a left coset.

If the group G is abelian, then we have $a*h = h*a$, or simply $ah = ha \, \forall \, h \in H$; therefore, the right coset Ha will be equal to the corresponding left coset aH.

4.9.1 Properties of Cosets

Let H be a subgroup of G, and $a, b \in G$, then

I. $a \in aH$
II. $aH = H$ if and only if $a \in H$
III. $aH = bH$ or $aH \cap bH = \phi$
IV. $aH = bH$ if and only if $a^{-1}b \in H$

Similar results hold good for right cosets.

Proof

I. Here, $a = ae \in aH$, e is the identity element of G.

II. If e is the identity in G and so is in H, then

$$aH = H \implies ae \in H$$

i.e.,

$$aH = H \implies a \in H \tag{4.1}$$

Again, if $a \in H$ and $h \in H$, then $a \in H \implies ah \in H \, \forall h \in H$.
So, $aH \subset H$. On the other hand, $a \in H \implies a^{-1}H$, H being a subgroup of G, satisfies group axioms. Then,

$$\begin{aligned} & a^{-1}h \in H \quad \forall h \in H \quad \text{(closure law in } H\text{)} \\ \implies & a(a^{-1}h) \in aH \quad \forall h \in H \quad \text{(closure law in } H\text{)} \\ \implies & h \in aH \quad \forall h \in H \end{aligned}$$

Therefore, $H \subset aH$. Thus, $aH \subset H$ and $H \subset aH \implies aH = H$
Lastly,

$$a \in H \implies aH = H \tag{4.2}$$

From Eqs (4.1) and (4.2), $aH = H \iff a \in H$.

III. Let H be a subgroup of a group G. Let aH and bH be two of its left cosets. Assume that $aH \cap bH \neq \phi$. Also, let c be the common element of the two cosets. Then, we may write $c = ah$ and $c = bh'$ for $h, h' \in H$. Therefore, $ah = bh'$, yielding $a = bh'h^{-1}$. Since H is a subgroup, we have $h'h^{-1} \in H$. Let $h'h^{-1} = h''$, so that $a = bh''$.
Hence, $aH = (bh'')H = b(h''H) = bH$, since $h''H = H$. Thus, the two left cosets aH and bH are identical if $aH \cap bH \neq \phi$. Thus, either $aH \cap bH = 0$, or $aH = bH$.

IV. Here,

$$\begin{aligned} aH = bH & \implies a^{-1}(aH) = a^{-1}(bH) \implies (a^{-1}a)H = (a^{-1}b)H \\ \implies & eH = (a^{-1}b)H \end{aligned}$$

e being the identity in G and so in H.

$$\implies H = (a^{-1}b)H$$

Therefore,

$$aH = bH \implies a^{-1}b \in H \tag{4.3}$$

Again, if $a^{-1}b \in H$, then
$$bH = e(bH) = (aa^{-1})(bH) = a(a^{-1}b)H = aH \qquad (4.4)$$

From Eqs (4.3) and (4.4), $aH = bH \Leftrightarrow a^{-1}b \in H$.

4.10 NORMAL SUBGROUP

A subgroup H of a group G is called *normal subgroup* if $a^{-1}Ha \subseteq H$ for every $a \in G$. Equally, H is normal if $aH = Ha$ for every $a \in G$, i.e., if the right and left cosets coincide.

The importance and applicability of normal subgroup can be observed from the following theorems.

Theorem 4.22 Every subgroup of an abelian group is a normal subgroup.

Proof Let G be an abelian group and H be a subgroup of G. So, $g \in G$ and $h \in H$,
$$\Rightarrow \quad ghg^{-1} = hgg^{-1} \text{ (since } G \text{ is an abelian and } H \subseteq G\text{)}$$
$$= he = h$$
$$\Rightarrow \quad ghg^{-1} = h \in H \; \forall g \in G, h \in H$$

Hence, H is a normal subgroup.

Theorem 4.23 A subgroup H of a group G is normal if and only if $g^{-1}hg \in H$ for every $h \in H, g \in G$.

Proof Let H be a normal subgroup of G. Let $h \in H$ and $g \in G$. Then, $Hg = gH$, from definition of normal subgroup. Now, $hg \in Hg = gH$. So, $hg = gh_1$ for some $h_1 \in H$, i.e., $g^{-1}hg = g^{-1}gh_1 = h_1 \in H$. Conversely, let H be such that $g^{-1}hg \in H \; \forall h \in H, g \in G$. Consider $a \in G$, then for any $h \in H$, $a^{-1}ha \in H$.

Therefore, $ha = a(a^{-1}ha) \in aH$. Consequently, $Ha \subseteq aH$. Assume $b = a^{-1}$. Then, $b^{-1}hb \in H$. But, $b^{-1}hb = (a^{-1})^{-1}ha^{-1} = aha^{-1}$, which yields $aha^{-1} \in H$, so that $ah = (aha^{-1})a \in Ha$. Thus, $aH \subseteq Ha$. Hence, $aH = Ha$.

The above theorem shows that a subgroup H of a group G can be defined as a normal subgroup if $g^{-1}hg \in H \; \forall h \in H, g \in G$.

Theorem 4.24 Let H be a normal subgroup of a group G. Then the cosets of H form a group under coset multiplication
$$(aH)(bH) = abH$$
This group is called *quotient group* (or, *factor group*) and is denoted by G/H.

Proof Coset multiplication is well defined, since
$$(aH)(bH) = a(Hb)H = a(bH)H = ab(HH) = abH$$
(since H is normal, i.e., $Hb = bH$ and $HH = H$).

Coset multiplication is associative because G is associative. H is the identity element of G/H, since
$$(aH)H = a(HH) = aH$$
and
$$H(aH) = (Ha)H = (aH)H = H$$

G/H has an identity element $H = eH$, for

$$(aH)(eH) = (ae)H = aH \quad \text{and} \quad (eH)(aH) = (ea)H = aH$$

Also, $a^{-1}H$ is the inverse of aH since

$$(a^{-1}H)(aH) = (a^{-1}a)H = eH = H \quad \text{and} \quad (aH)(a^{-1}H) = (aa^{-1})H = eH = H$$

Thus, G/H is a group under coset multiplication.

If the group operation be addition, then the right coset of H in G, generated by a, is defined as

$$H + a = \{h + a : h \in H\}$$

and the left coset is

$$a + H = \{a + h : h \in H\}$$

If H is a subgroup of a group G, the number of distinct left or right cosets of H in G is called the *index* of H in G and is denoted by $[G:H]$ or by $i_G(H)$.

Corollary 4.5 If G is a finite group and H is a normal subgroup of G, then

$$O(G/H) = O(G)/O(H) = i_G(H)$$

For example If $H_3 = \{\ldots, -6, -3, 0, 3, 6, \ldots\}$, then H_3 is a normal subgroup of Z (integers) under addition.

The cosets of H in Z are $H_3 = \{\ldots, -6, -3, 0, 3, 6, 9, \ldots\}$, $1 + H_3 = \{\ldots, -5, -2, 1, 4, 7, 10, \ldots\}$, and $2 + H_3 = \{\ldots, -4, -1, 2, 5, 8, \ldots\}$. Therefore, $G/H = \{H_3, 1 + H_3, 2 + H_3\}$. Here, Z/H_3 forms a group under addition.

EXAMPLE 4.37 Consider the group Z of integers under addition and the subgroup $H = \{\ldots, -10, -5, 0, 5, 10, \ldots\}$ considering of the multiple of 5. Find (a) the cosets of H in Z and (b) the index of H in Z.

Solution

(a) There are five distinct [left] cosets of H in Z, as follows:

$$0 + H = \{\ldots, -10, -5, 0, 5, 10, \ldots\}$$
$$1 + H = \{\ldots, -9, -4, 1, 6, 11, \ldots\}$$
$$2 + H = \{\ldots, -8, -3, 2, 7, 12, \ldots\}$$
$$3 + H = \{\ldots, -7, -2, 3, 8, 13, \ldots\}$$
$$4 + H = \{\ldots, -6, -1, 4, 9, 14, \ldots\}$$

Any other coset $n + H$ coincides with one of the above cosets.

(b) Although Z and H both are infinite, the index of H in Z is finite. Specifically, the number of cosets is $[Z:H] = 5$.

EXAMPLE 4.38 If H is a subgroup of G such that $x^2 \in H$ for every $x \in G$, then prove that H is a normal subgroup of G.

Solution For any $g \in G$, $h \in H$, $(gh)^2 \in H$ and $g^{-2} \in H$. Since H is a subgroup, $h^{-1} g^{-2} \in H$ and simultaneously,

$$(gh)^2 h^{-1} g^{-2} \in H \Rightarrow ghghh^{-1} g^{-2} \in H \Rightarrow ghg^{-1} \in H$$

Hence, H is a normal subgroup of G.

 EXAMPLE 4.39 If G be an abelian group with identity e, then prove that all elements x of G satisfying the equation $x^2 = e$ form a subgroup H of G.

Solution Let $H = \{x : x^2 = e\}$. Now, $x^2 = e \Rightarrow x = x^{-1}$.

Therefore, if $x \in H$, then x^{-1} also belongs to H. Moreover, $e^2 = e$. Thus, the identity element of G also belongs to H.

Let $x, y \in H$. Then, since G is abelian, we have

$$xy = yx = y^{-1}x^{-1} \text{ (since } x^{-1} = x \text{ and } y^{-1} = y)$$
$$= (xy)^{-1}$$

Thus, $(xy)^2 = e$. Hence, $xy \in H$ and H is a subgroup of G.

 EXAMPLE 4.40 Let $(G, *)$ be a group and $(H, *)$ be a normal subgroup. Show that $(G/H, \Theta)$ is also a group.

Solution Let $\phi : G \to G/H$ be defined by $\phi(x) = Hx \forall x \in G$. Assume $y \in G$. Then,

$$\forall x, y \in G, \phi(x*y) = Hx*y = (Hx)\Theta(Hy)$$

So, ϕ is a homomorphism from G to G/H. Also, $\forall Hx \in G/H$, there exists $x \in G$ such that $\phi(x) = Hx$. Thus $(G/H, \Theta)$ is a group.

Theorem 4.25 For any two subgroups H and K of a group G, the following hold:
 (i) $H \cap K$ is a subgroup of G
 (ii) If H is normal in G, then $H \cap K$ is normal in K
 (iii) If H and K both are normal in G, then $H \cap K$ is normal in G

Proof
 (i) Since, an identity, $e \in H \cap K$, $H \cap K$ is non-empty. Now, $a, b \in H \cap K \Rightarrow a, b \in H$ and $a, b \in K$. Then, $ab^{-1} \in H$ and $ab^{-1} \in K \Rightarrow ab^{-1} \in H \cap K$.
 Hence, $H \cap K$ is a subgroup of G.
 (ii) Let $x \in K$ and $a \in H \cap K$. Then, $x^{-1}ax \in K$, since $x, a \in K$. Further, $x^{-1}ax \in H$, since H is normal and $a \in H$. Consequently, $x^{-1}ax \in H \cap K \forall x \in K$ and $a \in H \cap K$.
 Hence, $H \cap K$ is a normal subgroup of K.
 (iii) Left to the reader.

4.11 LAGRANGE'S THEOREM

If G is a finite group and H a subgroup, then the order of H is a divisor of the order of G.

Proof Let the order of H be m and G be n, i.e., $O(H) = m$ and $O(G) = n$, respectively. Given G is finite and $H \subset G$, so H is also finite.

Now, G is partitioned into finite number of distinct right cosets of H in G, say, k right cosets, namely $H, Ha_1, Ha_2, \ldots, Ha_{k-1}$. The order of each coset is also the order of H. Therefore,

$$O(Ha_1) = O(Ha_2) = \cdots = O(Ha_{k-1}) = O(H) = m$$

Also,
$$O(G) = O(H) + O(Ha_1) + \cdots + O(Ha_{k-1})$$
i.e.,
$$n = m + m + \cdots + m, \text{ k times} = km$$
Thus, m divides n, i.e., the order of H divides the order of G.

Note The converse of Lagrange's theorem need not be true, i.e., if m divides $O(G)$, then G need not have a subgroup of the order m.

Corollary 4.6 If G is a finite group and $a \in G$, then $O(a)$ divides $O(G)$.

Proof Let us consider the cyclic subgroup generated by a. Let $O(a) = m$. So, $(a) = \{a, a^2, \ldots, a^m = e\}$. Therefore, $O(a) = m$ divides $O(G)$, i.e., $O(a)$ divides the order of G.

Corollary 4.7 If G is a finite group and $a \in G$, then $a^{O(G)} = e$.

Proof Let $O(a) = m$ and $O(G) = n$. Also, m is the least positive integer such that $a^m = e$. Again, by Corollary 4.6, m divides n. So, $n = mq$, for some integer q.
Thus,
$$a^{O(G)} = a^n = a^{mq} = (a^m)^q = e^q = e$$

EXAMPLE 4.41 Let G be a group of the order p, where p is a prime. Find all subgroups of G.

Solution By Lagrange's theorem, the order of a subgroup H of G divides the order of G. Hence $|H| = 1$ or p. Thus, $\{e\}$ and G, itself are the only subgroups of G.

EXAMPLE 4.42 Let G be a finite group and H be a subgroup of G. For $a \in G$, define $aH = \{ah : h \in H\}$. Then,
(a) Show that $|aH| = |H|$.
(b) Show that for every pair of elements $a, b \in$ either $aH = bH$ or aH and bH which are disjoint.
(c) Use the above to analyse that the order of H must divide the order of G.

Solution

(a) Since H is a subgroup of G, for any $a \in G$, therefore, $a \equiv a|H|$. Taking modulus on both sides,
$$|a| = |aH| \Rightarrow H|a| = \{h*a : h \in H\} \Rightarrow |aH| = |H|$$
hence proved.

(b) Suppose aH and bH are not disjoint, then we have to prove that $aH = bH$. $aH \cap bH \neq \phi \Rightarrow$ there exists $h_1, h_2 \in H$ such that
$$ah_1 = bh_2 \text{ (a common element)} \tag{4.5}$$
Now, let us prove that $aH \subseteq bH$.
Let $x \in aH$. Then, $x = ah$ for some $h \in H$. From Eq. (4.5),
$$a = bh_2h_1^{-1} \text{ (post multiplying by } h_1^{-1})$$
So,
$$x = (bh_2h_1^{-1})h = b(h_2h_1^{-1}h) \text{ (by associative law)}$$

But, $h_2 h_1^{-1} h \in H$ by closure property of a subgroup. So,

$$x \in bH, \text{ i.e., } x \in aH \Rightarrow x \in bH \Rightarrow aH \subseteq bH \qquad (4.6)$$

Similarly, we can prove that

$$bH \subseteq aH \qquad (4.7)$$

From Eqs (4.6) and (4.7), $aH = bH$.

(c) The number of left cosets of H in G is called the index of H in G. From Lagrange's theorem, we have the index K of G given by

$$K = |G|/|H|$$

It immediately follows from Lagrange's theorem that any group of prime order has only trivial subgroups, while a group of the order n which has divisions other than 1 and n has proper subgroups.

Group theory possesses an important application in coding theory. The coding theory is used for introducing redundant information in transmitting data that help in detecting and in correcting errors. These are presented in detail in the next article.

4.12 GROUP CODES

Error detection and error correction techniques play an important role in the design of computer systems. Most of the systems today contain telephone and communication lines which disturb the transmitted messages by the presence of noise. The noise may be due to the lightning, folds in a magnetic tape, etc. Peripheral equipment associated with such systems is by far the most unreliable component of these systems, and both error detection and error correction are frequently performed.

Structure in the design of error-correcting codes is important. First, it makes easy in finding the properties of a code; second, and more importantly, it makes to realize the hardware of such practical codes. Algebraic structures are the basis of the most important codes which have been designed.

A communication process may take place in a variety of ways, for instance, by making a telephone call, sending a message by a telephone or a letter, using a sign language, etc. In all such cases, the process involves the flow of some information carrying commodity can vary from music to speech to electricity to water to a sequence of binary digits.

An ideal communication system can be represented by at least three essential parts, as shown in Figure 4.5, namely

1. Transmitter, sender, or source
2. Channel or storage medium
3. Receiver

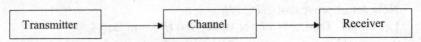

FIGURE 4.5 An ideal communication system

The channel conveys the message sent by the transmitter to the receiver. In practical situations, a system may consist of a number of transmitters and receivers, such as in a telephone exchange or a large computer centre. In a computer centre, there are good numbers of computers connected by communication lines. Such a system will consist of many users that can function both as transmitters and receivers of information.

In any case, the important task of a communication system designer is to minimize the losses due to noise and to recover in some optimal manner the original commodity when it is corrupted by the presence of noise. A device that can be used to improve the efficiency of communication channel is an *encoder*. This encoder transforms the incoming messages to detect the presence of noise on the transformed messages. The use of an encoder requires that a *decoder* be employed to transform the encoded messages into their original form that is acceptable to the receiver. It may be possible not only to detect the distortion due to the noise in the channel but also to rectify the message by using a proper encoder and decoder, and thus transmit the message in a perfect manner. The model of the communication system, given in Figure 4.5, can now be modified to include the encoder and decoder, and to show the presence of noise in the channel. Such a model is shown in Figure 4.6.

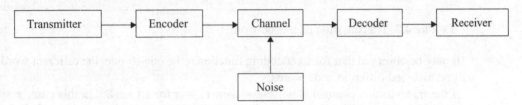

FIGURE 4.6 Modification of the communication system by encoder and decoder

Now, we shall discuss the communication channels in respect to the symbols from a specified set, called the *alphabet of the communication language*.

4.12.1 Coding of Binary Information

The basic unit of information is called a *message*. In fact, this unit is a finite sequence of characters from a finite alphabet. We shall select as our alphabet the set $B = \{0, 1\}$. Actually, the information is sent by a sequence of 0s and 1s, 0 and 1 is called a digit. Also, the basic unit of information, called a *word*, is a sequence of digits. The number of digits in the word is called the length of the word.

For example 1100 is of length 4. Also, 00101 is of length 5. That is, every character or symbol is represented in binary form.

Now, from the set B we can write $B^2 = B \times B = \{(0, 0), (0, 1), (1, 0), (1, 1)\}$. Also, we express this as $B^2 = \{00, 01, 10, 11\}$. This set is a binary code of words of length 2. Actually, $(B^2, +)$ is a group having identity $0 = 00 \cdots 00$ and every element is its own inverse. Similarly, $B^m = B \times B \times \cdots \times B$ (m factors) is a group under the operation $+$ having identity and its inverse as B^2.

A word is transmitted by sending its digits, one after the other across a binary channel. Figure 4.7 shows the basic process of sending a word from one point to another point over a transmission channel. An element $x \in B^m$ is sent through the transmission channel and is received as an element $x_t \in B^m$. In actual practice, the transmission channel may suffer disturbances (or noise) due to weather interference, electrical problems, etc., as a result the receiver can make an uncorrectable and undetectable mistake like the interchange of 0 or 1 in a particular code, for instance, the code 000 may be changed to 100 or 010 or 001; thus, $x = x_t$. Hamming developed a code by introducing redundant (extra) digits. This method is useful in detecting and correcting errors.

To reduce the transmission of information problem, let us proceed as follows. First of all, let us choose an integer $n > m$ and a one-to-one function $e : B^m \to B^n$. The function e is called an (m, n) *encoding function*. It can be thought of as a means of representing every word in B^m as a word in B^n. If $b \in B^m$, then $e(b)$ is called the code word denoting b. Through the conception of Hamming, the additional 0s and 1s can provide the means to detect or correct errors produced in the transmission channel.

We now transmit the code word by means of a transmission channel. Then each code word $x = e(b)$ is received as the word x_t in B^n. This situation is described in Figure 4.7.

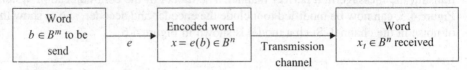

FIGURE 4.7 Transmission of a code word

It may be observed that for an encoding function to be one-to-one, the different words in B^m will be assigned different code words.

If the transmission channel is noiseless, then $x_t = x$ for all $x \in B^n$. In this case, $x = e(b)$ is received for each $b \in B^m$ and since e is a known function, b may be identified.

In general, errors in transmission do occur. We will say that the code word $x = e(b)$ has been transmitted with k or *fewer errors* if x and x_t differ in at least 1 but no more than k positions.

Let $e : B^m \to B^n$ be an (m, n) encoding function. We say that e detects k or fewer errors if whenever $x = e(b)$ is transmitted with k or fewer errors, then x_t is not a code word (thus $x_t \neq x$ and therefore it could not have been correctly transmitted). For $x \in B^n$, the number of 1s in x is called the *weight* of x and is denoted by $|x|$.

For example The weight of each of the following words in B^4,

(i) $x = 0100$ (ii) $x = 1110$ (iii) $x = 0000$ (iv) $x = 1111$

is

(i) $|x| = 1$ (ii) $|x| = 3$ (iii) $|x| = 0$ (iv) $|x| = = 4$

In a message of n digits long, m digits ($m < n$) are used to represent the information part of the message and the remaining $k = n - m$ digits are used to detect and correct errors. The later digits are called parity checks. We will discuss the same here.

4.12.2 Parity and Generator Matrices

The encoding function $e : B^m \to B^{m+1}$ is called the *parity* $(m, m + 1)$ *check code*. If $b = b_1 b_2 \cdots b_m \in B^m$, define

$$e(b) = b_1 b_2 \cdots b_m b_{m+1},$$

where

$$b_{m+1} = \begin{cases} 0, & \text{if } |b| \text{ is even} \\ 1, & \text{if } |b| \text{ is odd} \end{cases}$$

Here, b_{m+1} is zero if and only if the number of 1s in b is an even number. It means that every code word $e(b)$ has even weight. A single error in the transmission of a code word will change the received word to a word of odd weight and therefore can be detected. In the same way, we see that any odd number of errors can be detected.

For example To show the behaviour of the encoding function, let us assume $m = 3$. Then

$$e(000) = 0000$$
$$e(001) = 0011$$
$$e(010) = 0101$$
$$e(011) = 0110 \quad \text{(code with redundant digit)}$$
$$e(100) = 1001$$
$$e(101) = 1010$$
$$e(110) = 1100$$
$$e(111) = 1111$$

Let us consider the word $b = 000$. Then $x = e(b) = e(000) = 0000$. If a single error occurs in 0000, then the combination must be 1000 or 0100 or 0010 or 0001. But it is not available for any of the other codes, i.e., an odd number of errors (at least one) is occurred.

Hence by including redundant digit, we are able to detect and correct a single error appearing in one of the codes.

It should be noted that if the received word has even weight, then we cannot conclude that the code word is transmitted correctly, since this encoding function does not detect an even number of errors. Irrespective of this limitation, the parity check code is widely used.

Note If there exists more than one error in the code, then we have to supply more than one extra digit.

Now, in the present part generator matrices will be discussed.

The encoding and decoding functions can be expressed by matrices over B. Through one of the matrices one can locate the nearest code word for a received word.

Let

$$G = \begin{bmatrix} 1 & 0 & 0 & 1 & 1 & 0 \\ 0 & 1 & 0 & 0 & 1 & 1 \\ 0 & 0 & 1 & 1 & 0 & 1 \end{bmatrix}$$

be a 3×6 matrix over B. Then $G = [I_3/A]$ where

$$I_3 = \begin{bmatrix} 1 & 0 & 0 \\ 0 & 1 & 0 \\ 0 & 0 & 1 \end{bmatrix} \quad \text{and} \quad A = \begin{bmatrix} 1 & 1 & 0 \\ 0 & 1 & 1 \\ 1 & 0 & 1 \end{bmatrix}$$

The partitioned matrix G is called a *generator matrix*.

Let $e: B^3 \to B^6$ be an encoding function defined by $e(b) = bG \in B^6$. Now, $B^3 = \{000, 100, 010, 001, 110, 101, 011, 111\}$

$$e(000) = (000) \begin{bmatrix} 1 & 0 & 0 & 1 & 1 & 0 \\ 0 & 1 & 0 & 0 & 1 & 1 \\ 0 & 0 & 1 & 1 & 0 & 1 \end{bmatrix} = (0\ 0\ 0\ 0\ 0\ 0)$$

$$e(100) = (100) \begin{bmatrix} 1 & 0 & 0 & 1 & 1 & 0 \\ 0 & 1 & 0 & 0 & 1 & 1 \\ 0 & 0 & 1 & 1 & 0 & 1 \end{bmatrix} = (1\ 0\ 0\ 1\ 1\ 0)$$

In a similar way, we can calculate all other elements of B^3 and write the set of code words as

$$C = \{000000, 100110, 010011, 001101, 110101, 101011, 011110, 111000\} \subset B^6$$

We can recapture the message by simply dropping the last three digits of the code word. If the word u is transformed to the word v and $v \in C$, we cannot detect error though there may be an error. We can calculate the distances of words in C and find the minimum distance as 3. Hence, we can detect 2 or less errors and correct single error.

Let $b = b_1 b_2 b_3 \in B^3$ and $e(b) = b_1 b_2 b_3 b_4 b_5 b_6 \in B^6$. Then we have

$$e(b) = (b_1 b_2 b_3) \begin{bmatrix} 1 & 0 & 0 & 1 & 1 & 0 \\ 0 & 1 & 0 & 0 & 1 & 1 \\ 0 & 0 & 1 & 1 & 0 & 1 \end{bmatrix} = (b_1 b_2 b_3 (b_1 + b_3)(b_1 + b_2)(b_2 + b_3))$$

Then,

$$b_1 b_2 b_3 b_4 b_5 b_6 = b_1 b_2 b_3 (b_1 + b_3)(b_1 + b_2)(b_2 + b_3)$$

i.e.,

$$b_4 = b_1 + b_3 \quad b_5 = b_1 + b_2 \quad b_6 = b_2 + b_3$$

These equations are called parity check equations. Since b_4, b_5, and b_6 are either 0s or 1s and $0 = -0 \pmod 2, 1 = -1 \pmod 2$, we can write the above equations as follows:

$$b_1 + b_3 + b_4 = 0 \quad b_1 + b_2 + b_5 = 0 \quad b_2 + b_3 + b_6 = 0$$

Thus, it can be seen that

$$\begin{bmatrix} 1 & 0 & 1 & 1 & 0 & 0 \\ 1 & 1 & 0 & 0 & 1 & 0 \\ 0 & 1 & 1 & 0 & 0 & 1 \end{bmatrix} \begin{bmatrix} b_1 \\ b_2 \\ b_3 \\ b_4 \\ b_5 \\ b_6 \end{bmatrix} = \begin{bmatrix} b_1 + b_3 + b_4 \\ b_1 + b_2 + b_5 \\ b_2 + b_3 + b_6 \end{bmatrix} = \begin{bmatrix} 0 \\ 0 \\ 0 \end{bmatrix}$$

i.e.,

$$H(e(b))^T = \begin{bmatrix} 0 \\ 0 \\ 0 \end{bmatrix}$$

where $H = [A^T/I_3$ and $[e(b)]^T$ is the transpose of $e(b)$. Also, H is called the *parity check matrix* associated with the generator matrix G.

Now, $q = q_1q_2\ldots q_6 \in B^6$ is a code word if and only if

$$Hq^T = \begin{bmatrix} 0 \\ 0 \\ 0 \end{bmatrix}$$

Assume that we receive a word $q = 110110$. We observe that $q \notin C$, i.e., q is not a code word. Now,

$$Hq^T = \begin{bmatrix} 1 & 0 & 1 & 1 & 0 & 0 \\ 1 & 1 & 0 & 0 & 1 & 0 \\ 0 & 1 & 1 & 0 & 0 & 1 \end{bmatrix} \begin{bmatrix} 1 \\ 1 \\ 0 \\ 1 \\ 1 \\ 0 \end{bmatrix} = \begin{bmatrix} 0 \\ 1 \\ 1 \end{bmatrix}$$

which is the same as second column of H. Hence, we can say that an error is occurred in the second component of q. By changing the second component, we get 100110 and by dropping the last three digits, we can conclude that the message sent is 100.

Suppose if Hq^T is not one of the columns of H, but can be obtained as the sum of the two columns of H, we cannot correct multiple errors. Hence, if the parity check matrix H has no column of 0s and no two columns of H are the same, then it is helpful in correcting single error.

4.12.3 Decoding and Error Correction

An onto function $D: B^n \to B^m$ is called an (n, m) *decoding function* associated with e, if $D(y) = x_t \in B^m$, and is such that when the transmission channel has no noise then $x_t = x$, i.e., $D \circ e = I$, where I is the identity function on B^m. The decoding function decodes properly received words correctly but the decoding of improperly received words may or may not be correct.

For example Let us consider the even parity check given by $e: B^3 \to B^4$. Let $D: B^4 \to B^3$ be the decoding function.

Code (x)	000	001	010	011	100	101	110	111
$y = e(x)$	0000	0011	0101	0110	1001	1010	1100	1110
$D(y) = (D \circ e)(x)$	000	001	010	011	100	101	110	111

Let C denote a code. Let the word be transmitted into a word v. If $v \in C$, we cannot detect error, even if there exists no error. A code can detect all combinations of k or fewer errors if and only if the minimum distance of $x, y \in C$ is at least $k + 1$. It can correct all combinations of k or fewer errors if and only if the minimum distance is at least $2k + 1$.

For example If the minimum distance of a code is 5, it can detect 4 or less errors and correct 2 or less errors.

4.13 ALGEBRAIC SYSTEMS WITH TWO BINARY OPERATIONS

The algebraic systems with one binary operation, which we have studied so far, are not adequate enough to describe the system of real numbers. Therefore, we will consider here an abstract algebraic system called a *ring*, which is a special case of a group on which an additional binary operation satisfying certain properties could be defined. Other algebraic systems with two binary operations can be obtained by implementing further restrictions on rings.

4.13.1 Rings

An algebraic system $(R, +, \bullet)$ is called a *ring* if the binary operations $+$ and $\bullet$ on R satisfying the following axioms:

[R_1] For any $a, b, c \in R, (a + b) + c = a + (b + c)$
[R_2] For any $a, b \in R, a + b = b + a$
[R_3] There exists an element $0 \in R$, called the *zero element*, such that $a + 0 = 0 + a = a$ for every $a \in R$
[R_4] For each $a \in R$, there exists an element $-a \in R$, called the *negative* of a, such that $a + (-a) = (-a) + a = 0$
[R_5] For any $a, b, c \in R, (a \bullet b) \bullet c = a \bullet (b \bullet c)$
[R_6] For any $a, b, c \in R$, (i) $a \bullet (b + c) = a \bullet b + a \bullet c$ (ii) $(b + c) \bullet a = b \bullet a + c \bullet a$

It may be observed that the axioms [R_1] – [R_4] may be summarized by saying that R is an abelian group under addition.

Briefly, it may be said that an algebraic system $(R, +, \bullet)$ is called a ring if
(i) $(R, +)$ is an abelian group
(ii) $(R, \bullet)$ is a semigroup
(iii) the operation $\bullet$ is distributive over $+$

Depending upon the structure of $(R, \bullet)$, various special cases of rings will be defined.

If $(R, \bullet)$ is commutative, then the ring $(R, +, \bullet)$ is called a *commutative ring*. It follows that a ring $(R, +, \bullet)$ is commutative if and only if the semigroup $(R, \bullet)$ is commutative. Similarly, if $(R, \bullet)$ is a monoid, then $(R, +, \bullet)$ is called a *ring with identity*. In a ring, an element $e \in R$ is called an unit (identity) element if $ea = a = ae \; \forall a \in R$. An unit element of a ring $(R, +, \bullet)$, if it exists, is an element of the semigroup $(R, \bullet)$. The unit of a ring (if it exists) is generally denoted by 1. A ring $(R, +, \bullet)$ is called a ring with unity, if it possesses a unit element.

For example

(i) The set Z (of integers) under usual operations of addition and multiplication is a commutative ring with unity 1. The units in Z are only 1 and -1.
(ii) The set M of 2×2 matrices with integer or real entries is a non-commutative ring with unity.

We cannot show that $(R, \bullet)$ is a group, for a group with more than one element cannot have a zero element. We, therefore, investigate that $(R - \{0\}, \bullet)$ is closed with respect to the operation $\bullet$. If it is closed, then we have for any $a, b \in R$ such that $a \neq 0$ and $b \neq 0, a \bullet b \neq 0$; and we call $(R, +, \bullet)$ a *ring without divisors of zero*. Also, it can be shown that in a ring without divisors of zero, $a \bullet b = 0 \Rightarrow a = 0$ or $b = 0$.

4.13.2 Elementary Properties of a Ring

If $(R, +, \bullet)$ is a ring, then for any $a, b, c \in R$ the following properties are satisfied, where 0 is the additive identity and $-a$ denotes the additive inverse of an element $a \in R$:

I. $a \cdot 0 = 0 \cdot a = 0$
II. $a \cdot (-b) = (-a) \cdot b = -(a \cdot b)$
III. $(-a) \cdot (-b) = a \cdot b$
IV. $a \cdot (b - c) = a \cdot b - a \cdot c$ and $(b - c) \cdot a = b \cdot a - c \cdot a$

Proof

I. We have
$$a \cdot 0 + a \cdot a = a \cdot (0 + a) \quad \text{(right distributive law)}$$
$$= a \cdot a$$
$$= 0 + a \cdot a$$

Hence,
$$a \cdot 0 = 0 \quad \text{(right cancellation law)}$$

Similarly,
$$0 \cdot a = 0 \quad \text{(left cancellation law)}$$

II. We have
$$a \cdot (-b + b) = a \cdot (-b) + a \cdot b \quad \text{(right distributive law)}$$
i.e.,
$$a \cdot 0 = a \cdot (-b) + a \cdot b \quad [(-b) \text{ is the additive inverse of } b]$$
$$\Rightarrow \quad 0 = a \cdot (-b) + a \cdot b \quad \text{[from property (I)]}$$

Thus,
$$(-a) \cdot b = -(a \cdot b)$$

Similarly, $a \cdot (-b) = -(a \cdot b)$.

III. Left to the reader.
IV. Left to the reader.

Note The units of Z_m, the ring of integers modulo m can be developed as follows:
If a is a unit in Z_m, then $a^{-1}a \equiv 1 (\bmod m)$, or, in Z,
$$a^{-1}a = 1 + rm \quad \text{or} \quad a^{-1}a - rm = 1$$

This shows that any common divisor of a and m must divide 1, i.e., a and m are relatively prime.

EXAMPLE 4.43 Let $(R, +, \bullet)$ be a ring. The operation $\bullet$ is defined such that $ab = a \bullet b + b \bullet a$. Show that $(R, +, \bullet)$ is a commutative ring.

Solution Let $a, b \in R$, then $ab = a \bullet b + b \bullet a = b \bullet a + a \bullet b = ba$. Thus, $(R, +, \bullet)$ is a commutative ring.

EXAMPLE 4.44 Consider the ring $Z_{10} = \{0, 1, 2, \ldots, 9\}$ of integers modulo 10.

(a) Find the unit of Z_{10}.
(b) Find $-3, -8,$ and 3^{-1}.
(c) Let $f(x) = 2x^2 + 4x + 4$. Find the roots of $f(x)$ over Z_{10}.

Solution

(a) It is known that the integers relatively prime to the modulus $m = 10$ are the units in Z_{10}. Hence, the units are 1, 3, 7, and 9.
(b) The element $-a$ implies that the element is such that $a + (-a) = 0 = (-a) + a$. Hence, $-3 = 7$, since $3 + 7 = 7 + 3 = 0$ in Z_{10}. Similarly, $-8 = 2$. Again, by a^{-1} in a ring R we mean that the element is such that $a \cdot a^{-1} = a^{-1} \cdot a = 1$. Hence, $3^{-1} = 7$ since $3 \cdot 7 = 7 \cdot 3 = 1$ in Z_{10}.
(c) Substitute each of the 10 elements of Z_{10} into $f(x)$ to check which elements yield 0. We have

$$f(0) = 4, f(2) = 0, f(4) = 2, f(6) = 0, f(8) = 4$$
$$f(1) = 0, f(3) = 4, f(5) = 4, f(7) = 0, f(9) = 2$$

Thus the roots are 1, 2, 6, 7. (The present illustration shows that a polynomial of degree n can have more than n roots over an arbitrary ring. However, this is not possible if the ring is a field.)

A group is endowed with one binary operation; on the other hand, a ring is endowed with two binary operations connected by some interrelations.

4.13.3 Special Kinds of Rings

In this section, we will discuss different structures such as integral domain and field with illustrations of their own kind, respectively.

4.13.4 Integral Domain

A ring $(R, +, \bullet)$ is called an *integral domain* if it is commutative with identity and without divisors of zero.

For example
(i) The ring of integers $(Z, +, \bullet)$ is an integral domain, because it is commutative ring with unity and for any two integers $a, b, a \bullet b = 0 \Rightarrow a = 0$ or $b = 0$.
(ii) The ring of real numbers $(R, +, \bullet)$ is an integral domain.
(iii) The ring of even integers $(2Z, +, \bullet)$ is not an integral domain, for it does not contain the unit element, although it is without zero divisors.

EXAMPLE 4.45 Show that $S = \{a + \sqrt{2}b : a, b \in I\}$ is an integral domain.

Solution It can be verified that it is a commutative ring with unity element $1 + \sqrt{2} \cdot 0 = 1$. It is also a ring without zero divisors. Again, we have

$$(a + \sqrt{2}b)(c + \sqrt{2}d) = 0 \Rightarrow (ac + 2bd)\sqrt{2}(bc + ad) = 0$$

So,
$$ac + 2bd = 0 \qquad (4.8)$$
$$bc + ad = 0 \qquad (4.9)$$

From Eqs (4.8) and (4.9), we have $a = 0, b = 0$ or $c = 0$, and $d = 0$, i.e., either
$$a + \sqrt{2}b = 0 \quad \text{or} \quad c + \sqrt{2}d = 0$$

Therefore, S is an integral domain.

4.13.5 Field

A ring $(R, +, \bullet)$ which contains more than one element (with identity) such that every non-zero element of R has a multiplicative inverse in R is called a *field*.

Thus, a structure $(R, +, \bullet)$ is a field if

(i) $(R, +)$ is an abelian group
(ii) $(R', \bullet)$ is a commutative group, where $R' (= R - \{0\})$ is the set of non-zero elements of R
(iii) the distributive laws

$$a \bullet (b + c) = a \bullet b + a \bullet c \quad \text{and} \quad (b + c) \bullet a = b \bullet a + c \bullet a$$

hold good for any $a, b, c \in R$

For example

(i) The ring of rational numbers $(Q, +, \bullet)$ is a field, since it is commutative with identity element unity and each non-zero element has multiplicative inverse.
(ii) The set R of all real numbers is a field.
(iii) The set C of all complex numbers is a field.

An important characteristic of a field F is that if $x, y \in F$, then
$$x + y \in F, x - y \in F, xy \in F, xy^{-1} \in F$$

Hence, a field F is closed with respect to the four arithmetical operations, namely addition, subtraction, multiplication, and division.

Theorem 4.26 Every field is an integral domain.

Proof Since a field F is a commutative ring with unity, hence in order to show that every field is an integral domain, we have to show that a field has no zero divisors.

Let $a, b \in F$ and $a \neq 0$ such that $ab = 0$. Again, since $a \neq 0$, a^{-1} exists and we have
$$ab = 0 \Rightarrow a^{-1}(ab) = a^{-1}0 \Rightarrow (a^{-1}a)b = 0 \Rightarrow 1b = 0 \Rightarrow b = 0$$

Similarly, let $ab = 0$ and $b \neq 0$. Since, $b \neq 0$, b^{-1} exists, we have
$$ab = 0 \Rightarrow (ab)b^{-1} = 0b^{-1} \Rightarrow a(bb^{-1}) = 0 \Rightarrow a1 = 0 \Rightarrow a = 0$$

Thus, in a field $ab = 0 \Rightarrow a = 0$ or $b = 0$. Therefore, a field has no zero divisors. Hence, every field is an integral domain.

The converse is not true, i.e., every integral domain is not a field.

For example The ring of integers is an integral domain but it is not a field. The only invertible elements of the ring of integers are 1 and -1.

EXAMPLE 4.46 Show that a field F is an integral domain which has no zero divisors.

Solution If $ab = 0$ and $a \neq 0$, then $b = 1 \cdot b = (a^{-1}a) \cdot b = a^{-1} \cdot (ab) = a^{-1} \cdot 0 = 0$.

4.14 SUBRING

Let $(R, +, \bullet)$ be a ring and A be a non-empty subset of R. If $(A, +, \bullet)$ is itself a ring under the operations of addition and multiplication in R, then $(A, +, \bullet)$ is called a *subring* of $(R, +, \bullet)$.

Thus, $(A, +, \bullet)$ is a subring of $(R, +, \bullet)$ if

(i) $A \neq \phi$
(ii) $(A, +)$ is a subgroup of $(R, +)$
(iii) A is closed under multiplication, i.e., $a \bullet b \in A$ for all $a, b \in A$

Two subrings ($\{0\}, +, \bullet$) and ($R, +, \bullet$) of the ring ($R, +, \bullet$) are called *improper* or *trivial subring* of R. Any subring other than these two subrings is called a *proper* or *non-trivial subring*.

For example
(i) The ring of integers is a subring of the ring of rational numbers which in turn is a subring of the ring of real numbers.
(ii) The ring of even integers is a subring of the ring of integers.

The conditions for a subring will be discussed in the following theorem.

Theorem 4.27 The necessary and sufficient conditions for a non-empty subset S of a ring R to be a subring of R are

(i) $a \in S, b \in S \Rightarrow a - b \in S$
(ii) $a \in S, b \in S \Rightarrow ab \in S$

Proof

Necessary condition Let $(S, +, \bullet)$ be a subring of $(R, +, \bullet)$. Since S is the subring of R, then S is itself a ring and $(S, +)$ is the abelian group.

Now, $a, b \in S \Rightarrow a \in S$ and $-b \in S$, i.e., $a + (-b) \in S \Rightarrow a - b \in S$.

Also, since S is a ring, hence S is closed with respect to multiplication.

Hence,

$$a \in S, b \in S \Rightarrow ab \in S$$

Sufficient condition Let S be a non-empty subset of R, and S satisfy the following conditions:

(i) $a \in S, b \in S \Rightarrow a - b \in S$
(ii) $a \in S, b \in S \Rightarrow ab \in S$

From condition (i), $a \in S, a \in S \Rightarrow a - a \in S \Rightarrow 0 \in S$. Hence, zero is the additive identity.
Again, from condition (i), $0 \in S, a \in S \Rightarrow 0 - a \in S \Rightarrow -a \in S$. Hence, $-a$ is the additive inverse of a.

So, every element of S possesses additive inverse.

Now, $b \in S \Rightarrow -b \in S$; then, from condition (i), $a \in S, -b \in S \Rightarrow a - (-b) \in S \Rightarrow a + b \in S$.

Hence, addition is closed in S.

Since R satisfies associative and commutative laws with respect to addition, hence S is also associative and commutative with respect to addition.

From condition (ii), S is closed with respect to multiplication. Since R is associative with respect to multiplication, hence S is also associative with respect to multiplication.

Since R satisfies left and right distribution laws, hence S also holds left and right distribution laws.

Thus, S is a ring. Since S is the subset of R, then S is the subring of R.

Theorem 4.28 The intersection of two subrings of a ring is a subring of R.

Proof Let $(S, +, \bullet)$ and $(T, +, \bullet)$ be two subrings of a ring R. Since $(S, +)$ and $(T, +)$ are subgroups of $(R, +)$ both S and T contain the zero of the ring R (or, group $(R, +)$). Hence, $S \cap T \neq \phi$. Since the intersection of two subgroups of a group is a group; therefore, $(S \cap T, +)$ is a subgroup of $(R, +)$. Again,

$$a, b \in S \cap T \Rightarrow a, b \in S \text{ and } a, b \in T$$
$$\Rightarrow ab \in S \text{ and } ab \in T$$

as S and T are subrings

$$\Rightarrow ab \in S \cap T$$

Hence, $S \cap T$ is a subring of R.

Note The sum of two subrings may not be a subring; for instance, the multiple of 3 and multiple of 4 are subrings of the ring of integers. The sum of these two subrings contains the elements 3 and 4, but not their sum $3 + 4 = 7$. Consequently, the sum is not a subring.

4.14.1 Ideal

Let A be a non-empty subset of a ring R such that A is an additive subgroup of R, i.e., $a - b \in A$, whenever $a, b \in A$. Then,

(i) A is a *left ideal* of R iff $ra \in R$ for each $a \in A$ and $r \in R$.
(ii) A is a *right ideal* of R iff $ar \in R$ for each $a \in A$ and $r \in R$.
(iii) A is an *ideal* of R iff it is both a left ideal of R and a right ideal of R.

Thus, if the ring is commutative, the concepts of left, right, and two-sided ideals coincide.

For example

(i) The subring of even integers is an ideal of ring of integers.
(ii) The set $\{nx : x \in Z\}$ is an ideal of the ring of integers, n being any fixed integer.
(iii) Consider the matrix ring $M_2(R)$ of a ring R. Then in $M_2(R)$,
   (a) the subring

$$A = \left\{ \begin{pmatrix} a & b \\ 0 & 0 \end{pmatrix} : a, b \in R \right\}$$

is a right ideal but not a left ideal of M. To check that A is a right ideal, let us proceed as follows:

Let
$$\begin{pmatrix} a & b \\ 0 & 0 \end{pmatrix} \in A \text{ and } \begin{pmatrix} x & y \\ c & d \end{pmatrix} \in M_2(R)$$

Then
$$\begin{pmatrix} a & b \\ 0 & 0 \end{pmatrix}\begin{pmatrix} x & y \\ c & d \end{pmatrix} = \begin{pmatrix} ax+bc & ay+bd \\ 0 & 0 \end{pmatrix} \in A$$

But A is not a left ideal, for
$$\begin{pmatrix} 0 & 1 \\ 1 & 0 \end{pmatrix} \in A, \quad \begin{pmatrix} 0 & 0 \\ 1 & 0 \end{pmatrix} \in M_2(R)$$

then
$$\begin{pmatrix} 0 & 0 \\ 1 & 0 \end{pmatrix}\begin{pmatrix} 0 & 1 \\ 0 & 0 \end{pmatrix} = \begin{pmatrix} 0 & 0 \\ 0 & 1 \end{pmatrix} \notin A$$

(b) the subring
$$B = \left\{ \begin{pmatrix} a & 0 \\ b & 0 \end{pmatrix} : a, b \in R \right\}$$

is a left ideal but not a right ideal.

Let
$$\begin{pmatrix} a & 0 \\ b & 0 \end{pmatrix} \in B \text{ and } \begin{pmatrix} x & y \\ c & d \end{pmatrix} \in M_2(R)$$

Then
$$\begin{pmatrix} x & y \\ c & d \end{pmatrix}\begin{pmatrix} a & 0 \\ b & 0 \end{pmatrix} = \begin{pmatrix} ax+by & 0 \\ ac+bd & 0 \end{pmatrix} \in B$$

But
$$\begin{pmatrix} 1 & 0 \\ 0 & 0 \end{pmatrix} \in B \text{ and } \begin{pmatrix} 1 & 1 \\ 0 & 0 \end{pmatrix} \in M_2(R)$$

then
$$\begin{pmatrix} 1 & 0 \\ 0 & 0 \end{pmatrix}\begin{pmatrix} 0 & 1 \\ 0 & 0 \end{pmatrix} = \begin{pmatrix} 0 & 1 \\ 0 & 0 \end{pmatrix} \notin B$$

(c) the subring
$$C = \left\{ \begin{pmatrix} a & 0 \\ 0 & 0 \end{pmatrix} : a \in R \right\}$$

is neither a left ideal nor a right ideal.
An ideal S of a ring R is a subring of R but every subring of R is not an ideal.

For example The set S of integers is a subring of the ring of rational numbers but S is not an ideal of R. Because the product of a rational and an integer is not always an integer, i.e.,

$$2 \in S, (1/6) \in R \Rightarrow 2 \times (1/6) = 1/3 \notin S$$

EXAMPLE 4.47 Let J and K be ideals in a ring. Show that $J \cap K$ is an ideal in R.

Solution Since J and K are ideals, $0 \in J$ and $0 \in K$. Hence, $0 \in J \cap K$. Now, let $a, b \in J \cap K$ and let $r \in R$. Then $a, b \in J$ and $a, b \in K$. Since J and K are ideals,

$$a - b, ra, ae \in J \quad \text{and} \quad a - b, ra, ar \in K$$

Hence, $a - b, ra, ar \in J \cap K$. Therefore, $J \cap K$ is an ideal.

EXAMPLE 4.48 Let J be an ideal in a ring with an identity element 1. Show that
(i) if $1 \in J$, then $J = R$
(ii) if any unit $u \in J$, then $J = R$

Solution

(i) If $1 \in J$, then for any $r \in R$ we have $r \cdot 1 \in J$. Hence $J = R$.
(ii) If $u \in J$, then $u^{-1} \cdot u \in J$, or $1 \in J$. Hence $J = R$, by part (i).

4.14.2 Quotient Ring

Let $(R, +)$ be a ring on an ideal I of R. A ring R is an additive abelian group and, hence, $(I, +)$ is a subgroup of the abelian group $(R, +)$. Again, every subgroup of an abelian group is normal. So, I is a normal subgroup of R and, hence, the quotient group $(R/I, +)$ is defined under normal addition of cosets, i.e., $(a + I) + (b + I) = a + b + I$ for all $a, b \in R$. In this group, the identity is the trivial coset I and the inverse of $a + I$ is $-a + I$ for all $a \in R$.

Given an ideal I of a ring $(R, +, \bullet)$, the normal addition and multiplication of cosets, namely

$$(a + I) + (b + I) = a + b + I \quad \text{and} \quad (a + I)(b + I) = ab + I \text{ for all } a, b \in R$$

make $(R/I, +, \bullet)$ a ring.

Given an ideal I of a ring, the ring R/I is so also.

EXAMPLE 4.49 Let I be an ideal in a commutative ring R. Then show that R/I is a ring.

Solution

$$(a + I)(b + I) = ab + aI + Ib + II = ab + I = ba + I = (b + I)(a + I)$$

$\Rightarrow$ R/I is a ring.

EXAMPLE 4.50 Let I be an ideal in a ring with unity element 1, and suppose that $1 \notin I$. Then show that $1 + I$ is a unity element for R/I.

Solution For any coset $a + I$, we have

$$(a + I)(1 + I) = a.1 + I = a + I$$

and

$$(1 + I)(a + I) = 1.a + I = a + I$$

Thus, $1 + I$ is a unity element in R/I.

4.14.3 Morphisms of Rings

A mapping of a ring R to a ring S is called a of *morphism of ring* if for any $a, b \in R$

$$f(ab) = f(a)f(b)$$

Again, a mapping from R to S is called *homomorphism of ring* or simply *homomorphism* if for any $a, b \in R$

$$f(a + b) = f(a) + f(b) \quad \text{and} \quad f(ab) = f(a)f(b)$$

A homomorphism $f: R \to S$ is called an *epimorphism* if f is onto. It is called a *monomorphism* if it is one-to-one and *isomorphism* if it is both one-to-one and onto. A homomorphism f of a ring R into itself is called an *endomorphism*. An endomorphism is called an *automorphism* if it is an isomorphism.

For example

(i) Let R be a ring and let $I: R \to S$ be the identity function. Then I is a homomorphism of ring which is one-to-one and onto, and hence is an isomorphism.
(ii) Let S be a subring of R. Define $f: R \to S$ by $f(x) = x$ for all $x \in S$. Then, for any $x, y \in S$,

$$f(x + y) = x + y = f(x) + f(y) \quad \text{and} \quad f(xy) = xy = f(x)f(y)$$

Thus, f is a homomorphism of ring from S to R.

4.14.4 Properties of Homomorphism of Ring

Let f be a homomorphism from a ring R to a ring S. Let A be a subring and B an ideal of S.

I. For any $r \in R$ and any positive integer n,

$$f(nr) = nf(r) \quad \text{and} \quad f(r^n) = (f(r))^n$$

II. $f(A) = \{f(a): a \in A\}$ is a subring of S.
III. If A is an ideal and f is onto on S, then $f(A)$ is an ideal.
IV. If f is an isomorphism from R onto S, then f^{-1} is an isomorphism from S onto R.

EXAMPLE 4.51 Consider two groups G and G' where $G = (Z, +)$ and $G' = \{Z^m: m = 0, \ldots, n\}$. Let $\phi: Z \to \{Z^m, m = 0, \ldots, n\}$ be defined by $\phi(m) = 2^m$, where $m \in Z$. Prove that ϕ is homomorphism.

Solution We have $\phi(m) = 2^m$, where $m \in Z$. Then,

$$\phi(m + r) = 2^{m+r} = 2^m \cdot 2^r = \phi(m) \cdot \varphi(r)$$

Hence, ϕ is homomorphism.

EXAMPLE 4.52 Let R and R' be the rings, defined by $R = 2Z$, multiples of 3. Prove that R is not isomorphic to R'.

Solution Let $f: R \to R'$ be a ring homomorphism. Then $f(2) = 3k$ for some integer k. Since f is a homomorphism,

$$f(4) = f(2+2) = f(2) + f(2) = 3k + 3k = 6k$$

Further,

$$f(4) = f(2 \times 2) = f(2) \times f(2) = (3k) \times (3k) = 9k^2$$

Thus, $9k^2 = 6k$ and, since k is an integral, $k = 0$. Hence, $f(2) = 0$. But $f(0) = 0$. Thus, f is not an isomorphism.

SUMMARY

The study of algebraic structures is important in computer science, not only because of the variety of applications of computing techniques that involve massive algebra, namely operational research, numerical methods, and relational databases, but also because the way a computer operates, or is formally described by means of programming languages, requires an algebraic perspective. In this chapter, we have investigated various kinds of algebraic systems which are of significance in their own right, and which are also needed later in the text. The main types of structure examined are groups, semigroups, permutation groups, rings, fields, and group codes.

EXERCISES

1. Let the binary operation '$*$' be defined by $x * y = x^y$ for all $x, y \in N$. Show that the operation '$*$' on N is neither commutative nor associative.
2. Investigate the following subsets of the positive integers N as a closed one under the operation of multiplication:
   (i) $A = \{0, 1\}$
   (ii) $B = \{1, 2\}$
   (iii) $C = \{x : x \text{ is prime}\}$
   (iv) $D = \{x : x \text{ is even}, 2, 4, 6, \ldots\}$
3. Investigate the binary operation $+$ defined on $(Z, +)$ as a group.
4. Consider $a \in R$ as a constant real number. Assume $G = \{a^n : n \in Z\}$. Prove that G is an abelian group under usual multiplication.
5. Show that the set of four transformations $f_1, f_2, f_3,$ and f_4 on the set of complex numbers be defined by $f_1(z) = z, f_2(z) = -z, f_3(z) = 1/z,$ and $f_4(z) = -1/z$ form a finite abelian group with respect to the binary operation $\circ$ as the composition of product of two functions.
6. Let G be the set of all 2×2 non-singular matrices over R and the operation $\bullet$ be the matrix multiplication. Show that $(G, \bullet)$ is a non-abelian group.
7. Assume $G = \{(a, b) : a, b \in R, a \neq 0\}$. Consider a binary operation '$*$', defined on $(G, *)$, by $(a, b) * (c, d) = (ac, bc + d)$ for all $(a, b), (c, d) \in G$. Show that $(G, *)$ is a group.
8. If every element of a group $(G, *)$ be its own inverse, then prove that it is an abelian group.
9. Let S be a semigroup with identity e, and let b and b' be inverses of a. Show that $b = b'$, i.e., the inverses are unique if they exist.
10. Find the order of every element in the multiplicative group $G = \{a, a^2, a^3, a^4, a^5, a^6 = e\}$.
11. Show that every subgroup of an abelian group is normal.
12. Let $S = \{1, 2, 3\}$. Also, assume

$$f = \begin{pmatrix} 1 & 2 & 3 \\ 2 & 3 & 1 \end{pmatrix} \text{ and } g = \begin{pmatrix} 1 & 2 & 3 \\ 3 & 2 & 1 \end{pmatrix}$$

as two permutations on S. Compute $f \circ g$.

13. If
$$f = \begin{pmatrix} 1 & 2 & 3 & 4 & 5 & 6 & 7 & 8 \\ 1 & 2 & 3 & 4 & 5 & 6 & 8 & 7 \end{pmatrix}$$
Compute f^2 and find its order.

14. Investigate the oddness and evenness of the following permutations:

(i) $\begin{pmatrix} 1 & 2 & 3 & 4 & 5 & 6 & 7 & 8 \\ 7 & 3 & 1 & 8 & 5 & 6 & 2 & 4 \end{pmatrix}$

(ii) $\begin{pmatrix} 1 & 2 & 3 & 4 & 5 & 6 & 7 & 8 \\ 1 & 4 & 3 & 6 & 5 & 8 & 7 & 2 \end{pmatrix}$

15. Investigate the following permutation as odd or even
$$p = \begin{pmatrix} a & b & c & d & e & f & g & h \\ b & c & a & e & f & d & h & g \end{pmatrix}$$

16. Express the permutation
$$\begin{pmatrix} 1 & 2 & 3 & 4 & 5 & 6 \\ 6 & 5 & 2 & 4 & 3 & 1 \end{pmatrix}$$
as a product of transposition.

17. Find the inverse of the permutation
$$\begin{pmatrix} 1 & 2 & 3 & 4 & 5 & 6 \\ 3 & 1 & 5 & 4 & 6 & 2 \end{pmatrix}$$

18. If for a group $G, f: G \to G$ is given by $f(x) = x^2$, $x \in G$ and f is homomorphism, show that G is abelian.

19. Find the generators and the number of generators of a cyclic group of order 8.

20. Let $(G, *)$ be a group and $(N, *)$ be a normal subgroup. Show that the system $(G/N, *)$ is also a group. (*known as quotient group or factor group*).

21. Let $(G, *)$ be a group of order 2, where $G = \{e, a\}$. Find $(G \times G, \circ)$, the direct product of $(G, *)$ with itself.

22. If G be an abelian group with identity e, then prove that all elements, x of G satisfying the equation $x^2 = e$, form a subgroup H of G.

23. Let G be $(Z, +)$, a group. Consider $H = \{3n : n \in Z\}$, prove that H is a subgroup of Z.

24. Consider the semigroup $(I, +)$. Let $f(x) = x^2 - 2x - 3$ and also let R be a relation on I defined by aRb iff $f(a) = f(b)$. Show that R is not a congruence relation.

25. Prove that every subgroup of a cyclic group G is cyclic.

26. Show that the mapping ϕ, defined as follows, is an automorphism; in which
   (i) G, the group of integers under addition and $\phi(x) = -x$.
   (ii) G, the group of positive real numbers under multiplication and $\phi(x) = x^2$.

27. Let $(R, +, \times)$ be a ring. The operation $\otimes$ is defined by $a \otimes b = a \times b + b \times a$. Show that $(R, +, \times)$ is a commutative ring.

28. Consider a rang $(R, +, *)$ defined by $a * a = a$. Investigate the ring as a commutative one.

5 Matrix Algebra

> **LEARNING OBJECTIVES**
>
> After reading this chapter you will be able to understand the following:
> - Description of several types of matrices from the grass roots of the topic
> - Operations on matrices with an exposure to algorithms in respect to matrix multiplication
> - Difference between matrix and determinant
> - A variety of typical square matrices such as orthogonal matrix, unitary matrix, and so on
> - The adjoint of a matrix, inverse of a matrix, and rank of a matrix
> - Boolean matrix or a zero-one matrix which are often used to represent discrete structures
> - Elementary row transformation on a matrix
> - Gaussian elimination method to find rank
> - Existence and uniqueness of solutions

5.1 INTRODUCTION

Matrix algebra plays an important and powerful role in quantitative analysis of management decisions in several disciplines such as production, marketing, finance, economics, computer science, discrete mathematics, network analysis, Markov models, input–output models, and some statistical models. All these models are built by establishing a system of linear equations. These equations represent the problems to be solved. Simultaneous linear equations involving more than three variables cannot be solved by using 'ordinary algebra'. Actually, real-world business problems involve more than three variables. In such cases, matrices are used to represent a complex system of equations and large quantities of data in a compact form. Once the system of equations is expressed in matrix form, they can be solved by using a computer.

Matrices are useful because they enable us to consider an array of numbers as a single object, denote it by a single symbol, and perform operations with these symbols in a precise form.

The definition of matrices and related concepts are discussed in this chapter, together with a discussion of basic algebraic operations for matrices, such as addition, multiplication, and scalar multiplication. We also define here special types of matrices and the operation of transposition of a matrix. The remaining part centres on linear system of equations. This includes inverse of a matrix, the important role of a rank and the basic existence and uniqueness problem for solutions.

5.2 DEFINITION OF A MATRIX

A rectangular array of entries arranged in m rows and n columns is called a *matrix* of order m by n, written as $m \times n$ matrix. Such array is enclosed between [ ] or ( ). A matrix is usually denoted by a boldface capital letter, or by a typical element enclosed within brackets, for example, **A** or $[a_{ij}]$ respectively. Here, a_{ij} represents the elements which lie in the ith row and the jth column of a matrix A. Matrix A can also be written as

$$A = [a_{ij}]_{m \times n}, \quad i \leq i \leq m, \quad i \leq j \leq n$$

i.e., A is of order $m \times n$ matrix.

In general, an $m \times n$ matrix A may be written as

$$A = \begin{pmatrix} a_{11} & a_{12} & a_{13} & \cdots & a_{1n} \\ a_{21} & a_{22} & a_{23} & \cdots & a_{2n} \\ \cdots & \cdots & \cdots & \cdots & \cdots \\ a_{i1} & a_{i2} & a_{i3} & \cdots & a_{in} \\ \cdots & \cdots & \cdots & \cdots & \cdots \\ a_{m1} & a_{m2} & a_{m3} & \cdots & a_{mn} \end{pmatrix} = [a_{ij}]_{m \times n}$$

Thus, the ith row consists of the entries

$$a_{i1}, a_{i2}, \ldots, a_{in}$$

while the jth column consists of the entries

$$a_{1j}, a_{2j}, \ldots, a_{mj}$$

Consider the matrix

$$A = \begin{pmatrix} 3 & 2 & 5 \\ 1 & 4 & 7 \end{pmatrix}$$

Here, the element in the 1st row and 2nd column is 2. So, one can write $a_{12} = 2$. Similarly, $a_{11} = 3$, $a_{12} = 2$, $a_{13} = 5$, $a_{21} = 1$, $a_{22} = 4$, and, $a_{23} = 7$.

Actually, a matrix of order $m \times n$ possesses mn elements. Such an array does not have a numerical value, it is to be regarded only as an ordered arrangement of elements. Thus, a matrix is a system of numbers without any numerical value. For example, 3 is a number whereas [3] is a matrix. In respect to the word *ordered*, the matrix [2 3 4] is different from the matrix [3 2 4].

5.3 TYPES OF MATRICES

Let

$$A = [a_{ij}]_{m \times n} \qquad (5.1)$$

be the general matrix.

5.3.1 Rectangular and Square Matrices

If $m \neq n$ in Eq. (5.1), then the matrix A is a *rectangular* matrix of order $m \times n$. Also, if $m = n$ in Eq. (5.1), then the matrix A is a *square* matrix of order n.

If $A = [a_{ij}]_{n \times n}$ (square matrix) then the diagonal consisting of the elements $a_{ij} \forall i = j$ ($\forall$ implies 'for all') is called *principal* or *leading diagonal*, and the elements are known as *principal diagonal elements*. The sum of principal diagonal elements of a square matrix is known as *trace* of the matrix. Thus,

$$\text{Trace of } A = \sum_{i=1}^{n} a_{ij} = a_{11} + a_{22} + \cdots + a_{nn}$$

For example

$$\begin{pmatrix} 1 & 2 & 3 \\ 3 & 6 & 7 \end{pmatrix} \text{ and } \begin{pmatrix} 3 & 2 & 5 & 1 \\ 1 & 4 & 7 & 4 \\ 3 & 1 & -2 & 6 \end{pmatrix}$$

are *rectangular* matrices of orders 2×3 and 3×4, respectively.
Also,

$$[5], \quad \begin{pmatrix} 2 & 3 \\ 5 & 0 \end{pmatrix}, \text{ and } \begin{pmatrix} 3 & 2 & 5 \\ 0 & 6 & -1 \\ 7 & 5 & 2 \end{pmatrix}$$

are square matrices of orders 1, 2, and 3, respectively. Also, for the square matrices of order 2 and 3, trace $= 2 + 0 = 2$ and $3 + 6 + 2 = 11$, respectively.

5.3.2 Row Matrix or a Row Vector

If $m = 1$ in Eq. (5.1) then the matrix A is a $1 \times n$ matrix (i.e., a matrix having only one row and any finite number of columns), called a *row matrix* or a *row vector*.

For example [1 5 3 4] is a row matrix of order 1×4 and [3] is a row matrix of order 1×1.

5.3.3 Column Matrix or a Column Vector

If $n = 1$ in Eq. (5.1) then the matrix A is an $m \times 1$ matrix (i.e., a matrix having only one column and any finite number of rows), called a *column matrix* or a *column vector*.

For example

$$\begin{pmatrix} 3 \\ 2 \\ 1 \end{pmatrix}$$

is a column matrix of order 3×1 and

$$\begin{pmatrix} 5 \\ -3 \\ 4 \\ 0 \end{pmatrix}$$

is a column matrix of order 4×1.

5.3.4 Zero or Null Matrix

If $a_{ij} = 0 \ \forall i, j$, i.e., a matrix, whose elements are all zero, is called a *zero matrix* or a *null matrix* of order $m \times n$ and is denoted by $0_{m \times n}$.

For example

$$[0 \quad 0]$$

is a zero matrix of order 1×2 and is written as $0_{1 \times 2}$.

$$\begin{pmatrix} 0 \\ 0 \end{pmatrix}$$

is a null matrix of order 2×1 and is represented as $0_{2 \times 1}$.

$$\begin{pmatrix} 0 & 0 & 0 & 0 \\ 0 & 0 & 0 & 0 \\ 0 & 0 & 0 & 0 \end{pmatrix}$$

is a null matrix of order 3×4 and is denoted by $0_{3 \times 4}$.

5.3.5 Diagonal Elements of a Matrix

The elements a_{ij} of the matrix A in Eq. (5.1), for which $i = j$, are called the *diagonal elements* of A.

For example The diagonal elements of $A = [a_{ij}]_{m \times n}$ are $a_{11}, a_{22}, a_{33}, \ldots, a_{nn}$. The line, along which the diagonal elements lie, is called the *diagonal* of the matrix.

5.3.6 Diagonal Matrix

If $m = n$, $a_{ij} = 0 \ \forall i \neq j$, and $a_{ij} \neq 0 \ \forall i = j$ in Eq. (5.1) (i.e., a square matrix having non-zero elements on the main diagonal), then the matrix A is called a *diagonal matrix* of order n. A diagonal matrix of order n with diagonal elements $a_{11}, a_{22}, \ldots, a_{nn}$ is denoted by $A = \text{diag} [a_{11}, a_{22}, \ldots, a_{nn}]$.

For example

$$\begin{pmatrix} 5 & 0 \\ 0 & 2 \end{pmatrix}$$

is a diagonal matrix of order 2×2 and is written as diag [5 2].
Also,

$$\begin{pmatrix} 3 & 0 & 0 \\ 0 & 2 & 0 \\ 0 & 0 & 6 \end{pmatrix}$$

is a diagonal matrix of order 3×3 and one can write it as diag [3 2 6].

5.3.7 Scalar Matrix

If $m = n$, $a_{ij} = 0 \: \forall \: i \neq j$, and $a_{ij} = k \: \forall \: i = j$ in Eq. (5.1), where k is some scalar (i.e., a square matrix in which every non-diagonal element is zero and all diagonal elements are equal), then the matrix A is called a *scalar matrix* of order n.

For example

$$\begin{pmatrix} 2 & 0 \\ 0 & 2 \end{pmatrix}$$

is a scalar matrix of order 2×2. Also,

$$\begin{pmatrix} 4 & 0 & 0 \\ 0 & 4 & 0 \\ 0 & 0 & 4 \end{pmatrix}$$

is a scalar of order 3.

5.3.8 Unit Matrix or Identity Matrix

If $m = n, a_{ij} = 0 \: \forall \: i \neq j$, and $a_{ij} = 1, \: \forall \: i = j$ in Eq. (5.1) (i.e., a square matrix in which every non-diagonal element is zero and every diagonal element is 1), then the matrix A is called a *unit matrix* or *identity matrix* of order n and is generally denoted by I_n.

For example

$$I_1 = [1], \quad I_2 = \begin{pmatrix} 1 & 0 \\ 0 & 1 \end{pmatrix}, \quad \text{and} \quad I_3 = \begin{pmatrix} 1 & 0 & 0 \\ 0 & 1 & 0 \\ 0 & 0 & 1 \end{pmatrix}$$

are unit or identity matrices of order 1, 2, and 3, respectively.

5.3.9 Comparable Matrices

Two matrices $A = [a_{ij}]$ and $B = [b_{ij}]$ are called comparable matrices if they are of the same order, i.e., they possess the same number of rows and columns. In other words, the matrices $[a_{ij}]_{m \times n}$ and $[b_{ij}]_{p \times q}$ are *comparable* if $m = p$ and $n = q$.

For example The matrices

$$\begin{pmatrix} 5 & 0 & -2 \\ 3 & 2 & -7 \end{pmatrix} \quad \text{and} \quad \begin{pmatrix} 4 & -3 & -6 \\ -1 & 0 & 4 \end{pmatrix}$$

are comparable because both are of the order 2×3.

5.3.10 Equal Matrices

Two matrices $A = [a_{ij}]$ and $B = [b_{ij}]$ are said to be *equal*, written as $A = B$, if they are of the same order and their corresponding elements are equal. In other words, the matrices $[a_{ij}]_{m \times n}$ and $[b_{ij}]_{p \times q}$ are equal if (i) $m = p$, (ii) $n = q$, and (iii) $a_{ij} = b_{ij} \: \forall \: i, j$.

For example

$$A = \begin{pmatrix} a_{11} & a_{12} \\ a_{21} & a_{22} \end{pmatrix} = B = \begin{pmatrix} 4 & 0 \\ 3 & -1 \end{pmatrix}$$

say, if and only if

$$a_{11} = 4, \quad a_{12} = 0$$
$$a_{21} = 3, \quad a_{22} = -1$$

Matrices of the same order that are all different (not equal) are, for instance,

$$\begin{pmatrix} 3 & \frac{1}{4} \\ 0 & -7 \end{pmatrix}, \begin{pmatrix} 3 & -7 \\ 0 & \frac{1}{4} \end{pmatrix}, \begin{pmatrix} 3 & \frac{1}{4} \\ 0 & 7 \end{pmatrix}, \begin{pmatrix} 3 & 1 \\ 0 & -7 \end{pmatrix}$$

5.3.11 Upper Triangular Matrix

If $a_{ij} = 0 \; \forall \, i > j$ in Eq. (5.1), (i.e., all elements below the main diagonal are zero), then the matrix A is said to be *upper triangular matrix*.

For example

$$A = \begin{pmatrix} 1 & 2 & 6 & 5 \\ 0 & 3 & 4 & 7 \\ 0 & 0 & 1 & 6 \\ 0 & 0 & 0 & 8 \end{pmatrix}$$

is an upper triangular matrix of order 4.

Note The matrix

$$\begin{pmatrix} 1 & 2 & 0 \\ 0 & 3 & 1 \\ 0 & 0 & 2 \end{pmatrix}$$

is an upper triangular matrix of order 3.

5.3.12 Lower Triangular Matrix

If $a_{ij} = 0 \; \forall \, i < j$ in Eq. (5.1), (i.e., all elements below the main diagonal are zero), then the matrix A is said to be *lower triangular matrix*.

For example

$$A = \begin{pmatrix} 1 & 0 & 0 \\ 4 & 2 & 0 \\ 5 & 3 & 6 \end{pmatrix}$$

is a lower triangular matrix of order 3.

Note The matrix

$$\begin{pmatrix} 1 & 0 & 0 \\ 2 & 1 & 0 \\ 0 & 2 & 3 \end{pmatrix}$$

is a lower triangular matrix of order 3.

Note

(i) Triangular matrix need not be a square.
(ii) A square matrix which is both upper and lower triangular is a diagonal matrix.
(iii) If $m = n, a_{ij} = 0 \,\forall\, i \le j$ in Eq. (5.1) then A is strictly lower triangular matrix and if $m = n, a_{ij} = 0 \,\forall\, i \ge j$ in Eq. (5.1) then A is strictly upper triangular matrix.

5.4 OPERATIONS ON MATRICES

There are several operations that can be performed on matrices. These are described below.

5.4.1 Addition of Matrices

If $A = [a_{ij}]_{m \times n}$ and $B = [b_{ij}]_{m \times n}$ are two matrices of same order, then their sum $A + B$ is a matrix of order $m \times n$, obtained by adding the corresponding elements of A and B.

Thus, $A + B = [a_{ij}]_{m \times n} + [b_{ij}]_{m \times n} = [a_{ij} + b_{ij}]_{m \times n}$.

Note For two matrices A and B, the sum $A + B$ exists only when A and B are comparable, that is having same order.

For example

(i) Let

$$A = \begin{pmatrix} 5 & 0 & -2 \\ 3 & 2 & -2 \end{pmatrix} \text{ and } B = \begin{pmatrix} 4 & -3 & -6 \\ -1 & 0 & 4 \end{pmatrix}$$

then A and B are matrices of same order. Hence, $A + B$ is defined and

$$A + B = \begin{bmatrix} 5+4 & 0-3 & -2-6 \\ 3-1 & 2+0 & -2+4 \end{bmatrix} = \begin{bmatrix} 9 & -3 & -8 \\ 2 & 2 & 2 \end{bmatrix}$$

(ii) Let

$$A = \begin{pmatrix} 1 & 2 \\ 4 & 0 \end{pmatrix} \text{ and } B = \begin{pmatrix} 2 & 3 & -6 \\ -4 & 0 & 9 \end{pmatrix}$$

Here, A and B are matrices of order 2×2 and 2×3, respectively, i.e., they are of different orders. Hence $A + B$ is not defined.

5.4.2 Subtraction of Matrices

If A and B are two matrices of the same order then their difference is given by $A - B = A + (-B)$, where the matrix $(-B)$ is the negative of the matrix B.

For example Consider

$$A = \begin{pmatrix} 2 & 4 \\ 3 & 2 \end{pmatrix}, \quad B = \begin{pmatrix} 1 & 3 \\ -2 & 5 \end{pmatrix}$$

and the negative of the matrix B as

$$(-B) = \begin{pmatrix} -1 & -3 \\ 2 & -5 \end{pmatrix}$$

then,

$$A - B = A + (-B) = \begin{pmatrix} 2 & 4 \\ 3 & 2 \end{pmatrix} + \begin{pmatrix} -1 & -3 \\ 2 & -5 \end{pmatrix}$$

$$= \begin{pmatrix} 2 + (-1) & 4 + (-3) \\ 3 + 2 & 2 + (-5) \end{pmatrix} = \begin{pmatrix} 1 & -3 \\ 5 & -3 \end{pmatrix}$$

5.4.3 Scalar Multiple of a Matrix

The product of an $m \times n$ matrix $A = [a_{ij}]$ and a scalar c (any number c) is the $m \times n$ matrix cA, i.e., $cA = [ca_{ij}]$ (obtained by multiplying each entry in A by c).

For example If

$$A = \begin{pmatrix} 2.7 & -1.8 \\ 0 & 0.9 \\ 9.0 & -4.5 \end{pmatrix}$$

then

$$-A = \begin{pmatrix} -2.7 & 1.8 \\ 0 & -0.9 \\ -9.0 & 4.5 \end{pmatrix}$$

Also,

$$\frac{10}{9} A = \begin{pmatrix} 3 & -2 \\ 0 & 1 \\ 10 & -5 \end{pmatrix}, \quad 0A = \begin{pmatrix} 0 & 0 \\ 0 & 0 \\ 0 & 0 \end{pmatrix}$$

5.4.4 Multiplication of Matrices

Let A be an $m \times n$ matrix and B an $n \times p$ matrix, i.e., the number of columns of A is equal to the number of rows of B. Then the multiplication of matrices is possible.

To obtain (i, j)th element of AB matrix, one should follow the steps given below:

Step I Consider each entry in the ith row of A and in the jth column of B.
Step II Multiply the entries element-wise.
Step III Add all these n products.

Thus, if

$$A = [a_{ik}]_{m \times n}, \quad 1 \leq i \leq m, \quad 1 \leq k \leq n$$

and
$$B = [b_{kj}]_{n \times p}, \quad 1 \leq k \leq n, \quad 1 \leq j \leq p$$
then the jth row of A is $[a_{i1}, a_{i2}, \ldots, a_{in}]$ and the jth column of B is

$$\begin{pmatrix} b_{ij} \\ b_{2j} \\ \vdots \\ \vdots \\ b_{nj} \end{pmatrix}$$

Thus, the product AB is given by
$$AB = [c_{ij}]_{m \times p}, \text{ where } c_{ij} = a_{i1} \cdot b_{1j} + a_{i2} b_{2j} + \cdots + a_{in} b_{nj}$$
$$= \sum_{k=1}^{n} a_{ik} b_{kj}, \quad 1 \leq i \leq m, \quad 1 \leq j \leq p$$

Note

(i) Two matrices A and B are said to be conformable for product AB if the number of columns in A is same as the number of rows in B.
(ii) If A and B are two matrices such that the product AB exists, then the product BA may or may not exist.
(iii) If the product AB exists, then the matrix A is known as the *pre-factor* (pre-multiplier) and the matrix B is known as the *post-factor* (post-multiplier).

For example Let

$$A = \begin{pmatrix} a_{11} & a_{12} & a_{13} \\ a_{21} & a_{22} & a_{23} \end{pmatrix} \text{ and } B = \begin{pmatrix} b_{11} \\ b_{21} \\ b_{32} \end{pmatrix}$$

Here, A is a 2×3 matrix and B is a 3×1 matrix, i.e., the number of columns in A equals to the number of rows in B. So, AB is defined and it is a 2×1 matrix, i.e.,

$$AB = \begin{pmatrix} a_{11} & a_{12} & a_{13} \\ a_{21} & a_{22} & a_{23} \end{pmatrix}_{2 \times 3} \begin{pmatrix} b_{11} \\ b_{21} \\ b_{32} \end{pmatrix}_{3 \times 1} = \begin{pmatrix} a_{11}b_{11} + a_{12}b_{21} + a_{13}b_{31} \\ a_{21}b_{11} + a_{22}b_{21} + a_{23}b_{31} \end{pmatrix}_{2 \times 1}$$

Here, BA is not defined, because the number of column in B is not the same as the number of rows in A.

5.4.5 Properties of Matrix Multiplication

I. **Matrix multiplication is not commutative:** Assume $A = [a_{ij}]_{m \times n}$ and $B = [b_{ij}]_{n \times p}$. Then only the product AB exists, however BA does not. Again, suppose $A = [a_{ij}]_{m \times n}$ and $B = [b_{ij}]_{n \times m}$. Then both AB and BA exist, but not equal, since they are of different order (size), such as $m \times n$ and $n \times m$. Even when both AB and BA exist and are matrices of same order, it is not necessary that $AB = BA$.

The necessary condition for commutative law to hold good is that both AB and BA should exist and are to be of the same order. However, these conditions are not sufficient for the commutative law.

For example

(i) If A is a 2×3 matrix and B is a 3×2 matrix, then AB as well as BA exists. But AB is a 2×2 matrix while BA is a 3×3 matrix. Further, if AB and BA both exist and their orders are same and as such they may not be equal.

(ii) If
$$A = \begin{pmatrix} 1 & 0 \\ 0 & -1 \end{pmatrix} \text{ and } B = \begin{pmatrix} 0 & 1 \\ 1 & 0 \end{pmatrix}$$

then both AB and BA are defined and each one is a 2×2 matrix.

But,
$$AB = \begin{pmatrix} 0 & 1 \\ -1 & 0 \end{pmatrix} \text{ and } BA = \begin{pmatrix} 0 & -1 \\ 1 & 0 \end{pmatrix} \Rightarrow AB \neq BA$$

(iii) Consider
$$A = \begin{pmatrix} 4 & 3 \\ 7 & 2 \\ 9 & 0 \end{pmatrix} \text{ and } B = \begin{pmatrix} 2 & 5 \\ 1 & 6 \end{pmatrix}$$

$$AB = \begin{pmatrix} 4.2 + 3.1 & 4.5 + 3.6 \\ 7.2 + 2.1 & 7.5 + 2.6 \\ 9.2 + 0.1 & 9.5 + 0.6 \end{pmatrix} = \begin{pmatrix} 11 & 38 \\ 16 & 47 \\ 18 & 45 \end{pmatrix}$$

Here, A is 3×2 and B is 2×2, so that AB yields 3×2, whereas BA is not defined.

(iv) Multiplication of a matrix and a column vector
$$\begin{pmatrix} 4 & 2 \\ 1 & 8 \end{pmatrix} \begin{pmatrix} 3 \\ 5 \end{pmatrix} = \begin{pmatrix} 12 + 10 \\ 3 + 40 \end{pmatrix} = \begin{pmatrix} 22 \\ 43 \end{pmatrix}$$

whereas
$$\begin{pmatrix} 3 \\ 5 \end{pmatrix} \begin{pmatrix} 4 & 2 \\ 1 & 8 \end{pmatrix}$$

is undefined.

(v) Product of row and column vectors
$$[3 \ 6 \ 1] \begin{pmatrix} 1 \\ 2 \\ 4 \end{pmatrix} = [19], \quad \begin{pmatrix} 1 \\ 2 \\ 4 \end{pmatrix} [3 \ 6 \ 1] = \begin{pmatrix} 3 & 6 & 1 \\ 6 & 12 & 2 \\ 12 & 24 & 4 \end{pmatrix}$$

(vi) Matrix multiplication is not commutative for square matrices also
$$\begin{pmatrix} 9 & 3 \\ -2 & 0 \end{pmatrix} \begin{pmatrix} 1 & -4 \\ 2 & 5 \end{pmatrix} = \begin{pmatrix} 9.1 + 3.2 & 9.(-4) + 3.5 \\ -2.1 + 0.2 & (-2).(-4) + 0.5 \end{pmatrix} = \begin{pmatrix} 15 & -21 \\ -2 & 8 \end{pmatrix}$$

whereas
$$\begin{pmatrix} 1 & -4 \\ 2 & 5 \end{pmatrix} \begin{pmatrix} 9 & 3 \\ -2 & 0 \end{pmatrix} = \begin{pmatrix} 1.9 + (-4).(-2) & 1.3 + (-4).0 \\ 2.9 + 5.(-2) & 2.3 + 5.0 \end{pmatrix} = \begin{pmatrix} 17 & 3 \\ 8 & 6 \end{pmatrix}$$

Thus,
$$\begin{pmatrix} 9 & 3 \\ -2 & 0 \end{pmatrix} \begin{pmatrix} 1 & -4 \\ 2 & 5 \end{pmatrix} \neq \begin{pmatrix} 1 & -4 \\ 2 & 5 \end{pmatrix} \begin{pmatrix} 9 & 3 \\ -2 & 0 \end{pmatrix}$$

Note

(i) If $AB = -BA$, then the matrices A and B are said to be anti-commute.

(ii) The product of two diagonal matrices of order n is another diagonal matrix of the same order and commute always.

II. Matrix multiplication is associative: If A, B, and C are any three matrices conformable for multiplication, then $(AB)C = A(BC)$.

Proof Let
$$A = [a_{ij}]_{m \times n}, \quad B = [b_{ij}]_{n \times p}, \quad \text{and} \quad C = [c_{ij}]_{p \times q}$$

then,
$$(AB)C = \left[\left(\sum_{k=1}^{n} a_{ik} \sum_{s=1}^{n} b_{ks}\right) \sum_{s=1}^{n} c_{sj}\right]_{m \times q} = \left[\sum_{k=1}^{n} a_{ik}\left(\sum_{s=1}^{n} b_{ks} c_{sj}\right)\right]_{m \times q} = A(BC)$$

Hence, $(AB)C = A(BC)$.

III. Matrix multiplication is distributive with respect to addition of matrices:

(a) If A, B, and C are matrices of the type $m \times n$, $n \times p$, and $n \times p$, respectively, then $A(B + C) = AB + AC$.

(b) If A, B, and C are matrices of the type $m \times n$, $m \times n$, and $n \times p$, respectively, then $(A + B)C = AC + BC$.

Proof Let
$$A = [a_{ij}]_{m \times n}, \quad B = [b_{jk}]_{n \times p}, \quad \text{and} \quad C = [c_{jk}]_{n \times p}$$

where $i = 1, 2, 3, \ldots, m$; $j = 1, 2, 3, \ldots, n$; and $k = 1, 2, 3, \ldots, p$. Since A is an $m \times n$ matrix and $B + C$ is an $n \times p$ matrix; so, $A(B + C)$ is an $m \times p$ matrix. Again, since each one of AB and AC is an $m \times p$ matrix, so $(AB + BC)$ is also an $m \times p$ matrix. Thus, $A(B + C)$ and $(AB + AC)$ are comparable matrices. Now, (i, k)th element of $A(B + C)$

$$= \sum_{j=1}^{n} a_{ij}(b_{jk} + c_{jk}) = \sum_{j=1}^{n} (a_{ij} b_{jk} + a_{ij} c_{jk})$$

$$= \sum_{j=1}^{n} a_{ij} b_{jk} + \sum_{j=1}^{n} a_{ij} c_{jk}$$

$= (i, k)$th element of $AB + (i, k)$th element of AC

$= (i, k)$th element of $(AB + AC)$

$\therefore A(B + C) = AB + AC$

Similarly, the result $(A + B)C = (AC + BC)$ may be proved.

IV. **The product of two non-zero matrices can be a zero matrix:** If $AB = 0$, it does not imply that either A or $B = 0$, even if $A \neq 0$ and $B \neq 0$, still AB may be a null matrix.

For example

(i) Let

$$A = \begin{pmatrix} 3 & 2 & 2 \\ 5 & -2 & 6 \end{pmatrix} \text{ and } B = \begin{pmatrix} 4 & -6 \\ -2 & 3 \\ -4 & 6 \end{pmatrix}$$

Then

$$AB = \begin{pmatrix} 3 & 2 & 2 \\ 5 & -2 & 6 \end{pmatrix} \begin{pmatrix} 4 & -6 \\ -2 & 3 \\ -4 & 6 \end{pmatrix} = \begin{pmatrix} 12 - 4 - 8 & -18 + 6 + 12 \\ 20 + 4 - 24 & -30 - 6 + 36 \end{pmatrix}$$

$$= \begin{pmatrix} 0 & 0 \\ 0 & 0 \end{pmatrix}$$

Thus, $AB = 0$, neither $A = 0$ nor $B = 0$.

(ii) Consider

$$A = \begin{pmatrix} 1 & -1 \\ -1 & 1 \end{pmatrix} \text{ and } B = \begin{pmatrix} 2 & 2 \\ 2 & 2 \end{pmatrix}$$

Clearly, $A \neq 0$ and $B \neq 0$
But,

$$AB = \begin{pmatrix} 1 & -1 \\ -1 & 1 \end{pmatrix} \begin{pmatrix} 2 & 2 \\ 2 & 2 \end{pmatrix} = \begin{pmatrix} 0 & 0 \\ 0 & 0 \end{pmatrix} = 0$$

V. **Multiplication of a matrix by a unit matrix (i.e., existence of identity):** If A is a matrix of order $m \times n$, then there exists unit matrices of order m and n, such that $AI_n = A = I_m A$. Here, I_n is called *right multiplicative identity* and I_m is called *left multiplicative identity*. If A is a square matrix of order n, then there exists unit matrix of order n such that $IA = A = AI$ and I is called *multiplicative identity*. It illustrates the existence of identity.

For example If

$$A = \begin{pmatrix} 3 & -1 \\ 2 & -2 \end{pmatrix} \text{ and } I = \begin{pmatrix} 1 & 0 \\ 0 & 1 \end{pmatrix}$$

then

$$AI = \begin{pmatrix} 3 & -1 \\ 2 & -2 \end{pmatrix} \begin{pmatrix} 1 & 0 \\ 0 & 1 \end{pmatrix} = \begin{pmatrix} 3 & -1 \\ 2 & -1 \end{pmatrix} = A$$

Similarly, $IA = A$ may be shown.

VI. **Cancellation law in matrix multiplication:** Cancellation law does not hold well in matrix multiplication. If A, B, and C are three matrices such that AB and AC are defined, then $AB = AC$ does not imply $B = C$.

For example If

$$A = \begin{pmatrix} 0 & 4 \\ 0 & 0 \end{pmatrix}, \quad B = \begin{pmatrix} 5 & 0 \\ 0 & 0 \end{pmatrix}, \quad \text{and} \quad C = \begin{pmatrix} 6 & 0 \\ 0 & 0 \end{pmatrix}$$

Then,

$$AB = \begin{pmatrix} 0 & 0 \\ 0 & 0 \end{pmatrix} \quad \text{and} \quad AC = \begin{pmatrix} 0 & 0 \\ 0 & 0 \end{pmatrix}$$

i.e., $AB = AC$ but $B \neq C$.

Here, we will just construct a model of algorithm for matrix multiplication.

The definition of the product of two matrices leads to an algorithm that computes the product of two matrices. Suppose that $C = [c_{ij}]$ is the $m \times n$ matrix that is the product of the $m \times k$ matrix $A = [a_{ij}]$ and the $k \times n$ matrix $B = [b_{ij}]$. The algorithm is presented below.

ALGORITHM Matrix Multiplication
procedure matrix multiplication (A, B: matrices)
for $i := 1$ to m
for $j := 1$ to n
begin
    $c_{ij} := 0$
    for $q := 1$ to k
        $c_{ij} := c_{ij} + a_{iq}b_{qj}$
end
{$C = [c_{ij}]$ is the product of A and B}

5.4.6 Positive Integral Powers of Matrices

Let A be an $m \times n$ matrix. The product AA (by definition of matrix multiplication) is defined only when $n = m$, i.e., AA is defined, if A is a square matrix. The product AA is written as A^2. In other words, if A is an n-rowed square matrix, then write $A^2 = A \cdot A$, $A^3 = A \cdot A \cdot A = A^2 \cdot A$, and so on.

In general, $A^n = A \cdot A \cdot A \cdot \cdots n$ times, $n \in N$.

Note The above operation can be stated as the multiplication of a matrix by itself.

5.4.7 Sub-Matrix

The matrix obtained on deleting any number of rows and columns of the given matrix A is called the *sub-matrix* of A.

For example If

$$A = \begin{bmatrix} 1 & 2 & 3 & 4 \\ 7 & 2 & 1 & 0 \end{bmatrix}$$

then
$$B = \begin{bmatrix} 1 & 3 \\ 7 & 1 \end{bmatrix}$$
and
$$C = \begin{bmatrix} 2 & 3 & 4 \\ 2 & 1 & 0 \end{bmatrix}$$
are sub-matrices of A.

Note The determinants of square sub-matrices are called minors of the matrix.

(Application of matrix multiplication to allied problems is given below.)

EXAMPLE 5.1 A manufacturer is investigating which of the three methods he should use in producing three goods A, B, and C. The amount of each item produced by each method is shown in the matrix:

$$\begin{array}{c} \\ \text{Method I} \\ \text{Method II} \\ \text{Method III} \end{array} \begin{array}{ccc} A & B & C \end{array} \\ \begin{bmatrix} 4 & 8 & 2 \\ 5 & 7 & 1 \\ 5 & 3 & 9 \end{bmatrix}$$

The vector [10 4 6] represents the profit per unit for goods A, B, and C in that order. Using matrix multiplication, find that method which maximizes the total profit.

Solution Let

$$X = \begin{bmatrix} 4 & 8 & 2 \\ 5 & 7 & 1 \\ 5 & 3 & 9 \end{bmatrix} \begin{array}{l} \text{Method I} \\ \text{Method II} \\ \text{Method III} \end{array} \quad Y = \begin{bmatrix} 10 \\ 4 \\ 6 \end{bmatrix} \begin{array}{l} A \\ B \\ C \end{array}$$

Total profits under the three methods are given by

$$XY = \begin{array}{l} \text{Method I} \\ \text{Method II} \\ \text{Method III} \end{array} \begin{bmatrix} 4 & 8 & 2 \\ 5 & 7 & 1 \\ 5 & 3 & 9 \end{bmatrix} \times \begin{bmatrix} 10 \\ 4 \\ 6 \end{bmatrix} \begin{array}{l} A \\ B \\ C \end{array}$$

$$\text{Profits} = \begin{bmatrix} 84 \\ 84 \\ 116 \end{bmatrix} \begin{array}{l} \text{Method I} \\ \text{Method II} \\ \text{Method III} \end{array}$$

Thus, method III maximizes the total profit.

EXAMPLE 5.2 There are two families X and Y. In family X, there are 2 men, 3 women, and 1 child and 1 man, 1 woman, and 2 children are in family Y. The recommended daily requirement of calories is men: 2400, woman: 1900, children: 1800 and for proteins is men: 55 g, woman: 45 g, children: 33 g. Represent the above information by matrices. Using matrix multiplication, calculate the total requirements of calories and proteins for each of the two families.

Solution Let men, women, and child be denoted by M, W, and C, respectively. The family matrix is

$$F = \begin{matrix} & M & W & C & \\ & \begin{bmatrix} 2 & 3 & 1 \\ 1 & 1 & 2 \end{bmatrix} & \begin{matrix} X \\ Y \end{matrix} \end{matrix}$$

The matrix indicating the daily requirements of calories and proteins for different members of the family is

$$R = \begin{bmatrix} 2400 & 55 \\ 1900 & 45 \\ 1800 & 33 \end{bmatrix} \begin{matrix} M \\ W \\ C \end{matrix}$$

Now, the total requirement of calories and proteins for each of the two families is given by the matrix product:

$$FR = \begin{bmatrix} 2 & 3 & 1 \\ 1 & 1 & 2 \end{bmatrix} \times \begin{bmatrix} 2400 & 55 \\ 1900 & 45 \\ 1800 & 33 \end{bmatrix}$$

$$= \begin{bmatrix} 4800 + 5700 + 1800 & 110 + 135 + 33 \\ 2400 + 1900 + 3600 & 55 + 45 + 66 \end{bmatrix}$$

$$= \begin{bmatrix} 12300 & 278 \\ 7900 & 166 \end{bmatrix} \begin{matrix} X \\ Y \end{matrix}$$

Thus, family X requires 12,300 calories and 278 g proteins and family Y requires 7900 calories and 166 g proteins.

5.4.8 Partition of Matrices

If the matrices are sub-divided into rectangular blocks by drawing lines parallel to the rows and columns of the matrix, then the process of sub-division is called the *partitioning of matrices*.

For example

$$A = \begin{bmatrix} 1 & 2 & 3 \\ 4 & 5 & 6 \\ 7 & 8 & 9 \end{bmatrix} \quad \text{and} \quad B = \begin{bmatrix} 10 & 11 & 12 \\ 13 & 14 & 15 \\ 16 & 17 & 18 \end{bmatrix}$$

5.5 RELATED MATRICES

5.5.1 Transpose of a Matrix

If A is an $m \times n$ matrix, then the matrix obtained by interchanging the rows and columns of A is called the *transpose* of A. Transpose of the matrix A is denoted by A^T or A'. Thus, if $A = [a_{ij}]_{m \times n}$, then $A^T = [a_{ij}]_{n \times m}$.

For example

$$A = \begin{pmatrix} a_{11} & a_{12} & a_{13} \\ a_{21} & a_{22} & a_{23} \\ a_{31} & a_{32} & a_{33} \end{pmatrix}, \text{ then } A^T = \begin{pmatrix} a_{11} & a_{21} & a_{31} \\ a_{12} & a_{22} & a_{32} \\ a_{13} & a_{32} & a_{33} \end{pmatrix}$$

Also, $(A^T)^T = A$, i.e., the transpose of A^T coincides with A.

5.5.2 Symmetric and Skew-Symmetric Matrices

A square matrix A is said to be *symmetric* if $A^T = A$ and *skew-symmetric* if $A^T = -A$ (or, $A = -A^T$). Thus, a square matrix $A = [a_{ij}]_{n \times n}$ is said to be symmetric if $a_{ij} = a_{ij} \ \forall \ i, j$ and skew-symmetric if $a_{ij} = -a_{ji} \ \forall \ i, j$.

For example The matrix

$$A = \begin{pmatrix} a & b & c \\ b & d & e \\ c & e & f \end{pmatrix}$$

is symmetric matrix as $A^T = A$. The matrix

$$\begin{pmatrix} 0 & a & b \\ -a & 0 & c \\ -b & -c & 0 \end{pmatrix}$$

is skew-symmetric since $A^T = -A$.

Note If A is symmetric matrix of order n, then the number of independent elements $= \frac{1}{2}n(n+1)$. Again, if A is a skew-symmetric matrix of order n, then the number of independent elements $= \frac{1}{2}n(n-1)$.

5.5.3 Complex Matrix

If each or a few elements of a matrix be complex numbers, then the matrix is called a *complex matrix*. A complex matrix A can be expressed in the form $P + iQ$, where P and Q are real matrices.

For example

$$A = \begin{pmatrix} 4+5i & 2 \\ 1-i & 1 \end{pmatrix} = \begin{pmatrix} 4 & 2 \\ 1 & 0 \end{pmatrix} + i\begin{pmatrix} 5 & 0 \\ -1 & 2 \end{pmatrix} \equiv P + iQ$$

is a complex matrix.

5.5.4 Conjugate of a Matrix

If $A = [a_{ij}]_{m \times n}$, then the matrix obtained by replacing each element of A by its complex conjugate is called the *conjugate matrix* of A and is denoted by $\bar{A}$. Thus, $A = [a_{ij}]_{m \times n}$ gives $\bar{A} = [\bar{a}_{ij}]_{m \times n}$, where $\bar{a}_{ij}$ is the complex conjugate of a_{ij}.

For example If

$$A = \begin{pmatrix} 1+i & 2 & 1-i \\ 2-i & -1-i & -2 \end{pmatrix} \text{ then } \overline{A} = \begin{pmatrix} 1-i & 2 & 1+i \\ 1+i & -1+i & -2 \end{pmatrix}$$

Also, if $\overline{(\overline{A})} = A$, i.e., the conjugate of $\overline{A}$ coincides with A.

5.5.5 Conjugate Transpose of a Matrix

The conjugate of the transpose of a matrix A is called the *conjugate transpose* of A and is denoted by A^*. Thus, if $A = [a_{ij}]_{m \times n}$ then $A^* = \overline{(A^T)}$.

For example If

$$A = \begin{pmatrix} 1-i & 2+i \\ 1+i & 2-i \end{pmatrix},$$

then

$$A^T = \begin{pmatrix} 1-i & 1+i \\ 2+i & 2-i \end{pmatrix}$$

Consequently,

$$\overline{(A^T)} = \begin{pmatrix} 1+i & 1-i \\ 2-i & 2+i \end{pmatrix} = A^*$$

Similarly, it can be shown that

$$(\overline{A})^T = \begin{pmatrix} 1+i & 1-i \\ 2-i & 2+i \end{pmatrix} = A^*$$

Thus,

$$\overline{(A^T)} = (\overline{A})^T$$

i.e., the conjugate of the transpose of A coincides with the transpose of the conjugate of A. Also, the conjugate transpose of a matrix A coincides with itself, i.e.,

$$(A^*)^* = A$$

5.5.6 Hermitian and Skew-Hermitian Matrices

A square matrix $A = [a_{ij}]_{m \times n}$ is said to be *Hermitian* if $a_{ij} = \overline{a}_{ji} \, \forall \, i, j$ and is said to be *skew-Hermitian* if $a_{ij} = -\overline{a}_{ji}$. Thus, a square matrix A is Hermitian if $A^* = A$ and skew-Hermitian if $A^* = -A$ (or, $A = -A^*$). It may be noted here that the principal diagonal elements of a Hermitian matrix are purely real (since when $j = i, a_{ii} = \overline{a}_{ii} \Rightarrow a_{ii}$ is real and the principal diagonal elements of skew-Hermitian are purely imaginary or zero).

For example

$$\begin{pmatrix} 2 & 3-i & 4+3i \\ 3+i & 5 & 6i \\ 4-3i & -6i & 1 \end{pmatrix} \text{ and } \begin{pmatrix} -2i & 4+3i & -3-i \\ -4+3i & 0 & 6i \\ -3+i & -6i & 4i \end{pmatrix}$$

are Hermitian and skew-Hermitian matrices.

Corollary 5.1 Every square matrix can be uniquely expressed as the sum of a Hermitian matrix and a skew-Hermitian matrix.

Proof Let A be the given square matrix, then

$$A = \frac{1}{2}(A + A^*) + \frac{1}{2}(A - A^*)$$

Let

$$B = \frac{1}{2}(A + A^*) \text{ and } C = \frac{1}{2}(A - A^*)$$

Now,

$$B^* = \left[\frac{1}{2}(A + A^*)\right]^* = \frac{1}{2}(A + A^*) = \frac{1}{2}\left[A^* + (A^*)^*\right]$$

$$= \frac{1}{2}\left[A^* + A\right] = B$$

$$\Rightarrow B = \frac{1}{2}(A + A^*)$$

is a Hermitian matrix. Again,

$$C^* = \left[\frac{1}{2}(A - A^*)\right]^* = \frac{1}{2}\left[A^* - (A^*)^*\right] = \frac{1}{2}\left[A^* - A\right]$$

$$= \frac{1}{2}(A - A^*) = -C$$

$$\Rightarrow C = \frac{1}{2}(A - A^*)$$

is a skew-Hermitian matrix.

Hence, A can be expressed as the sum of a Hermitian and a skew-Hermitian.

To prove the uniqueness, assume that P is a Hermitian matrix and Q is a skew-Hermitian matrix such that

$$A = P + Q$$

Now,

$$A^* = (P + Q)^* = P^* + Q^* = P - Q$$

Thus,

$$P = \frac{1}{2}(A + A^*) \text{ and } Q = \frac{1}{2}(A - A^*)$$

which shows that there exists one and only way of expressing A as the sum of a Hermitian and skew-Hermitian matrix.

EXAMPLE 5.3 If both A and B are skew-symmetric matrices of the same order such that $AB = BA$, then show that AB is symmetric.

Solution If both A and B are skew-symmetric matrices, then
$$A = -A^T \text{ and } B = -B^T \quad (5.2)$$

Also, given that $AB = BA = (-B^T)(-A^T)$, from Eq. (5.2), i.e., $B^T A^T = (AB)^T$ or, $AB = (AB)^T$, i.e., AB is a symmetric matrix. Hence, proved.

EXAMPLE 5.4 If A is any square matrix, show that AA^T is symmetric matrix.

Solution We have $(AA^T)^T = $ transpose of $AA^T = (A^T)^T A^T = AA^T$, i.e.,

$$AA^T = (AA^T)^T$$

Hence, AA^T is a symmetric matrix, by definition.

EXAMPLE 5.5 If A and B are symmetric matrices, prove that (BAB) is also symmetric.

Solution Given that A and B are symmetric
$$\Rightarrow A^T = A \text{ and } B^T = B$$

Now,
$$(BAB)^T = B^T A^T B^T = BAB$$

Thus, (BAB) is symmetric.

EXAMPLE 5.6 Show that a Hermitian matrix is normal.

Solution Let A be a Hermitian matrix. Then, $A = A^*$. Thus, $AA^* = AA = A^*A$. So, A is normal. For the definition of normal matrices refer to Section 5.12.2.

EXAMPLE 5.7 If

$$A = \begin{bmatrix} 1 & 2 & -1 \\ 3 & 0 & 2 \\ 4 & 5 & 0 \end{bmatrix} \text{ and } B = \begin{bmatrix} 1 & 0 & 0 \\ 2 & 1 & 0 \\ 0 & 1 & 3 \end{bmatrix}$$

show that

$$(AB)^T = B^T A^T$$

Solution Here

$$AB = \begin{bmatrix} 1 & 2 & -1 \\ 3 & 0 & 2 \\ 4 & 5 & 0 \end{bmatrix} \begin{bmatrix} 1 & 0 & 0 \\ 2 & 1 & 0 \\ 0 & 1 & 3 \end{bmatrix}$$

i.e.,

$$AB = \begin{bmatrix} 1(1)+2(2)-1(0) & 1(0)+2(1)-1(1) & 1(0)+2(0)-1(3) \\ 3(1)+0(2)+2(0) & 3(0)+0(1)+2(1) & 3(0)+0(0)+2(3) \\ 4(1)+5(2)+0(0) & 4(0)+5(1)+0(1) & 4(0)+5(0)+0(3) \end{bmatrix} = \begin{bmatrix} 5 & 1 & -3 \\ 3 & 2 & 6 \\ 14 & 5 & 0 \end{bmatrix}$$

and

$$(AB)^T = \begin{bmatrix} 5 & 3 & 14 \\ 1 & 2 & 5 \\ -3 & 6 & 0 \end{bmatrix} \tag{5.3}$$

Now,

$$B^T = \begin{bmatrix} 1 & 2 & 0 \\ 0 & 1 & 1 \\ 0 & 0 & 3 \end{bmatrix} \text{ and } A^T = \begin{bmatrix} 1 & 3 & 4 \\ 2 & 0 & 5 \\ -1 & 2 & 0 \end{bmatrix}$$

Hence,

$$B^T A^T = \begin{bmatrix} 1 & 2 & 0 \\ 0 & 1 & 1 \\ 0 & 0 & 3 \end{bmatrix} \begin{bmatrix} 1 & 3 & 4 \\ 2 & 0 & 5 \\ -1 & 2 & 0 \end{bmatrix}$$

$$= \begin{bmatrix} 1(1)+2(2)+0(-1) & 1(3)+2(0)+0(2) & 1(4)+2(5)+0(0) \\ 0(1)+1(2)+1(-1) & 0(3)+1(0)+1(2) & 0(4)+1(5)+1(0) \\ 0(1)+0(2)+3(-1) & 0(3)+0(0)+3(2) & 0(4)+0(5)+3(0) \end{bmatrix}$$

i.e.,

$$B^T A^T = \begin{bmatrix} 5 & 3 & 14 \\ 1 & 2 & 5 \\ -3 & 6 & 0 \end{bmatrix} \tag{5.4}$$

From Eqs (5.3) and (5.4), $(AB)^T = B^T A^T$.

5.6 DETERMINANT OF A MATRIX

Let A be a square matrix. A determinant exists for a square matrix only. It yields a pure number. *Determinant of a square matrix A* may be denoted by det A or $|A|$ or Δ. It plays an important role in solving a system of linear equations.

Let us now define the determinant of matrices (square) of different order as follows.
Determinant of square matrix A of order 1, i.e., $A = [a]$ can be expressed as

$$\det A = |A| = a$$

Define the determinant of a 2 × 2 matrix (i.e., of order 2)

$$A = \begin{pmatrix} a_{11} & a_{12} \\ a_{21} & a_{22} \end{pmatrix}$$

by

$$|A| = a_{11} a_{22} - a_{21} a_{12}$$

It may also be written as
$$\begin{vmatrix} a_{11} & a_{12} \\ a_{21} & a_{22} \end{vmatrix} = a_{11}a_{22} - a_{21}a_{12}$$

However, for defining the determinant of a matrix of order 3 (or determinant of order 3), one should first define minors and co-factors of elements of the determinant.

5.6.1 Minor and Co-Factor

Let $A = [a_{ij}]_{m \times n}$. The *minor* of an element a_{ij} of determinant of a matrix A is the determinant formed by suppressing the ith row and the jth column in which the element a_{ij} exists. The minor of the element a_{ij} is denoted by M_{ij}. It may be noted that the minor of an element of a determinant of order n is a determinant of order $(n-1)$. The *co-factor* of an element a_{ij} denoted by A_{ij} is defined as $A_{ij} = (-1)^{i+j} M_{ij}$.

For example The minor and co-factor of the elements a_{11}, a_{22}, and a_{32} of the determinant,

$$\Delta = \begin{vmatrix} a_{11} & a_{12} & a_{13} \\ a_{21} & a_{22} & a_{23} \\ a_{31} & a_{32} & a_{33} \end{vmatrix}$$

of which the matrix is

$$A = \begin{pmatrix} a_{11} & a_{12} & a_{13} \\ a_{21} & a_{22} & a_{23} \\ a_{31} & a_{32} & a_{33} \end{pmatrix}$$

can be obtained as follows:

$$M_{11} \text{ (minor of } a_{11}) = \begin{vmatrix} a_{22} & a_{23} \\ a_{32} & a_{33} \end{vmatrix} = a_{22}a_{33} - a_{23}a_{32}$$

$$A_{11} \text{ (co-factor of } a_{11}) = (-1)^{1+1} M_{11} = +(a_{22}a_{33} - a_{23}a_{32})$$

$$M_{22} \text{ (minor of } a_{22}) = \begin{vmatrix} a_{11} & a_{13} \\ a_{31} & a_{33} \end{vmatrix} = a_{11}a_{33} - a_{13}a_{31}$$

$$A_{22} \text{ (co-factor of } a_{22}) = (-1)^{2+2} M_{22} = +(a_{11}a_{33} - a_{13}a_{31})$$

$$M_{32} \text{ (minor of } a_{32}) = \begin{vmatrix} a_{11} & a_{13} \\ a_{21} & a_{23} \end{vmatrix} = a_{11}a_{23} - a_{13}a_{21}$$

$$A_{32} \text{ (co-factor of } a_{32}) = (-1)^{3+2} M_{32} = -(a_{11}a_{23} - a_{13}a_{21})$$

5.6.2 Expansion of the Determinant (Δ)

The determinant (Δ) of a matrix A can be expanded as the sum of the products of elements of any row (or column) by their corresponding co-factors. Thus, the determinant of the matrix A when expanded along the first row is given by

$$\begin{aligned} \Delta &= a_{11}A_{11} + a_{12}A_{12} + a_{13}A_{13} \\ &= a_{11} \begin{vmatrix} a_{22} & a_{23} \\ a_{32} & a_{33} \end{vmatrix} - a_{12} \begin{vmatrix} a_{21} & a_{23} \\ a_{31} & a_{33} \end{vmatrix} + a_{13} \begin{vmatrix} a_{21} & a_{22} \\ a_{31} & a_{32} \end{vmatrix} \\ &= a_{11}(a_{22}a_{33} - a_{23}a_{32}) - a_{12}(a_{21}a_{33} - a_{23}a_{31}) + a_{13}(a_{21}a_{32} - a_{22}a_{31}) \end{aligned}$$

Thus,

$$\Delta = \sum_{j=1}^{3}(-1)^{i+j}a_{ij}M_{ij} = \sum_{j=1}^{3}a_{ij}\{(-1)^{i+j}M_{ij}\} = \sum_{j=1}^{3}a_{ij}A_{ij} = a_{i1}A_{i1} + a_{i2}A_{i2} + a_{i3}A_{i3}$$

for either $i = 1$ or $i = 2$ or $i = 3$, i.e., the determinant (Δ) is expanded along ith row. It can be verified that the value of the determinant of the matrix A will remain the same if one expands it along any row or column.

5.6.3 Difference between a Matrix and a Determinant

A matrix is an arrangement of numbers in which the number of rows may not be equal to the number of columns. On the other hand, a determinant must possess equal number of rows and columns.

A matrix defines representation without any fixed numerical value. However, a determinant possesses a fixed value. The value of the determinant of order 2 can be evaluated by the rule,

$$\begin{vmatrix} a_{11} & a_{12} \\ a_{21} & a_{22} \end{vmatrix} = a_{11}a_{22} - a_{21}a_{12}$$

EXAMPLE 5.8 Prove that $A^3 - 4A^2 - 3A + 11I = 0$, where A is given by

$$A = \begin{bmatrix} 1 & 3 & 2 \\ 2 & 0 & -1 \\ 1 & 2 & 3 \end{bmatrix}$$

and I is the unit matrix of order 3.

Solution

$$A^2 = \begin{bmatrix} 1 & 3 & 2 \\ 2 & 0 & -1 \\ 1 & 2 & 3 \end{bmatrix} \begin{bmatrix} 1 & 3 & 2 \\ 2 & 0 & -1 \\ 1 & 2 & 3 \end{bmatrix};$$

$$= \begin{bmatrix} 1(1) + 3(2) + 2(2) & 1(3) + 3(0) + 2(2) & 1(2) + 3(-1) + 2(3) \\ 2(1) + 0(2) - 1(1) & 2(3) + 0(0) - 1(2) & 2(2) + 0(-1) - 1(3) \\ 1(1) + 2(2) + 3(1) & 1(3) + 2(0) + 3(2) & 1(2) + 2(-1) + 3(3) \end{bmatrix}$$

$$= \begin{bmatrix} 9 & 7 & 5 \\ 1 & 4 & 1 \\ 8 & 9 & 9 \end{bmatrix}$$

Then $A^3 = A^2 \cdot A = \begin{bmatrix} 9 & 7 & 5 \\ 1 & 4 & 1 \\ 8 & 9 & 9 \end{bmatrix} \begin{bmatrix} 1 & 3 & 2 \\ 2 & 0 & -1 \\ 1 & 2 & 3 \end{bmatrix}$

$$= \begin{bmatrix} 9(1) + 7(2) + 5(1) & 9(3) + 7(0) + 5(2) & 9(2) + 7(-1) + 5(3) \\ 1(1) + 4(2) + 1(1) & 1(3) + 4(0) + 1(2) & 1(2) + 4(-1) - 1(3) \\ 8(1) + 9(2) + 9(1) & 8(3) + 9(0) + 9(2) & 8(2) + 9(-1) + 3(3) \end{bmatrix}$$

$$= \begin{bmatrix} 28 & 37 & 26 \\ 10 & 5 & 1 \\ 35 & 42 & 34 \end{bmatrix}$$

Hence,
$$A^3 - 4A^2 - 3A + 11I$$

$$= \begin{bmatrix} 28 & 37 & 26 \\ 10 & 5 & 1 \\ 35 & 42 & 34 \end{bmatrix} - \begin{bmatrix} 36 & 28 & 20 \\ 4 & 16 & 4 \\ 32 & 36 & 36 \end{bmatrix} - \begin{bmatrix} 3 & 9 & 6 \\ 6 & 0 & -3 \\ 3 & 6 & 9 \end{bmatrix} + \begin{bmatrix} 11 & 0 & 0 \\ 0 & 11 & 0 \\ 0 & 0 & 11 \end{bmatrix}$$

$$= \begin{bmatrix} 0 & 0 & 0 \\ 0 & 0 & 0 \\ 0 & 0 & 0 \end{bmatrix} = 0$$

EXAMPLE 5.9 Show that the matrix
$$A = \begin{bmatrix} 2 & 3 \\ 1 & 2 \end{bmatrix}$$
satisfies the equation $A^2 - 4A + I = 0$ and hence find A^{-1}.

Solution
$$A^2 = A \cdot A = \begin{bmatrix} 2 & 3 \\ 1 & 2 \end{bmatrix}\begin{bmatrix} 2 & 3 \\ 1 & 2 \end{bmatrix} = \begin{bmatrix} 2(2)+3(1) & 2(3)+3(2) \\ 1(2)+2(1) & 1(3)+2(2) \end{bmatrix} = \begin{bmatrix} 7 & 12 \\ 4 & 7 \end{bmatrix}$$

Hence,
$$A^2 - 4A + I = \begin{bmatrix} 7 & 12 \\ 4 & 7 \end{bmatrix} - 4\begin{bmatrix} 2 & 3 \\ 1 & 2 \end{bmatrix} + \begin{bmatrix} 1 & 0 \\ 0 & 1 \end{bmatrix} = \begin{bmatrix} 0 & 0 \\ 0 & 0 \end{bmatrix} = 0$$

i.e.,
$$A^2 - 4A + I = 0 \Rightarrow A \cdot A - 4A = -I$$
$$\Rightarrow A(AA^{-1}) - 4AA^{-1} = -IA^{-1} \Rightarrow AI - 4I = -A^{-1}$$
$$\Rightarrow A^{-1} = 4I - A = \begin{bmatrix} 4 & 0 \\ 0 & 4 \end{bmatrix} - \begin{bmatrix} 2 & 3 \\ 1 & 2 \end{bmatrix} = \begin{bmatrix} 2 & -3 \\ -1 & 2 \end{bmatrix}$$

Hence,
$$A^{-1} = \begin{bmatrix} 2 & -3 \\ -1 & 2 \end{bmatrix}$$

5.7 TYPICAL SQUARE MATRICES

There are five types of square matrices. A brief description of each is provided below.

5.7.1 Orthogonal Matrix

A square matrix A is said to be orthogonal if $AA^T = A^TA = I$.
It may be observed from the definition that

(i) A^T is an orthogonal matrix
(ii) $|A| = \pm 1$.

EXAMPLE 5.10 Show that
$$\begin{bmatrix} \cos\theta & 0 & \sin\theta \\ 0 & 1 & 0 \\ -\sin\theta & 0 & \cos\theta \end{bmatrix}$$
is orthogonal. Determine the value of $|A|$.

Solution From the given matrix,
$$A^T = \begin{bmatrix} \cos\theta & 0 & -\sin\theta \\ 0 & 1 & 0 \\ \sin\theta & 0 & \cos\theta \end{bmatrix}$$

Then,
$$AA^T = \begin{bmatrix} \cos\theta & 0 & \sin\theta \\ 0 & 1 & 0 \\ -\sin\theta & 0 & \cos\theta \end{bmatrix} \begin{bmatrix} \cos\theta & 0 & -\sin\theta \\ 0 & 1 & 0 \\ \sin\theta & 0 & \cos\theta \end{bmatrix}$$
$$= \begin{bmatrix} \cos^2\theta + \sin^2\theta & 0 & 0 \\ 0 & 1 & 0 \\ 0 & 0 & \sin^2\theta + \cos^2\theta \end{bmatrix} = \begin{bmatrix} 1 & 0 & 0 \\ 0 & 1 & 0 \\ 0 & 0 & 1 \end{bmatrix}$$
$$= I$$

Hence, A is an orthogonal matrix.

Now,
$$|A| = \begin{vmatrix} \cos\theta & 0 & \sin\theta \\ 0 & 1 & 0 \\ -\sin\theta & 0 & \cos\theta \end{vmatrix} = 1 \begin{vmatrix} \cos\theta & \sin\theta \\ -\sin\theta & \cos\theta \end{vmatrix}$$
$$= \cos^2\theta + \sin^2\theta = 1$$

Thus,
$$|A| = 1$$

5.7.2 Unitary Matrix

A square matrix A is called a *unitary matrix* if $AA^* = I = A^*A$, where A^* is the transposed conjugate of A and I is the unit matrix. From the definition, it is clear that if A is unitary then A^T and A^{-1} are also unitary.

For example The matrix
$$\frac{1}{2}\begin{bmatrix} 1+i & -1+i \\ 1+i & 1-i \end{bmatrix}$$
is unitary. Let
$$A = \frac{1}{2}\begin{bmatrix} 1+i & 1+i \\ -1+i & 1-i \end{bmatrix}$$

Since
$$A^* = (A^T)$$
So,
$$A^* = \frac{1}{2}\begin{bmatrix} 1-i & 1-i \\ -1-i & 1+i \end{bmatrix}$$
Now,
$$A^*A = \frac{1}{2}\begin{bmatrix} 1-i & 1-i \\ -1-i & 1+i \end{bmatrix} \cdot \frac{1}{2}\begin{bmatrix} 1+i & -1+i \\ 1+i & 1-i \end{bmatrix}$$
$$= \frac{1}{4}\begin{bmatrix} 1-i^2+1-i^2 & (1-i)^2+(1-i)^2 \\ -(1+i)^2+(1+i)^2 & 1-i^2+1-i^2 \end{bmatrix} = \frac{1}{4}\begin{bmatrix} 4 & 0 \\ 0 & 4 \end{bmatrix}$$
$$= \begin{bmatrix} 1 & 0 \\ 0 & 1 \end{bmatrix} = I$$

Similarly, $AA^* = I$. Hence, the matrix A is unitary.

5.7.3 Involutory Matrix

A square matrix A is known as involutory if $A^2 = I$.

For example Let
$$A = \begin{bmatrix} 0 & 1 \\ 1 & 0 \end{bmatrix}$$
Then
$$A^2 = \begin{bmatrix} 0 & 1 \\ 1 & 0 \end{bmatrix}\begin{bmatrix} 0 & 1 \\ 1 & 0 \end{bmatrix} = \begin{bmatrix} 1 & 0 \\ 0 & 1 \end{bmatrix} = I$$
Thus,
$$A = \begin{bmatrix} 0 & 1 \\ 1 & 0 \end{bmatrix}$$
is an involutory matrix.

5.7.4 Idempotent Matrix

A square matrix A is called *idempotent* if $A^2 = A$.

For example Assume
$$A = \begin{bmatrix} 2 & -2 & -4 \\ -1 & 3 & 4 \\ 1 & -2 & -3 \end{bmatrix}$$
Then,
$$A^2 = \begin{bmatrix} 2 & -2 & -4 \\ -1 & 3 & 4 \\ 1 & -2 & -3 \end{bmatrix}\begin{bmatrix} 2 & -2 & -4 \\ -1 & 3 & 4 \\ 1 & -2 & -3 \end{bmatrix} = \begin{bmatrix} 2 & -2 & -4 \\ -1 & 3 & 4 \\ 1 & -2 & -3 \end{bmatrix} = A$$

Hence, A is an idempotent matrix.

5.7.5 Nilpotent Matrix

A square matrix A is called a *nilpotent matrix* if there exists a +ve integer n such that $A^n = 0$. If n is the least positive integer such that $A^n = 0$, then n is called the *index* or *order* of the nilpotent matrix A.

For example The matrix

$$A = \begin{bmatrix} 0 & 0 \\ 2 & 0 \end{bmatrix}$$

is a nilpotent matrix, since

$$\begin{bmatrix} 0 & 0 \\ 2 & 0 \end{bmatrix}^2 = 0$$

and as such, here 2 is the index or order of the matrix A.

5.8 ADJOINT AND INVERSE OF A MATRIX

To describe adjoint and inverse of a matrix, the following definitions are necessary.

5.8.1 Singular and Non-Singular Matrices

A matrix $A = [a_{ij}]_{n \times n}$ is said to be *non-singular* if $|A| \neq 0$. If $|A| = 0$, then the matrix A is said to be *singular*.

For example The determinant of the matrix

$$\begin{bmatrix} 2 & 2 \\ 3 & 3 \end{bmatrix}$$

is

$$\begin{vmatrix} 2 & 2 \\ 3 & 3 \end{vmatrix} = 0$$

hence, it is a singular matrix. On the other hand, the determinant of the matrix

$$\begin{bmatrix} 4 & -3 \\ 2 & 1 \end{bmatrix}$$

is

$$\begin{vmatrix} 4 & -3 \\ 2 & 1 \end{vmatrix} = 10 \neq 0$$

hence, it is a non-singular matrix.

5.8.2 Adjoint of a Square Matrix

The *adjoint* of a square matrix $A = [a_{ij}]$, of order n, is defined as the transpose of the matrix $[A_{ij}]$ where A_{ij} is the co-factor of the element a_{ij}. Adjoint of the matrix A is denoted by adj A.

Thus, if $A = [a_{ij}]_{n \times n}$ then adj $\mathbf{A} = [\mathbf{A}_{ij}]_{n \times n}^{\mathrm{T}}$, where $\mathbf{A}_{ij}$ is the co-factor of a_{ij}. In other way of representataion, if

$$A = \begin{bmatrix} a_{11} & a_{12} & \cdots & a_{1n} \\ a_{21} & a_{22} & \cdots & a_{2n} \\ \cdots & \cdots & \cdots & \cdots \\ a_{n1} & a_{n2} & \cdots & a_{nn} \end{bmatrix}$$

then the matrix formed by the co-factors of the elements in $|A|$ is

$$\begin{bmatrix} A_{11} & A_{12} & \cdots & A_{1n} \\ A_{21} & A_{22} & \cdots & A_{2n} \\ \cdots & \cdots & \cdots & \cdots \\ A_{n1} & A_{n2} & \cdots & A_{nn} \end{bmatrix}$$

The transpose of this matrix is

$$\operatorname{adj} A = \begin{bmatrix} A_{11} & A_{21} & \cdots & A_{n1} \\ A_{12} & A_{22} & \cdots & A_{n2} \\ \cdots & \cdots & \cdots & \cdots \\ A_{1n} & A_{2n} & \cdots & A_{nn} \end{bmatrix}$$

5.8.3 Properties of Adjoint of a Matrix

The following are the properties of adjoint matrix:

I. $A \,(\operatorname{adj} A) = |A| I = (\operatorname{adj} A) A$, iff $|A| \neq *0$

II. $A \left\{ \dfrac{(\operatorname{adj} A)}{|A|} \right\} = I = \left\{ \dfrac{(\operatorname{adj} A)}{|A|} \right\} A$, iff $|A| \neq 0$

III. $\operatorname{adj}(AB) = (\operatorname{adj} A)(\operatorname{adj} B)$, iff $|A| \neq 0, |B| \neq 0$

The proofs of each one of above are given below.

I. Since the orders of A and adj A are the same, therefore

$$\operatorname{adj} A (A) = \begin{bmatrix} a_{11} & a_{12} & \cdots & a_{1n} \\ a_{21} & a_{22} & \cdots & a_{2n} \\ \cdots & \cdots & \cdots & \cdots \\ a_{n1} & a_{n2} & \cdots & a_{nn} \end{bmatrix} \begin{bmatrix} A_{11} & A_{21} & \cdots & A_{n1} \\ A_{12} & A_{22} & \cdots & A_{n2} \\ \cdots & \cdots & \cdots & \cdots \\ A_{1n} & A_{2n} & \cdots & A_{nn} \end{bmatrix}$$

The elements in the ith row and the jth column of $A \cdot (\operatorname{adj} A)$ is the scalar product of the elements of the ith row of A and the corresponding elements of the jth column of adj A, which is

$$a_{i1} A_{j1} + a_{i2} A_{j2} + \cdots + a_{in} A_{jn} = \sum \begin{cases} |A|, & \text{when } i = j \\ 0, & \text{when } i \neq j \end{cases}$$

This shows that each diagonal element of $A(\operatorname{adj} A)$ is $|A|$ and each one of its non-diagonal elements is 0.

Therefore,

$$A \cdot (\text{adj } A) = \begin{bmatrix} |A| & 0 & \cdots & 0 \\ 0 & |A| & \cdots & 0 \\ \cdots & \cdots & \cdots & \cdots \\ 0 & 0 & \cdots & |A| \end{bmatrix}$$

$$= |A| \begin{bmatrix} 1 & 0 & \cdots & 0 \\ 0 & 1 & \cdots & 0 \\ \cdots & \cdots & \cdots & \cdots \\ 0 & 0 & \cdots & 1 \end{bmatrix}$$

Similarly, it may be shown that $(\text{adj } A) A = |A| I$.
Hence, $A (\text{adj } A) = |A| I = (\text{adj } A) A$.

II. Let A be an invertible square matrix of order n, then there exists a square matrix B such that

$$AB = I = BA$$

Consider,

$$AB = I$$

i.e.,

$$|AB| = |I| = 1$$

$$\Rightarrow |A| |B| = I \Rightarrow |A| \neq 0 \text{ and } |B| \neq 0$$

In particular, $|A| \neq 0$, i.e., A is non-singular.
Conversely, let A be non-singular, i.e., $|A| \neq 0$.
Then,

$$A (\text{adj } A) = |A| I = (\text{adj } A) A$$

$$\Rightarrow A \left[\frac{1}{|A|} (\text{adj } A) \right] = I = \left[\frac{1}{|A|} (\text{adj } A) \right] A$$

Hence, proved.

III. Let A and B be invertible square matrices, each of order n. Then,

$$|A| \neq 0 \text{ and } |B| \neq 0$$

So,

$$|AB| = |A| |B| \neq 0$$

Therefore, AB is an invertible square matrix of order n.
Also,

$$(AB)(\text{adj } AB)$$
$$= (AB) \cdot (\text{adj } B \cdot \text{adj } A)$$
$$= A (B \cdot \text{adj } B) (\text{adj } A) \qquad \text{(By associative law)}$$
$$= A(|B| \cdot I) \cdot (\text{adj } A) \qquad \text{(since } B \cdot \text{adj } B = |B| I\text{)}$$
$$= (A I)|B| (\text{adj } A)$$

$$= |B| \cdot |A| \, I \qquad \text{(since } A \text{ adj } A = |A| \, I\text{)}$$
$$= |A| \cdot |B| \, I$$
$$= |AB| \cdot I \qquad \text{(since } |A| \cdot |B| = |AB|\text{)}$$

Thus,
$$(AB) \cdot (\text{adj } AB) = (AB) \cdot (\text{adj } B \cdot \text{adj } A) = |AB| \, I$$
Hence,
$$(AB) \cdot (\text{adj } AB) = (AB) \cdot (\text{adj } B \cdot \text{adj } A)$$

But AB being invertible, by cancellation law, we finally obtain $(\text{adj } AB) = (\text{adj } B)(\text{adj } A)$.

5.9 INVERSE OF A MATRIX

If corresponding to a matrix A, there exists matrix B such that $AB = BA = I$, then B is called the *inverse* of A and is denoted by A^{-1}. Thus, by definition
$$AA^{-1} = A^{-1}A = I$$

Here, both the matrices A and A^{-1} should be square matrices of the same order. For this reason, rectangular matrices cannot have an inverse.

Finally, it follows from the property II of adjoint matrix, for a square matrix A, inverse of A can be expressed as
$$A^{-1} = \frac{\text{adj } A}{|A|}, \quad |A| \neq 0$$

5.9.1 Properties of Inverse of a Matrix

I. The necessary and sufficient condition for a square matrix A to possess the inverse is that A is to be non-singular, i.e., $|A| \neq 0$.

Proof

Case I The condition is necessary

Let A be a square matrix whose inverse is B, then $AB = I$. Implementing determinant on both sides,
$$|AB| = |I|$$
$$\Rightarrow |A| \, |B| = I$$
which is permissible if neither of $|A|$ and $|B|$ is zero, i.e., if the matrices A and B are non-singular, i.e., if $|A| \neq 0$.

Case II The condition is sufficient

Let $|A| \neq 0$ and there exists a matrix B such that
$$B = \frac{\text{adj } A}{|A|}$$

Then
$$AB = A \cdot \frac{\text{adj } A}{|A|} = \frac{1}{|A|} (A \text{ adj } A)$$
$$= \frac{1}{|A|} \cdot |A| I = I$$

Similarly,
$$BA = I$$
i.e.,
$$AB = BA = I$$

$\Rightarrow$ A possesses an inverse or A is invertible.

II. **The inverse of a matrix, if exists, is unique.**

Proof Let a matrix A possess inverses as B and C
Then,
$$AB = BA = I \quad \text{and} \quad AC = CA = I$$

Now,
$$CAB = C(AB) = CI = C$$

Also,
$$CAB = (CA)B = IB = B$$

So,
$$CAB = B = C$$

$\Rightarrow$ the inverse is unique.

III. If A and B are two non-singular matrices of the same order, then $(AB)^{-1} = B^{-1}A^{-1}$, i.e., inverse of a product of two matrices is the product of their inverses in the reverse order.

Proof Since the matrix multiplication is associative, so, by definition
$$(B^{-1}A^{-1})(AB) = B^{-1}(A^{-1}A)B = B^{-1}IB = B^{-1}B = I$$
and
$$(AB)(B^{-1}A^{-1}) = A(BB^{-1})A^{-1} = AIA^{-1} = AA^{-1} = I$$

Thus,
$$(B^{-1}A^{-1})(AB) = (AB)(B^{-1}A^{-1}) = I$$

$\Rightarrow$
$$(AB)^{-1} = B^{-1}A^{-1}$$

which is also known as the *reversal law* for inverses.

Note In general, if $A_1, A_2, \ldots, A_n$ are non-singular matrices, then
$$(A_1, A_2, \ldots, A_n)^{-1} = A_n^{-1} \cdots A_2^{-1} A_1^{-1}$$

IV. If A is non-singular, then

(a) $(A^{-1})^{-1} = A$ (b) $(A^{-1})^T = (A^T)^{-1}$

i.e., transposition and inverse are commutative.

Proof

(a) It is known that $AA^{-1} = I$. Taking inverse on both the sides,
$$(AA^{-1})^{-1} = I^{-1}$$
$$\Rightarrow (A^{-1})^{-1}A^{-1} = I = AA^{-1}$$
$$\Rightarrow (A^{-1})^{-1} = A \;(\because A^{-1} \neq 0)$$

(b) It is known that $AA^{-1} = A^{-1}$. Taking transpose of the above equation
$$(A^{-1})^T A^T = A^T (A^{-1})^T = I^T$$
$$\Rightarrow (A^{-1})^T A^T = A^T (A^{-1})^T = I$$
$$\Rightarrow (A^{-1})^T \text{ is the inverse of } A^T$$
$$\Rightarrow (A^{-1})^T = (A^T)^{-1}$$

V. A^{-1} is an orthogonal matrix.
The proof is left to the reader.

EXAMPLE 5.11 If the product of two non-zero square matrices is a zero matrix, then prove that both of them are singular matrices.

Solution Let A and B be two non-zero $n \times n$ matrices. Given that $AB = O$, where O is a zero matrix. Suppose that B is a non-singular matrix, then B^{-1} exists. Then,
$$AB = O \Rightarrow (AB)B^{-1} = OB^{-1}$$
post-multiplying both sides by B^{-1}
$$\Rightarrow A(BB^{-1}) = O, \text{ by associative law of multiplication}$$
$$\Rightarrow AI = O \quad (\text{since } BB^{-1} = I)$$
$$\Rightarrow A = O$$

which is against hypothesis as A is a non-zero matrix.
Hence, B is a singular matrix. In a similar manner, we can prove that A is also a singular matrix.

EXAMPLE 5.12 If A is a non-singular matrix, then prove that $AB = AC \Rightarrow B = C$, where B and C are square matrices of the same order.

Solution Since A is a non-singular matrix, so A^{-1} exists. Now,
$$AB = AC \Rightarrow A^{-1}(AB) = A^{-1}(AC)$$
pre-multiplying both sides by A^{-1}
$$\Rightarrow (A^{-1}A)B = (A^{-1}A)C, \text{ by associative law of multiplication}$$
$$\Rightarrow IB = IC \;(\text{since } A^{-1}A = I)$$
$$\Rightarrow B = C \;(\text{since } IB = B \text{ etc.})$$
Hence, proved.

EXAMPLE 5.13 Find the adjoint of

(i) $\begin{bmatrix} 4 & 2 \\ -1 & 3 \end{bmatrix}$ (ii) $\begin{bmatrix} 1 & 2 & 3 \\ 2 & -4 & 5 \\ 6 & 1 & 0 \end{bmatrix}$

Solution

(i) Let
$$A = \begin{bmatrix} 4 & 2 \\ -1 & 3 \end{bmatrix}$$

Then
$$\det(A) = |A| = \begin{vmatrix} 4 & 2 \\ -1 & 3 \end{vmatrix} \equiv \begin{vmatrix} a_{11} & a_{12} \\ a_{21} & a_{22} \end{vmatrix}$$

The co-factors of the elements of det (A) are

$A_{11} = 3$, $A_{12} = -(-1) = 1$, $A_{21} = 2$, $A_{22} = 4$

$$\text{adj } A = \text{transpose of } \begin{bmatrix} A_{11} & A_{12} \\ A_{21} & A_{22} \end{bmatrix} = \begin{bmatrix} A_{11} & A_{21} \\ A_{12} & A_{22} \end{bmatrix}$$

Thus,
$$\text{adj } A = \begin{bmatrix} 3 & -2 \\ 1 & 4 \end{bmatrix}$$

(ii) Let
$$A = \begin{bmatrix} 1 & 2 & 3 \\ 2 & -4 & 5 \\ 6 & 1 & 0 \end{bmatrix}$$

Then,
$$\det(A) = \begin{vmatrix} 1 & 2 & 3 \\ 2 & -4 & 5 \\ 6 & 1 & 0 \end{vmatrix} \equiv \begin{vmatrix} a_{11} & a_{12} & a_{13} \\ a_{21} & a_{22} & a_{23} \\ a_{31} & a_{32} & a_{33} \end{vmatrix}$$

The co-factors of the elements of det (A) are

$A_{11} = \begin{vmatrix} -4 & 5 \\ 1 & 0 \end{vmatrix} = -5$ $A_{12} = -\begin{vmatrix} 2 & 5 \\ 6 & 0 \end{vmatrix} = 30$ $A_{13} = \begin{vmatrix} 2 & -4 \\ 6 & 1 \end{vmatrix} = 26$

$A_{21} = \begin{vmatrix} 2 & 3 \\ 1 & 0 \end{vmatrix} = 22$ $A_{32} = \begin{vmatrix} 1 & 3 \\ 6 & 0 \end{vmatrix} = 18$ $A_{23} = -\begin{vmatrix} 1 & 2 \\ 6 & 1 \end{vmatrix} = 11$

$A_{31} = \begin{vmatrix} 2 & 3 \\ -4 & 5 \end{vmatrix} = 22$ $A_{32} = -\begin{vmatrix} 1 & 3 \\ 2 & 5 \end{vmatrix} = 1$ $A_{33} = \begin{vmatrix} 1 & 2 \\ 2 & -4 \end{vmatrix} = -8$

$$\text{adj } A = \text{transpose of } \begin{bmatrix} A_{11} & A_{12} & A_{13} \\ A_{21} & A_{22} & A_{23} \\ A_{31} & A_{32} & A_{33} \end{bmatrix} = \begin{bmatrix} A_{11} & A_{12} & A_{13} \\ A_{21} & A_{22} & A_{23} \\ A_{31} & A_{32} & A_{33} \end{bmatrix}$$

$$\text{adj } A = \begin{bmatrix} -5 & 3 & 22 \\ 30 & -18 & 1 \\ 26 & 11 & -8 \end{bmatrix}$$

EXAMPLE 5.14 If
$$A = \begin{bmatrix} 2 & 3 \\ 4 & 8 \end{bmatrix}$$
verify that $A(\text{adj } A) = (\text{adj } A)A = \det(A)I$

Solution
$$\det(A) = \begin{vmatrix} 2 & 3 \\ 4 & 8 \end{vmatrix} \equiv \begin{vmatrix} a_{11} & a_{12} \\ a_{21} & a_{22} \end{vmatrix}$$

i.e.,
$$\det(A) = 4$$

The co-factors of the elements of det (A) are
$$A_{11} = 8, \quad A_{12} = -4, \quad A_{21} = -3, \quad A_{22} = 2$$

$$\text{adj } A = \text{transpose of } \begin{bmatrix} A_{11} & A_{12} \\ A_{21} & A_{22} \end{bmatrix} = \begin{bmatrix} A_{11} & A_{21} \\ A_{12} & A_{22} \end{bmatrix}$$

$$\text{adj } A = \begin{bmatrix} 8 & -3 \\ -4 & 2 \end{bmatrix}$$

Now,
$$A(\text{adj } A) = \begin{bmatrix} 2 & 3 \\ 4 & 8 \end{bmatrix} \begin{bmatrix} 8 & -3 \\ -4 & 2 \end{bmatrix} = \begin{bmatrix} 2 \times 8 + 3 \times (-4) & 2 \times (-3) + 2 \times 2 \\ 4 \times 8 + 8 \times (-4) & 4 \times (-3) + 8 \times 2 \end{bmatrix}$$

$$= \begin{bmatrix} 4 & 0 \\ 0 & 0 \end{bmatrix} = 4 \begin{bmatrix} 1 & 0 \\ 0 & 1 \end{bmatrix} = \det(A) I$$

Again,
$$(\text{adj } A)A = \begin{bmatrix} 8 & -3 \\ -4 & 2 \end{bmatrix} \times \begin{bmatrix} 2 & 3 \\ 4 & 8 \end{bmatrix} = \begin{bmatrix} 8 \times 2 + (-3) \times 4 & 8 \times 3 + (-3) \times 8 \\ (-4) \times 2 + 2 \times 4 & (-4) \times 3 + 2 \times 8 \end{bmatrix}$$

$$= \begin{bmatrix} 4 & 0 \\ 0 & 4 \end{bmatrix} = 4 \begin{bmatrix} 1 & 0 \\ 0 & 1 \end{bmatrix} = \det(A) I$$

Hence, $A(\text{adj } A) = (\text{adj } A)A = \det(A)I$.

 EXAMPLE 5.15 Find the inverse of

(i) $\begin{bmatrix} 2 & 1 \\ 0 & 1 \end{bmatrix}$ (ii) $\begin{bmatrix} 1 & 1 & 3 \\ 1 & 3 & -3 \\ -2 & -4 & -4 \end{bmatrix}$

Solution

(i) $|A| = \begin{vmatrix} 2 & 1 \\ 0 & 1 \end{vmatrix} = 2 \neq 0$

i.e., A is non-singular and therefore A^{-1} exists.

Now, the co-factors of the elements of $|A|$ are

$$A_{11} = 1 \quad A_{12} = 0 \quad A_{21} = -1 \quad A_{22} = 2$$

$$\therefore A^{-1} = \frac{1}{|A|} \text{adj } A = \frac{1}{|A|} \begin{bmatrix} A_{11} & A_{21} \\ A_{12} & A_{22} \end{bmatrix}$$

i.e.,

$$A^{-1} = \frac{1}{2}\begin{bmatrix} 1 & -1 \\ 0 & 2 \end{bmatrix} = \begin{bmatrix} \frac{1}{2} & -\frac{1}{2} \\ 0 & 1 \end{bmatrix}$$

(ii) Suppose

$$A = \begin{bmatrix} a_{11} & a_{12} & a_{13} \\ a_{21} & a_{22} & a_{23} \\ a_{31} & a_{32} & a_{33} \end{bmatrix} \equiv \begin{bmatrix} 1 & 1 & 3 \\ 1 & 3 & -3 \\ -2 & -4 & -4 \end{bmatrix}$$

Then,

$$|A| = \begin{bmatrix} 1 & 1 & 3 \\ 1 & 3 & -3 \\ -2 & -4 & -4 \end{bmatrix} \begin{matrix} R_2 \to R_2 - R_1 \\ R_3 \to R_3 + 2R_1 \end{matrix} = \begin{vmatrix} 1 & 1 & 3 \\ 0 & 2 & -6 \\ 0 & -2 & 2 \end{vmatrix}$$

expand C_1

$$= \begin{vmatrix} 2 & -6 \\ -2 & 2 \end{vmatrix} = -8 \neq 0$$

$\Rightarrow$ A is non-singular, i.e., inverse of the matrix A exists. Now, the co-factors of the elements of $|A|$ are

$$A_{11} = \begin{vmatrix} 3 & -3 \\ -4 & -4 \end{vmatrix} = -24 \quad A_{12} = -\begin{vmatrix} 1 & -3 \\ -2 & -4 \end{vmatrix} = 10 \quad A_{13} = \begin{vmatrix} 1 & 3 \\ -2 & -4 \end{vmatrix} = 2$$

$$A_{21} = -\begin{vmatrix} 1 & 3 \\ -4 & -4 \end{vmatrix} = -8 \quad A_{22} = -\begin{vmatrix} 1 & 3 \\ -2 & -4 \end{vmatrix} = 2 \quad A_{23} = -\begin{vmatrix} 1 & 1 \\ -2 & -4 \end{vmatrix} = 2$$

$$A_{31} = -\begin{vmatrix} 1 & 3 \\ 3 & -3 \end{vmatrix} = -12 \quad A_{32} = -\begin{vmatrix} 1 & 3 \\ 1 & -3 \end{vmatrix} = 6 \quad A_{33} = \begin{vmatrix} 1 & 1 \\ 1 & 3 \end{vmatrix} = 2$$

$$\text{adj } A = \text{transpose of } \begin{bmatrix} a_{11} & a_{12} & a_{13} \\ a_{21} & a_{22} & a_{23} \\ a_{31} & a_{32} & a_{33} \end{bmatrix} = \begin{bmatrix} a_{11} & a_{21} & a_{31} \\ a_{12} & a_{22} & a_{32} \\ a_{13} & a_{23} & a_{33} \end{bmatrix}$$

$$= \begin{bmatrix} 3 & 1 & \frac{3}{2} \\ -\frac{5}{4} & -\frac{1}{4} & -\frac{3}{4} \\ -\frac{1}{4} & -\frac{1}{4} & -\frac{1}{4} \end{bmatrix}$$

Hence,

$$A^{-1} = \frac{1}{|A|} \text{adj } A = -\frac{1}{8}\begin{bmatrix} -24 & -8 & -12 \\ 10 & 2 & 6 \\ 2 & 2 & 2 \end{bmatrix} = \begin{bmatrix} 3 & 1 & \frac{3}{2} \\ -\frac{5}{4} & -\frac{1}{4} & -\frac{3}{4} \\ -\frac{1}{4} & -\frac{1}{4} & -\frac{1}{4} \end{bmatrix}$$

5.10 RANK OF A MATRIX

A matrix A of order $m \times n$ is said to be a matrix of rank r if
 (i) there exists at least one non-zero minor of A of order r, and
 (ii) each minor of A of order $(r + 1)$ vanishes.
 In other words, the rank of a matrix is the order of the largest non-vanishing minors of the matrix. It follows that for a non-singular matrix of order n, the rank is n. For a null matrix, the rank is zero. If only one element of a matrix is non-zero, then its rank will be 1. In general, rank of a matrix A is represented by $\rho(A)$.

Note

 (i) If a matrix A possesses a non-zero minor of order r, then its rank $\rho(A) \geq r$.
 (ii) If all minors of A of order $(r + 1)$ vanishes, then its rank $\rho(A) \leq r$.
 (iii) If A is an $m \times n$ matrix and $m \leq n$, then its rank $\rho(A) \leq m$.
 (iv) If A is an $m \times n$ matrix and $m \geq n$, then its rank $\rho(A) \leq n$.
 (v) The rank of the unit matrix n, I_n is n, i.e., $\rho(I_n) = n$.
 (vi) If A is a diagonal matrix of order n, or upper (or lower) triangular matrix with non-zero diagonal elements, then $\rho(A) = n$.

For example

(i) $\begin{bmatrix} 0 & 0 \\ 0 & 0 \end{bmatrix}$, $\rho(A) = 0$

since all elements are zero.

(ii) $A = \begin{bmatrix} 0 & 0 \\ 0 & 3 \end{bmatrix}$, $\rho(A) = 1$

since

$$|A| = 0$$

and there exists at least one non-zero element as a minor.

(iii) $A = \begin{bmatrix} 2 & 0 \\ 0 & -3 \end{bmatrix}$

Here,
$|A| = -6 - 0 = -6 \neq 0$ and $\rho(A) = 2$

(iv) $A = \begin{bmatrix} 3 & 1 \\ 2 & 4 \end{bmatrix}$

Here,
$$|A| = 10 \neq 0 \text{ and } \rho(A) = 2$$

(v) $A = \begin{bmatrix} 2 & 1 \\ 4 & 2 \end{bmatrix}$, $\rho(A) = 1$

since $|A| = 0$ and there exists at least one non-vanishing element.

(vi) $A = \begin{bmatrix} 1 & 2 & 3 \\ 4 & 5 & 6 \\ 3 & 2 & 1 \end{bmatrix}$

Here,
$$|A| = 0$$
and
$$\begin{bmatrix} 1 & 2 \\ 4 & 5 \end{bmatrix} \neq 0$$

so $\rho(A) = 2$.

(vii) $A = \begin{bmatrix} 1 & 1 & 1 \\ 1 & 2 & 3 \\ 1 & 4 & 9 \end{bmatrix}$

Here, $|A| \neq 0$, hence $\rho(A) = 3$.

5.10.1 Elementary Transformations (Operations) of a Matrix

The *elementary transformations* may be performed by the following operations:
(i) Interchange of any two rows (or columns) denoted by $R_i \leftrightarrow R_j$ (or $C_i \leftrightarrow C_j$).
(ii) Multiplication of every element of a row (or a column) by a non-zero number k denoted by $R_i \to kR_i$ (or $C_i \to kC_i$).
(iii) Addition of the elements of one row (or column) with the corresponding elements of another row (or column), after multiplication by a non-zero number k denoted by $R_i \to R_i + kR_j$ (or $C_i \to C_i + kC_j$).

Elementary transformations are known as *row transformations* or *column transformations* in accordance with its application to rows or columns, respectively. It may also be noted that the rank of a matrix is invariant under elementary transformations.

EXAMPLE 5.16 Let A and B be two matrices of order $n \times n$ such that $A \neq 0, B \neq 0$, and $AB = 0$. Show that both A and B have a rank less than n.

Solution Suppose rank $(A) = n$. Since A is of type $n \times n$, rank $(A) = n$ implies that $|A| \neq 0$. So, A^{-1} exists. Now, $AB = 0 \Rightarrow A^{-1}(AB) = 0 \Rightarrow (A^{-1}A)B = 0 \Rightarrow IB = 0 \Rightarrow B = 0$ which contracticts $B \neq 0$. Thus, rank $(A) < n$. Similarly, we can show that rank $(B) < n$.

EXAMPLE 5.17 Under what condition the rank of the following matrix A is 3? Is it possible for the rank to be 1?

$$A = \begin{bmatrix} 2 & 4 & 2 \\ 3 & 1 & 2 \\ 1 & 0 & x \end{bmatrix}$$

Solution If the rank of the matrix A is 3, then the minor of order 3 of A should be non-zero, i.e.,

$$\begin{vmatrix} 2 & 4 & 2 \\ 3 & 1 & 2 \\ 1 & 0 & x \end{vmatrix} \neq 0$$

which is the required condition.

The rank of A cannot be 1 as at least one minor of order 1 of A, i.e., one element of A is zero.

EXAMPLE 5.18 Find the rank of the matrix

$$\begin{bmatrix} 1 & 2 & 3 \\ 1 & 4 & 2 \\ 2 & 6 & 5 \end{bmatrix}$$

Solution

Let

$$A = \begin{bmatrix} 1 & 2 & 3 \\ 1 & 4 & 2 \\ 2 & 6 & 5 \end{bmatrix}$$

Then using elementary row operations [i.e., $R_3 \to R_3 - (R_1 + R_2)$], the above matrix reduces to

$$A = \begin{bmatrix} 1 & 2 & 3 \\ 1 & 4 & 2 \\ 0 & 0 & 0 \end{bmatrix}$$

in which the number of non-zero rows is 2, i.e., $\rho(A) = 2$.

Aliter Let

$$A = \begin{bmatrix} 1 & 2 & 3 \\ 1 & 4 & 2 \\ 2 & 6 & 5 \end{bmatrix}$$

which is a 3×3 matrix. So, $\rho(A) \leq 3$.

Now,

$$A = \begin{bmatrix} 1 & 2 & 3 \\ 1 & 4 & 2 \\ 2 & 6 & 5 \end{bmatrix} \sim \begin{bmatrix} 1 & 2 & 3 \\ 0 & 2 & -1 \\ 0 & 2 & -1 \end{bmatrix} \begin{bmatrix} R_2 \to R_2 - R_1 \\ R_3 \to R_3 - 2R_1 \end{bmatrix}$$

$$\sim \begin{bmatrix} 1 & 2 & 3 \\ 0 & 2 & -1 \\ 0 & 0 & 0 \end{bmatrix} R_3 \to R_3 - R_2$$

Here, only the third order is zero, but the second order

$$\begin{vmatrix} 1 & 2 \\ 0 & 2 \end{vmatrix} = 2 \neq 0$$

Hence, $\rho(A) = 2$, i.e., the rank of the given matrix is 2.

EXAMPLE 5.19 Find the rank of an $m \times n$ matrix, every element of which is unity.

Solution Let an $m \times n$ matrix be

$$A = \begin{bmatrix} 1 & 1 & 1 & \cdots & \cdots & 1 \\ 1 & 1 & 1 & \cdots & \cdots & 1 \\ \cdots & \cdots & \cdots & \cdots & \cdots & \cdots \\ 1 & 1 & 1 & \cdots & \cdots & 1 \end{bmatrix}$$

Now, we find that every square sub-matrix of A higher than 1×1 will be a matrix, each element of which is unity, and therefore, the value of the determinant will always be zero, since its rows and columns are identical. But the square sub-matrices of order 1×1 are $[1]$, and the determinant of these are $|1| \neq 0$. Hence, the rank of $A = 1$. Hence, proved.

5.11 BOOLEAN MATRIX OR A ZERO-ONE MATRIX

A matrix with entries 0 and 1 subject to the Boolean operations $\vee$ and $\wedge$ is known as *Boolean matrix* or a *zero-one matrix*. Here, the arithmetic is based on the operational symbols $\vee$ and $\wedge$, operating on pairs of bits as

$$b_1 \wedge b_2 = \begin{cases} 1, & \text{if } b_1 = b_2 = 1 \\ 0, & \text{otherwise} \end{cases}$$

$$b_1 \vee b_2 = \begin{cases} 1, & \text{if } b_1 = 1 \text{ or } b_2 = 1 \\ 0, & \text{otherwise} \end{cases}$$

Boolean matrices are useful in the representation of discrete structures in respect to relation and graph theory.

5.11.1 Operations on Zero-One Matrices

Let $A = [a_{ij}]$ and $B = [b_{ij}]$ be $m \times n$ zero-one matrices. Then the *join* of A and B, expressed as $a_{ij} \vee b_{ij}$ with (i, j)th entry, is a Boolean matrix. This join is denoted by $A \vee B$. Also, the *meet* of A and B, represented as $a_{ij} \wedge b_{ij}$ with (i, j)th entry, is a Boolean matrix. The meet is denoted by $A \wedge B$.

EXAMPLE 5.20 Find the join and meet of the zero-one matrices

$$A = \begin{bmatrix} 1 & 1 & 0 \\ 1 & 0 & 1 \end{bmatrix}, \quad B = \begin{bmatrix} 0 & 1 & 0 \\ 1 & 0 & 1 \end{bmatrix}$$

Solution The join of A and B is

$$A \vee B = \begin{bmatrix} 1 \vee 0 & 1 \vee 0 & 0 \vee 0 \\ 0 \vee 1 & 0 \vee 0 & 1 \vee 1 \end{bmatrix} = \begin{bmatrix} 1 & 1 & 0 \\ 1 & 0 & 1 \end{bmatrix}$$

Also, the meet of A and B is

$$A \wedge B = \begin{bmatrix} 1 \wedge 0 & 1 \wedge 1 & 0 \wedge 0 \\ 0 \wedge 1 & 0 \wedge 0 & 1 \wedge 1 \end{bmatrix} = \begin{bmatrix} 0 & 1 & 0 \\ 0 & 0 & 1 \end{bmatrix}$$

5.11.2 Boolean Product of Matrices

Let $A = [a_{ij}]$ be an $m \times p$ zero-one matrix and $B = [b_{ij}]$ be a $p \times n$ zero-one matrix. Then the *Boolean product* of A and B, denoted by $A \otimes B$, is defined as

$$A \otimes B = (a_{i1} \wedge b_{1j}) \vee (a_{i2} \wedge b_{2j}) \vee \cdots \vee (a_{in} \wedge b_{nj})$$

which is an $m \times n$ matrix.

Note The Boolean product of two matrices is similar to the ordinary product of the matrices except the addition is replaced with the operation $\vee$ and the multiplication by $\wedge$. However, the computation of $A \otimes B$ is illustrated in the following example.

EXAMPLE 5.21 Determine the Boolean product of the following matrices:

$$A = \begin{bmatrix} 0 & 1 \\ 1 & 0 \\ 1 & 1 \end{bmatrix}, \quad B = \begin{bmatrix} 1 & 0 & 1 \\ 0 & 1 & 0 \end{bmatrix}$$

Solution The Boolean product $A \otimes B$ is determined as

$$A \otimes B = \begin{bmatrix} (0 \wedge 1) \vee (1 \wedge 0) & (0 \wedge 0) \vee (1 \wedge 1) & (0 \wedge 1) \vee (1 \wedge 0) \\ (1 \wedge 1) \vee (0 \wedge 0) & (1 \wedge 0) \vee (0 \wedge 1) & (1 \wedge 1) \vee (0 \wedge 0) \\ (1 \wedge 1) \vee (1 \wedge 0) & (1 \wedge 0) \vee (1 \wedge 1) & (1 \wedge 1) \vee (1 \wedge 0) \end{bmatrix}$$

$$= \begin{bmatrix} 0 \vee 0 & 0 \vee 1 & 0 \vee 0 \\ 1 \vee 0 & 0 \vee 0 & 1 \vee 0 \\ 1 \vee 1 & 0 \vee 1 & 1 \vee 0 \end{bmatrix}$$

$$= \begin{bmatrix} 0 & 1 & 0 \\ 1 & 0 & 1 \\ 1 & 1 & 1 \end{bmatrix}$$

5.12 ELEMENTARY ROW OPERATION ON A MATRIX

We can perform elementary row operations on a matrix as described below.

5.12.1 Echelon Matrix (Row-Reduced Echelon Form)

A matrix $A = [a_{ij}]$ is an *echelon matrix* or is said to be in *echelon form*, if the number of zeros preceding the first non-zero entry (known as *distinguished elements*) of a row increases row by row until only zero rows remain.

A matrix is said to be in row-reduced echelon form if the distinguished elements are unity and are the only non-zero entry irrespective of the zero elements in their respective columns.

For example The matrix

$$\begin{bmatrix} 0 & 1 & 1 & 0 & 1 \\ 0 & 0 & 0 & 1 & 0 \\ 0 & 0 & 0 & 0 & 0 \end{bmatrix}$$

illustrates the row-reduced echelon form, whereas the matrix

$$\begin{bmatrix} 0 & 0 & 0 \\ 1 & 1 & 0 \\ 0 & 0 & 1 \end{bmatrix}$$

do not come under this category. Similarly, the column-reduced echelon form can also be defined. It follows from the definition that the rank of a matrix is equal to the number of non-zero rows in its row-reduced echelon form.

EXAMPLE 5.22 Reduce the following matrix in the row-reduced echelon form:

$$\begin{bmatrix} 2 & 4 & -2 & 2 \\ 1 & 2 & -3 & 0 \\ 3 & 6 & -4 & 3 \end{bmatrix}$$

Solution Let

$$A = \begin{bmatrix} 2 & 4 & -2 & 2 \\ 1 & 2 & -3 & 0 \\ 3 & 6 & -4 & 3 \end{bmatrix} \sim \begin{bmatrix} 1 & 2 & -3 & 0 \\ 2 & 4 & -2 & 2 \\ 3 & 6 & -4 & 3 \end{bmatrix} \text{ Interchange } R_1 \text{ and } R_2$$

$$\sim \begin{bmatrix} 1 & 2 & -3 & 0 \\ 0 & 0 & 4 & 2 \\ 0 & 0 & 5 & 3 \end{bmatrix} R_2 \rightarrow R_2 - 2R_1, R_3 \rightarrow R_3 - 3R_1$$

$$\sim \begin{bmatrix} 1 & 2 & -3 & 0 \\ 0 & 0 & 1 & \frac{1}{2} \\ 0 & 0 & 5 & 3 \end{bmatrix} R_2 \rightarrow \frac{1}{4} R_2$$

$$\sim \begin{bmatrix} 1 & 2 & -3 & 0 \\ 0 & 0 & 1 & \frac{1}{2} \\ 0 & 0 & 1 & 1 \end{bmatrix} R_3 \rightarrow R_3 - 4R_2$$

$$\sim \begin{bmatrix} 1 & 2 & -3 & 0 \\ 0 & 0 & 1 & \frac{1}{2} \\ 0 & 0 & 0 & \frac{1}{2} \end{bmatrix} R_3 \rightarrow R_3 - R_2$$

Since the number of non-zero is 3, so, $\rho(A) = 3$, i.e., the rank of the given matrix is 3.

5.12.2 Normal Form of a Matrix

Every non-zero $m \times n$ matrix A of rank $r(\rho > 0)$ can be reduced by elementary transformation to any of the forms

$$\begin{bmatrix} I_r & 0 \\ 0 & 0 \end{bmatrix}, \quad (I_r, 0), \quad \begin{bmatrix} I_r \\ 0 \end{bmatrix}, \quad (I_r)$$

is called *normal* or *canonical form* of matrix A. The form

$$\begin{bmatrix} I_r & 0 \\ 0 & 0 \end{bmatrix}$$

is called the *first canonical form* of the matrix A.

5.12.3 Procedure of Reduction of a Matrix A to Its Normal Form

Let

$$A_{m \times n} = I_{m \times m} A_{m \times n} I_{n \times n} \tag{5.5}$$

Performing elementary row operations on A [Eq. (5.5)] and on the pre-factor $I_{m \times n}$, and applying elementary column operations on A and on the post-factor $I_{n \times n}$, the matrix A [on the left-hand side (LHS) of Eq. (5.5)] is reduced to its normal form. Then $I_{m \times m}$ reduces to $P_{m \times m}$ and $I_{n \times n}$ reduces to $Q_{n \times n}$, yielding $N = PAQ$, where P and Q are non-singular matrices.

Thus for any matrix A of rank r, there exists non-singular matrices P and Q such that

$$PAQ = N = \begin{bmatrix} I_r & 0 \\ 0 & 0 \end{bmatrix}$$

EXAMPLE 5.23 Reduce the matrix

$$A = \begin{bmatrix} 1 & -1 & -1 \\ 1 & 1 & 1 \\ 3 & 1 & 1 \end{bmatrix}$$

to the normal form as PAQ. Hence, find the rank of A.

Solution Consider $A_{3 \times 3} = I_{3 \times 3} A_{3 \times 3} I_{3 \times 3}$

i.e.,

$$\begin{bmatrix} 1 & -1 & -1 \\ 1 & 1 & 1 \\ 3 & 1 & 1 \end{bmatrix} = \begin{bmatrix} 1 & 0 & 0 \\ 0 & 1 & 0 \\ 0 & 0 & 1 \end{bmatrix} A \begin{bmatrix} 1 & 0 & 0 \\ 0 & 1 & 0 \\ 0 & 0 & 1 \end{bmatrix}$$

Every elementary row (column) transformation will be affected by pre-multiplication (post-multiplication) of A.

Operate $C_2 \to C_1 + C_2$
$C_3 \to C_1 + C_3$
Post

$$\begin{bmatrix} 1 & 0 & 0 \\ 1 & 2 & 2 \\ 3 & 4 & 4 \end{bmatrix} = \begin{bmatrix} 1 & 0 & 0 \\ 0 & 1 & 0 \\ 0 & 0 & 1 \end{bmatrix} A \begin{bmatrix} 1 & 1 & 1 \\ 0 & 1 & 0 \\ 0 & 0 & 1 \end{bmatrix}$$

Operate $R_2 \to R_2 + R_1$
$R_3 \to R_3 + 3R_3$
Pre

$$\begin{bmatrix} 1 & 0 & 0 \\ 0 & 2 & 2 \\ 0 & 4 & 4 \end{bmatrix} = \begin{bmatrix} 1 & 0 & 0 \\ -1 & 1 & 0 \\ -3 & 0 & 1 \end{bmatrix} A \begin{bmatrix} 1 & 1 & 1 \\ 0 & 1 & 0 \\ 0 & 0 & 1 \end{bmatrix}$$

Operate $R_2 \to 1/2\, R_2$
$R_3 \to 1/4\, R_3$
Pre

$$\begin{bmatrix} 1 & 0 & 0 \\ 0 & 1 & 1 \\ 0 & 1 & 1 \end{bmatrix} = \begin{bmatrix} 1 & 0 & 0 \\ -\dfrac{1}{2} & \dfrac{1}{2} & 0 \\ -\dfrac{3}{4} & 0 & \dfrac{1}{4} \end{bmatrix} A \begin{bmatrix} 1 & 1 & 1 \\ 0 & 1 & 0 \\ 0 & 0 & 1 \end{bmatrix}$$

Operate $R_3 \to R_3 - R_2$
Pre

$$\begin{bmatrix} 1 & 0 & 0 \\ 0 & 1 & 1 \\ 0 & 0 & 0 \end{bmatrix} = \begin{bmatrix} 1 & 0 & 0 \\ -\dfrac{1}{2} & \dfrac{1}{2} & 0 \\ -\dfrac{1}{4} & -\dfrac{1}{2} & \dfrac{1}{4} \end{bmatrix} A \begin{bmatrix} 1 & 1 & 1 \\ 0 & 1 & 0 \\ 0 & 0 & 1 \end{bmatrix}$$

Operate $C_3 \to C_3 - C_2$
Post

$$\begin{bmatrix} 1 & 0 & 0 \\ 0 & 1 & 0 \\ 0 & 0 & 0 \end{bmatrix} = \begin{bmatrix} 1 & 0 & 0 \\ -\dfrac{1}{2} & \dfrac{1}{2} & 0 \\ -\dfrac{1}{4} & -\dfrac{1}{2} & \dfrac{1}{4} \end{bmatrix} A \begin{bmatrix} 1 & 1 & 1 \\ 0 & 1 & -1 \\ 0 & 0 & 1 \end{bmatrix}$$

Thus, the LHS of above equation is in the normal form

$$\begin{bmatrix} I_2 & 0 \\ 0 & 0 \end{bmatrix}$$

Hence,

$$P_{3 \times 3} \begin{bmatrix} 1 & 0 & 0 \\ -\dfrac{1}{2} & \dfrac{1}{2} & 0 \\ -\dfrac{1}{4} & -\dfrac{1}{2} & \dfrac{1}{4} \end{bmatrix} \text{ and } Q_{3 \times 3} = \begin{bmatrix} 1 & 1 & 1 \\ 0 & 1 & -1 \\ 0 & 0 & 1 \end{bmatrix}$$

and the rank of $A = 2$.

5.13 SOLUTION OF LINEAR ALGEBRAIC EQUATIONS

A system of m linear algebraic equations in n unknowns $x_1, x_2, \ldots, x_n$ is a set of equations of the form

$$\begin{aligned} a_{11} x_1 + a_{12} x_2 + \cdots + a_{1n} x_n &= b_1 \\ a_{21} x_1 + a_{22} x_2 + \cdots + a_{2n} x_n &= b_2 \\ \cdots \quad \cdots \quad \cdots \quad \cdots \quad \cdots& \\ \cdots \quad \cdots \quad \cdots \quad \cdots \quad \cdots& \\ a_{m1} x_1 + a_{m2} x_2 + \cdots + a_{mn} x_n &= b_n \end{aligned} \quad (5.6)$$

In matrix notation, the above system (5.6) can be expressed as

$$Ax = b \quad (5.7)$$

where

$$A = \begin{bmatrix} a_{11} & a_{12} & \cdots & a_{1n} \\ a_{21} & a_{22} & \cdots & a_{2n} \\ \cdots & \cdots & \cdots & \cdots \\ \cdots & \cdots & \cdots & \cdots \\ a_{m1} & a_{m2} & \cdots & a_{mn} \end{bmatrix}, \quad x = \begin{bmatrix} x_1 \\ x_2 \\ \cdots \\ \cdots \\ x_n \end{bmatrix}, \text{ and } b = \begin{bmatrix} b_1 \\ b_2 \\ \cdots \\ \cdots \\ b_n \end{bmatrix}$$

in which A is called the *coefficient matrix*, x is the *column matrix of unknowns*, and b is the *column matrix of constants*.

If A is a non-singular matrix, i.e., $|A| \neq 0$, then its inverse A^{-1} exists. Multiplying both sides of $Ax = b$ by A^{-1}, one obtains

$$A^{-1}Ax = A^{-1}b$$
$$\Rightarrow \quad (A^{-1}A)x = A^{-1}b \quad \Rightarrow Ix = A^{-1}b$$
$$\Rightarrow \quad x = A^{-1}b$$

which is the required solution of the given set of equations. It may be noted here that when $|A| = 0$, i.e., A^{-1} does not exist, the method fails.

5.13.1 Linear Homogenous Equations ($Ax = 0$)

For a given set of equations, if $b = 0$, i.e., if $b_1, b_2, \ldots, b_n = 0$, then the matrix equation $Ax = b$ reduces to $Ax = 0$. Such a system of equations is called a *linear homogenous* equations.

Working rule for finding the solutions Reduce the coefficient matrix A to echelon form by applying elementary row transformations only. As a result, the rank of the matrix will be known.

Let the matrix A be of order $m \times n$ and its rank be $\rho(A) = r$.

If $r < m$, then $m - r$ equations will be eliminated. The given system of m equations will thus be replaced by an equivalent system of r equations. Solving these r equations one can express the values of some r unknowns in terms of the remaining $n - r$ unknowns. These $n - r$ unknowns can be given any arbitrary chosen values.

Conditions for existence of the solutions

(i) If $\rho(A) = n$, the number of unknowns, then $x = 0$ possesses always a solution, which is the *null* solution or the *trivial* solution. Thus, a homogenous system is always consistent.
(ii) If $\rho(A) < n$, the number of unknowns, the system possesses infinite number of solutions.

5.13.2 Linear Non-Homogenous Equations ($Ax = b$)

In a given set of equations, if all b_i are not zero, i.e., at least one b_i, or one of $b_1, b_2, \ldots, b_n$ is non-zero, then $b \neq 0$, i.e., $Ax = b$. This type of system of equations is called a *linear non-homogenous* equations.

To determine the solution of the system of linear non-homogenous equations one needs to define that augmented matrix, denoted by $[A:b]$.

The augmented matrix is the matrix in which the elements of A and b are written side by side.

5.13.3 Consistent and Inconsistent Equations

If the above system has a solution, then the equations are said to be *consistent* otherwise the equations are said to be *inconsistent*.

Note

(i) The system of equation $Ax = b$ is consistent, i.e., possesses a solution, iff the coefficient matrix A and the augmented matrix $[A:b]$ are of the same rank.

(ii) If A be an n-rowed non-singular matrix, x be an $n \times 1$ matrix, and b be an $n \times 1$ matrix, then the system of equation $Ax = b$ has a unique solution.

Working rule for determining the solutions Let the coefficient matrix A be of the type $m \times n$. Write the augmented matrix $[A:b]$ and reduce it to an echelon form by applying only elementary row transformation on it. This echelon form will give the ranks of the augmented matrix $[A:b]$ and the coefficient matrix A.

The following different cases arise:

Case I $\rho(A) < \rho[A:b]$

In this case, the equation $Ax = b$ is inconsistent, i.e., they have no solution.

Case II $\rho(A) = \rho[A:b] = r$, say

In this case, the equation $Ax = b$ is consistent.

If $\rho(A) < m$, then in the process of reducing the matrix $[A:b]$ to echelon form, $m - r$ equations will be eliminated. The given system of m equations will then be replaced by an equivalent system of r equations. From these r equations one shall be able to express the values of some r unknowns in terms of the remaining $n - r$ unknowns. These $n - r$ unknowns can be given any arbitrary chosen values.

If $\rho(A) = n$, then there will be a unique solution.

If $\rho(A) < n$, then $n - r$ variables can be assigned arbitrary values. So, in this case there will be an infinite number of solutions.

If $m < n$, then $r \leq m < n$. Thus, in this case $n - r > 0$. So, when the number of equations is less than the number of unknowns, the equations will always have an infinite number of solutions, provided they are consistent.

EXAMPLE 5.24 Test for the consistency of the following set of equations and solve it:

$$x + y + z = 0, \quad 2x - 5y + 3z = 0, \quad 2x + 5y + 7z = 0$$

Solution The given set of equations can be written in the matrix form as

$$\begin{bmatrix} 1 & 1 & 1 \\ 2 & -5 & 3 \\ 2 & 5 & 7 \end{bmatrix} \begin{bmatrix} x \\ y \\ z \end{bmatrix} = \begin{bmatrix} 0 \\ 0 \\ 0 \end{bmatrix}$$

which is of the type $Ax = 0$. Here,

$$A = \begin{bmatrix} 1 & 1 & 1 \\ 2 & -5 & 3 \\ 2 & 5 & 7 \end{bmatrix}$$

$$A \sim \begin{bmatrix} 1 & 1 & 1 \\ 0 & -7 & 1 \\ 0 & 3 & 5 \end{bmatrix} \begin{array}{l} \text{Operate } R_2 \to R_2 - 2R_1 \\ R_3 \to R_3 - 2R_1, \end{array}$$

$$A \sim \begin{bmatrix} 1 & 1 & 1 \\ 0 & 1 & -11 \\ 0 & 3 & 5 \end{bmatrix} \text{Operate } R_2 \to -(R_2 + 2R_3)$$

$$A \sim \begin{bmatrix} 1 & 1 & 12 \\ 0 & 1 & -11 \\ 0 & 0 & 1 \end{bmatrix} \begin{array}{l} \text{Operate } R_1 \to R_1 - 2R_2 \\ R_3 \to \left(\dfrac{1}{38} R_3 - 3R_2\right) \end{array}$$

$$A \sim \begin{bmatrix} 1 & 0 & 0 \\ 0 & 1 & 0 \\ 0 & 0 & 1 \end{bmatrix} \begin{array}{l} \text{Operate } R_1 \to R_1 - 12R_3 \\ R_2 \to R_2 + 11R_3 \end{array}$$

$\Rightarrow \rho(A) = 3 =$ number of unknowns. Thus, the system is consistent and possesses only a trivial solution as $x = 0, y = 0$, and $z = 0$.

EXAMPLE 5.25 Check for consistency the system of equations
$$x + y + z = 9, \quad 2x - y + z = 8, \quad 5x + y - 3z = 2$$

Solution Let

$$A = \begin{bmatrix} 1 & 1 & 1 \\ 2 & -1 & 1 \\ 5 & 1 & -3 \end{bmatrix}$$

Then
$$|A| = 20 \neq 0$$

So, rank $(A) = 3$.
Augmented matrix,

$$[A:b] = \begin{bmatrix} 1 & 1 & 1 & : & 9 \\ 2 & -1 & 1 & : & 8 \\ 5 & 1 & -3 & : & 2 \end{bmatrix}$$

Rank $[A:b]$ cannot be 4, since $[A:b]$ is of type 3×4.
Rank $[A:b] = 3$, since $[A:b]$ contains the 3-rowed minor.

$$\begin{vmatrix} 1 & 1 & 1 \\ 2 & -1 & 1 \\ 5 & 1 & -3 \end{vmatrix} \neq 0$$

Thus, rank $[A:b] =$ rank $(A) = 3$. So, the given system is consistent.

EXAMPLE 5.26 Test for the consistency of the following system of equations and hence solve:
$$x + y + z = 3, \quad x + 2y + 3z = 4, \quad x + 4y + 9z = 6$$

Solution In a matrix notation, the given system of equations can be written as

$$\begin{bmatrix} 1 & 1 & 1 \\ 1 & 2 & 3 \\ 1 & 4 & 9 \end{bmatrix} \begin{bmatrix} x \\ y \\ z \end{bmatrix} = \begin{bmatrix} 3 \\ 4 \\ 6 \end{bmatrix}$$

which is of the type $Ax = b$.

The augmented matrix of the system is then

$$[A:b] = \begin{bmatrix} 1 & 1 & 1 & : & 3 \\ 1 & 2 & 3 & : & 4 \\ 1 & 4 & 9 & : & 6 \end{bmatrix} \quad \text{Operate } R_2 \to R_2 - 2R_1$$
$$R_3 \to R_3 - 2R_1$$

$$\sim \begin{bmatrix} 1 & 1 & 1 & : & 3 \\ 0 & 1 & 2 & : & 1 \\ 0 & 0 & 8 & : & 0 \end{bmatrix} \quad \text{Operate } R_3 \to R_3 - 3R_2$$

$$\sim \begin{bmatrix} 1 & 1 & 1 & : & 3 \\ 0 & 1 & 2 & : & 1 \\ 0 & 0 & 2 & : & 0 \end{bmatrix} \quad \text{Operate } R_3 \to 1/2\, R_3$$

$$\sim \begin{bmatrix} 1 & 1 & 1 & : & 3 \\ 0 & 1 & 2 & : & 1 \\ 0 & 0 & 2 & : & 0 \end{bmatrix} \quad \text{Operate } R_1 \to R_1 - R_3$$
$$R_2 \to R_2 - 2R_3$$

$$\sim \begin{bmatrix} 1 & 1 & 1 & : & 3 \\ 0 & 1 & 2 & : & 1 \\ 0 & 0 & 1 & : & 0 \end{bmatrix} \quad \text{Operate } R_1 \to R_1 - R_2$$

$$\sim \begin{bmatrix} 1 & 0 & 0 & : & 2 \\ 0 & 1 & 0 & : & 1 \\ 0 & 0 & 1 & : & 0 \end{bmatrix}$$

Thus,

$$[A:b] = \begin{bmatrix} 1 & 0 & 0 & : & 2 \\ 0 & 1 & 0 & : & 1 \\ 0 & 0 & 1 & : & 0 \end{bmatrix} \quad \text{and } A \sim \begin{bmatrix} 1 & 0 & 0 \\ 0 & 1 & 0 \\ 0 & 0 & 1 \end{bmatrix}$$

Here, the rank of the augmented matrix is 3, i.e., $\rho[A:b] = 3$ and rank of the given matrix is also 3, i.e., $\rho(A) = 3$. Hence, $\rho[A:b] = 3$, which is equal to the number of unknowns. Therefore, the given system of equations is consistent and possesses a unique solution.

Finally, the given system of equations in matrix notation reduces to
$$\begin{bmatrix} 1 & 0 & 0 \\ 0 & 1 & 0 \\ 0 & 0 & 1 \end{bmatrix} \begin{bmatrix} x \\ y \\ z \end{bmatrix} = \begin{bmatrix} 2 \\ 1 \\ 0 \end{bmatrix}$$
$\Rightarrow x = 2, y = 1, z = 0$, which are the required solutions of the given system.

5.13.4 Gaussian Elimination (Direct Method)

The Gaussian elimination method is an elementary elimination method for solving linear systems. The method reduces the system of equations to an equivalent upper triangular system which can be solved by *back substitution*. It is a systematic elimination process, a method of great importance, that works in practice and is reasonable in respect to computing time and storage demand.

We first illustrate the method by considering two-variable linear system of equations such as
$$3x + 2y = 3$$
$$6x + 5y = 9$$
To solve the system we multiply the first equation by 2 and subtract it from the second, obtaining
$$3x + 2y = 3$$
$$-y = -3$$
The solution now follows by back substitution,
$$y = 3, \quad x = \frac{(3 - 2y)}{3} = \frac{(3 - 6)}{3} = -1$$

Since a linear system is completely determined by an augmented matrix, the elimination process can be done by merely considering the matrices. To link this correspondence, we shall write system of equations and augmented matrices side by side.

To generalize the method to a certain extent (for simplicity), we describe by considering a system of three equations as follows.

Let the system be
$$a_{11}x_1 + a_{12}x_2 + a_{13}x_3 = b_1$$
$$a_{21}x_1 + a_{22}x_2 + a_{23}x_3 = b_2 \qquad (5.8)$$
$$a_{31}x_1 + a_{32}x_2 + a_{33}x_3 = b_3$$

We first form the augmented matrix of the system (5.8), viz.,
$$\begin{bmatrix} a_{11} & a_{12} & a_{13} : b_1 \\ a_{21} & a_{22} & a_{23} : b_2 \\ a_{31} & a_{32} & a_{33} : b_3 \end{bmatrix} \qquad (5.9)$$

To eliminate x_1 from the second equation, we multiply the first equation by
$$-\frac{a_{21}}{a_{11}}, \quad a_{11} \neq 0$$

and add it to the second equation. Similarly, to eliminate x_1 from the third equation, we multiply the first equation by

$$-\frac{a_{31}}{a_{11}}$$

and add it to the third. This procedure can be shown thus

$$\begin{matrix} -\dfrac{a_{21}}{a_{11}} \\ \\ -\dfrac{a_{31}}{a_{11}} \end{matrix} \begin{bmatrix} a_{11} & a_{21} & a_{13} & : & b_1 \\ a_{21} & a_{22} & a_{23} & : & b_2 \\ a_{31} & a_{32} & a_{33} & : & b_3 \end{bmatrix} \qquad (5.10)$$

where

$$-\frac{a_{21}}{a_{11}} \quad \text{and} \quad -\frac{a_{31}}{a_{11}}$$

are called the *multipliers* for the *first* stage of elimination. The first equation in Eq. (5.8) is called the *pivotal* equation and a_{11} is called the first *pivot*. At the end of the first stage, the augmented matrix Eq. (5.10) becomes

$$\begin{matrix} \\ -\dfrac{a'_{32}}{a'_{22}} \end{matrix} \begin{bmatrix} a_{11} & a_{12} & a_{13} & : & b_1 \\ 0 & a'_{22} & a'_{23} & : & b'_2 \\ 0 & a'_{32} & a'_{33} & : & b'_3 \end{bmatrix} \qquad (5.11)$$

where $a'_{22}, a'_{23}, \ldots$, are all modified elements. Again, a'_{22} is the new pivot and the multiplier is $-a'_{32}/a'_{22}$. At the end of the second stage, we have the upper triangular system

$$\begin{bmatrix} a_{11} & a_{12} & a_{13} & : & b_1 \\ 0 & a'_{22} & a'_{23} & : & b'_2 \\ 0 & 0 & a''_{33} & : & b''_3 \end{bmatrix} \qquad (5.12)$$

from which the values of x_1, x_2, and x_3 can be obtained by back substitution.

It is seen that the method fails if one of the elements a_{11}, a'_{22}, or a''_{33} vanishes. In such a case, the method can be modified by re-arranging the rows so that the pivot is non-zero. This procedure is called *partial pivoting* and can be implemented in a computer. If this is not the case, then the matrix is singular and the equations have no solution. Importantly, *the number of non-zero diagonal elements in Eq. (5.12) will represent the rank of the original matrix.*

Gaussian elimination method is an efficient numerical method and can be implemented on high-speed digital computer. It should be noted that the numerical error in numerical procedure can also be controlled by this method. Therefore, the Gaussian elimination method is of much popular one. The following example will illustrate the method.

EXAMPLE 5.27 Solve the following system:

$$2x + y + z = 10$$
$$3x + 2y + 3z = 18$$
$$x + 4y + 9z = 16$$

Solution In the first step, the multipliers are $-3/2$ and $-1/2$. We multiply the first equation by $-3/2$ and $-1/2$ and add it to the second and third equations, respectively, to obtain the equations

$$\frac{1}{2}y + \frac{3}{2}z = 3$$
$$\frac{7}{2}y + \frac{17}{2}z = 11$$

The augmented matrix therefore becomes

$$\begin{bmatrix} 2 & 1 & 1 & : 10 \\ 0 & \frac{1}{2} & \frac{3}{2} & : 3 \\ 0 & \frac{7}{2} & \frac{17}{2} & : 11 \end{bmatrix}$$

At the second step, we eliminate y from the third equation by multiplying the second equation by -7 and adding it to the third. The resulting system will be upper triangular

$$2x + y + z = 10$$
$$\frac{1}{2}y + \frac{3}{2}z = 3$$
$$-2z = -10$$

which gives another augmented matrix

$$\begin{bmatrix} 2 & 1 & 1 & : 10 \\ 0 & \frac{1}{2} & \frac{3}{2} & : 3 \\ 0 & 0 & -2 & : -10 \end{bmatrix}$$

By back substitution we obtain the solution
$$x = 7, \quad y = -9, \quad \text{and} \quad z = 5$$

Note It can be observed from the last augmented matrix that there exists three non-zero elements in the diagonal, so the rank of the matrix evolved from the system of equations is 3.

5.14 EIGENVALUES AND EIGENVECTORS

Algebraic *eigenvalue* problems are the most important problems in connection with matrices.

Assume $A = [a_{jk}]$ as a given $n \times n$ matrix and consider the vector equation

$$Ax = \lambda x \tag{5.13}$$

where x is an unknown vector and λ an unknown scalar. Here, the purpose is to determine both.

A value of λ for which Eq. (5.13) possesses a solution $x \neq 0$ is called an *eigenvalue* or *characteristic value* of the matrix A. The corresponding solutions $x \neq 0$ of Eq. (5.13) are called *eigenvectors* or *characteristic vectors* of A for *eigenvalue* λ. Also, the set of eigenvalues is called the *spectrum* of A.

The set of all eigenvectors corresponding to an *eigenvalue* of A, together with 0, forms a *vector space*, called the *eigen space* of A for this *eigenvalue*.

The problem of determining the eigenvalues and eigenvectors of a matrix is called an *eigenvalue problem*, or more precisely an *algebraic eigenvalue problem* (since there are other eigenvalue problems dealing with a differential equation or an integral equation).

5.14.1 Determination of Eigenvalues and Eigenvectors

EXAMPLE 5.28 Find the eigenvalues and eigenvectors of the matrix

$$A = \begin{bmatrix} 8 & -4 \\ 2 & 2 \end{bmatrix}$$

Solution

(a) *To find the eigenvalues* The matrix A can be expressed in the form of Eq. (5.7) as

(i) $Ax = \begin{bmatrix} 8 & -4 \\ 2 & 2 \end{bmatrix} \begin{bmatrix} x_1 \\ x_2 \end{bmatrix} = \lambda \begin{bmatrix} x_1 \\ x_2 \end{bmatrix}$

which can be written in components,

(ii) $8x_1 - 4x_2 = \lambda x_1$
$2x_1 + 2x_2 = \lambda x_2$

Transferring the terms on the right side to the left

(iii) $(8 - \lambda)x_1 + (-4)x_2 = 0$
$2x_1 + (2 - \lambda)x_2 = 0$

The above equations can be written in matrix notation,

(iv) $(A - \lambda I)x = 0$ (here λx may be converted into $\lambda I x$.) Property (iv) is a homogenous linear system, which possesses a non-trivial solution $x \neq 0$ (an eigenvector of A) iff its coefficient as a determinantal form vanishes, i.e.,

$$D(\lambda) = \det|A - \lambda I| = \begin{vmatrix} 8 - \lambda & -4 \\ 2 & 2 - \lambda \end{vmatrix} = 0$$

$\Rightarrow \quad (8-\lambda)(2-\lambda)-(-4)2 = 0$
$\Rightarrow \quad \lambda^2 - 10\lambda + 24 = 0 \quad \Rightarrow \quad (\lambda-4)(\lambda-6) = 0$

which are the two distinct eigenvalues of A, i.e., $\lambda_1 = 4$ and $\lambda_2 = 6$. Here, $D(\lambda)$ is called the *characteristic determinant*. Expansion of this determinant gives the *characteristic polynomial*. Also, $D(\lambda) = 0$ is called *characteristic equation* of A.

(b) *To find the eigenvectors* Eigenvector of A corresponding to eigenvalue $\lambda_1 = 4$. The equations of the components are [from property (iii)]

$4x_1 - 4x_2 = 0$
$2x_1 - 2x_2 = 0$

$\Rightarrow x_1 = x_2$, which yields an eigenvector corresponding to an eigenvalue, $\lambda_1 = 4$. Assume $x_2 = 1$, then the eigenvector is

$$x_1 = \begin{bmatrix} 1 \\ 1 \end{bmatrix}$$

To verify the solution, let us proceed as follows:

$$Ax_1 = \begin{bmatrix} 8 & -4 \\ 2 & 2 \end{bmatrix} \begin{bmatrix} 1 \\ 1 \end{bmatrix} = \begin{bmatrix} 4 \\ 4 \end{bmatrix} = 4 \begin{bmatrix} 1 \\ 1 \end{bmatrix} = 4x_1$$

Eigenvector of A corresponding to eigenvalue $\lambda_2 = 6$.

Distributing $\lambda_2 = 6$ in property (iii), the components are obtained as

$$2x_1 - 4x_2 = 0$$
$$2x_1 - 4x_2 = 0$$

$\Rightarrow x_2 = \frac{1}{2}x_1$. If one assumes $x_1 = 2$, then $x_2 = 1$. Thus an eigenvector of A corresponding to $\lambda_2 = 6$ is

$$x_2 = \begin{bmatrix} 2 \\ 1 \end{bmatrix}$$

5.14.2 Linear Transformations

Following the above illustrations the general case of linear transformation can be described as follows.

Equation (5.13) acts as the linear transformation $\gamma = Ax$, which transforms a column vector x into a column vector γ through a square matrix A. Thus, if a vector x is transformed into a scalar multiple of the same vector, i.e., x is transformed to λx, then

$$\gamma = \lambda x = Ax \Rightarrow Ax = \lambda x = \lambda Ix$$
$$\Rightarrow (A - \lambda I)x = 0 \qquad (5.14)$$

Here,

$$A = \begin{bmatrix} a_{11} & a_{12} & \cdots & a_{1n} \\ a_{21} & a_{22} & \cdots & a_{2n} \\ \cdots & \cdots & \cdots & \cdots \\ \cdots & \cdots & \cdots & \cdots \\ a_{n1} & a_{n2} & \cdots & a_{nn} \end{bmatrix}, \quad X = \begin{bmatrix} x_1 \\ x_2 \\ \cdot \\ \cdot \\ x_n \end{bmatrix}, \quad \gamma = \begin{bmatrix} y_1 \\ y_2 \\ \cdots \\ \cdots \\ y_n \end{bmatrix}$$

Then Eq. (5.14) can be expressed, in general, as

$$\begin{aligned} (a_{11} - \lambda)x_1 + a_{12}x_2 + \cdots + a_{1n}x_n &= 0 \\ a_{21}x_1 + (a_{22} - \lambda)x_2 + \cdots + a_{2n}x_n &= 0 \\ \cdots \quad \cdots \quad \cdots \quad \cdots \quad \cdots & \\ \cdots \quad \cdots \quad \cdots \quad \cdots \quad \cdots & \\ a_{n1}x_1 + a_{n2}x_2 + \cdots + (a_{nn} - \lambda)x_n &= 0 \end{aligned} \qquad (5.15)$$

The homogenous linear system of equations possesses a non-trivial solution if and only if the determinant of the coefficients is zero, i.e.,

$$D(\lambda) = \det(A - \lambda I) = \begin{vmatrix} a_{11} - \lambda & a_{12} & \cdots & a_{1n} \\ a_{21} & a_{22} - \lambda & \cdots & a_{2n} \\ \cdots & \cdots & \cdots & \cdots \\ a_{n1} & a_{n2} & \cdots & a_{nn} - \lambda \end{vmatrix} = 0 \qquad (5.16)$$

where $D(\lambda)$ is called the *characteristic determinant*. Equation (5.16) is called the *characteristic equation* of the matrix A. The roots of the characteristic equation of A are called *characteristic roots* or *latent roots* or eigenvalues of A. By expanding $D(\lambda)$, one can obtain a polynomial of nth degree in λ, which is called the *characteristic polynomial* of A.

5.14.3 Properties of Eigenvalues and Eigenvectors

I. **Eigenvalues:** The eigenvalues of a square matrix A are the roots of the characteristic Eq. (5.12) of A. Hence, an $n \times n$ matrix has at least one eigenvalue and at most n numerically different eigenvalues.

II. **Eigenvectors:** If x is an eigenvector of a matrix A corresponding to an eigenvalue λ, so is kx with arbitrary $k \neq 0$ [since $Ax = \lambda x \Rightarrow k(Ax) = A(kx) = \lambda(kx)$].

III. **Real and complex eigenvalues:** If A is real, its eigenvalues are real or complex conjugates in pairs.

IV. **Determinant:** Determinant of $A = |A|$ = product of eigenvalues = $\lambda_1 \cdot \lambda_2 \cdots \lambda_n$.

V. **Singular matrix:** If at least one eigenvalue is zero, then $|A|$ = product of eigenvalues
$$= \lambda_1, \lambda_2, \ldots, \lambda_n = 0, \text{ i.e., } A \text{ is singular.}$$

VI. **Non-singular matrix:** If all the eigenvalues are non-zero, then $|A|$ = product of the eigenvalues $\neq 0$, i.e., A is non-singular.

VII. **Transpose:** A and A^T possess same eigenvalues. Since the diagonal elements in the determinants of A and A^T are same, the determinant

$$|A - \lambda I| \quad \text{and} \quad |A^T - \lambda I|$$

are equal; hence, possess the same eigenvalues

$$(|A| = |A^T|, \quad |A - \lambda I| = |(A - \lambda I)^T| = |A^T - (\lambda I)^T| = |A^T - \lambda I|).$$

VIII. **Inverse:** A^{-1} exists if 0 is not an eigenvalue of A.
Eigenvalues of A^{-1} are

$$\frac{1}{\lambda_1}, \frac{1}{\lambda_2}, \ldots, \frac{1}{\lambda_n}$$

i.e., the reciprocals of the eigenvalues of A.

$$Ax = \lambda x, \text{ pre-multiply by } A^{-1},$$
$$\Rightarrow \qquad A^{-1}Ax = A^{-1}\lambda x = \lambda A^{-1}x$$
$$\Rightarrow \qquad x = \lambda A^{-1}x \quad \Rightarrow A^{-1}x = \frac{1}{\lambda}x$$

IX. **Scalar multiples:** kA has eigenvalues $k\lambda$.

$$(A - \lambda I)x = 0$$

Multiplying both the sides by k, $k(A - \lambda I)x = 0$
Characteristic equation is

$$|k(A - \lambda I)| = 0$$
$$\Rightarrow \qquad |kA - k\lambda I| = 0$$

thus kA has eigenvalues $k\lambda$.

 EXAMPLE 5.29 Find the eigenvalues and eigenvectors of

$$A = \begin{bmatrix} -5 & 2 \\ 2 & -2 \end{bmatrix}$$

Solution The eigenvalues are the roots of the characteristic equation

$$\begin{bmatrix} -5-\lambda & 2 \\ 2 & -2-\lambda \end{bmatrix} = 0$$

$$\Rightarrow (-5-\lambda)(-2-\lambda) - 4 = 0, \quad \Rightarrow \lambda^2 + 7\lambda + 6 = 0 \Rightarrow (\lambda+6)(\lambda+1) = 0$$

The two distinct eigenvalues are $\lambda = -6, -1$.
Eigenvector corresponding to eigenvalue $\lambda = -6$.

$$(A - \lambda I)x = 0$$

$$\Rightarrow \begin{bmatrix} -5+6 & 2 \\ 2 & -2+6 \end{bmatrix} \begin{bmatrix} x_1 \\ x_2 \end{bmatrix} = 0 \quad \Rightarrow \begin{bmatrix} 1 & 2 \\ 2 & 4 \end{bmatrix} \begin{bmatrix} x_1 \\ x_2 \end{bmatrix} = 0$$

Thus,
$$x_1 + 2x_2 = 0, \quad 2x_1 + 4x_2 = 0$$

and its solution is $x_1 = 2x_2$, which yields an eigenvector corresponding to an eigenvalue $\lambda_1 = -6$. Assume $x_2 = -1$, then the eigenvector is

$$x_1 = \begin{bmatrix} 2 \\ -1 \end{bmatrix}$$

Now, corresponding to $\lambda_2 = -1$,

$$\begin{bmatrix} -5+1 & 2 \\ 2 & -2+1 \end{bmatrix} \begin{bmatrix} x_1 \\ x_2 \end{bmatrix} = 0 \quad \Rightarrow \begin{bmatrix} -4 & 2 \\ 2 & -1 \end{bmatrix} \begin{bmatrix} x_1 \\ x_2 \end{bmatrix} = 0$$

i.e.,
$$-4x_1 + 2x_2 = 0 \quad \text{and} \quad 2x_1 - x_2 = 0$$

$$\Rightarrow \quad x_2 = 2x_1$$

If one assumes $x_1 = 1$, then $x_2 = 2$. Thus, an eigenvector of A corresponding to $\lambda_2 = -1$ is

$$x_2 = \begin{bmatrix} 1 \\ 2 \end{bmatrix}$$

EXAMPLE 5.30 Find the eigenvalues and eigenvectors of

$$A^T = \begin{bmatrix} -5 & 2 \\ 2 & -2 \end{bmatrix}$$

Solution Characteristic equation is

$$\begin{bmatrix} -5-\lambda & 2 \\ 2 & -2-\lambda \end{bmatrix} = 0$$

i.e., $\lambda^2 + 7\lambda + 6 = 0$, same as the characteristic equation of A. Thus, the eigenvalues of A and A^T are same and the eigenvalues are $\lambda_1 = -6, \lambda_2 = -1$.

For $\lambda_1 = -6$,

$$\begin{bmatrix} -5+6 & 2 \\ 2 & -2+6 \end{bmatrix} \begin{bmatrix} x_1 \\ x_2 \end{bmatrix} = 0 \quad \Rightarrow \begin{bmatrix} 1 & 2 \\ 2 & 4 \end{bmatrix} \begin{bmatrix} x_1 \\ x_2 \end{bmatrix} = 0$$

Hence,
$$x_1 + 2x_2 = 0, \quad 2x_1 + 4x_2 = 0$$

$$\Rightarrow \quad x_1 = -2x_2$$

Assuming $x_2 = 1$, the eigenvector corresponding to an eigenvalue $\lambda_1 = -6$ is

$$x_1 = \begin{bmatrix} -2 \\ 1 \end{bmatrix}$$

Again, for the eigenvalue $\lambda_2 = -1$,

$$\begin{bmatrix} -5+1 & 2 \\ 2 & -2+1 \end{bmatrix} \begin{bmatrix} x_1 \\ x_2 \end{bmatrix} = 0 \Rightarrow \begin{bmatrix} -4 & 2 \\ 2 & -1 \end{bmatrix} \begin{bmatrix} x_1 \\ x_2 \end{bmatrix} = 0$$

i.e.,

$$-4x_1 + 2x_2 = 0 \quad \text{and} \quad 2x_1 - x_2 = 0$$
$$\Rightarrow x_2 = 2x_1$$

If $x_1 = -1$, then the eigenvector corresponding to $\lambda_2 = -1$, is

$$x_2 = \begin{bmatrix} -1 \\ -2 \end{bmatrix}$$

Thus, the above examples illustrate that the eigenvectors corresponding to the same eigenvalues of A and A^T are not the same.

EXAMPLE 5.31 Find the eigenvalues and eigenvectors of the matrix

$$A = \begin{bmatrix} 6 & -2 & 2 \\ -2 & 3 & -1 \\ 2 & -1 & 3 \end{bmatrix}$$

Solution The characteristic equation of A is

$$|A - \lambda I| = 0$$

i.e.,

$$\begin{bmatrix} 6-\lambda & -2 & 2 \\ -2 & 3-\lambda & -1 \\ 2 & -1 & 3 \end{bmatrix} = 0$$

Expanding along the first row

$$(6-\lambda)[(3-\lambda)^2 - 1] + 2[-2(3-\lambda) + 2] + 2[2 - 2(3-\lambda)] = 0$$
$$\Rightarrow \lambda^3 - 12\lambda^2 + 36\lambda - 32 = 0 \quad \Rightarrow (\lambda-2)(\lambda-2)(\lambda-8) = 0$$

i.e., $\lambda = 2, 2, 8$, which are the eigenvalues of the matrix A.
Corresponding to $\lambda = 2$, the eigenvectors are given by

$$[A - 2I]x = 0$$

i.e.,

$$\begin{bmatrix} 4 & -2 & 2 \\ -2 & 1 & -1 \\ 2 & -1 & 1 \end{bmatrix} \begin{bmatrix} x_1 \\ x_2 \\ x_3 \end{bmatrix} = 0$$

The above system of equations reduces to a single equation

$$2x_1 - x_2 + x_3 = 0$$

i.e., the rank of the coefficient matrix is 1, less than the number of unknowns by 2. So, independent characteristic vector can be obtained by specifying arbitrary values of 2 of the quantitative x_1, x_2, x_3.

Choosing $x_2 = 0$, $2x_1 = -x_3$ or $x_1/1 = x_3/-2$. Hence, $x_1 = x_2/0 = x_3/-2$. This gives eigenvectors as $(1, 0, -2)$.

Next, choose $x_3 = 0$, $2x_1 = x_2$, i.e., $x_1/1 = x_2/2$. Hence, $x_1/1 = x_2/2 = x_3/0$. This yields the other independent eigenvector as $(1, 2, 0)$. Any other eigenvector corresponding to $\lambda = 2$ will be a linear combination of these two eigenvectors.

Again, corresponding to $\lambda = 8$, the components are

$$-2x_1 - 2x_2 + 2x_3 = 0$$
$$-2x_1 - 5x_2 + x_3 = 0$$
$$2x_1 - x_2 - 5x_3 = 0$$

Solving any two of the equations,

$$\frac{x_1}{2} = \frac{x_2}{-1} = \frac{x_3}{1}$$

and the eigenvectors are $(2, -1, 1)$.

5.15 CAYLEY–HAMILTON THEOREM

Every square matrix satisfies its characteristic equation. Thus, if

$$|A - \lambda I| = a_0\lambda^n + a_1\lambda^{n-1} + a_2\lambda^{n-2} + \cdots + a_{n-1}\lambda + a_n = 0$$

is the characteristic equation of A, then

$$a_0 A^n + a_1 A^{n-1} + a_2 A^{n-2} + \cdots + a_{n-1} A + a_n I = 0 \tag{5.17}$$

5.15.1 Inverse of a Matrix

Cayley–Hamilton theorem may be used to obtain A^{-1}, if A is non-singular (i.e., $|A| \neq 0$). Equation (5.17) can be re-written as

$$-a_n I = a_0 A^n + a_1 A^{n-1} + a_2 A^{n-2} + \cdots + a_{n-1} A$$

Multiplying both sides by A^{-1} and transposing,

$$A^{-1} = -\frac{1}{a_n}[a_0 A^{n-1} + a_1 A^{n-2} + \cdots + a_{n-1} I] \tag{5.18}$$

EXAMPLE 5.32 Verify Cayley–Hamilton theorem for the matrix

$$A = \begin{bmatrix} 1 & 2 \\ 2 & -1 \end{bmatrix}$$

and using it determine A^{-1}.

Solution The characteristic equation is

$$|A - \lambda I| = \begin{bmatrix} 1-\lambda & 2 \\ 2 & -1-\lambda \end{bmatrix} = 0$$

$$\Rightarrow (\lambda - 1)(1+\lambda) - 4 = 0 \quad \Rightarrow \lambda^2 - 5 = 0$$

Now,

$$A^2 = A \cdot A = \begin{bmatrix} 1 & 2 \\ 2 & -1 \end{bmatrix} \cdot \begin{bmatrix} 1 & 2 \\ 2 & -1 \end{bmatrix} = \begin{bmatrix} 5 & 0 \\ 0 & 5 \end{bmatrix} = 5I$$

$$\Rightarrow A^2 - 5I = 0 \qquad (5.19)$$

Thus, A satisfies the characteristic equation.
To find A^{-1}, multiply both the sides of Eq. (5.19) by A^{-1}, i.e.,

$$A^{-1} \cdot A^2 - 5A^{-1}I = 0$$
$$\Rightarrow A - 5A^{-1} = 0$$
$$\Rightarrow A - 1 = \frac{1}{5}A = \frac{1}{5}\begin{bmatrix} 1 & 2 \\ 2 & -1 \end{bmatrix}$$

Thus,

$$A^{-1} = \frac{1}{5}\begin{bmatrix} 1 & 2 \\ 2 & -1 \end{bmatrix}$$

SUMMARY

Matrices are used throughout discrete mathematics to express relationships between elements in sets. Matrices are useful in tackling an array of many numbers as a single object, denote it by a single symbol, and perform calculations with these symbols in a very compact form. Matrix arithmetic operations are imperative in model of communication network and transportation system.

The definitions of matrices and related concepts are presented with a discussion of two algebraic operations for matrices: addition and scalar multiplication. We have defined typical square matrices, namely orthogonal matrix, involutory matrix, nilpotent matrix, and adjoint and inverse of a matrix. We have also demonstrated the evaluation of rank of a matrix with a number of illustrations. We then have continued to discuss row-reduced echelon form and normal form of a matrix. This chapter is also centred around linear system of equations accompanied with eigenvalues and eigenvectors. Importantly, this includes Gaussian elimination, and the basic existence and uniqueness problem for solutions. A discussion on the Cayley–Hamilton theorem has also been provided.

EXERCISES

1. If A and B are two matrices such that $AB = O$, then show by means of example that it does not necessarily mean either $A = 0$ or $B = 0$, where 0 stands for the null matrix.

2. If a matrix A satisfies a relation $A^2 + A - I = 0$, prove that A^{-1} exists and $A^{-1} = I + A$, being an identity matrix.

3. Let

$$A = \begin{bmatrix} 1 & 2 & 0 \\ 3 & -1 & 4 \end{bmatrix}$$

Find (i) AA^T and (ii) A^TA.

4. Let $f(x) = 2x^3 - 4x + 5$ and $g(x) = x^2 + 2x - 11$. Given the matrix

$$A = \begin{bmatrix} 1 & 2 \\ 4 & -3 \end{bmatrix}$$

Determine (i) $f(A)$; (ii) $g(A)$.

5. Let A be an invertible matrix with inverse B. In other words $AB = BA = I$. Show that the inverse matrix B is unique.

6. Let A and B be invertible matrices of the same order. Show that AB is invertible and that $(AB)^{-1} = B^{-1}A^{-1}$.

7. Show that the matrix
$$\begin{bmatrix} 2 & 5 & -7 \\ -9 & 12 & 4 \\ 15 & -13 & 6 \end{bmatrix}$$
can be expressed as a sum of a lower triangular matrix and an upper triangular matrix with zero leading diagonal.

8. Investigate whether the following matrices are orthogonal or not.

(i) $\begin{bmatrix} \cos\theta & 0 & \sin\theta \\ 0 & 1 & 0 \\ \sin\theta & 0 & \cos\theta \end{bmatrix}$ (ii) $\dfrac{1}{3}\begin{bmatrix} 1 & -2 & 2 \\ -2 & 1 & 2 \\ -2 & -2 & -1 \end{bmatrix}$

9. Find the inverse of the following matrices:

(i) $A = \begin{bmatrix} 1 & -2 & 2 \\ 2 & -3 & 6 \\ 1 & 1 & 7 \end{bmatrix}$

(ii) $B = \begin{bmatrix} 1 & 3 & -4 \\ 1 & 5 & -1 \\ 3 & 13 & -6 \end{bmatrix}$

10. Show that the matrix
$$A = \begin{bmatrix} -5 & -8 & 0 \\ 3 & 5 & 0 \\ 1 & 2 & -1 \end{bmatrix}$$
is involutory.

11. Show that the matrix
$$A = \begin{bmatrix} 1 & 1 & 3 \\ 5 & 2 & 6 \\ -2 & -1 & -3 \end{bmatrix}$$
is nilpotent matrix of order 3.

12. Determine the rank of the following matrices:

(i) $\begin{bmatrix} 4 & 2 & 3 \\ 8 & 4 & 6 \\ -2 & -1 & -1.5 \end{bmatrix}$

(ii) $\begin{bmatrix} 1 & 2 & 3 \\ 1 & 4 & 2 \\ 2 & 6 & 5 \end{bmatrix}$

(iii) $\begin{bmatrix} 0 & 1 & -3 & -1 \\ 0 & 0 & 1 & 1 \\ 3 & 1 & 0 & 2 \\ 1 & 1 & -2 & 0 \end{bmatrix}$

(iv) $\begin{bmatrix} 1 & 2 & -2 & 3 \\ 2 & 5 & -4 & 6 \\ -1 & -3 & 2 & -2 \\ 2 & 4 & -1 & 6 \end{bmatrix}$

13. Find the values of p such that the rank of
$$A = \begin{bmatrix} 1 & 1 & -1 & 0 \\ 4 & 4 & -3 & 1 \\ p & 2 & 2 & 2 \\ 9 & 9 & p & 3 \end{bmatrix}$$
is 3.

14. Interchange the rows in each matrix to obtain an echelon matrix.

(i) $\begin{bmatrix} 0 & 1 & -3 & 4 & 6 \\ 4 & 0 & 2 & 5 & -3 \\ 0 & 0 & 7 & -2 & 8 \end{bmatrix}$

(ii) $\begin{bmatrix} 0 & 0 & 0 & 0 & 0 \\ 1 & 2 & 3 & 4 & 5 \\ 0 & 0 & 5 & -4 & 7 \end{bmatrix}$

15. Find the Echelon form and row canonical (or, row reduced) Echelon form of the following matrices.

(i) $\begin{bmatrix} 1 & 2 & -5 \\ -4 & 1 & -6 \\ 6 & 3 & -4 \end{bmatrix}$ (ii) $\begin{bmatrix} 1 & 3 & -1 & 2 \\ 0 & 11 & -5 & 3 \\ 2 & -5 & 3 & 1 \\ 4 & 1 & 1 & 5 \end{bmatrix}$

16. Find the non-singular matrices P and Q such that PAQ is in the normal form for the following matrices:

(i) $\begin{bmatrix} 1 & 1 & 2 \\ 1 & 2 & 3 \\ 0 & -1 & -1 \end{bmatrix}$

(ii) $\begin{bmatrix} 1 & 3 & 3 & -1 \\ 1 & 4 & 5 & 1 \\ 1 & 5 & 4 & 3 \end{bmatrix}$

17. Find P and Q such that the normal form of
$$A = \begin{bmatrix} 1 & -1 & -1 \\ 1 & 1 & 1 \\ 3 & 1 & 1 \end{bmatrix}$$
is PAQ. Hence, find the rank of A.

18. Let
$$A = \begin{bmatrix} 1 & 0 & 0 \\ 0 & 0 & 1 \\ 1 & 1 & 0 \end{bmatrix} \text{ and } B = \begin{bmatrix} 0 & 1 & 1 \\ 1 & 0 & 0 \\ 0 & 1 & 0 \end{bmatrix}$$
be Boolean matrices. Determine the Boolean products AB, BA, and A^2.

19. Find the values of λ such that the system of homogeneous equations
$$2x + y + 2z = 0, \quad x + y + 3z = 0,$$
$$4x + 3y + \lambda z = 0 \text{ possitive (1) trivial}$$
solutions non-trival solution. Find the non-trival solution

20. Determine the values of λ and μ for which the following system of equations possesses (i) no solution, (ii) unique solution, and (iii) infinitely many solutions:
    (a) $x + y + z = 6$
    $x + 2y + 3z = 10$
    $x + 2y + \lambda z = \mu$
    (b) $2x + 3y + 5z = 9$
    $7x + 3y + -2z = 8$
    $2x + 3y + \lambda z = \mu$

21. Test for the consistency of the following set of equations and solve:
    (i) $x + 2y + 3z = 6$, $2x + 3y = 11$,
    $4x + y - 5z = -3$
    (ii) $5x + 3y + 7z = 4$,
    $3x + 26y + 2z = 9$,
    $7x + 2y + 10z = 5$
    (iii) $4x - 2y + 6z = 8$,
    $x + y - 3z = -1$,
    $15x - 3y + 9z = 21$
    (iv) $x + y + z = 6$, $x - y + 2z = 5$,
    $3x + y + z = 8$, $2x - 2y + 3z = 7$

22. Show that if $\lambda \neq -5$ the system of equations
    $3x - y + 4z = 3$, $x + 2y - 3z = -2$,
    $6x + 5y + \lambda z = -3$
    possesses unique solution. If $\lambda = -5$, show that the equations are consistent. Determine the solutions in each case.

23. Solve the following system using the augmented matrix M:
    $x - 2y + 4z = 2$, $2x - 3y + 5z = 3$,
    $3x - 4y + 6z = 7$

24. Find the eigenvalues and eigenvectors of the matrices.
    (i) $\begin{bmatrix} 3 & 1 & 4 \\ 0 & 2 & 6 \\ 0 & 0 & 5 \end{bmatrix}$ (ii) $\begin{bmatrix} 1 & 1 & 3 \\ 1 & 5 & 1 \\ 3 & 1 & 1 \end{bmatrix}$
    (iii) $\begin{bmatrix} 3 & -1 & 1 \\ -1 & 5 & -1 \\ 1 & -1 & 3 \end{bmatrix}$

25. Find the sum and product of the eigenvalues of the matrix
$$\begin{bmatrix} 7 & 2 & 2 \\ -6 & -1 & 2 \\ 6 & 2 & -1 \end{bmatrix}$$
without evaluating classically (eigenvalues).

26. If
$$A = \begin{bmatrix} 2 & 2 & 2 \\ 0 & 5 & 2 \\ 0 & 0 & 4 \end{bmatrix}$$
determine the eigenvalues of A, A^2, A^{-1}, and A^{100}.

27. Find the characteristic equation of the matrix
$$A = \begin{bmatrix} 2 & 1 & 1 \\ -1 & 2 & -1 \\ 1 & -1 & 2 \end{bmatrix}$$
Show that the equation is satisfied by A.

28. If
$$A = \begin{bmatrix} 1 & -2 & 3 \\ 2 & 3 & -1 \\ -3 & 1 & 1 \end{bmatrix}$$
obtain the characteristic polynomial for the matrix and show how this can be used to find A^{-1}.

29. Use the Cayley–Hamilton theorem to find A^{-1} of the matrix
$$A = \begin{bmatrix} 1 & 2 & 3 \\ 2 & 4 & 5 \\ 3 & 5 & 6 \end{bmatrix}$$

30. Verify Cayley–Hamilton theorem for the matrix
$$A = \begin{bmatrix} 11 & -4 & -7 \\ 7 & -2 & -5 \\ 10 & -4 & -6 \end{bmatrix}$$
Hence, compute A^{-1} if possible.

6 Order, Relation, and Lattices

LEARNING OBJECTIVES

After reading this chapter you will be able to understand the following:
- Definitions of partially ordered set and their properties
- Construction of Hasse diagram and topological sorting
- Ordering of the letters in the alphabet on a set constructed from a partial ordering on the set, known as *lexicographic order*
- About lattices which play an important role in multilevel security policy used in government and military systems, and in Boolean algebra
- Some special class of lattices, namely complete lattice, bounded lattice, modular lattice, and finally homomorphism

6.1 INTRODUCTION

We have clearly defined various types of relations on a set in Chapter 1. In this chapter, we discuss relations associated with the idea of 'ordering'. The notion of ordering on a set of data implies the order in which the items are arranged in a sequence. With this the conception grows up, what is called a linear ordering (because the elements are thought of as being arranged in some sort of 'line') or a total ordering (in which two elements in the set are comparable). Often we have only partial information: sometimes, given a and b, we can definitely assert that a precedes b; sometimes, not. In other way, the ordering may be inherently partial, in which case there would be no reason to place either a or b first. Thus, a partial ordering is obtained, which is the most general kind of ordering we shall consider here.

Order and relationships occur in several areas of mathematics and computer science. The relationships and ordering finally tend to the concepts of lattices which will be discussed in this chapter. This chapter also describes some of the important properties of lattices and their applications in respect to homomorphism and isomorphism. These structures will be of much use to students towards the understanding of the concepts of Boolean algebra and switching theory.

6.2 PARTIALLY ORDERED SET

Let R be a relation on a set S. Then the relation R is called a *partial order* if R is reflexive, antisymmetric, and transitive. The set S with the partial order R is said to be a *partially ordered set*, or POSET and is denoted by (S, R) or simply by S.

For example

(i) Let S be any collection of sets. The relation $\subseteq$ of set inclusion is a partial order of S. In particular, $A \subseteq A$ for any set S; if $A \subseteq B$ and $B \subseteq A$ then $A = B$; and if $A \subseteq B$ and $B \subseteq C$ then $A \subseteq C$, so $(S, \subseteq)$ is a poset.

(ii) The set of natural numbers, N, forms a poset with respect to the relation $\leq$. Specifically, $a \leq a$; if $a \leq b$ and $b \leq a$, then $a = b$; and finally, if $a \leq b$ and $b \leq c$, then $a \leq c$ for all $a, b, c \in N$.

(iii) The relation of divisibility ($a\, R\, b$ if and only if $a|b$, '|' signifies divisibility) is not a partial order of the set Z of integers. In particular, the relation is not antisymmetric, for example, $3|-3$ and $-3|3$, but $3 \neq -3$.

(iv) The relation of divisibility $a\, R\, b$ iff $a|b$ is a partial order on Z^+, since, $a|a$ for every $a \in Z^+$; if $a|b$ and $b|a$, then $a = b$; if $a|b$ and $b|c$, then $a|c$ for every $a, b, c \in Z^+$.

(v) The relation $<$ on Z^+ is not a partial order because it is not reflexive.

The most familiar partial orders, called the *usual order*, are the relations $\leq$ and $\geq$ on the positive integers Z or, more generally, on any subset of the real numbers R. So, while thinking in general of a partial order R on a set S, the symbols $\leq$ and $\geq$ (does not necessarily mean 'less than or equal to' or 'greater than or equal to') will be used frequently for R.

If R be a partial order on a set S and if R^{-1} be the inverse relation of R, then R^{-1} is also a partial order. Thus, the poset (S, R^{-1}) can also be called the dual of (S, R). Whenever $(S, \leq)$ is a poset, then $(S, \leq^{-1})$ is also a poset. However, the symbol $\geq$ can be used for $\leq^{-1}$ and the poset $(S, \geq)$ is called the *dual poset* of $(S, \leq)$.

6.2.1 Comparability of Elements

Let $(S, \leq)$ be a poset and $a, b \in S$. The elements a and b are said to be *comparable* if $a \leq b$ or $b \leq a$.

When a and b are elements of the poset $(S, \leq)$, it is not necessary that either $a \leq b$ or $b \leq a$. For instance, in the relation of divisibility on Z, 3 is not related to 4 and 4 is not related to 3, since neither $3 \dagger 4$ nor $4 \dagger 3$ ('$\dagger$' signifies non-divisibility).

In a partially ordered set, every pair of elements need not be comparable, for example, the elements 2 and 5 are not comparable, since $2 \dagger 5$ and $5 \dagger 2$ ('$\dagger$' signifies non-divisibility) or $2 \| 5$ and $5 \| 2$ ('$\|$' denotes non-comparability). Thus, by the word 'partial' in partially ordered set, one should recognize that some elements may not be comparable. Hence, in general, the elements a and b are non-comparable if neither $a \leq b$ nor $b \leq a$, or $a \| b$ and $b \| a$.

EXAMPLE 6.1 If $S = (1, 2, 3, 5, 6, 10, 15)$ is ordered by divisibility, find all the comparable and non-comparable pairs of elements of S.

Solution The comparable pairs of elements of S are

$$(1, 2), (1, 3), (1, 5), (1, 6), (1, 10), (1, 15)$$

(2, 6), (2, 10)
(3, 6), (3, 15)
(5, 15)

The non-comparable pairs of elements of S are

(2, 3), (2, 5), (2, 15)
(3, 5), (3, 10), (5, 6), (6, 10), (6, 15), (10, 15)

EXAMPLE 6.2 Let Z be the set of positive integers. Prove that the relation 'divisibility' and 'integral multiplicity' are partial orderings on Z.

Solution Let us consider a and b as positive integers. Then we say that 'a divides b' represented as a/b iff there exists an integer c such that $ac = b$ (or b is an integral multiple of a). Again, for any $a, b \in x$, set $a \leq b \Leftrightarrow a/b$ and $ac = b$.

(i) Reflexive: Since $a.1 = a, a/a$ and hence $a \leq a$.
(ii) Antisymmetric: Suppose $a \leq b$ and $b \leq a$. Then there exists $c, d \in Z$ such that $ac = b$ and $bd = a$. Since $a = bd = acd \Rightarrow cd = 1, c = 1$ and $d = 1$. Thus, $a = b$.
(iii) Transitive: Assume for any $a, b, c \in Z$, $a \leq b, b \leq c$. Then there exists $x, y \in Z$ such that $ax = b$ and $by = c$. Now, $axy = by = c$. Also, since $x, y \in Z$, a/c. Thus, $a \leq c$.

Hence, $(Z, \leq)$ is a partial order.

6.2.2 Linearly Ordered Set

Let $(S, \leq)$ be a poset. The set S is called *linearly ordered set* or *totally ordered set*, if every pairs of elements in a poset S is comparable, i.e., if for every $a, b \in S$, there exists $a \leq b$ or $b \leq a$ ('$\leq$' is called linear ordering or total ordering on S). Also, the set S, here, can be named as a *chain*. A chain is said to be *antichain* if no two distinct elements of the set are related.

Note The word 'partial' in partially ordered set means that some of the elements of the set may not be comparable.

For example

(i) Let Z be the set of positive integers ordered by divisibility. The elements 3 and 6 are comparable, since $3|6$; similarly, the elements 3 and 9 are comparable, since $3|9$.
(ii) The set of positive integers Z^+, with the usual order $\leq$ is a linearly ordered set.

EXAMPLE 6.3 If the set $Z = (1, 2, 3, \ldots,)$ is ordered by divisibility, then investigate each of the following subsets of Z for linearly ordered subset.

(a) (2, 4, 8)   (b) (3, 6, 9, 11)   (c) (1)   (d) (2, 4, 6, 8, 10)

Solution

(a) Here, the pair of elements (2, 4), (2, 8), (4, 8), with respect to the relation divisibility, is comparable, i.e., the given subset is linearly ordered.
(b) Since the pair (3, 11) is not comparable, the subset is not linearly ordered.
(c) The subset is linearly ordered, since the set containing one element is always linearly ordered.

(d) The subset is not linearly ordered, since every pair of elements is not comparable, i.e., neither 4|6 nor 6|4, or 4|6 and 6|4.

EXAMPLE 6.4 Let $B = \{a, b, c, d, e\}$ be ordered alphabetically for letters. Draw the diagram of B. What is your observation from the diagram?

Solution The diagram is shown in Figure 6.1. From the diagram it is observed that there exists a single path starting with a and ending at e, since B is linearly ordered by the given relation.

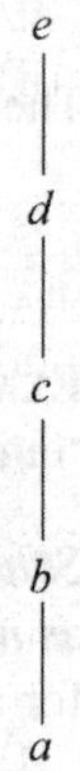

Theorem 6.1 If $(S, \leq)$ and $(T, \leq)$ are posets, then $(S \times T, \leq)$ is a poset, with partial order $\leq$ defined by

$$(a, b) \leq (a', b') \text{ if } a \leq a' \text{ in } S \text{ and } b \leq b' \text{ in } T$$

Proof Given that $a \leq a'$ in S and $b \leq b'$ in T. Then $(a, b) \in S \times T$ $\Rightarrow (a, b) \leq (a, b)$. Hence, $\leq$ satisfies the reflexive property in $S \times T$.

FIGURE 6.1 Diagram of the set B

Again, assume $(a, b) \leq (a', b')$ and $(a', b') \leq (a, b)$, where $a, a' \in S$ and $b, b' \in T$. Then

$$a \leq a' \quad \text{and} \quad a' \leq a \text{ in } A$$

and

$$b \leq b' \quad \text{and} \quad b' \leq b \text{ in } B$$

Since S and T are posets, the antisymmetric property of the partial orders in S and T gives

$$a = a' \quad \text{and} \quad b = b'$$

Hence, $\leq$ satisfies the antisymmetric property in $S \times T$.
Lastly, assume that

$$(a, b) \leq (a', b') \text{ and } (a', b') \leq (a'', b'')$$

where

$$a, a', a'' \in S \quad \text{and} \quad b, b', b'' \in T$$

Then,

$$a \leq a' \quad \text{and} \quad a' \leq a'' \Rightarrow a \leq a''$$

by the transitive property of the partial order in S. Similarly,

$$b \leq b' \quad \text{and} \quad b' \leq b'' \Rightarrow b \leq b''$$

by the transitive property of the partial order in T. Hence,

$$(a, b) \leq (a'', b'')$$

Thus, the transitive property holds for the partial order in $S \times T$. Hence, $(S \times T, \leq)$ is a poset.

Note

(i) The above theorem is useful in constructing a new poset from given posets.

(ii) The partial order $\leq$ defined on the Cartesian product $S \times T$ (as above) is known as the *product partial order*.

EXAMPLE 6.5 Assume $N \times N$ is given as product order, where N has usual order $\leq$. Insert the correct symbol, $<$ and $>$, between each of the following pairs of elements of $N \times N$:

(i) (4, 6)–(4, 2) (ii) (1, 3)–(1, 7) (iii) (7, 9)–(4, 1)

Solution Here $(a, b) \leq (a', b')$ provided $a \leq a'$ and $b \leq b'$. Hence, $(a, b) < (a', b')$ if $a < a'$ and $b \leq b'$ or if $a \leq a'$ and $b < b'$.

(i) $>$; since $4 \geq 4$ and $6 > 2$
(ii) $<$; since $1 \leq 1$ and $3 < 7$
(iii) $>$; since $7 > 4$ and $9 > 1$

6.2.3 Cover of an Element

Let $(S, \leq)$ be a poset. We say that an element $y \in S$ *covers* an element $x \in S$ if $x < y$ and there is no element $z \in S$ such that $x < z < y$. If y covers x then a line is drawn between the elements x and y such that $x < y$.

The set of pairs (x, y) such that y covers x is called the *covering relation* of $(S, \leq)$.

For example The covering relations of the partial ordering $\{(a, b): \text{'}a \text{ divides } b\text{'}\}$ on $\{1, 2, 3, 4, 6, 12\}$ are the following sets:

$$(2, 4), (2, 6), (2, 12), (3, 6), (3, 12), (4, 12)$$

6.3 HASSE DIAGRAM

A partial ordering $\leq$ on set S can be represented by means of a diagram known as *Hasse diagram* of $(S, \leq)$. The diagram acts as a very useful tool in describing completely the associated partial order. So, it is also known as ordering diagram.

Suppose S is a finite partially ordered set. Then the order of S is completely known once one knows all pairs a, b in S such that $a \ll b$ (i.e., b is a cover of a, or, b is an immediate successor of a), i.e., if one knows the relation '$\ll$' on S. This follows from the fact that $a < b$ iff $a \ll b$ or there exists elements $a_1, a_2, \ldots, a_m$ in S such that

$$a \ll a_1 \ll a_2 \ll \cdots \ll a_m \ll b$$

The Hasse diagram of a finite partially ordered set S, is the directed graph (*given in Section 1.19*), can be drawn following the procedure given below:

(i) Since a partial order is reflexive, so each vertex of S must be related to itself (i.e., every vertex in the digraph acts as a cycle, defined as a path that begins and ends at the same vertex, of length 1). Consequently, for convenience all such cycles are deleted in Hasse diagram. Hence, the digraph shown in Figure 6.2(i) is to be drawn like Figure 6.2(ii).

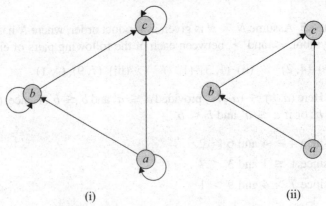

FIGURE 6.2 Digraphs

(ii) Since a partial order is transitive, thus, if $a \leq b$ and $b \leq c$, it follows that $a \leq c$. In this case, the edge from a to c is omitted, but the edges from a to b and b to c are drawn. For instance, the digraph depicted in Figure 6.3(i) is to be drawn as shown in Figure 6.3(ii).

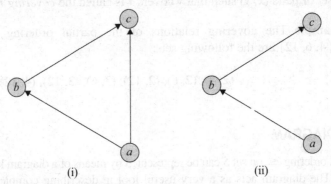

FIGURE 6.3 Digraphs

(iii) If a vertex a is connected to vertex b, i.e., aRb, then vertex b appears to be an *immediate successor* of a, i.e., the movement upwards indicates succession. Thus, the arrows may be omitted.

(iv) Finally, the vertices in Hasse diagram are denoted by dots rather than circles. Thus, the diagram shown in Figure 6.4 gives the ultimate form of the digraph, as shown in Figure 6.2(i).

The resulting diagram of a partial order, much simpler than its digraph, is known as the *Hasse diagram* of the partial order of the poset $(S, \leq)$.

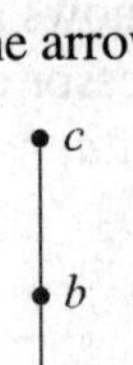

FIGURE 6.4 Hasse diagram of partial order of $(S, \leq)$

Theorem 6.2 The Hasse diagram for a given set possesses different Hasse diagrams.

Proof We will prove the theorem using an illustration.

Consider the set $A = \{a, b\}$, the relation of inclusion $\leq$ on $S(A)$ as a partial ordering. The Hasse diagram of $(S(A), \leq)$ is depicted in Figure 6.5.

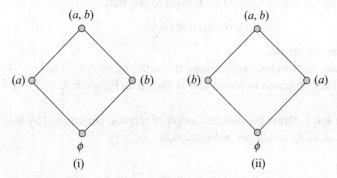

FIGURE 6.5 Hasse diagram of $(S(A), \leq)$

If $(A, \leq)$ is a poset, the Hasse diagram of $(A, \leq)$ can be formed by revolving the Hasse diagram of $(A, \leq)$ through π^c or $180°$, so that the point at the top becomes the point at the bottom. However, some of the Hasse diagrams possess a unique point which lies at the top in comparison to other points in the diagram. Thus, for a given set different Hasse diagrams can be obtained.

Corollary 6.1 The Hasse diagram represented by the partially ordered set or the poset $(A, \leq)$ is not unique.

EXAMPLE 6.6 Let R be the relation on the set $S = (a, b, c, d)$. Draw the directed graph and the Hasse diagram of R.

Solution The relation $\leq$ on the set S is given by

$$R = \{(a, b), (a, c), (a, d), (b, c), (b, d), (c, d), (a, a), (b, b), (c, c), (d, d)\}$$

The directed graph of the relation R is shown in Figure 6.6.

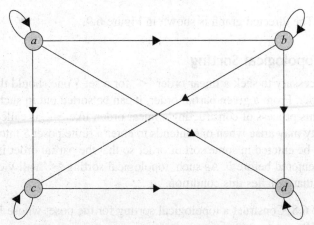

FIGURE 6.6 Directed graph of the relation R

To draw the Hasse diagram of partial order of the poset $(S, \leq)$ use the following norms:

1. Omit all edges implied by reflexive property,

    $(a, a), (b, b), (c, c),$ and (d, d)

2. Neglect all edges implied by transitive property,

    $(a, d), (b, d),$ and (a, c)

3. Delete the arrows.
4. Replace the circles, representing the vertices, by dots. Thus, the Hasse diagram is drawn and is shown in Figure 6.7.

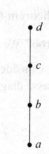

FIGURE 6.7 Hasse diagram of the poset $(S, \leq)$

EXAMPLE 6.7 Draw the directed graph of relation determined by the Hasse diagram on the set $S = \{1, 4, 6, 8\}$ as shown in Figure 6.8.

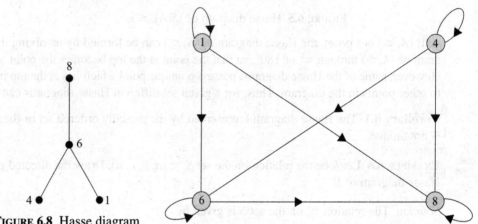

FIGURE 6.8 Hasse diagram on the set $S = \{1, 4, 6, 8\}$

FIGURE 6.9 Directed graph on the set $S = \{1, 4, 6, 8\}$

Solution The directed graph is shown in Figure 6.9.

6.3.1 Topological Sorting

If it is necessary to seek a linear order '<' for a set S one should think of a poset $(S, \leq)$ with partial order. From a given partial order, it can be sorted out in such a way that if $a \leq b$, then $a < b$. This process of constructing a linear order, like '<', is called *topological sorting*. This complexity may arise when one intends to enter a finite poset S into a computer. The elements of S must be entered in some sort of order so that the partial order is maintained, i.e., if $a \leq b$ then a is entered before b. As such, topological sorting '<' will yield an order of entry of the elements that satisfies this condition.

EXAMPLE 6.8 Construct a topological sorting for the poset whose Hasse diagram is shown in Figure 6.10.

Solution The partial order $<$ whose Hasse diagram is depicted in Figure 6.11(a) is a linear order. Also, every pair exists in the order $<$, so that $<$ is a topological sorting. Solutions of other two are shown in Figures 6.11(b) and (c).

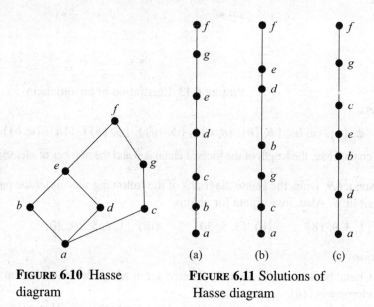

FIGURE 6.10 Hasse diagram

FIGURE 6.11 Solutions of Hasse diagram

6.3.2 Chain

Let $(A, \leq)$ be a poset. A subset of A is called a *chain* if every two elements in the subset are related.

For example

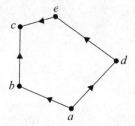

FIGURE 6.12 Illustration of a chain

It is observed from Figure 6.12 that $\{a, b, c, e\}$, $\{a, b, c\}$, $\{a, d, e\}$, and $\{a\}$ are chains.

If A is itself a chain then the poset $(A, \leq)$ is called a totally ordered (linearly ordered) set.

6.3.3 Antichain

A subset of A is called an *antichain* if there exists no relation between two distinct elements in the subset.

For example

Let $A = \{a, b\}$, then the subset $\{\{a\}, \{b\}\}$, in Figure 6.13, is an antichain. However, the following

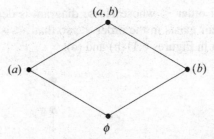

FIGURE 6.13 Illustration of an antichain

subsets

$\phi, \{a\}, \{a, b\}, \{\phi, \{b\}, \{a, b\}\}, \{\phi, \{a\}\}, \{\phi, \{b\}\}, \{\{a\}, \{a, b\}\}, \{\{b\}, \{a, b\}\}$

are a chain. Here, the length of the longest chain is 3 and the number of elements in the antichain is 2.

 EXAMPLE 6.9 Draw the Hasse diagrams of the following sets under the partial ordering relation 'divisibility'. Also, investigate for chains.

(i) {1, 3, 9, 18}   (ii) {3, 5, 30}   (iii) {1, 2, 5, 10, 20}

Solution

(i) Chain: because this is a totally ordered set in which the least element is (1) and the greatest element is (18).
(ii) Not a chain: because there exists no least element in the Hasse diagram.

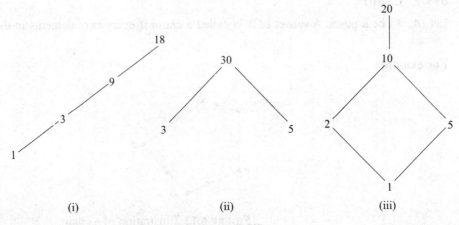

(i)   (ii)   (iii)

FIGURE 6.14 Hasse diagrams of sets (i), (ii), and (iii)

(iii) Chain: here the least element is (1) and the greatest element is (20) (see Figure 6.14).

6.4 ISOMORPHISM

Let $(S, \leq)$ and $(S', \leq')$ be two posets. Consider $f: S \to S'$ as a one-to-one correspondence between S and S'. The function f is called an *isomorphism* from $(S, \leq)$ to $(S', \leq')$ if, for any a and $b \in S$,

$$a \leq b \quad \text{iff} \quad f(a) \leq' f(b)$$

also, if $f: S \rightarrow S'$ is an isomorphism, then it can be said that $(S, \leq)$ and $(S', \leq')$ are *isomorphic posets*.

EXAMPLE 6.10 Let S be the set of positive integers (Z^+), and let $\leq$ be the usual partial order on S. Let S' be the set of positive even integers, and let $\leq'$ be the usual partial order on S'. Show that the function $f: S \rightarrow S'$ satisfying the property

$$f(a) = 2a$$

is an isomorphism from $(S, \leq)$ to $(S, \leq')$.

Solution If $f(a) = f(b)$, then $2a = 2b \Rightarrow a = b$, i.e., f is one-to-one. Next, Dom(f) = S, i.e., f is defined. Finally, if $c \in S'$, then $c = 2a$ for some $a \in Z^+$; so, $c = f(a)$. This shows that f is onto which implies that f is a one-to-one correspondence. Lastly, if a and b are elements of S, then $a \leq b$ iff $2a \leq 2b$, i.e., $f(a) \leq f(b)$. Thus, f is an isomorphism.

Consider two finite posets as $(S, \leq)$ and $(S', \leq')$. Let $f: S \rightarrow S'$ be a one-to-one correspondence. Also, assume H as the Hasse diagram of $(S, \leq)$.

Now, if f is an isomorphism and each label a of H is replaced by $f(a)$, then H will become a Hasse diagram of $(S', \leq')$. Conversely, if H becomes a Hasse diagram for $(S', \leq')$, whenever each label a is replaced by $f(a)$, then f is an isomorphism.

Thus, it may be concluded that two finite isomorphic posets must possess the same Hasse diagrams.

The name 'isomorphism' is thus justified, since isomorphic posets possess the same (*iso*) shape (*morph*) as shown by their Hasse diagrams.

6.4.1 Isomorphic Ordered Sets

Let S and T be two partially ordered sets. A one-to-one (injective) function $f: S \rightarrow T$ is called an *isomorphic mapping* from S into T if f preserves the order relation, i.e., if the following conditions hold good for any pair a and b in S

(i) If $a < b$, then $f(a) < f(b)$

(ii) If $a \| b$, then $f(a) \| (b)$

Nevertheless, for f to be an isomorphic mapping, the condition (i) is essential (or needed).

Two-ordered sets S and T are said to be isomorphic, expressed as

$$S \approx T$$

if there exists a one-to-one correspondence (bijective mapping) $f: S \rightarrow T$, which preserves the order relation, that is an isomorphic mapping.

EXAMPLE 6.11 Given $S = (\{1, 2, 4, 8, 16\}, |)$ and assume $T = \{a, b, c, d, e\}$ as isomorphic to S. Suppose the following function f be an isomorphic mapping from S onto T, written as

$$f = \{(1, e), (2, d), (4, b), (8, c), (16, a)\}$$

Draw the Hasse diagram of T.

Solution The isomorphic mapping preserves the order of the set S and is one-to-one and onto. Thus, the mapping can be observed as relabelling of the vertices in the Hasse diagram of set S. The Hasse diagrams for both S and T are shown in Figure 6.15.

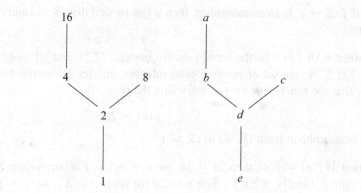

FIGURE 6.15 Hasse diagrams of the sets S and T

6.5 LEXICOGRAPHIC ORDERING

Another partial order on $A \times B$ [where $(A, \leq)$ and $(B, \leq)$ are partially ordered sets], denoted by $<$, can be defined as follows:

$$(a, b) < (a^*, b^*) \text{ if } a < a^* \text{ or if } a = a^* \text{ and } b < b^*$$

This ordering is called *lexicographic ordering* (or *dictionary ordering*). In the ordering of elements, the first coordinate dominates except in case of ties. However, in case of equality of coordinates we should move to the second coordinate and so on. Consequently, the lexicographic ordering can be extended as follows.

Let $(A_1, \leq), (A_2, \leq), \ldots, (A_n, \leq)$ denote partially ordered sets. Define a partial order $<$ on

$$A_1 \times A_2 \times \cdots \times A_n$$

in the following fashion

$$(a_1, a_2, \ldots, a_n) < (a_1^*, a_2^*, \ldots, a_n^*)$$

iff

$$a_1 < a_1^*$$

or

$$a_1 = a_1^* \text{ and } a_2 < a_2^*$$

or

$$a_1 = a_1^*, \quad a_2 = a_2^*, \quad \text{and} \quad a_3 < a_3^*$$

or

$$a_1 = a_1^*, \quad a_2 = a_2^*, \ldots, a_{n-1} = a_{n-1}^* \quad \text{and} \quad a_n \leq a_n^*$$

The order in which the words in an English dictionary appear is an example of lexicographic ordering. Lexicographic ordering is used in sorting character data on a computer.

For example Let $A = (a, b, c, \ldots, z)$ be linearly ordered in the usual way ($a \leq b, b \leq c, \ldots, y \leq z$). Then, the set $A^n = A \times A \times \cdots \times A$ (n factors) can be recognized with a set of all words having length n. The lexicographic ordering on A^n has the property that if $w_1 < w_2$ ($w_1, w_2 \in S^n$) then w_1 would precede w_2 in the dictionary listing. Thus, card < cart, loss < lost, park < part, salt < seat, mark < mast.

Lexicographic order can be extended to posets. If A is a partially ordered set and A^* denotes the set of all finite sequences of elements of A, then the lexicographic order to A^* can be extended as follows:

Let

$$u = a_1, a_2, \ldots, a_m \text{ and } v = b_1, b_2, \ldots, b_n \in S^*$$

with $m \leq n$. Then $u < v$ if

$$(a_1, a_2, \ldots, a_m) < (b_1, b_2, \ldots, b_n)$$

in

$$A^n = A \times A \times A \times \cdots \times A \ (n \text{ factors})$$

under the lexicographic ordering of A.

For example Let $A = (a, b, c, \ldots, z)$ Let R be a simple ordering on A denoted by $\leq$, where $(a \leq b \leq c \leq \cdots \leq z)$.

Assume $S = A \cup A^2 \cup A^3$. Then S consists of strings (words) of three or fewer than three letters from A. Let $<$ be the lexicographic ordering on S, then

$$be < bet, leg < let, peg < pet, \ldots$$

EXAMPLE 6.12 Investigate the lexicographic ordering, constructed from the relation $\leq$ on Z, of the following:

(a) $(2, 6) < (5, 9)$ (b) $(1, 3, 5, 7) < (1, 3, 7, 9)$

Solution
(a) Here, $2 < 5$, which implies that $(2, 6) < (5, 9)$.
(b) In this case, the first two elements are equal; so move to the next elements in which $5 < 7$. Thus, it follows that $(1, 3, 5, 7) < (1, 3, 7, 9)$.

EXAMPLE 6.13 Suppose $N = \{1, 2, 3, \ldots\}$ and $A = \{a, b, c, \ldots, z\}$ are ordered with the usual orders. Consider the following subset of $N \times A$:

$$S = \{(2, a), (1, c), (2, c), (4, b), (4, z), (3, b)\}$$

Assuming $N \times A$ is ordered lexicographically, arrange the elements of S in the appropriate order.

Solution In a lexicographic order, we order by the first element unless there is a tie, in which case we order by the second element. Thus the appropriate order for these elements is

$$(1, c), (2, a), (2, c), (3, b), (4, b), (4, z)$$

6.6 EXTREMAL ELEMENTS OF POSETS

Consider a poset $(S, \leq)$ with partial order $\leq$.

Maximal element An element $a \in S$ is called a *maximal element* of S if there exists no element c in S such that $a < c$.

Minimal element An element $b \in S$ is called a *minimal element* of A if there exists no element c in S such that $c < b$.

Thus, it follows that if $(S, \leq)$ is a poset and $(S, \geq)$ is its dual poset, an element $a \in S$ is a maximal element of $(A, \geq)$ iff a is a minimal element of $(A, \leq)$. In the same way, a is a minimal element of $(A, \geq)$ iff it is a maximal element of $(A, \leq)$.

Note There can be more than one maximal or more than one minimal element.

For example Let S be the poset of non-negative real numbers with partial order $\leq$. Then 0 is a minimal element of S. There exits no maximal elements of S.

EXAMPLE 6.14 Find all the maximal and minimal elements of the poset S whose Hasse diagram is depicted in Figure 6.16.

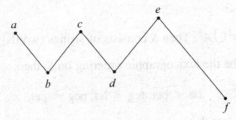

FIGURE 6.16 Hasse diagram of the poset $(S, \leq)$

Solution There are three maximal elements and three minimal elements of S.

The maximal elements are a, c, e and b, d, f are the minimal one.

EXAMPLE 6.15 Consider the poset A shown in Figure 6.17. Determine the maximal and minimal elements in the poset.

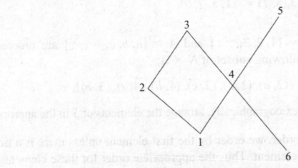

FIGURE 6.17 Hasse diagram of the poset $(A, \leq)$

Solution Here, the maximal elements are 3 and 5. Also, the minimal elements are 1 and 6.

EXAMPLE 6.16 Let $S = (2, 3, 4, 6, 8, 24, 48)$ with partial ordering of divisibility. Determine all the maximal and minimal elements of S.

Solution Here the maximal element is 48. The minimal elements are 2 and 3.

6.6.1 Greatest and Least Elements

An element $a \in S$ is called a *greatest element* of S if $x \leq a$ for all $x \in S$.
An element $a \in S$ is called a *least element* of S if $a \leq x$ for all $x \in S$.

The greatest element of a poset S is denoted by 1, known as a unit element.
Also, the least element of a poset S is denoted by 0, called the *zero element*.

For example

(i) Let S be the poset of non-negative real numbers with partial order $\leq$. Then 0 is a least element, but there is no greatest element.
(ii) The poset Z with the partial order $\leq$ possesses neither a least nor a greatest element.

Theorem 6.3 A poset possesses at most one greatest element and at most one least element.

Proof Let x and y be two greatest elements of a poset S. Now, since y is a greatest element, so $x \leq y$.

Again, as x is a greatest element, so $y \leq x$ which implies that $x = y$ (antisymmetric property). Hence, it can be concluded that if poset possesses a greatest element, then there exists only one such element.

Using similar arguments, one can prove that if the poset has a least element, then there exists only one such element.

EXAMPLE 6.17 Determine the greatest and least elements of the poset whose Hasse diagrams are shown in Figure 6.18, if they exist

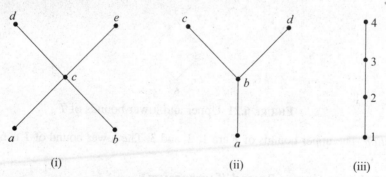

FIGURE 6.18 Hasse diagrams

Solution

(i) The poset possesses neither greatest nor least element.
(ii) The poset has no greatest element but possesses a as the least element.
(iii) The poset contains 4 as the greatest element and 1 as the least element.

6.6.2 Upper and Lower Bounds

Let $(S, \leq)$ be a partially ordered set and let T be a subset of S. An element $m \in S$ is called an *upper bound* of T if $n \leq m$ for all $n \in T$. Similarly, an element $m \in S$ is called a *lower bound* of T if $m \leq n$ for all $n \in T$.

 EXAMPLE 6.18 Let the poset $S = (1, 2, 3, 4, 5, 6)$ be ordered as shown in Figure 6.19. Also, let $T = (3, 4)$. Determine the upper and lower bounds of T.

Solution The upper bound of T is 4, 5, and 6 because every element of T is '$\leq$' 4, 5, and 6.

The lower bound of T is 3, 2, and 1 because 3, 2, and 1 are '$\leq$' every element of T.

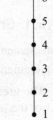

FIGURE 6.19 Ordered poset $(S, \leq)$

 EXAMPLE 6.19 Let the poset $S = (1, 2, 3, 4, 5, 6, 7, 8)$ be ordered as shown in Figure 6.20. Also, let $T = (3, 4, 5)$. Find the upper and lower bounds of T.

Solution The upper bound of T is 6 because every element of T is '$\leq$' 6. The number 8 is not an upper bound of T, since $4 \in T$, but 8 does not.

The lower bound of T is 1, 2, and 3 because 1 and 2 are '$\leq$' every element of T. The number 7 is not a lower bound of T, 3, $4 \in T$, but 7 does not.

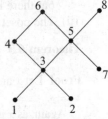

FIGURE 6.20 Ordered poset $(S, \leq)$

 EXAMPLE 6.20 Let $S = (1, 2, 3, 4, 5, 6)$ be ordered as shown in Figure 6.21. If $T = (4, 5)$, then find the upper and lower bounds of T.

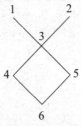

FIGURE 6.21 Upper and lower bounds of T

Solution The upper bounds of T are 1, 2, and 3. The lower bound of T is 6.

6.6.3 Least Upper Bound (Supremum)

Let S be a partially ordered set (or poset) and T be a subset of S. An element $m \in S$ is called a *least upper bound* (LUB) of T if m is an upper bound of T and $m \leq m'$ whenever m' is an upper bound of T. Thus, $m = $ LUB (T) if $n \leq m$ for all $n \in T$, and if whenever $m' \in S$ is also an upper bound of T, then $m \leq m'$.

An LUB of a partially ordered set, if it exists, is unique.

For example Let $S = (a, b, c, d, e, f, g, h)$ denote a partially ordered set, whose Hasse diagram is shown in Figure 6.22.

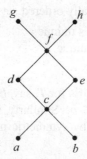

FIGURE 6.22 Hasse diagram of the poset $(S, \leq)$

If $T = (c, d, e)$, then f, g, h are upper bounds of T. But the element f is the LUB, i.e., LUB (T) or, LUB $(c, d, e) = f$.

6.6.4 Greatest Lower Bound (Infimum)

Let S be a partially ordered set and T denote a subset of S. An element $m \in S$ is called a *greatest lower bound* (GLB) of T if m is a lower bound of T and $m' \leq m$ whenever m' is a lower bound of T. Thus, $m =$ GLB (T) if $m \leq n$ for all $n \in T$, and if whenever $m' \in S$ is also a lower bound of T, then $m' \leq m$.

The GLB of a poset, if it exists, is unique. Hence, it can be concluded that upper bounds in $(S, \leq)$ correspond to lower bounds in $(S, \leq)$ (for the same set of elements), and lower bounds in $(S, \leq)$ correspond to upper bounds in $(S, \leq)$. Similar statements can be made for GLBs and LUBs.

For example Consider the poset $S = (1, 2, 3, 4, 5, 6, 7, 8)$, whose Hasse diagram is shown in Figure 6.23. Also, assume $T = (3, 4, 5)$.

Here, the elements 1, 2, 3 are lower bounds of T; and 3 is the GLB, i.e., GLB $(1, 2, 3) = 3$.

The LUB and the GLB of subset of T are also called the *supremum* and *infimum* of the subset T.

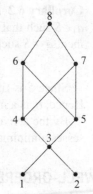

FIGURE 6.23 Hasse diagram of the poset $(S, \leq)$

Note

(i) The least upper bound of set is abbreviated as LUB or Sup and the greatest lower bound as GLB or Inf.
(ii) Let Z^+ be the set of positive integers and Z be ordered by divisibility. If a and b are two elements of Z, then

$$\text{Inf}(a, b) = GCD(a, b)$$

and

$$\text{Sup}(a, b) = LCM(a, b)$$

(iii) If $(S, \leq)$ is a poset, then its dual $(S, \geq)$ is also a poset. The least member of $(S, \leq)$ is the greatest member in $(S, \geq)$ relative to $\geq$ and vice versa. Similarly, the GLB of S with respect to the relation $\leq$ is the GLB of S with respect to the relation $\leq$ and vice versa.

(iv) GLB of a and b is called the meet or product of a and b, and LUB of a and b is called the join or sum of a and b, where $a, b \in Z$.

Theorem 6.4 Let $(S, \leq)$ be a partially ordered set and T be a subset of S. Then
 (i) The LUB of set S, if it exists, is unique.
 (ii) The GLB of S, if it exists, is unique.

Proof
 (i) If possible, let there be two LUBs for T, say, b_1 and b_2. Now, b_2 is supremum and b_1 is an upper bound of T, implies $b_2 \leq b_1$. Similarly, b_1 is supremum and b_2 is an upper bound of T, implies $b_1 \leq b_2$. $T \subseteq S$, so by symmetric property, $b_2 \leq b_1, b_1 \leq b_2$ implies $b_1 = b_2$. Hence, LUB of T is unique.
 (ii) Left to the reader as practice.

Theorem 6.5 Let $(S, \leq)$ be a poset. A subset T of S possesses at most one LUB and at most one GLB.

Proof The proof is similar to the proof of the theorem on greatest element and least element, and is left to the reader.

EXAMPLE 6.21 Prove that a finite partial-ordered set has (i) at most one greatest element and (ii) at most one least element.

Solution Assume a and b are greatest elements of $(S, \leq)$. Since a is the greatest element, we have $b \leq a$. Also, since b is the greatest element, we have $a \leq b$. Thus, $b \leq a$ and $a \leq b$. Since $\leq$ is antisymmetric for partial-ordered set, it follows that $a = b$. Thus there cannot be two different greatest elements of $(S, \leq)$, if it exists. Hence, a finite partial-ordered set has at most one greatest element.

Corollary 6.2 Let $a \in S$. Assume a is not the maximal element. Then one can find an element $a_1 \in S$ such that $a < a_1$. Again, if a_1 is not a maximal element of S, then one can find an element $a_2 \in S$ such that $a_1 < a_2$. Continuing this argument, one obtains a chain

$$S < a_1 < a_2 < a_3 < \cdots < a_{i-1} < a_i$$

Since S is finite, this chain cannot be extended and for any $b \in S$, one cannot find $a_i < b$. Hence, a_i is a maximal element of $(S, \leq)$.

By the above argument, the dual poset $(S, \geq)$ has a maximal element such that $(S, \leq)$ possesses a minimal element.

6.7 WELL-ORDERED SET

A set S with an ordering relation is known as *well-ordered set* if every non-empty subset A of the set S possesses a least element.

For example The poset $(Z^+, \leq)$ is a well-ordered set but the poset $(Z, \leq)$ is not a *well-ordered set*, since the subset of the negative integers does not contain a minimal element. Also, the ordered set represented by

$$S = \{A; B\} = \{1, 3, 5, \ldots; 2, 4, 6, \ldots\}$$

is well-ordered.

Note

(i) A well-ordered set is linearly ordered if $a, b \in S$ are comparable.
(ii) Every subset of a well-ordered set is well-ordered.

Theorem 6.6 Let A be a well-ordered set and B a subset of A, and let the function $f: A \to B$ be a similarity mapping from A into B. Then, for every a in A, we have $a \leq f(a)$.

Proof Let $X = \{x: f(x) < x\}$. If X is empty, the theorem is true. Suppose $X \neq \phi$. Then, since A is well-ordered, X possesses an element x_0 such that $x_0 \leq x$. It may be noted that $x_0 \in X$ implies $f(x_0) < x_0$. Since f is a similarity mapping,

$$f(x_0) < x_0$$

implies

$$f(f(x_0)) < f(x_0)$$

Consequently,

$$f(x_0)$$

also belongs to X. But

$$f(x_0) < x_0 \quad \text{and} \quad f(x_0) \in X$$

contradicts the fact that $x_0 \in X$. Hence, the original assumption that $X \neq \phi$ leads to a contradiction. Therefore, X is empty and the theorem is true.

EXAMPLE 6.22 Find a similarity mapping from N into its subset $A = \{1, 3, 5, 7, \ldots\}$.

Solution For each element x in N, define $f(x) = 2x - 1$. This mapping preserves the order of the positive integers N.

6.8 CONSISTENT ENUMERATIONS

Let S be a finite partially ordered set. It is required to assign positive integers to the elements of S in such a way that the order is preserved, i.e., to enumerate them in some order $a, b, c \ldots$. That is, a function is to be sought as

$$f: S \to N$$

so that if $a < b$ then $f(a) < f(b)$. Such a function is called a *consistent enumeration* of S. The fact of this enumeration lies in the following theorem.

Theorem 6.7 If S is a finite poset with n elements, then there exists a consistent enumeration $f: S \to \{1, 2, 3, \ldots, n\}$.

Proof The theorem will be proved by mathematical induction. Let there be n number of elements in S. Assume $n = 1$ and $S = \{s\}$, say. Then $f(s) = 1$ forms a consistent enumeration of S. Again, suppose $n > 1$ and the theorem holds good for posets with fewer than n elements.

Let $a \in S$ be a minimal element. Assume $T = S\{a\}$. Then T is a finite poset with $n - 1$ elements and therefore, by induction, T yields consistent enumeration, say,

$$g: T \to \{1, 2, 3, \ldots, n - 1\}$$

Let us define
$$f: S \rightarrow \{1, 2, 3, \ldots, n\}$$
by
$$f(x) = \begin{cases} 1, & \text{if } x = a \\ g(x) + 1, & \text{if } x \neq a \end{cases}$$

Then, f is the required consistent enumeration.

 EXAMPLE 6.23 Let $S = \{a, b, c, d, e\}$ be ordered as in Figure 6.24. Find all possible consistent enumerations of $f: S \rightarrow \{1, 2, 3, 4, 5\}$.

Solution Here, a is the only minimal element, i.e., $f(a) = 1$. Also, e is the only maximal element here, i.e., $f(e) = 5$. Again, $f(b) = 2$, since b is the only successor of a. But the choices for e and d are $f(c) = 3$ and $f(d) = 4$ or vice versa. Thus, two enumerations are possible as given below:
(i) $f(a) = 1, f(b) = 2, f(c) = 3, f(d) = 4, f(e) = 5$
(ii) $f(a) = 1, f(b) = 2, f(c) = 4, f(d) = 3, f(e) = 5$

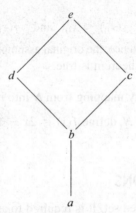

FIGURE 6.24 Ordered diagram of S

6.9 LATTICES

A lattice is a partially ordered set $(L, \leq)$ in which every pair of elements $a, b \; \varepsilon \; L$ possesses a GLB or infimum and an LUB or supremum.

The GLB of a subset $(a, b) \subseteq L$ is denoted by $a \wedge b$ (meet of a and b) and the LUB by $a \vee b$ (join of a and b). It may be noted here that a lattice is a mathematical structure with symbols '$\wedge$' and '$\vee$' as binary operations. It may be denoted by $(L, \wedge, \vee)$. Lattices have many special properties. Moreover, lattices are used in several applications such as models of information flow and play a key role in Boolean algebra.

For example

(i) Let Z^+ denote the set of all positive integers and R denote the relation 'division' in Z^+. Let S_n be the set of all divisors of n (a positive integer); assume $n = 6$, $S_6 = (1, 2, 3, 6)$ and for $n = 24$, $S_{24} = (1, 2, 3, 4, 6, 8, 12, 24)$. Then the lattices are (S_6, R), (S_8, R), and (S_{24}, R).

(ii) Let n be a positive integer and S_n be the set of all division of $n \cdots S_n$. If $n = 30$, $S_{30} = (1, 2, 3, 5, 6, 10, 15, 30)$. Let R denote the relation 'division' as defined in example (i). Then (S_{30}, R) is a lattice, shown in Figure 6.25.

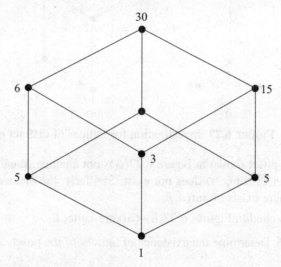

FIGURE 6.25 A lattice (S_{30}, R)

(iii) Let S and $L = S$ be a non-empty set and $L = P(S)$ be its power set, i.e., $(L, \leq)$ is a partially ordered set. If a and b are two elements of L, then $a \wedge b = a \cap b$ and $a \vee b = a \cup b$. Thus, $(L, \leq)$ is a lattice.
(iv) Let Z^+ be the set of positive integers. For any $a, b \in Z^+$, $a \leq b$ if $a|b$, then $(Z^+, \wedge, \vee)$ is a lattice in which $a \wedge b = $ GCD (a, b) (greatest common divisor of a and b) and $a \vee b = $ LCM (a, b) (least common multiple of a and b).

EXAMPLE 6.24 Assume a partial order set $A = \{(2, 4, 6, 8), |\}$. Then, investigate the following:

(i) Every pair of elements in the poset has a GLB.
(ii) Every pair of elements in the poset has an LUB.
(iii) The given poset is not a lattice
for true or false.

Solution The Hasse diagram of the given poset is shown in Figure 6.26.

(i) True; since

GLB $(2, 4) = 2$ GLB $(4, 6) = 2$
GLB $(2, 6) = 2$ GLB $(4, 8) = 4$
GLB $(2, 8) = 2$ GLB $(6, 8) = 2$

(ii) False; since the LUB of $(6, 8)$ does not exist.
(iii) False; the given poset is not a lattice.

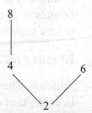

FIGURE 6.26 Hasse diagram of the poset $A = \{(2, 4, 6, 8), |\}$

EXAMPLE 6.25 Investigate the posets shown in Figure 6.27 as lattices.

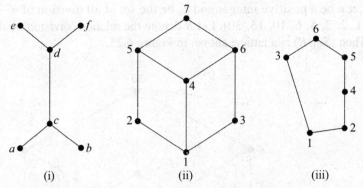

FIGURE 6.27 Investigation for lattices of distinct posets

Solution The poset shown in Figure 6.27(i) is not a lattice, since the elements e and f have no upper bound, i.e., $\sup(e, f)$ does not exist. Similarly, the elements a and b possess no lower bound, i.e., there exists no $\inf(a, b)$.

The posets cited in Figures 6.27(ii)–(iii) are lattices.

 EXAMPLE 6.26 Determine the existence of lattices of the posets shown in Figure 6.28.

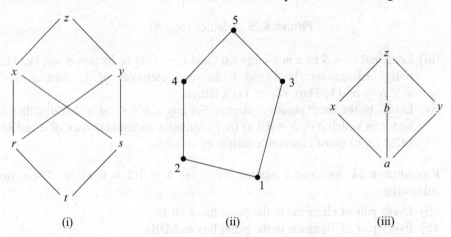

FIGURE 6.28 Test for lattices

Solution The poset given in Figure 6.28(i) is not a lattice because (x, y) possesses three lower bounds r, s, and t, but there is no existence of $\inf(x, y)$. Also, there exists no $\sup(r, s)$. The posets shown in Figures 6.28(ii)–(iii) are lattices.

 EXAMPLE 6.27 Investigate the posets shown in Figure 6.29 as lattices.

Solution The posets represented by the Hasse diagrams in (i) and (iii) are both lattices because in each poset every pair of elements has a both LUB and a GLB, as the reader can easily verify. On the other hand, the poset with the Hasse diagram shown in (ii) is not a lattice, because the elements b and c have no upper bound. It may be noted here that each of the elements d, e, and f is an upper bound, but none of these three elements precedes the other two with respect to the ordering of this poset.

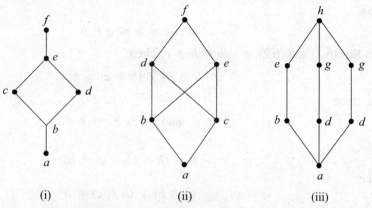

FIGURE 6.29 Investigation for lattices

 EXAMPLE 6.28 Determine the existence of lattices of the posets shown in Figure 6.30.

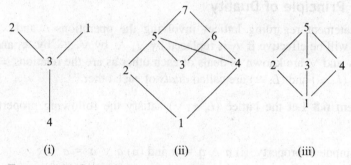

FIGURE 6.30 Investigation for lattices

Solution
(i) The given poset is not a lattice since the elements 1 and 2 have no upper bounds and hence LUB (1, 2) does not exist.
(ii) The given poset is not a lattice since the elements 5 and 6 have two lower bounds 3 and 4, but neither precedes the other; hence, GLB (5, 6) does not exist.
(iii) The given poset is a lattice.

 EXAMPLE 6.29 Prove that in a lattice $(L, \leq)$, for any $a, b, c \in L$, if $a \leq b \leq c$ then

$$a \vee b = b \wedge c$$

and

$$(a \wedge b) \vee (b \wedge c) = (a \vee b) \wedge (a \vee c) = b$$

Solution Since $a \leq b$ and $a \leq c$, then $a \leq b \wedge c$.
Again,

$$b \leq b \quad \text{and} \quad b \leq c \Rightarrow b \leq b \wedge c$$

Now,

$$a \leq b \wedge c \quad \text{and} \quad b \leq b \wedge c \Rightarrow a \vee b \leq b \wedge c \tag{6.1}$$

Again,
$$b \wedge c \leq b \leq a \wedge b \quad (6.2)$$

From Eqs (6.1) and (6.2), $a \vee b = b \wedge c$. Then,
$$(a \wedge b) \vee (b \wedge c) \leq b$$

Also, since
$$b \leq c \text{ and } b \leq b \Rightarrow b \leq b \wedge c$$

Thus,
$$b \leq (b \wedge c) \vee (a \wedge b)$$

Hence,
$$(a \wedge b) \vee (b \wedge c) = b$$

Similarly, it can be proved that
$$(a \vee b) \wedge (a \vee c) = b$$

6.9.1 Principle of Duality

Any statement regarding lattices involving the operations $\wedge$ and $\vee$, and the relations $\leq$ and $\geq$ will be effective if $\wedge$ is replaced by $\vee$, $\wedge$ by $\vee$, $\leq$ by $\geq$, and $\geq$ by $\leq$. The operations $\wedge$ and $\vee$ are known as duals of each other as are the relations $\leq$ and $\geq$. Similarly, the lattices $(L, \leq)$ and $(L, \geq)$ are called *duals* of each other.

Theorem 6.8 Let the lattice $(L, \wedge, \vee)$ satisfy the following properties (for any elements a, b, c):

I. Idempotent property: (i) $a \wedge a = a$ and (ii) $a \vee a = a$
II. Commutative property: (i) $a \wedge b = b \wedge a$ and (ii) $a \vee b = b \vee a$
III. Associative property: (i) $(a \wedge b) \wedge c = a \wedge (b \wedge c)$ and (ii) $(a \vee b) \vee c = a \vee (b \vee c)$
IV. Absorption property: (i) $a \wedge (a \vee b) = a$ and (ii) $a \vee (a \wedge b) = a$

Proof

I. Since $a \leq a$, so a is a lower bound of (a, a). Also, if b is a lower bound and if $a \leq b$, then $b \leq a$ and $a \leq b$. By the definition of antisymmetry, $a = b$. So, a is the GLB of (a). Thus, $a \wedge a = a$. By the principle of duality, it can be said that $a \vee a = a$.

II. Let $x = a \wedge b = $ GLB (a, b). Since $(a, b) = (b, a)$, where $(a, b) \subseteq L$, GLB $(b, a) = x$. So, $b \wedge a = x$. Hence, $x = a \wedge b = b \wedge a$. Dually, $a \vee b = b \vee a$ follows.

III. Let
$$x = a \wedge (b \wedge c) \text{ and } y = (a \wedge b) \wedge c$$

Now,
$$x = a \wedge (b \wedge c) \Rightarrow x \leq a, x \leq b \wedge c$$
$$\Rightarrow x \leq a, x \leq b, x \leq c$$
$$\Rightarrow x \leq a \wedge b, x \leq c$$
$$\Rightarrow x \leq (a \wedge b) \wedge c = y$$

Similarly, $y \leq x$ follows. By antisymmetry, $x = y$.

Hence,
$$a \wedge (b \wedge c) = (a \wedge b) \vee c$$

Dually,
$$a \vee (b \vee c) = (a \vee b) \vee c$$

follows.

IV. By definition, for any $a \in L$, $a \leq a$ and $a \leq a \vee b$. So, $a \leq a \wedge (a \vee b)$. But
$$a \wedge (a \vee b) \leq a$$

Hence,
$$a \wedge (a \vee b) = a$$

Dually,
$$a \vee (a \wedge b) = a$$

follows.

Theorem 6.9 Let $(L, \leq)$ be a lattice in which $\wedge$ and $\vee$ denote the operations of meet and join, respectively. For any $a, b \in L$,
$$a \leq b \Leftrightarrow a \wedge b = a \Leftrightarrow a \vee b = b$$

Proof Let $a \leq a$. Since, $a \leq b$, it implies that $a \leq a \wedge b$; but from the definition of $\wedge$, $a \wedge b \leq a$.

Thus, $a \leq b \Rightarrow a \wedge b = a$. On the other hand, let $a \wedge b = a$, which is possible only if $a \leq b$, i.e., $a \wedge b = a \Rightarrow a \leq b$. Hence, $a \leq b \Rightarrow a \wedge b = a$ and $a \wedge b = a \Rightarrow a \leq b$. Combining, $a \leq b \Leftrightarrow a \wedge b = a$.

Again, assume $a \wedge b = a$. Then,
$$b \vee (a \wedge b) = b \vee (a \vee a) \vee b = b \vee a \vee b = (a \vee b) \vee b = a \vee (b \vee b) = a \vee b$$

Also,
$$b \vee (a \wedge b) = b \vee (b \wedge a) \quad \text{(commutative property)}$$
$$= b \quad \text{(absorption property)}$$

Hence, $a \vee b = b$.

Similarly, by assuming $a \vee b = b$, one can show that $a \wedge b = a$. Hence, $a \leq b \Leftrightarrow a \wedge b = a \Leftrightarrow a \vee b = b$.

EXAMPLE 6.30 If $(L, \leq)$ is a lattice, then show that $(L, \geq)$ is also a lattice.

Solution Let $a, b \in L$ be any elements. Since $(L, \leq)$ is a lattice, so GLB (a, b) and LUB (a, b) exist in $(L, \leq)$. Then, LUB $(a, b) = a \vee b$

$$\Rightarrow a \leq a \vee b, \quad b \leq a \vee b$$
$$\Rightarrow a \vee b \geq a, \quad a \vee b \geq b$$

$\Rightarrow a \vee b$ is the lower bound of a, b in $(L, \geq)$. Now, it is required to show that $a \vee b$ is the GLB of a and b in $(L, \geq)$.

Let $c \in L$ be another lower bound of a and b in $(L, \geq)$. Then

$$c \geq a, \quad c \geq b$$
$$\Rightarrow a \leq c, \quad b \leq c$$
$$\Rightarrow c \text{ is the upper bound of } a \text{ and } b$$

Since $(L, \leq)$ is the lattice, so LUB of a and b exists, i.e., $a \vee b$ exists in $(L, \leq)$. Then

$$a \vee b \leq c \quad \text{or} \quad c \geq a \vee b$$

$\Rightarrow a \vee b$ is the GLB of a and b in $(L, \geq)$. Similarly, it can be shown that the LUB of a and b exists in $(L, \geq)$. Thus, if $(L, \leq)$ is a lattice, then $(L, \geq)$ is also a lattice.

6.9.2 Isotonocity Property

Theorem 6.10 Let $(L, \geq)$ be a lattice. Then, for any $a, b, c \in L$,

$$a \wedge b \leq a \wedge c \text{ and } b \leq c \Rightarrow a \vee b \leq a \vee c$$

Proof Assume $b^* = a \wedge b$ and $c^* = a \wedge c$. But, from previous theorem,

$$b^* \leq c^* \Leftrightarrow b^* \wedge c^* = b^* \quad \text{[property (i)]}$$

i.e.,

$$(a \wedge b) \wedge (a \wedge c) = (a \wedge a) \wedge (b \wedge c) \quad \text{(associative property)}$$
$$= a \wedge (b \wedge c)$$
$$= a \wedge b \quad \text{(since } b \leq c \Leftrightarrow b \wedge c = b)$$

Hence, $a \wedge b \leq a \wedge c$, by property (i).
Similarly, it can be proved that $a \vee b \leq a \vee c$.

Corollary 6.3 For any a, b, c, d in a lattice $(L, \leq)$ if $a \leq b$ and $c \leq d$, then $a \vee c \leq b \vee d$ and $a \wedge c \leq b \wedge d$.

Proof As $a \leq b$, so $a \vee c \leq b \vee c$. Also, as $c \leq d$, so $b \vee c \leq b \vee d$.

Thus, it follows that

$$a \vee c \leq b \vee d \quad \text{(transitivity property)}$$

Similarly, it can be proved that $a \wedge c \leq b \wedge d$.

Theorem 6.11 With respect to the lattice $(L, \leq)$, the following inequalities hold good for any $a, b, c \in L$:

I. Distributive inequalities

   (i) $a \vee (b \wedge c) \leq (a \vee b) \wedge (a \vee c)$
   (ii) $(a \wedge b) \vee (a \wedge c) \leq a \wedge (b \vee c)$

II. Modular inequalities
  (i) $a \leq c \Leftrightarrow a \vee (b \wedge c) \leq (a \vee b) \wedge c$
  (ii) $a \geq c \Leftrightarrow a \wedge (b \vee c) \geq (a \wedge b) \vee c$

Proof

I. (i) Since $a \leq a \vee b$ and $a \leq a \vee c$,
$$a \leq (a \vee b) \wedge (a \vee c) \quad \text{[inequality (i)]}$$
Again, since
$$b \wedge c \leq b \leq a \vee b \quad \text{and} \quad b \wedge c \leq c \leq a \vee c$$
$$b \wedge c \leq (a \vee b) \wedge (a \vee c) \quad \text{[inequality (ii)]}$$

Thus, from the inequalities (i) and (ii), $a \vee (b \vee c) \leq (a \vee b) \wedge (a \vee c)$. Similarly, the inequality (ii) can be proved.

II. (i) Assume $a \leq c$. Then $a \vee c = c$. By distributive inequality,
$$a \vee (b \wedge c) \leq (a \vee b) \wedge (a \vee c)$$
Since $a \vee c = c$, so $a \vee (b \wedge c) \leq (a \vee b) \wedge c$.
Then,
$$a \leq a \vee (b \wedge c)$$
$$\leq (a \vee b) \wedge c$$
$$\leq c$$
$$\Rightarrow a \leq c$$

Hence, the proof.
Similarly, the inequality (ii) can be proved.

6.10 SUBLATTICES

Let $(L, \leq)$ be a lattice. A non-empty subset S of L is called a *sublattice* of L if $a \vee b \in S$ and $a \wedge b \in S$, whenever $a \in S$ and $b \in S$.

If S is a sublattice of L, then S is closed under the operations of '$\wedge$' and '$\vee$'.

For example

(i) Let Z^+ be the set of all positive integers and let R denote the relation 'diviion' in Z^+ such that for any $a, b \in Z^+$, aRb. If a divides b, then (Z^+, R) is a lattice in which $a \vee b = $ LCM of a and b and $a \wedge b = $ GCD of a and b.

(ii) Let n be a positive integer and S_n be the set of all divisions of n. If R denotes the relation as defined above [example (i)], then (S_n, D) is a sublattice of (Z^+, R).

(iii) Consider the lattice $L = (l, m, n, p, q, r, t)$ (Figure 6.31). The subset $S = (l, m, n, p)$ is a sublattice of L.

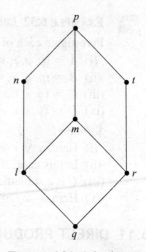

FIGURE 6.31 A lattice $L = (l, m, n, p, q, r, t)$

Theorem 6.12 If M is a sublattice of a distributive lattice L, then M is a distributive lattice.

Proof For a distributive lattice L,

$a \wedge (b \vee c) = (a \wedge b) \vee (a \wedge c)$ and $a \vee (b \wedge c) = (a \vee b) \wedge (a \vee c)$ for all $a, b, c \in L$.

Since M is closed, each element of M is also in L, the distributive laws hold for all elements in M.

Hence, M is a distributive lattice.

EXAMPLE 6.31 Let $L = (1, 2, 3, 4, 5)$ be the lattice shown in Figure 6.32. Find all the sublattices with three or more elements.

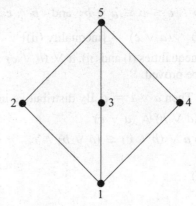

FIGURE 6.32 Lattice $L = (1, 2, 3, 4, 5)$

Solution All the sublattices with three or more elements are those whose supremum and infimum exist for every pair of elements. The pairs of elements are given below:

(a) $(1, 2, 3)$,   (b) $(1, 3, 5)$,   (c) $(1, 4, 5)$,   (d) $(1, 2, 3, 5)$
(e) $(1, 3, 4, 5)$,   (f) $(1, 2, 3, 4, 5)$, and   (g) $(1, 2, 4, 5)$

EXAMPLE 6.32 Let the lattice L be described in Figure 6.33.

Investigate each of the following as a sublattice of L:

(i) $A = (p, q, r, v)$
(ii) $B = (p, q, u, v)$
(iii) $C = (q, s, t, v)$
(iv) $D = (p, s, t, v)$

Solution
(i) Here, $q \vee r = s$ and $s \notin A$. So, A is not a sublattice.
(ii) In this case, B is a sublattice.
(iii) C is a sublattice.
(iv) Here, $s \wedge t = q$ and $q \notin D$. So, D is not a lattice.

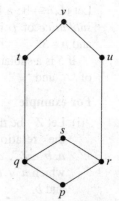

FIGURE 6.33 Lattice L

6.11 DIRECT PRODUCT OF LATTICES

Let $(L_1, \vee_1, \wedge_1)$ and $(L_2, \vee_2, \wedge_2)$ be two lattices. Then the direct product of lattices is denoted by $(L, \vee, \wedge)$, where $L = L_1 \times L_2$. Here, binary operations $\vee, \wedge$ are such that, for any (a_1, b_1) and (a_2, b_2) in L,

$$(a_1, b_1) \vee (a_2, b_2) = (a_1 \vee_1 a_2, b_1 \vee_1 b_2)$$

and
$$(a_1, b_1) \wedge_2 (a_2, b_2) = (a_1 \wedge_2 a_2, b_1 \wedge_2 b_2)$$

EXAMPLE 6.33 Let a lattice $(L, \leq)$ be defined in Figure 6.34, where $L = (1, 2)$. Evaluate the lattice $(L^2, \leq)$ in which $L^2 = L \times L$.

Solution The required lattice $(L^2, \leq)$ is described in Figure 6.35.

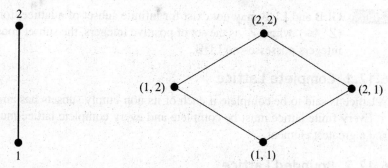

FIGURE 6.34 Lattice $(L, \leq)$ **FIGURE 6.35** Direct product of lattices $(L^2, \leq)$

EXAMPLE 6.34 Let $L_1 = (1, 2, 4)$ and $L_2 = (1, 3, 9)$. L_1 consists of divisor of 4 and L_2 consists of divisor of 9. Determine $L_1 \times L_2$.

Solution The lattice $(L_1 \times L_2, \leq)$ is depicted in Figure 6.36.

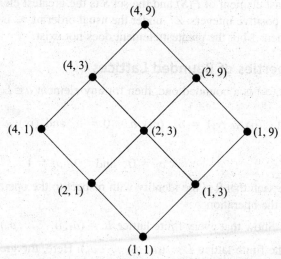

FIGURE 6.36 Product of two lattices $(L_1 \times L_2, \leq)$

6.12 SOME SPECIAL CLASS OF LATTICES

Let $(L, \wedge, \vee)$ be a lattice and $S \leq L$ be a finite subset of L, where $S = (a_1, a_2, \ldots, a_n)$. The GLB and LUB of S are defined as

$$\text{GLB } S = \bigwedge_{i=1}^{n} a_i \quad \text{and} \quad \text{LUB } S = \bigvee_{i=1}^{n} a_i$$

in which

$$\bigwedge_{i=1}^{n} a_i = a_1 \wedge a_2 \wedge \cdots \wedge a_n$$

and

$$\bigvee_{i=1}^{n} a_i = a_1 \vee a_2 \vee \cdots \vee a_n$$

Note GLB and LUB may not exist for infinite subset of a lattice, for instance, in the lattice $(Z^+, \leq)$, where Z^+ is the set of positive integers, the subset consisting of even-positive integers possesses no LUB.

6.12.1 Complete Lattice

A lattice is said to be complete if each of its non-empty subsets has an LUB and a GLB.

Every finite lattice must be complete and every complete lattice must have a least element and a greatest element.

6.12.2 Bounded Lattice

A lattice $(L, \leq)$ is called a bounded lattice if it has a greatest element denoted by 1 and a least element denoted by 0.

In a bounded lattice, 1 and 0 act as duals to each other.

For example
(i) The power set $P(S)$ of the set S under the operations '$\cap$' and '$\cup$' is a bounded lattice since ϕ is the least element of $P(S)$ and the set S is the greatest element of $P(S)$.
(ii) The set of positive integers, Z^+ under the usual order of $\leq$ is not a bounded lattice, since it has element 1 but the greatest element does not exist.

6.12.3 Properties of Bounded Lattices

If the lattice $(L, \leq)$ is a bounded one, then for any element $a \in L$, the following identities are satisfied:

(i) $a \vee 1 = 1$, (ii) $a \wedge 1 = a$, (iii) $a \vee 0 = a$, and (iv) $a \wedge 0 = 0$

Here,

$$\wedge a_i = 0 \quad \text{and} \quad \vee a_i = 1$$

Also, it may be seen that 0 is the identity with respect to the operation $\vee$ and 1 is the identity with respect to the operation $\wedge$.

Theorem 6.13 Show that every finite lattice $L = (a_1, a_2, \ldots, a_n)$ is bounded.

Proof Given the finite lattice $L = (a_1, a_2, \ldots, a_n)$. Here, the greatest element of lattice L is $a_1 \vee a_2 \vee \cdots \vee a_n$. Also, the least element of lattice L is $a_1 \wedge a_2 \wedge \cdots \wedge a_n$. Since the greatest and least elements exist for every finite lattice, hence L is bounded.

6.12.4 Distributive Lattice

A lattice $(L, \leq)$ is said to be distributive lattice if for any $a, b, c \in L$,

(i) $a \wedge (b \vee c) = (a \wedge b) \vee (a \wedge c)$
(ii) $a \vee (b \wedge c) = (a \vee b) \wedge (a \vee c)$

If the lattice L does not satisfy the above properties, it is called a non-distributive lattice.

For example

(i) The power set $P(S)$ of the sets under the operations of intersection ($\cap$) and union ($\cup$) is a distributive lattice. Since

$$a \cap (b \cup c) = (a \cap b) \cup (a \cap c)$$

and also,

$$a \cup (b \cap c) = (a \cup b) \cap (a \cup c)$$

for any elements $a, b, c \in P(S)$.

(ii) The lattice shown in Figure 6.37 is distributive. Since it satisfies the distributive properties for all ordered triplets from 1, 2, 3, and 4, such as

$$1 \wedge (2 \vee 3) = 1 \wedge 4 = 4$$

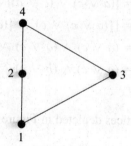

FIGURE 6.37 Distributive lattice

and

$$(1 \wedge 2) \vee (1 \wedge 3) = 2 \vee 3 = 4$$

Theorem 6.14 If $(L, \leq)$ is a distributive lattice, for any $a, b, c \in L$, then

$$a \vee b = a \vee c \quad \text{and} \quad a \wedge b = a \wedge c \Rightarrow b = c$$

Proof It is known that

$$b = b \vee (b \wedge a) \quad \text{(absorption property)}$$

then,

$$\begin{aligned}
b &= b \vee (a \wedge b) & \text{(commutative)} \\
&= b \vee (a \wedge c) & \text{(since } a \wedge b = a \wedge c\text{)} \\
&= (b \vee a) \wedge (b \vee c) & \text{(distributive)} \\
&= (a \vee b) \wedge (c \vee b) & \text{(commutative)} \\
&= (a \vee c) \wedge (c \vee b) & \text{(since } a \vee b = a \vee c\text{)} \\
&= (c \vee a) \wedge (c \vee b) & \text{(commutative)} \\
&= c \vee (a \wedge b) & \text{(distributive)}
\end{aligned}$$

$$= c \lor (a \land c) \quad \text{(since } a \land b = a \land c\text{)}$$
$$= c \lor (c \land a) \quad \text{(commutative)}$$
$$= c \quad \text{(absorption property)}$$

Hence, it is proved.

Theorem 6.15 In a distributive lattice $(L, \land, \lor)$,

$$(a \land b) \lor (b \land c) \lor (c \land a) = (a \lor b) \land (b \lor c) \land (c \lor a)$$

holds for all $a, b, c \in L$.

Proof Given that the lattice L is distributive. Using distributive property,

$$(a \land b) \lor (b \land c) \lor (c \land a) = [\{(a \lor b) \lor b\} \land \{(a \land b) \lor c\}] \lor (c \land a)$$
$$= [b \land \{(a \lor c) \land (b \lor c)\}] \lor (c \land a)$$
$$= [(a \lor c) \land \{b \land (b \lor c)\}] \lor (c \land a)$$
$$= [(a \lor c) \land b] \lor (c \land a)$$
$$= [(a \lor c) \lor (c \lor a)] \land [(b \lor (c \land a)]$$
$$= [\{(a \lor c) \lor c\} \land \{(a \lor c) \lor a\}] \land [(b \lor c) \land (b \lor a)]$$
$$= (a \lor c) \land (a \lor c) \land (b \lor c) \land (a \lor b)$$
$$= (a \lor c) \land (b \lor c) \land (a \lor b)$$

Hence, it is proved.

EXAMPLE 6.35 Show that the lattices depicted in Figure 6.38 are non-distributive.

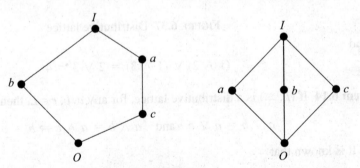

FIGURE 6.38 Non-distributive lattices

Solution

(A) From the distributive property (i),

$$a \land (b \lor c) = a \land I = a$$

and

$$(a \land b) \lor (a \land c) = 0 \lor c = c$$

So,

$$a \land (b \lor c) \neq (a \land b) \lor (a \land c).$$

Hence, the given lattice is not distributive.

(B) Using the distributive property,
$$a \wedge (b \vee c) = a \wedge I = a$$
Also,
$$(a \wedge b) \vee (a \vee c) = 0 \vee 0 = 0$$
Thus,
$$a \wedge (b \vee c) \neq (a \wedge b) \vee (a \wedge c)$$
Hence, the lattice is not distributive.

> **Note**
> (i) If one of the distributive property holds for all $a, b, c \in L$, then by duality principle, the other distributive property also holds for all $a, b, c \in L$.
> (ii) The distributive properties may be satisfied by some elements of lattice, but this does not fulfil that the lattice is distributive.

6.12.5 Modular Lattice

A lattice $(L, \wedge, \vee)$ is said to be a modular lattice if $a \vee (b \wedge c) = (a \vee b) \wedge c$, whenever $a \leq c$.

Theorem 6.16 Whenever $a \leq c$, $a \vee (b \wedge c) = (a \vee b) \wedge c$.

Proof If $a \leq c$, then $a \vee c = c$. Also, if L is distributive, then
$$a \vee (b \wedge c) = (a \vee b) \wedge (a \vee c) = (a \vee b) \wedge c \quad \text{(since } a \vee c = c\text{)}$$

> **Note**
> (i) Every distributive lattice is modular.
> (ii) Every modular lattice is not distributive (left to the reader for practice).

EXAMPLE 6.36 Show that the lattice described in Figure 6.39 is a non-distributive modular lattice.

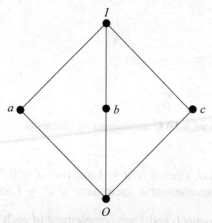

FIGURE 6.39 Non-distributive modular lattice

Solution From the distributive property of lattice,

$$a \wedge (b \vee c) = a \wedge I = a$$

and

$$(a \wedge b) \vee (a \wedge c) = 0 \vee 0 = 0$$

i.e.,

$$a \wedge (b \vee c) \neq (a \wedge b) \vee (a \wedge c)$$

Hence the result.

6.12.6 Complemented Lattices

Let the lattice $(L, \leq)$ be a bounded one. Then, an element $b \in L$ is said to be a *complement* of an element $a \in L$ if $a \vee b = 1$ and $a \wedge b = 1$ (denoting b by a').

Let $(L, \leq)$ be a bounded lattice with greatest element 1 and the least element 0. An element $a' \in L$ is called a complement of a if $a \vee a' = 1$ and $a \wedge a' = 0$.

From the definition of complement, if a' is a complement of a, then a is also a complement of a'. It is not necessary that an element a possesses a complement.

A lattice $(L, \leq)$ is called a *complemented* lattice if $(L, \leq)$ is bounded and each element $\in L$ possesses a complement. On the other hand, a bounded lattice is said to be complemented if every element has at least one complement in the lattice.

Note It is possible that $1' = 0$ and $0' = 1$.

 EXAMPLE 6.37 Determine the complement of a and c in Figure 6.40.

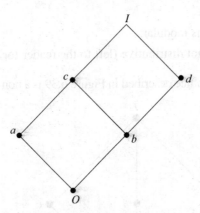

FIGURE 6.40 Complement of elements

Solution

The complement of a is d, since $a \vee d = 1$ and $a \wedge d = 0$. The complement of c does not exist. Since there exists no element c' such that $c \vee c' = 1$ and $c \wedge c' = 0$.

Theorem 6.17 The elements 0 and 1 are complement of each other.

Proof Assume that $c \neq 1$ is a complement of 0 and $c \in L$. Then, $0 \wedge c = 0$ and $0 \vee c = 1$. But $0 \vee c = c$ (property of bounded lattice) and $c \neq 1$, which is a contradiction.

Similarly, it can be shown that 0 is the only complement of 1.

EXAMPLE 6.38 Let L be a bounded lattice with lower bound 0 and upper bound I. Show that 0 and I are complements of each other.

Solution We have $0 \vee I = I \vee 0 = I$ and $0 \wedge I = I \wedge 0 = 0$. Thus 0 and I are complements of each other.

EXAMPLE 6.39 Prove that complemented distributive lattice establishes De Morgan's laws.

Solution To prove that

$$(a \wedge b)' = a' \vee b' \quad \text{and} \quad (a \vee b)' = (a' \wedge b')$$

Let us first consider

$$(a \wedge b)' \wedge (a' \vee b')$$

i.e.,

$$\begin{aligned}
(a \wedge b)' \wedge (a' \vee b') &= \{(a \wedge b) \wedge a'\} \vee \{(a \wedge b) \wedge b'\} \quad \text{(distributive)} \\
&= \{(b \wedge a) \wedge a'\} \vee \{(a \wedge b) \wedge b'\} \quad \text{(commutative)} \\
&= \{b \wedge (a \wedge a')\} \vee \{a \wedge (b \wedge b')\} \quad \text{(associative)} \\
&= (b \wedge 0) \vee (a \wedge 0) \\
&= 0 \vee 0 = 0
\end{aligned}$$

Next, consider

$$(a \wedge b) \wedge (a' \vee b')$$

i.e.,

$$\begin{aligned}
(a \wedge b) \wedge (a' \vee b') &= \{a \vee (a' \vee b')\} \wedge \{b \vee (a' \vee b')\} \quad \text{(distributive)} \\
&= \{a \vee (a' \vee b')\} \wedge \{b \vee (b' \vee a')\} \quad \text{(commutative)} \\
&= \{(a \vee a') \vee b'\} \wedge \{(b \vee b') \vee a'\} \quad \text{(associative)} \\
&= (1 \vee b') \wedge (1 \vee a') \\
&= 1 \wedge 1 = 1
\end{aligned}$$

Thus, $a' \vee b'$ is the complement of $a \wedge b$. Hence,

$$a' \vee b' = (a \wedge b)'$$

Similarly, it can be proved that

$$(a \vee b)' = a' \wedge b'$$

Theorem 6.18 In a bounded distributive lattice $(L, \leq)$, if an element has a complement then this complement is unique.

Solution Assume for some element $a \in L$, there exists two complements b and c in L. Then,

$$a \vee b = 1, \quad a \wedge b = 0 \quad \text{and} \quad a \vee c = 1, \quad a \wedge c = 0$$

Now,

$$\begin{aligned}
b &= b \wedge 1 = b \wedge (a \vee c) \\
&= (b \wedge a) \vee (b \vee c) \quad \text{(distributive)} \\
&= 0 \vee (b \vee c) \\
&= (a \wedge c) \vee (b \vee c) \\
&= (a \vee b) \wedge c \quad \text{(distributive)} \\
&= 1 \wedge c = c
\end{aligned}$$

Thus, $b = c$.

 EXAMPLE 6.40 In the bounded distributive lattice (Figure 6.41), show that every complement is unique, if it exists.

Solution From Figure 6.41, it is seen that the complement of a is b and vice versa. Also, the complement of 0 is 1 and vice versa. Thus, all the complements are unique.

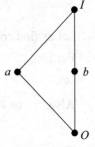 **EXAMPLE 6.41** Show that in a complemented lattice $(L, \leq)$,

$$a \leq b \Leftrightarrow a' \vee b = 1 \Leftrightarrow a \wedge b' = 0 \Leftrightarrow b' \leq a'$$

FIGURE 6.41 Complemented lattice

Solution Consider

$$\begin{aligned}
a \leq b &\Leftrightarrow a \wedge b = a \\
&\Leftrightarrow a' \vee a = a' \vee (a \wedge b) = 1 \\
&\Leftrightarrow (a' \vee a) \wedge (a' \vee b) = 1 \\
&\Leftrightarrow 1 \wedge (a' \vee b) = 1 \\
&\Leftrightarrow a' \vee b = 1
\end{aligned}$$

Again,

$$\begin{aligned}
a \leq b &\Leftrightarrow a \vee b = b \\
&\Leftrightarrow b \wedge b' = (a \vee b) \wedge b' = 0 \\
&\Leftrightarrow (a \wedge b') \vee (b \wedge b') = 1 \\
&\Leftrightarrow (a \wedge b') \vee 0 = 0 \\
&\Leftrightarrow a \wedge b' = 0
\end{aligned}$$

Finally,

$$\begin{aligned}
a \leq b &\Leftrightarrow a \vee b = b \\
&\Leftrightarrow (a \vee b)' = b'
\end{aligned}$$

$$\Leftrightarrow a' \wedge b' = b'$$
$$\Leftrightarrow a' \wedge a' = b'$$
$$\Leftrightarrow b' \leq a'$$

6.12.7 Isomorphic Lattices

Two lattices L_1 and L_2 are said to be *isomorphic* if there exists a one-to-one correspondence such that
$$f(a \wedge b) = f(a) \wedge f(b) \quad \text{and} \quad f(a \vee b) = f(a) \vee f(b)$$
for any elements $a, b, \in L$.

EXAMPLE 6.42 Investigate the lattices shown in Figure 6.42 as isomorphic.

Solution Consider the mapping
$$f = \{(a, 1), (b, 2), (c, 3), (d, 4)\}$$
For instance,
$$f(b \wedge c) = f(a) = 1$$
Also,
$$f(b) \wedge f(c) = 2 \wedge 3 = 1$$
Thus, the lattices shown in Figure 6.42 are isomorphic.

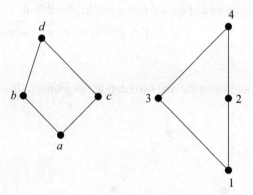

FIGURE 6.42 Isomorphic lattices

6.12.8 Join-Irreducible

Let $(L, \wedge, \vee)$ be a lattice. An element $a \in L$ is join-irreducible if it can be expressed as the join of two distinct elements of L, i.e., $a \in L$ is join-irreducible if
$$a = x \vee y \Rightarrow a = x \quad \text{or} \quad a = y$$
where $x, y \in L$.

EXAMPLE 6.43 Find the join-irreducible elements of the lattices shown in Figure 6.43.

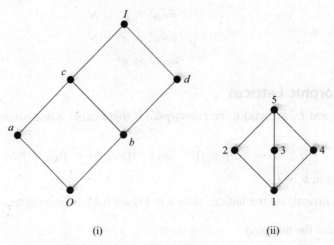

FIGURE 6.43 Join-irreducible elements

Solution

(a) The join-irreducible elements of Figure 6.43(i) are a, b, and d.

(b) The join-irreducible elements of Figure 6.43(ii) are 2, 3, and 4.

6.12.9 Meet-Irreducible

Let $(L, \wedge, \vee)$ be a lattice. An element $a \in L$ is *meet-irreducible* if it is defined as the meet of two distinct elements of L, i.e., $a \in L$ is meet-irreducible if

$$a = x \wedge y \Rightarrow a = x \quad \text{or} \quad a = y$$

where $x, y \in L$.

EXAMPLE 6.44 Determine the meet-irreducible elements of the lattices as shown in Figure 6.44.

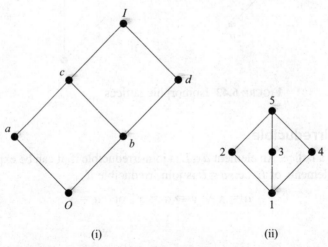

FIGURE 6.44 Meet-irreducible elements

Solution

(a) The meet-irreducible elements of Figure 6.44(i) are a, b, and c.
(b) The meet-irreducible elements of Figure 6.44(ii) are 2, 3, and 4.

6.13 LATTICE HOMOMORPHISM

Let L_1 and L_2 be two lattices. A mapping $f: L_1 \to L_2$ is called

I. Join-homomorphism if $f(x \vee y) = f(x) \vee f(y)$
II. Meet-homomorphism if $f(x \wedge y) = f(x) \wedge f(y)$
III. Order-homomorphism if $x \leq y$ then $f(x) \leq f(y)$

i.e., it defends the partial order and holds for all $x, y \in L$.

The mapping f is called a lattice homomorphism if it is both a join- and meet-homomorphism. If f is a homomorphism from L_1, then $f(L_1)$ is called a homomorphic image of L_1. It can be shown that every join- or meet-homomorphism is an order-homomorphism. But the converse is not true.

If a homomorphism f is bijective, i.e., one-to-one and onto, then f is called an *isomorphism*. If there exists an isomorphism from L_1 to L_2, then one can say that L_1 and L_2 are isomorphic. If f is an isomorphism from L to L, then it is called an *automorphism*.

EXAMPLE 6.45 Let L_1 be the lattice D_6 (divisor of 6) = $\{1, 2, 3, 6\}$ and let L_2 be the lattice $(P(S), \subseteq)$ where $S = \{a, b\}$. Show that the lattice L_1 and L_2 are isomorphic.

Solution The Hasse diagrams of L_1 and L_2 are constructed as follows (Figure 6.45).

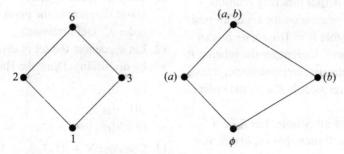

FIGURE 6.45 Isomorphic lattices

Let $f: L_1 \to L_2$ be defined as

$$f(1) = \Phi, \quad f_2(2) = \{a\}, \quad f(3) = \{b\}, \quad f(6) = \{a, b\}$$

Also, f is one-to-one function.

Now, consider $f(2 \wedge 3) = 1 = \{a\} \wedge \{b\} = \Phi$ and $f(2 \vee 3) = 6 = \{a\} \vee \{b\} = \{a, b\}$. Since the mapping is one-to-one and onto, f is isomorphic and the lattices L_1 and L_2 are isomorphic as well.

SUMMARY

Partially ordered structures occur widely in mathematics and computer science, and familiarity with the basic notions is essential. After giving some standard definitions and examples in partially ordered set we have presented the concepts of Hasse diagram, topological sorting, and isomorphisms followed by different parameters of posets. We have also discussed lattices, describing their use in analysing the semantics of programming languages. Consequently, we have discussed sublattices, some special class of lattices, and their properties as well as lattice homomorphism.

EXERCISES

1. Consider the relation of divisibility '|' of the set Z of integers. Is the relation an ordering of Z?
2. Consider $P(S)$ as the power set, i.e., the set of all subsets of a given set S. Then investigate $(P(S), \subseteq)$ as a partially ordered set, in which the symbol $\subseteq$ denotes the relation of set inclusion.
3. Give an example of a relation R which is both a partial ordering relation and an equivalence relation.
4. For a given set, $A = (a, b)$, assume the relation $R = \{(a, b)\} : a, b \in P(A), a \subseteq b\}$. Show that R is a partial ordering relation.
5. Let R be a binary relation on the set of all positive integers such that $R = \{(a, b): a-b$ is an odd positive integer$\}$. Investigate the relation R for reflexive, symmetric, antisymmetric, transitive, equivalence relation. Is R a partial order relation?
6. Let P be the set of all people. Let R be a binary relation on P such that $(a, b) \in R$, if a is a brother of b (neglect paternity brothers). Investigate the relation R for reflexive, symmetric, antisymmetric, transitive, equivalence relation. Is R a partial order relation.
7. If a relation R is transitive, then prove that its inverse relation R^{-1} is also transitive.
8. Let $A = \{1, 2, 3, 4\}$ and consider the relation $R = (1, 1), (1, 2), (1, 3), (2, 2), (2, 4), (3, 3), (3, 4), (1, 4), (4, 4)\}$. Show that R is a partial ordering and draw its Hasse diagram.
9. Draw the Hasse diagram of the following posets:
   (i) $(\{1, 2, 3, 4, 6, 9\}|)$
   (ii) $(\{3, 6, 12, 36, 72\}|)$
   (iii) $(\{2, 3, 4, 9, 12, 18\}|)$
   (iv) $(\{2, 3, 5, 30, 60, 120, 180, 360\}|)$
10. Let $A = \{1, 2, 3, 4, 12\}$ be defined by the partial order of divisibility on A, that is if a and $b \in A$, $a \leq b$ iff $a \,|\, b$. Draw the Hasse diagram of the poset $(A, \leq)$.
11. Let $S = \{1, 2, 3\}$ and $A = P(S)$. Draw the Hasse diagram of the poset with the partial order $\subseteq$ (set inclusion).
12. Let d_m denote the set of divisors of m ordered by divisibility. Draw the Hasse diagrams of
    (i) d_{15}
    (ii) d_{16}
    (iii) d_{17}
13. Consider $N = \{1, 2, 3, \ldots\}$ be ordered by divisibility. State whether each of the following subsets of N is linearly (totally) ordered:
    (i) $\{16, 4, 2\}$
    (ii) $\{3, 2, 15\}$
    (iii) $\{2, 4, 8, 12\}$
    (iv) $\{6\}$
    (v) $\{15, 5, 30\}$
14. Let $D_{100} = \{1, 2, 4, 5, 10, 20, 25, 50, 100\}$ whose all the elements are divisors of 100. Let

the relation ≤ be the relation | (divides) be a partial ordering on D_{100}.

(i) Determine the GLB of B where $B = \{10, 20\}$.
(ii) Determine the LUB of B where $B = \{10, 20\}$.
(iii) Determine the GLB of B where $B = \{5, 10, 20, 25\}$.
(iv) Determine the LUB of B where $B = \{5, 10, 20, 25\}$.

15. Find two incomparable elements in the following posets:
    (i) $(S\{0, 1, 2\}, \subseteq)$
    (ii) $(\{1, 2, 4, 6, 8\}, |)$

16. Find the dual of the following posets:
    (i) $(\{0, 1, 2\}, \leq)$
    (ii) $(Z^+, |)$
    (iii) $(Z, \geq)$
    (iv) $(S(Z), \subseteq)$

17. Find the lexicographic ordering of the following n-tuples:
    (i) (1, 1, 3) (1, 3, 1)
    (ii) (0, 1, 2, 3) (0, 1, 3, 2)
    (iii) (0, 1, 3, 4) (0, 1, 1, 5)
    (iv) (1, 0, 1, 0, 1) (0, 1, 1, 1, 0)

18. Find the lexicographic ordering of the bit strings: 0, 01, 11, 001, 010, 011, 0001, and 0101 based on the ordering 0 < 1.

19. Draw the Hasse diagram for the 'less than or equal to' relation on the set $A = \{0, 2, 5, 10, 11, 15\}$.

20. Draw the Hasse diagram for divisibility on the following sets:
    (i) $\{1, 2, 3, 4, 5, 6, 7, 8\}$
    (ii) $\{1, 2, 3, 5, 7, 11, 13\}$
    (iii) $\{1, 2, 3, 6, 12, 24, 36, 48\}$
    (iv) $\{1, 2, 4, 8, 16, 32, 64\}$

21. Suppose $X = \{1, 2, 6, 8, 12\}$ is ordered by divisibility and suppose $Y = \{a, b, c, d, e\}$ is isomorphic to X. The mapping from X in to Y is as follows:

    $f = \{(1, e), (2, d), (6, b)(8, c)(12, a)\}$

    Draw the Hasse diagram of Y.

22. Let $A = \{1, 2, 3, 4, 5\}$ be ordered as in the figure below. Find the number n of similarity mappings $f: A \to A$.

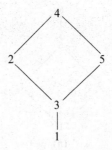

23. Let $S = \{a, b, c, d, e\}$ be ordered by the diagram in the figure below. Insert the correct symbol <, >, or || between each of the following pair of elements:
    (i) a–e, (ii) c–b, (iii) b–d, and (iv) a–d

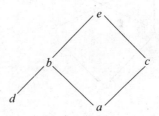

24. Consider the partially ordered set A in the figure below. Find all minimal and maximal elements of A.

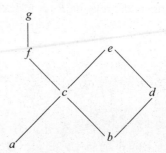

25. Consider the lattice L on the set D_m of divisors of a positive integer m where
$$a \vee b = \text{LCM}(a, b) \text{ and } a \wedge b = \text{GCD}(a, b)$$
Draw the diagram of the partial order induced by L for $m = 36$.

26. Which of the partially ordered sets shown in the figures below are lattices?

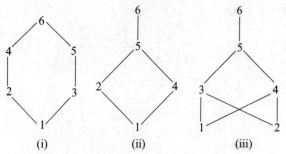

27. Consider the partially ordered set $S = \{a, b, c, d, e\}$ shown in the figure below. Find two subsets of S that are lattices with respect to the operation on S.

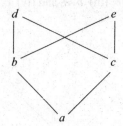

28. Are the two lattices shown in the figures below isomorphic?

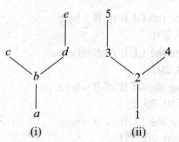

29. Are the two lattices shown in the figures below isomorphic?

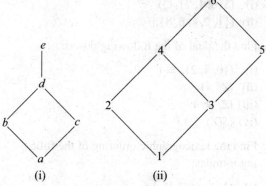

30. Let L be a bounded distributive lattice. Then prove that their complements are unique if they exist.

7 Boolean Algebra

LEARNING OBJECTIVES

After reading this chapter you will be able to understand the following:
- The laws of Boolean algebra
- The Boolean expressions containing the functions: AND, NOT, and XOR that can be used to specify a more complex behaviour
- The construction of truth tables
- The familiar and useful properties of Boolean algebra
- The expressions such as minterm and maxterm, switching networks, and Karnaugh maps (K-map)

7.1 INTRODUCTION

Mathematical rules are based on the defining limits that are placed on the numerical quantities. When we say that 1 + 1 = 2 or 3 + 4 = 7, we are implying the use of integer quantities: the same types of numbers we all learned to count in elementary education. People assume these to be self-evident rules of arithmetic, valid at all times and for all purposes.

Logic is much like mathematics in this respect: the so-called 'Laws' of logic depend on how we define a proposition. The Greek philosopher Aristotle founded a system of logic based on only two types of propositions: *true* and *false*. His bivalent (two-mode) definition of truth led to the four foundational laws of logic: the law of identity (A is A); the law of non-contradiction (A is not non-A); the law of the excluded middle (either A or non-A); and the law of rational inference. These so-called laws function within the scope of logic where a proposition is limited to one of two possible values, but may not apply in cases where propositions can hold values other than 'true' or 'false'. In fact, much work has been done and continues to be done on 'multi-valued' or *fuzzy* logic, where propositions may be true or false to a limited degree.

The English mathematician George Boole (1815–1864) sought to give symbolic form to Aristotle's system of logic. Boole wrote a treatise on the subject in 1854, titled *An Investigation of the Laws of Thought,* on which are founded the mathematical theories of logic and probabilities, which codified several rules of relationship between mathematical quantities limited to one of two possible values: true or false, 1 or 0. His mathematical system is known as *Boolean algebra*.

Two things about Boolean algebra make it a very important form of mathematics for practical applications. First, statements expressed in everyday language (such as 'If it rains, I will be home today') can be converted into mathematical expressions, using letters and numbers for example, $P =$ If it rains, $Q =$ I will be at home today. Second, those symbols generally have only one of two values. The statements above (P and Q), for example, are either true or false. That means they can be expressed in a binary system: true or false; yes or no; 0 or 1.

Binary mathematics is the number system most often used in computers. Computer systems consist of magnetic cores that can be switched on or switched off. The numbers 0 and 1 are used to represent the two possible states of a magnetic core. Boolean statements can be represented, then, by the numbers 0 and 1 and also by electrical systems that are either on or off. As a result, when engineers design circuitry for personal computers, pocket calculators, compact disk players, cellular telephones, and a host of other electronic products, they apply the principles of Boolean algebra.

In this chapter, basic operations on Boolean algebra, De Morgan's theorem, logic gate, sum of products, product of sums, normal form, expression of a Boolean function as a canonical form, simplification of Boolean expression by algebraic method, Boolean expression from logic and switching network, and Karnaugh map method for simplification of Boolean expression are discussed.

7.2 LAWS ON BOOLEAN ALGEBRA

To acquire knowledge on Boolean algebra, we need the definitions of literal and Boolean expression. In fact, a *literal* is a variable or its complement, whereas a *Boolean expression* consists of combinations of Boolean terms as well as Boolean operations. Further, a Boolean term consists of Boolean literals.

The basic operations used in Boolean algebra are logical addition/OR, logical multiplication/AND, and complementation. These are normally represented by the symbols '+', '·', and 'c' respectively. These operations are also called as *logic operations*. [It should be noted that we may use other symbols such as: ¬, ', and ¯ (bar) to represent complement. In this book, the symbol ' or ¯ is interchangeably used to represent complement.]

Basic Boolean laws are presented below. Note that every law has two expressions: (a) and (b). This is known as *duality*. These are obtained by changing every AND (·) to OR (+), every OR (+) to AND (·), and all 1s to 0s and vice versa. It has become conventional to drop the. (AND symbol), i.e., $A \cdot B$ is written as AB.

T1: Commutative law
(a) $x + y = y + x$
(b) $xy = yx$

T2: Associative law
(a) $(x + y) + z = x + (y + z)$
(b) $(xy) z = x (yz)$

T3: Distributive law
(a) $x(y + z) = xy + xz$
(b) $x + (yz) = (x + y)(x + z)$

T4: Identity law
(a) $x + x = x$ (b) $x \cdot x = x$

T5: Complement law
(a) $x' + x = 1$
(b) $x'x = 0$

T6: Redundance law
(i) (a) $x + xy = x$
   (b) $x(x + y) = x$
(ii) (a) $xy + xy' = x$
    (b) $(x + y)(x + y') = x$
(iii) (a) $x + x'y = x + y$
     (b) $x(x' + y) = xy$

T7: Uniqueness of complement law
(a) $0 + x = x$ (b) $0x = 0$

T8: Boundedness law
(a) $1 + x = 1$ (b) $1.x = x$

T9: Involution
$(x')' = x$

T10: De Morgan's theorem
(a) $(x + y)' = x'y'$
(b) $(xy)' = x' + y'$

Note If $E(x_1, x_2, \ldots, x_n)$ is a Boolean expression of n variables, and $E_d(x_1, x_2, \ldots, x_n)$ is its dual, then

$$E'(x_1, x_2, \ldots, x_n) = E_d(x'_1, x'_2, \ldots, x'_n)$$

7.3 TRUTH TABLES OF BOOLEAN OPERATIONS

Truth tables of some simple Boolean operations are shown in Table 7.1.

TABLE 7.1 Truth table of simple Boolean operations

OR operation			AND operation			NOT operation	
x	y	x OR y	x	y	x AND y	x	x'
F	F	F	F	F	F	F	T
F	T	T	F	T	F	T	F
T	F	T	T	F	F		
T	T	T	T	T	T		

The corresponding logic gates of the above simple Boolean operations are shown in Figure 7.1.

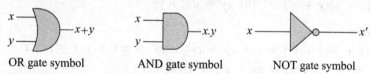

OR gate symbol   AND gate symbol   NOT gate symbol

FIGURE 7.1 Logic gates of simple Boolean operations

Apart from the above-mentioned simple operations, some other logic operators along with their respective gates are listed in Table 7.2.

 EXAMPLE 7.1 Prove that $A + A'B = A + B$.

Solution Algebraically

$$\begin{aligned}
A + A'B &= A1 + A'B & \text{[by T8 (b), } 1 \cdot A = A] \\
&= A(1 + B) + A'B & \text{[by T8 (a), } 1 = 1 + B] \\
&= A + AB + A'B & \text{[by T3 (a)]} \\
&= A + B(A + A') & \text{[by T3 (a)]} \\
&= A + B & \text{[by T5 (a)]}
\end{aligned}$$

TABLE 7.2 A summary on algebraic function and corresponding operators, and graphic symbols

Name	Algebraic function	Operator symbol	Graphic symbol	Comments
Exclusive OR	$f = x \cdot y' + x' \cdot y$	$f = x \oplus y$		x or y, but not both
NOR	$f = (x + y)$	$f = (x \downarrow y)$		Not OR
Equivalence	$f = x \cdot y + x' \cdot y'$	$f = (x \oplus y)'$		x equals y
NAND	$f = (x \cdot y)'$	$f = x \uparrow y$		Not AND

EXAMPLE 7.2 Prove that $(A + B')(C + B') = AC + B'$.
Solution $(A + B')(C + B') = AC + AB' + B'C + B'$ [expanding the terms and using T4]
$\qquad\qquad\qquad\qquad\quad = AC + AB' + B'$ [using T6 (a)]
$\qquad\qquad\qquad\qquad\quad = AC + B'$ [using T6 (a)]

EXAMPLE 7.3 Simplify the following Boolean expressions.
(a) $x(x' + y)$ (b) $x + x'y$

Solution
(a) $x(x' + y) = xx' + xy = 0 + xy = xy$
(b) $x + x'y = (x + x')(x + y) = 1(x + y) = x + y$

EXAMPLE 7.4 Simplify the following Boolean expressions.
(a) $(x + y)(x + y')$ (b) $xy + x'z + yz$

Solution
(a) $(x + y)(x + y') = x + xy + xy' + yy' = x(1 + y + y') = x$
(b) $xy + x'z + yz = xy + x'z + yz(x + x')$
$\qquad\qquad\qquad\; = xy + x'z + xyz + x'yz$
$\qquad\qquad\qquad\; = xy(1 + z) + x'z(1 + y) = xy + x'z$

EXAMPLE 7.5 Let '*' be defined as $X * Y = X' + Y$. Let $Z = X * Y$, find $Z * X$.
Solution Given,
$$Z = X * Y = X' + Y \qquad [\text{since } X * Y = X' + Y]$$
Thus,
$$Z * X = (X' + Y)' + X \qquad [\text{using } X * Y = X' + Y]$$
$$\qquad\; = (X')' \cdot Y' + X$$
$$\qquad\; = X \cdot Y' + X \qquad\qquad [\text{since } (X')' = X]$$
$$\qquad\; = X(Y' + 1) = X \qquad\qquad [\text{by T5 (b)}]$$

EXAMPLE 7.6 Simplify $ABC + ABD + BCD + A'CD + A'BD + ABC'D + A'BCD'$.

Solution

$ABC + ABD + BCD + A'CD + A'BD + ABC'D + A'BCD'$
$= (ABC + A'BCD') + (ABD + ABC'D) + BCD + A'CD + A'BD$
  [rearranging the terms]
$= BC(A + A'D') + ABD(1 + C') + BCD + A'CD + A'BD$
$= BC(A + A')(A + D') + ABD + BCD + A'CD + A'BD$
  [expanding first term by distributive law]
$= BC(A + D') + ABD + BCD + A'CD + A'BD$
$= ABC + BCD' + ABD + BCD + A'CD + A'BD$
$= ABC + A'CD + (BCD' + BCD) + (ABD + A'BD)$ [rearranging the terms]
$= ABC + A'CD + BC(D' + D) + (A + A')BD$
$= ABC + A'CD + BC + BD$
$= BC(A + 1) + A'CD + BD$
$= BC + A'CD + BD$

EXAMPLE 7.7 Simplify $A'B'C' + A'BC + A'BC'$.

Solution

$A'B'C' + A'BC + A'BC'$
$= A'B'C' + A'B(C + C')$
$= A'B'C' + A'B$
$= A'(B'C' + B)$
$= A'(B' + B)(C' + B)$ [using distributive law]
$= A'(C' + B) = A'C' + A'B$

EXAMPLE 7.8 Find the simplified Boolean function represented by the following truth table (Table 7.3).

TABLE 7.3 Truth table of $F(A, B, C)$

A	B	C	$F(A, B, C)$
0	0	0	1
0	0	1	0
0	1	0	0
0	1	1	1
1	0	0	1
1	0	1	1
1	1	0	1
1	1	1	0

Solution From the given truth table, we get

$F(A, B, C) = A'B'C' + A'BC + AB'C' + AB'C + ABC'$

It is simplified below as

$F(A, B, C) = A'B'C' + A'BC + AB'(C + C') + ABC'$ [grouping of terms]
$= A'B'C' + A'BC + AB' + ABC'$
$= A'B'C' + A'BC + A(B' + BC')$ [grouping of terms]
$= A'B'C' + A'BC + A(B' + B)(B' + C')$ [using distributive law]
$= A'B'C' + A'BC + A(B' + C')$
$= A'B'C' + A'BC + AB' + AC'$
$= (AC' + A'B'C') + A'BC + AB'$ [rearranging the terms]
$= C'(A + A'B') + A'BC + AB'$
$= C'(A + A')(A + B') + A'BC + AB'$
$= C'(A + B') + A'BC + AB'$
$= AC' + B'C' + A'BC + AB'$

EXAMPLE 7.9 Construct the truth table for the following Boolean function.

$F(A, B, C) = A'BC + ABC' + BC + A'C + A'B$

Solution First, let us simplify the expression before generating its truth table.

$F(A, B, C) = A'BC + ABC' + BC + A'C + A'B$
$= (A'BC + A'C + A'B) + (ABC' + BC)$ [reordering the terms]
$= A'(BC + C + B) + B(AC' + C)$ [grouping of terms]
$= A'(C(1 + B) + B) + B(C + C')(C + A)$ [using distributive law]
$= A'(C + B) + B(C + A)$
$= A'C + AB + BC + BA$
$= AB + A'C + BC$ [since $AB + BA = AB + AB = AB$]

The corresponding truth table is given in Table 7.4.

TABLE 7.4 Truth table of F

A	B	C	F(A, B, C)
0	0	0	0
0	0	1	1
0	1	0	0
0	1	1	1
1	0	0	0
1	0	1	0
1	1	0	1
1	1	1	1

7.4 UNIQUE FEATURES OF BOOLEAN ALGEBRA

The striking unique features of Boolean algebra, which distinguish it from other branches of algebra are

(a) Boolean algebra has only a finite set of elements, namely 1 and 0, which are usually called *true* and *false*. However, the ordinary algebra deals with a set of an infinite number of elements each of which may possess values other than 0 and 1.
(b) Boolean algebra does not provide operations such as *subtraction* and *division*.
(c) In ordinary algebra, there is no equivalent of the unary operations of complement.
(d) In Boolean algebra, the cancellation law does not hold, i.e., $A + B = A + C$ does not imply $B = C$.

7.5 MINTERM AND MAXTERM

Minterm A minterm of n variables is a product of n literals in which each variable appears exactly once in either true or complimented form, but not both. For example, the list of all the minterms of *two* variables x and y are $xy, x'y, xy', x'y'$. In general, n variables have 2^n minterms.

Maxterm A maxterm of n variables is a sum of n literals in which each variable appears exactly once in either true or complimented form, but not both. For example, the list of all the maxterms of *two* variables x and y are $x + y, x' + y, x + y', x' + y'$.

Clearly, n variables produce 2^n possible maxterms.

> **Note** Each maxterm (M_i) is the complement of its corresponding minterm (m'_i). In minterm, variable value 0 implies corresponding complement and 1 simply means the uncomplemented. The reverse is true for maxterms. The minterms and the respective maxterms for three variables are shown in Table 7.5.

TABLE 7.5 Minterms and the respective maxterms for three variables

Row number	x	y	z	Minterms	Notation	Maxterms	Notation
0	0	0	0	$x'y'z'$	m_0	$x + y + z$	M_0
1	0	0	1	$x'y'z$	m_1	$x + y + z'$	M_1
2	0	1	0	$x'yz'$	m_2	$x + y' + z$	M_2
3	0	1	1	$x'yz$	m_3	$x + y' + z'$	M_3
4	1	0	0	$xy'z'$	m_4	$x' + y + z$	M_4
5	1	0	1	$xy'z$	m_5	$x' + y + z'$	M_5
6	1	1	0	xyz'	m_6	$x' + y' + z$	M_6
7	1	1	1	xyz	m_7	$x' + y' + z'$	M_7

7.5.1 Boolean Expression in Sum of Products (SOP) and Product of Sums (POS) Form or Normal Form

A Boolean expression E is said to be in a *sum of products* (or *minterm*) *form* or the *disjunctive normal form* (DNF) if E consists of some products of variables (complemented or

uncomplemented), none of which is included in another. It is also called as *canonical sum of products* or *standard sum of products*. Some examples of this form are
(i) $xy + xy'z' + xyz'$
(ii) $a + ab' + a'b'c'$

On the other hand, a Boolean expression E is said to be in a *product of sums* (or *maxterm*) *form* or the *conjunctive normal form* (CNF) if E consists of several product of sum terms, such that none of the terms is included in other term. It is also called as *canonical product of sums* or *standard product of sums*.

Some examples of this form are
(i) $(x + y)(x + y' + z')$
(ii) $a(a + b')$

Note The *complement* of SOP = POS and the *complement* of POS = SOP.

7.6 BOOLEAN FUNCTION

A *Boolean function* is an algebraic expression formed with Boolean variables, the operators OR, AND, NOT, parentheses, and an equal sign. Some useful hints on Boolean function are presented below.

- When evaluating a Boolean function, we must consider the following order of computation in sequence: (1) parenthesis, (2) NOT, (3) AND, and (4) OR.
- Any Boolean function can be represented by a truth table. The number of rows in the table is 2^n, where n is the number of variables in the function. There are 2^{2^n} different Boolean functions for n binary variables.
- There are infinitely many algebraic expressions that specify a given Boolean function. It is important to find the simplest one.
- Any Boolean function can be transformed in a straightforward manner from an algebraic expression into a logic diagram composed of AND, OR, and NOT gates.
- The complement of any function F is F', which can be obtained by De Morgan's theorem: (1) take the dual of F and (2) complement each literal.
- A Boolean function F may be either in CNF or in DNF.

Consider the following example.

EXAMPLE 7.10 The Boolean function $F_1 = xy + xy'z + x'yz$ and its truth table is given in Table 7.6. Explain the function F_1.

Solution In this example, F_1 has eight literals (x, x', y, y', z, z'), one OR term (sum term), i.e., $xy + xy'z + x'yz$ in which three AND terms (product terms) are present.

EXAMPLE 7.11 Consider two functions F_1 and F_2, as shown in Table 7.7. Express them in the standard canonical forms.

Solution Sum of minterms:
$$F_1 = x'yz + xy'z + xyz' + xyz = m_3 + m_5 + m_6 + m_7 = \Sigma(3, 5, 6, 7)$$
$$F_2 = x'y'z' + xy'z' + xyz = m_1 + m_4 + m_7 = \Sigma(1, 4, 7)$$

TABLE 7.6 Truth table for F_1 and F_1'

Row number	x	y	z	F_1	F_1'
0	0	0	0	0	1
1	0	0	1	0	1
2	0	1	0	0	1
3	0	1	1	1	0
4	1	0	0	0	1
5	1	0	1	1	0
6	1	1	0	1	0
7	1	1	1	1	0

TABLE 7.7 Truth table for F_1, F_1', F_2, and F_2'

Row number	x	y	z	F_2	F_2'	F_1	F_1'
0	0	0	0	0	1	0	1
1	0	0	1	1	0	0	1
2	0	1	0	0	1	0	1
3	0	1	1	0	1	1	0
4	1	0	0	1	0	0	1
5	1	0	1	0	1	1	0
6	1	1	0	0	1	1	0
7	1	1	1	1	0	1	0

Product of maxterms:

$$F_1 = (x + y + z)(x + y + z')(x + y' + z')(x' + y + z)$$
$$= M_0 M_1 M_2 M_4 = \Pi(0, 1, 2, 4)$$
$$F_2 = (x + y + z)(x + y' + z)(x + y' + z')(x' + y + z')(x' + y' + z)$$
$$= M_0 M_2 M_3 M_5 M_6 = \Pi(0, 2, 3, 5, 6)$$

To convert from one canonical form to another, interchange Σ and Π, and list the numbers that were excluded from the original.

Now, $F_1 = \sum(3, 5, 6, 7)$. Thus, $F_1' = \sum(0, 1, 2, 4)$.

EXAMPLE 7.12 Convert the Boolean function $f(x, y) = x \cdot y' + x' \cdot y + x' \cdot y'$ to its CNF.

Solution Here, $f(x, y) = x \cdot y' + x' \cdot y + x' \cdot y' = m_2 + m_1 + m_0$. The complete DNF in two variables x and y is $F(x, y) = x \cdot y + x \cdot y' + x' \cdot y + x' \cdot y' = m_3 + m_2 + m_1 + m_0$. Thus, the *complement* of $f(x, y)$ is $f'(x, y) = F(x, y) - f(x, y) = x \cdot y = m_3$.

Again, $(f'(x, y))' = f(x, y)$ [by law of *involution*]. Clearly, $(f'(x, y))' = f(x, y) = (x \cdot y)' = x' + y' = M_3$, and it is the required CNF of the given DNF.

EXAMPLE 7.13 Convert the Boolean function $f(x, y, z) = (x' + y + z')(x' + y + z)(x + y' + z)$ to its DNF.

Solution Here, $f(x, y, z) = (x' + y + z')(x' + y + z)(x + y' + z) = M_5 M_4 M_2$.

Further, the complete CNF in three variables is

$$f(x, y, z) = (x + y + z)(x + y + z')(x + y' + z)(x' + y + z)$$
$$(x + y' + z')(x' + y + z')(x' + y' + z)(x' + y' + z')$$

Hence,

$$f'(x, y, z) = f(x, y, z) - f(x, y, z) = (x + y + z)(x + y + z')(x + y' + z)$$
$$(x' + y' + z)(x' + y' + z')$$

Thus,

$$(f'(x, y, z))' = [(x + y + z)(x + y + z')(x + y' + z)(x' + y' + z)$$
$$(x' + y' + z')]' = (M_0 \cdot M_1 \cdot M_3 \cdot M_6 \cdot M_7)'$$
$$= x'y'z' + x'y'z + x'yz' + xyz' + xyz = m_0 + m_1 + m_3 + m_6 + m_7$$

Note A Boolean function may be expressed in a non-standard form. For example, consider the function F_3 as follows.

$$F_3 = xy + z(t + w)$$

But it can be changed to standard form by using the distributive law to remove the parentheses.

$$F_3 = xy + z(t + w) = xy + zt + zw$$

EXAMPLE 7.14 Express $E = z(x' + y) + y'$ in its complete sum of products form.

Solution Here,

$$E = z(x' + y) + y'$$
$$= x'z + yz + y' = x'(y + y')z + (x + x')yz + (x + x')y'(z + z')$$
[since $(x + x') = (y + y') = (z + z') = 1$]
$$= xyz + xy'z + xy'z' + x'yz + x'y'z + x'y'z'$$

EXAMPLE 7.15 Express $E = (xy' + xz) + z'$.

Solution It is given that $E = (xy' + xz) + z'$
So,

$$E = (xy' + x)(xy' + z) + z' \qquad \text{[since } a + bc = (a + b)(a + c),$$
$$\text{i.e., distributive law]}$$

$$= (z' + xy' + x)(z' + xy' + z)$$
$$= (z' + x + x)(z' + y' + x)(z' + y' + z)(z' + x + z)$$

$$= (z' + x)(z' + y' + x)(z' + y' + z)(1 + x) \quad \text{[since } z' + z = 1\text{]}$$
$$= (z' + x)(z' + y' + x)(z' + y' + z) \quad \text{[since } 1 + x = 1\text{]}$$
$$= (z' + x + yy')(z' + y' + x)(z' + y' + z) \quad \text{[since } yy' = 0\text{]}$$
$$= (z' + x + y)(z' + x + y')(z' + y' + x)(z' + y' + z)$$

7.7 SWITCHING NETWORK FROM BOOLEAN EXPRESSION USING LOGIC GATES

The electric network can be designed from the simplified Boolean expression, using logic gates. For example, consider the following POS function using only NOR gates. It is assumed that you have all signal and signal complements.

$$f = (A + B)(A + \overline{C})(\overline{B} + C)$$

First, we implement function using AND and OR gates, noting where the inputs are inverted (Figure 7.2).

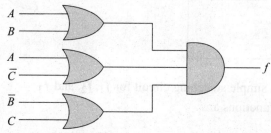

FIGURE 7.2 Switching network for f using AND and OR gates

Now, replace all gates with NOR gates or De Morgan equivalent NOR gates. Notice in Figure 7.3, by choosing NOR gates for implementing POS, *inverters* are required at the input only.

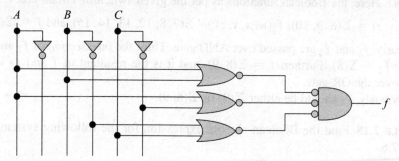

FIGURE 7.3 Switching network of f using NOR gates

EXAMPLE 7.16 Implement the following SOP function using only NAND gates. It is assumed that you have all signal and signal complements.

$$f = (A \cdot C) + (A \cdot \overline{B}) + (\overline{A} \cdot B \cdot \overline{C})$$

Solution First, we implement function using AND and OR gates, noting where the inputs are inverted (Figure 7.4).

Now, replace all gates with NAND gates or De Morgan equivalent NAND gates. Notice, by choosing NAND gates for implementing SOP, inverters are required at the input only.

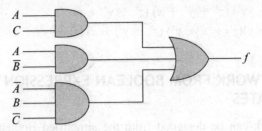

FIGURE 7.4 Logic diagram of f

 EXAMPLE 7.17 Consider the following switching circuit, as shown in Figure 7.5.

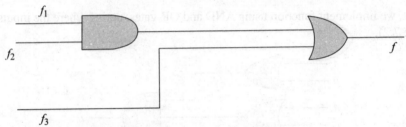

FIGURE 7.5 A simple switching circuit for f_1, f_2, and f_3

Here, the Boolean functions are

$$f_1(w, x, y, z) = \Sigma(8, 9, 10),\ f_2(w, x, y, z) = \Sigma(7, 8, 12, 13, 14, 15)\ \text{and}\ f = \Sigma(8, 9)$$

Find the Boolean function f_3.

Solution Here, the Boolean functions as per the given switching circuit are

$$f_1(w, x, y, z) = \Sigma(8, 9, 10),\ f_2(w, x, y, z) = \Sigma(7, 8, 12, 13, 14, 15),\ \text{and}\ f = \Sigma(8, 9)$$

Clearly, f_1 and f_2 are passed over AND gate. Thus, the *intersection* of f_1 and f_2 is simply $f = f_1 \cdot f_2 = \Sigma(8)$. Further, $f = \Sigma(8, 9)$, and it is the resultant of f and f_3 when they are passed over the OR gate.

Obviously, f_3 should be either $\Sigma(9)$ or $\Sigma(8, 9)$.

 EXAMPLE 7.18 Find the Boolean algebra expression for the following system, as shown in Figure 7.6.

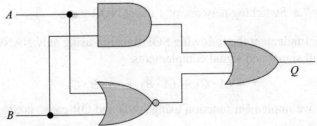

FIGURE 7.6 A simple logic circuit

Solution The system consists of an AND gate, a NOR gate, and finally an OR gate. The expression for the AND gate is $A.B$, and the expression for the NOR gate is $(A + B)'$. Both of these expressions have separate inputs to the OR gate which is defined as $(A + B)$. Thus, the final output expression is given, as shown in Figure 7.7.

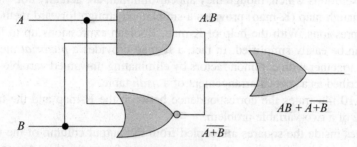

FIGURE 7.7 Simplified logic diagram

Thus, the output of the system is given as

$$Q = AB + \overline{A + B}$$

Further simplification of the above equation gives

$$Q = AB + A'B' = (A \text{ XOR } B)' = A \text{ XNOR } B$$

EXAMPLE 7.19 Minimize the number of logic gates in the following circuit.

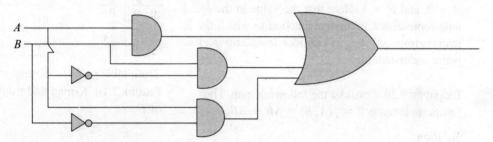

FIGURE 7.8 A redundant logic diagram

Solution The expression given by the logic circuit, as shown in Figure 7.8 is

$$\begin{aligned}
F &= AB + A'B + AB' \\
&= (A + A')B + AB' \quad &&\text{[grouping of terms]} \\
&= B + AB' \\
&= (B + A)(B + B') \quad &&\text{[using distributive law]} \\
&= B + A = A + B
\end{aligned}$$

The equivalent-minimized logic circuit needs just an OR gate, and is shown in Figure 7.9.

FIGURE 7.9 The equivalent-minimized logic diagram

7.8 KARNAUGH MAP

So far, we have seen that applying Boolean algebra can be awkward in order to simplify expressions. Apart from being laborious (and requiring to remember all the laws), the method can lead to solutions which, though they appear minimal, are actually not.

The Karnaugh map (K-map) provides a simple and straightforward method of minimizing Boolean expressions. With the help of K-map, Boolean expressions up to four and even six variables can be easily simplified. In fact, a K-map provides a *pictorial* method of grouping expressions together with common factors by eliminating unwanted variables. The K-map can also be described as a special arrangement of a *truth* table.

Figure 7.10 illustrates the correspondence between the K-map and the truth table for the general case of a two-variable problem.

The values inside the squares are copied from the output column of the truth table, therefore, there is one square in the map for every row in the truth table. Around the edge of the K-map are the values of the two input variable. A is along the top and B is down the left-hand side. The diagram explains this.

The values around the edge of the map can be thought of as coordinates. So, as an example, the square on the top right-hand corner of the map in the diagram has coordinates $A = 1$ and $B = 0$. This square corresponds to the row in the truth table where $A = 1$, $B = 0$, and $F = 1$. Note that the value in the F column represents a particular function to which the K-map corresponds. Let us consider *some examples* for better understanding.

A	B	F
0	0	a
0	1	b
1	0	c
1	1	d

Truth table of F

K-map of F

B \ A	0	1
0	a	b
1	c	d

A	B	F
0	0	0
0	1	1
1	0	1 ←
1	1	1

Truth table of F

K-map of F

B \ A	0	1
0	0	1 ←
1	1	1

FIGURE 7.10 K-map and truth table of F

EXAMPLE 7.20 Consider the following map. The function plotted is $Z = f(A, B) = A\overline{B} + AB$.

Solution

- Note that values of the input variables form the rows and columns. That is, the logic values of the variables A and B (with 1 denoting true form and 0 denoting false form) form the head of the rows and columns, respectively. (see Figure 7.11).
- Bear in mind that the above map is a one-dimensional type which can be used to simplify an expression in two variables.
- There is a two-dimensional map that can be used for up to four variables, and a three-dimensional map for up to six variables.

B \ A	0	1
0		(1)
1		(1)

FIGURE 7.11 K-map of F

Using algebraic simplification,

$Z = A\overline{B} + AB$

$Z = A(\overline{B} + B)$ [using complement law T5 (a)]

$Z = A$

Referring to the map above, the two adjacent 1s are grouped together. Through inspection, it can be seen that variable B has its true and false form within the group. This eliminates variable B, leaving only variable A which only has its true form. Therefore, the minimized answer is $Z = A$.

EXAMPLE 7.21 Consider the expression $Z = f(A, B) = \overline{A}\,\overline{B} + A\overline{B} + \overline{A}B$ plotted on the K-graph.

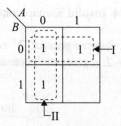

FIGURE 7.12 K-map of F

Solution Pairs of 1s are *grouped* as shown above, and the simplified answer is obtained by using the following steps:

- Two groups can be formed for the example given above, bearing in mind that the largest rectangular clusters that can be made consist of two 1s.
- A 1 can belong to more than one group.

Clearly, the first group, labelled I, consists of two 1s which correspond to $A = 0, B = 0$ and $A = 1, B = 0$. Put in another way, all squares in this example that correspond to the area of the map where $B = 0$ contains 1s, independent of the value of A. So, when $B = 0$ the output is 1. The expression of the output will contain the term $\overline{B}$. The second group, labelled II, corresponds to the area of the map where $A = 0$. The group can, therefore, be defined as $\overline{A}$. This implies that when $A = 0$, the output is 1. The output is, therefore, 1 whenever $B = 0$ and $A = 0$ (see Figure 7.12).

Hence, the simplified answer is $Z = \overline{A} + \overline{B}$.

7.8.1 Rules Used by K-Map for Simplification

The K-map uses the following rules for the simplification of expressions by *grouping* together adjacent cells containing *ones*.

- *Groups may not include any cell containing a zero* (see Figure 7.13).

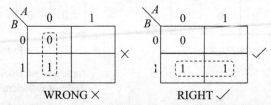

FIGURE 7.13 Incorrect K-map containing both 0 and 1

- *Groups may be horizontal or vertical, but not diagonal* (see Figure 7.14).

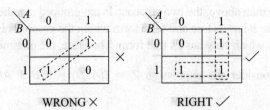

FIGURE 7.14 Invalid K-map containing diagonal elements

- *Groups must contain 1, 2, 4, 8, or, in general, 2^n cells. That is, if $n = 1$, a group will contain two 1s since $2^1 = 2$. If $n = 2$, a group will contain four 1s since $2^2 = 4$ (see Figure 7.15).*

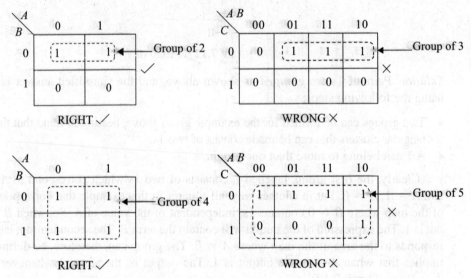

FIGURE 7.15 Incorrect K-map when group is not power of 2

- *Each group should be as large as possible* (see Figure 7.16).

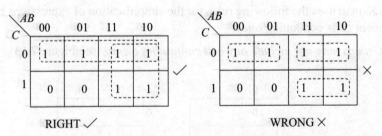

(note that no Boolean laws broken, but not sufficiently minimal)

FIGURE 7.16 K-map invalid when group is not the possible largest

- *Each cell containing a 1 must be in at least one group* (see Figure 7.17).

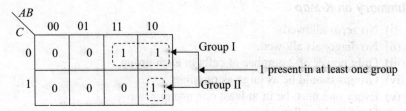

FIGURE 7.17 Invalid K-map

Note that in the above diagram, grouping is NOT optimal.

- *Groups may overlap* (see Figure 7.18).

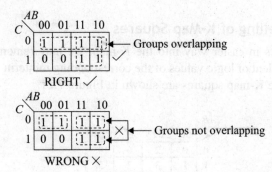

FIGURE 7.18 Valid K-map containing overlapping groups

- *Groups may wrap around the table. The leftmost cell in a row may be grouped with the rightmost cell and the top cell in a column may be grouped with the bottom cell* (see Figure 7.19).

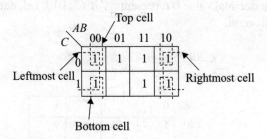

FIGURE 7.19 K-map with wrap around

- *There should be as few groups as possible, as long as this does not contradict any of the previous rules* (see Figure 7.20).

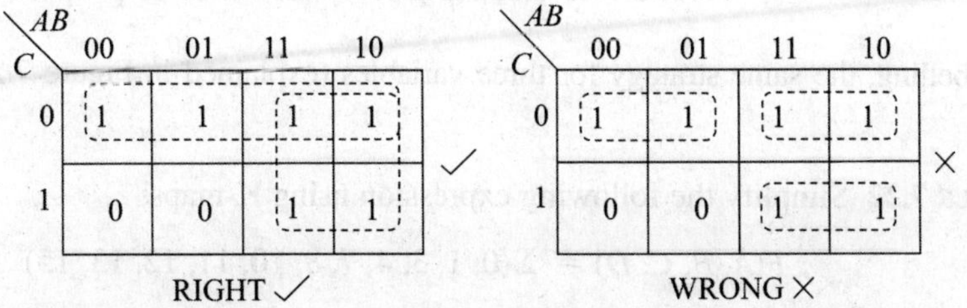

FIGURE 7.20 Valid K-map containing maximal possible groups

Summary on K-Map

(i) No zeros allowed.
(ii) No diagonals allowed.
(iii) Only power of 2 number of cells in each group.
(iv) Groups should be as large as possible.
(v) Every one must be in at least one group.
(vi) Overlapping allowed.
(vii) Wrap around allowed.
(viii) Fewest number of groups possible.
(ix) The much easier looking K-map becomes complicated and tedious while extending variable size to more than 6.

7.8.2 Labelling of K-Map Squares

For placing 1s in easier way into the K-map, we may remember the numbers equal to the binary equivalent of logic values of the corresponding minterm variables. The numbering of 3- and 4-variable K-map squares are shown in Figure 7.21.

C \ AB	00	01	11	10
0	0	2	6	4
1	1	3	7	5

FIGURE 7.21 Labelling of 3-variable K-map squares

Here, we take the binary values of AB rowwise, whereas C as columnwise. Clearly, binary value 000, i.e., equivalent decimal value 0 represents $A'B'C'$; 010, i.e., equivalent decimal value 2 represents $A'BC'$; and so on.

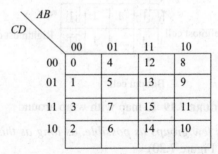

CD \ AB	00	01	11	10
00	0	4	12	8
01	1	5	13	9
11	3	7	15	11
10	2	6	14	10

FIGURE 7.22 Labelling of 4-variable K-map squares

In labelling, the same strategy for three variables (explained in Figure 7.22) is adopted here also.

EXAMPLE 7.22 Simplify the following expression using K-maps:
$$F(A, B, C, D) = \Sigma(0, 1, 3, 4, 7, 8, 10, 11, 12, 13, 15)$$

Solution The K-map is shown in Figure 7.23(a).

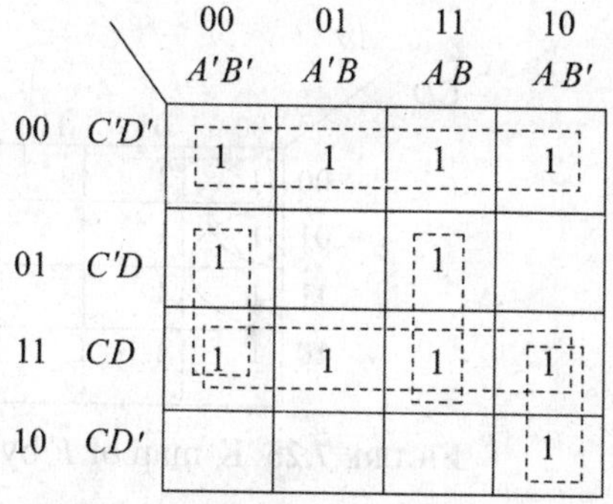

	00 $A'B'$	01 $A'B$	11 AB	10 AB'
00 $C'D'$	1	1	1	1
01 $C'D$	1		1	
11 CD	1	1	1	1
10 CD'				1

FIGURE 7.23(a) K-map of F

A is the most significant bit in the bitwise representation. This K-map gives the solution as $C'D' + CD + A'B'D + ABD + AB'C$. However, there can be several ways of minimizing this particular expression. Another possible solution using K-maps is shown in Figure 7.23(b).

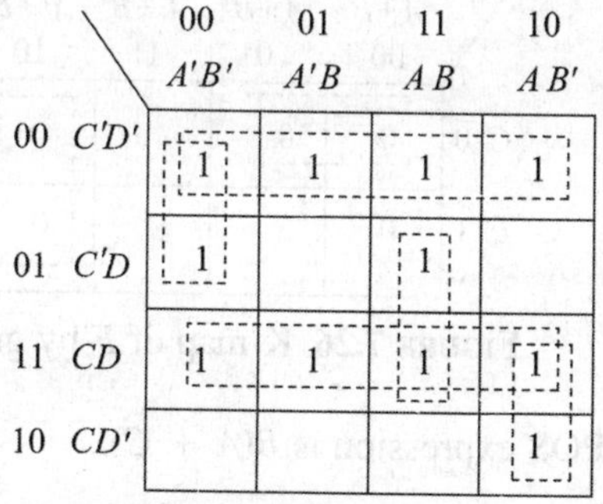

	00 $A'B'$	01 $A'B$	11 AB	10 AB'
00 $C'D'$	1	1	1	1
01 $C'D$	1		1	
11 CD	1	1	1	1
10 CD'				1

FIGURE 7.23(b) K-map of F

The above K-map gives the expression as $C'D' + CD + A'B'C' + ABD + AB'C$.

EXAMPLE 7.23 Simplify the following Boolean function:

$$F(A, B, C, D) = \Sigma(0, 1, 2, 3, 4, 5, 6, 7, 8, 9, 11)$$

Solution As per the given Boolean function, it is true that the K-map corresponding to function consists of label 1 at the columns whose decimal values are 0, 1, 2, 3, 4, 5, 6, 7, 8, 9, 11. Thus, the corresponding K-map table looks like the table shown in Figure 7.24.

CD \ AB	00	01	11	10
00	1	1		1
01	1	1		1
11	1	1		1
10	1	1		

FIGURE 7.24 K-map of F

Now, grouping the 1s by taking as many as possible, the marked K-map is shown in Figure 7.25 for simplification.

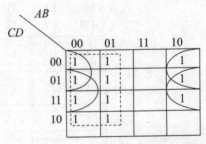

FIGURE 7.25 K-map of F by grouping 1s

Hence, the simplified function is $F = A' + B'C' + B'D$.

EXAMPLE 7.24 Simplify the following expression using K-maps (Figure 7.26):
$$F(A, B, C) = \Pi(0, 1, 2, 4, 5)$$

FIGURE 7.26 K-map of F by grouping 0s

Solution Simplified POS expression is $B(A + C)$.

EXAMPLE 7.25 Simplify the following expression using K-maps.
$$F(A, B, C, D, E) = \Sigma(0, 1, 2, 3, 4, 5, 6, 7, 8, 9, 12, 13, 16, 18, 19, 20, 21, 22, 23)$$

Solution This is a case of 5-variable K-map. Here, we may have groups of 16 as well as groups of 32. The K-map for this case is shown in Figure 7.27.

FIGURE 7.27 K-map of F

There are 4 octets and 1 quadrant. Simplified expression is

$$A'B' + B'D + B'C + A'D' + B'D'E'$$

EXAMPLE 7.26 Simplify the following expression using K-maps:

$G(A, B, C, D, E, F) = \Sigma(0, 1, 2, 3, 4, 5, 6, 7, 8, 9, 10, 11, 12, 13, 14, 15, 16, 17, 18, 19, 20, 21, 22, 23, 24, 25, 26, 27, 28, 29, 30, 31, 32, 33, 34, 35, 36, 37, 38, 39, 40, 41, 42, 43, 44, 45, 46, 47)$

	000 $D'E'F'$	001 $D'E'F$	011 $D'EF$	010 $D'EF'$	110 DEF'	111 DEF	101 $DE'F$	100 $DE'F'$
000 $A'B'C'$	1	1	1	1	1	1	1	1
001 $A'B'C$	1	1	1	1	1	1	1	1
011 $A'BC$	1	1	1	1	1	1	1	1
010 $A'BC'$	1	1	1	1	1	1	1	1
110 ABC'								
111 ABC								
101 $AB'C$	1	1	1	1	1	1	1	1
100 $AB'C'$	1	1	1	1	1	1	1	1

FIGURE 7.28 K-map of F

Solution This is a case of 6-variable K-map. Here, we may have groups of 32 as well as groups of 64. K-map is shown in Figure 7.28.

There are 2 sets of 32 elements each. Simplified expression is

$$G(A, B, C, D, E, F) = A' + B'$$

SUMMARY

Boolean algebra has a lot of applications in mathematics, logic, as well as in digital electronics. The chapter begins with an introduction to the fundamental concepts of logic and Boolean algebra and also explains the fundamental laws of Boolean algebra.

Then, the chapter moves on to the concepts of minterms, maxterms, SOP, and POS. A reasonable number of solved examples have been provided at suitable places to elaborate the concepts.

The chapter ends with the concepts of switching networks and K-maps. Solved problems related to minimization of switching networks using K-maps have been given.

EXERCISES

1. Define Boolean algebra. Write down the axioms of Boolean algebra.
2. State and prove De Morgan's theorem in both forms.
3. Complement the following expressions by De Morgan's theorem.
   - (i) $(A' + A')'$
   - (ii) $((A + B')C)'$
   - (iii) $((A' + C).(B + D')')$
   - (iv) $(A' + B'C)'$
4. Find sum of products for the following Boolean expressions.
   - (i) $F(x, y, z) = \Sigma(2, 3, 6, 7)$
   - (ii) $F(A, B, C, D) = \Sigma(7, 13, 14, 15)$
   - (iii) $F(A, B, C, D) = \Sigma(4, 6, 7, 15)$
5. Obtain the simplified expressions in sum of products for the following Boolean functions.
   - (a) $xy + x'y'z' + x'yz'$
   - (b) $A'B + BC' + B'C'$
   - (c) $a'b' + bc + a'bc'$
6. Obtain the simplified expressions in product of sums.
   - (i) $F(x, y, z) = \Pi(0, 1, 4, 5)$
   - (ii) $F(A, B, C, D) = \Pi(0, 1, 2, 3, 4, 10, 11)$
   - (iii) $F(A, B, C, D) = \Pi(1, 3, 5, 7, 13, 15)$
7. Use NAND gates to construct the following outputs.
   - (a) x'
   - (b) $x + y$
   - (c) xy
8. Express $A + BCD$, using NOR gates only.
9. Draw logic diagram to represent the Boolean expressions.
   - (a) $(x'y)' + x.y$
   - (b) $(xy)' + xyz$
   - (c) $(x' + y)'$
10. Draw a minimized logic circuit for the Boolean function
    $$F = AB'C + BC + AC' + A'BC + A'B'C$$
11. Draw a minimized logic circuit for the Boolean function represented by the following truth table.

A	B	C	F(A, B, C)
0	0	0	1
0	0	1	1
0	1	0	1
0	1	1	1
1	0	0	0
1	0	1	1
1	1	0	1
1	1	1	0

12. A Boolean function $F(A, B, C)$ generates an output of 1 if and only if even number of its input parameters have value as 1s, else the output is 0. Design an optimized logic circuit for this Boolean function.
13. Simplify the following Boolean expressions.
    - (i) $(A + B + AB)(A + C)$
    - (ii) $XY + XYZ + X'Y + XY'Z$
    - (iii) $XY + X'Y'Z' + YZ$
    - (iv) $XY' + Z + (X' + Y)Z'$
    - (v) $x(x' + y)$
    - (vi) $x + x'y$
    - (vii) $(x + y)(x + y')$
    - (viii) $xy + x'z + yz$
14. Express the following expressions in DNF.
    - (i) $F(A, B) = (A' + B')(A + B)$
    - (ii) $F(X, Y) = X + XY$
    - (iii) $F(X, Y, Z) = XY' + Z$
    - (iv) $F(A, B) = (A'B)'$
15. Find the output expression of the following gating network.

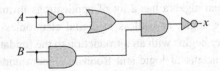

16. Solve the following equations by using K-map.
    (i) $F(w, x, y, z) = \Sigma(0, 1, 2, 7, 8, 11, 13, 15)$
    (ii) $F(A, B, C) = ABC + A'BC + BC + AB$

17. Design a minimized logic circuit for the following Boolean function.
    $$F(A, B, C) = \Sigma(0, 2, 4)$$

18. Design a minimized logic circuit for the following Boolean function.
    $$F(A, B, C, D) = \Sigma(9, 10, 11, 12, 13, 14, 15)$$

	00 C'D'	01 C'D	11 CD	10 CD'
00 A'B'				
01 A'B				
11 AB	1	1	1	1
10 AB'		1	1	1

19. Design a logic circuit using only one logic gate for the following Boolean function.
    $$F(A, B, C) = \Sigma(0, 1, 2, 3, 4, 5, 6)$$

20. Design a minimized logic circuit for the following Boolean expression.
    $$F(A, B, C, D, E) = A + BCD + A'B + ABCDE + A'CD' + B'CDE$$

21. Design a minimized logic circuit for the following Boolean function.
    $$F(A, B, C, D, E) = \Sigma(0, 1, 2, 3, 4, 5, 6, 7, 8, 9, 10, 11, 12, 13, 14, 15, 16, 17, 20, 21, 22, 23, 28, 29, 30, 31)$$

22. Design a minimized logic circuit for the following Boolean function.
    $$F(A, B, C, D, E) = \Sigma(0, 1, 2, 3, 8, 9, 10, 11, 12, 13, 14, 15, 16, 17, 18, 19, 24, 25, 26, 27, 28, 29, 30, 31)$$

23. Design a minimized logic circuit for the following Boolean function.
    $$F(A, B, C, D, E) = \Sigma(0, 1, 2, 3, 8, 9, 10, 11, 12, 13, 14, 15, 16, 17, 18, 19, 20, 21, 22, 23, 24, 25, 26, 27, 28, 29, 30, 31)$$

24. Design a minimized logic circuit for the following Boolean function.
    $$F(A, B, C, D, E) = \Sigma(4, 5, 6, 7, 12, 13, 14, 15, 16, 17, 18, 19, 20, 21, 22, 23, 24, 25, 26, 27, 28, 29, 30, 31)$$

25. Design a minimized logic circuit for the following Boolean function.
    $$F(A, B, C, D, E) = \Pi(4, 5, 6, 7, 8, 9, 10, 11, 12, 13, 14, 15, 16, 17, 18, 19, 20, 21, 22, 23, 24, 25, 26, 27, 28, 29, 30, 31)$$

26. Design a minimized logic circuit for the following Boolean function.
    $$f(A, B, C, D, E, F) = \Sigma(0, 1, 2, 3, 4, 5, 6, 7, 8, 9, 10, 11, 12, 13, 14, 15, 17, 25, 32, 33, 34, 35, 36, 37, 38, 39, 40, 41, 42, 43, 44, 45, 46, 47, 49, 57)$$

27. Design a minimized logic circuit for the following Boolean function.
    $$f(A, B, C, D, E, F) = \Pi(0, 1, 2, 3, 4, 5, 6, 7, 8, 9, 10, 11, 12, 13, 14, 15, 32, 33, 34, 35, 36, 37, 38, 39, 40, 41, 42, 43, 44, 45, 46, 47)$$

ANNEXURE 1

A.1.1 Quine-McCluskey Algorithm

Before discussing the algorithm and its objective, we should get some idea about the following terms.

In Boolean logic, an *implicant* is a 'covering' (sum term or product term) of one or more minterms in a sum of products (or maxterms in a product of sums) of a Boolean function. Formally, a product term P in a sum of products is an *implicant* of the Boolean function F if P implies F. More precisely, P implies F (and thus is an implicant of F) if F also takes the value 1 whenever P equals 1, where F is a Boolean function of n variables. P is a product term.

This means that $P \Rightarrow F$ with respect to the natural ordering of the Boolean space. For instance, the function $f(x, y, z, w) = xy + yz + w$ is implied such as xy by xyz, by $xyzw$, by w and many others; these are the implicants of f.

A *prime implicant* of a function is an implicant that cannot be covered by a more general (more reduced—meaning with fewer literals) implicant. W. V. Quine defined a *prime implicant* of Boolean function F to be an implicant that is minimal—that is, if the removal of any literal from P results in a non-implicant for F. *Essential prime implicants* are prime implicants that cover an output of the function that no combination of other prime implicants is able to cover.

Using the example above, one can easily see that while xy (and others) is a prime implicant, xyz and $xyzw$ are not. Clearly, from the latter, multiple literals can be removed to make it prime: x, y, and z can be removed, yielding w; alternatively, z and w can be removed, yielding xy. Finally, x and w can be removed, yielding yz.

The process of removing literals from a Boolean term is called *expanding* the term. Expanding by one literal doubles the number of input combinations for which the term is true (in binary Boolean algebra). Using the example function above, we may expand xyz to xy or to yz without changing the cover of f. The sum of all prime implicants of a Boolean function is called the *complete sum* of that function.

Further, in digital logic, a *don't-care* term is an input sequence (a series of bits) to a function that the designer does not care about, usually because that input would never happen, or because differences in that input would not result in any changes to the output. Obviously, 'don't-care' may refer to an unknown value in a multi-valued logic system, in which case it may also be called an X value. By considering these don't-care inputs, designers can potentially minimize their function much more so than if the don't-care inputs were taken to have an output of all 0 or all 1. Examples of don't-care terms are the binary values 1010 through 1111 (10 through 15 in decimal) for a function that takes a BCD value, because a BCD value never takes on values from 1010 to 1111. Don't-care terms are important to consider in minimizing, using *Karnaugh maps* and the *Quine–McCluskey algorithm*.

The Quine–McCluskey algorithm (or the method of prime implicants) is a method used for minimization of Boolean functions which was developed by W. V. Quine and Edward J. McCluskey. It is functionally identical to Karnaugh mapping, but the tabular form makes it more efficient for use in computer algorithms, and it also gives a deterministic way to check that the minimal form of a Boolean function has been reached. It is sometimes referred to as the *tabulation method*.

A.1.2 Quine–McCluskey Method

The method involves *two steps*:
Step 1 Finding all prime implicants of the function.
Step 2 Use those prime implicants in a *prime implicant chart* to find the essential prime implicants of the function, as well as other prime implicants that are necessary to cover the function.

Let us consider one example of how to use *Quine–McCluskey* algorithm to reduce Boolean expression.

$F(A, B, C, D) = \Sigma m(4, 5, 6, 8, 9, 10, 13) + d(0, 7, 15)$. The notation Σ says that function is the sum of *minterms* (*m*) listed, and the value of the function for the terms 0, 7, and 15 are *don't-care*(*d*).

The first step of Quine–McCluskey (QM) method is to list all of the minterms grouped by the number of 1s in the minterm's binary representation. This step includes the don't-care terms, too. We compare each implicant in adjacent groups to find patterns that *differ* by only a single bit which yields:

TABLE A.1 Grouping of minterms

Binary representation of the minterms	Size 2 implicants
0000	0-00
0100	-000
1000	
0101	010-
0110	01-0
1001	100-
1010	10-0
0111	
1101	01-1
1111	-101
	011-
	1-01
	-111
	11-1

Here, we are considering a total of 10 minterms whose decimal values are 0, 4, 5, 6, 7, 8, 9, 10, 13, and 15, respectively. The binary representations of the minterms are provided in the left column of Table A.1 and they are grouped according to the number of 1s. Here, grouping is meant about *adjacent* grouping. Further, 0-00 implies that the minterms 0 and 4 currently form a group, since $(0)_{10} = (0000)_2$ and $(4)_{10} = (0100)_2$, where subscripts 10 and 2 represent, respectively, the base for decimal and binary. Similarly, -000 means that (0, 8) makes another separate group, and so on. (Obviously, the implicants in the second column cover all of the implicants in the first/left column, and so the checkmarks are not shown.)

Replacing the group-wise comparison on the new set of terms, we get
The last column only has a single group, which ends this phase of the QM method. The remaining (unchecked) seven terms are 0-00, -000, 100-, 10-0, 1-01, 01- -, and -1-1 (see Table A.2).

TABLE A.2 Replacing group-wise comparison

Size 2 implicants	Size 4 implicants
0-00	01--
-000	01--
010- ✓	-1-1
01-0 ✓	-1-1
100- ✓	
10-0	
01-1 ✓	
-101 ✓	
011- ✓	
1-01	
-111 ✓	
11-1 ✓	

For the last phase, we build the table with the minterms in the top row, and the reduced terms (the prime implicants) in the left column. Note that the top row only includes the 'real' minterms, and the don't-cares are omitted. This is because we do not need to make sure that the don't-care entries get covered (see Table A.3).

TABLE A.3 Prime implicant chart

		4	5	6	8	9	10	13
0. 4	(0-00)	×						
0. 8	(-000)				×			
8. 9	(100-)				×	×		
8. 10	(10-0)				×		×	
9. 13	(1-01)					×		×
4. 5. 6. 7	(01--)	×	×	×				
5. 7. 13. 15	(-1-1)		×					×

Scanning the columns for minterms that are covered by only a single unique prime implicant. In this example, minterms 6 and 10 are each covered by only one prime implicant. Therefore, (01--) must be included in the final expression to cover minterm 6, and (10-0) must be included to cover minterm 10. By including these two prime implicants, we also end up covering 4, 5, and 8 (see Table A.4).

TABLE A.4 Deriving the final equation

		4	5	6	8	9	10	13	
0. 4	(0-00)	×							
0. 8	(-000)				×				
8. 9	(100-)				×	×			
8. 10	(10-0)	-|-	-|-	-|-	-×-	—	-×-	—	$A.\overline{B}.\overline{D}$
9. 13	(1-01)					×		×	$A.\overline{C}.D$
4. 5. 6. 7	(01--)	-×-	-×-	-×-	-|-	—	-|-	—	$\overline{A}.B$
5. 7. 13. 15	(-1-1)		×					×	

The only two remaining minterms that need to be covered are 9 and 13. We could choose to include (100-) to cover minterm 9 and (-1-1) to cover 13. Instead, we could choose to use (1-01) which simultaneously covers both. This leads to the final equation:

$$f(A, B, C, D) = \overline{A}.B + A.\overline{B}.\overline{D} + A.\overline{C}.d$$

Note also that in the end, the equation covered the don't-care minterm 7, but it does not cover 0 or 15.

ANNEXURE 2

A.2.1 Stone's Representation Theorem for Boolean Algebra

In mathematics, Stone's *representation theorem* for Boolean algebras states that every Boolean algebra is *isomorphic* to a field of sets. The theorem is fundamental to the deeper understanding of Boolean algebra that emerged in the first half of the twentieth century. The theorem was first proved by Stone (1936), and thus named in his honor. Stone was led to it by his study of the spectral theory of operators on a Hilbert space.

A.2.2 Stone Spaces

Each Boolean algebra B has an associated topological space, denoted as $S(B)$, called its *Stone space*. The points in $S(B)$ are the ultra filters on B, or equivalently the homeomorphisms from B to the two-element Boolean algebra. The topology on $S(X)$ is generated by a basis consisting of all sets of the form: $\{x$ in $S(X)$ mid b in $x\}$, where b is an element of B.

For any Boolean algebra B, $S(B)$ is a compact totally disconnected Hausdorff space; such spaces are called *Stone spaces*. Conversely, given any topological space X, the collection of subsets of X that are *clopen* (both closed and open) is a Boolean algebra.

A.2.3 Representation Theorem

A simple version of Stone's representation theorem states that any Boolean algebra B is isomorphic to the algebra of *clopen subsets* of its Stone space $S(B)$.

The full statement of the theorem uses the language of category theory; it states that there is a duality between the category of Boolean algebras and the category of Stone spaces. This duality means that in addition to the isomorphism between Boolean algebras and their Stone spaces, each homomorphism from a Boolean algebra *A* to a Boolean algebra *B* corresponds in a natural way to a continuous function from $S(B)$ to $S(A)$. In other words, there is a contravariant functor that gives equivalence between the categories. This was the first example of a non-trivial duality of categories.

The theorem is a special case of *Stone duality*, a more general framework for dualities between *topological spaces* and partially *ordered sets*.

Note The proof requires either the *axiom* of choice or a weakened form of it. Specifically, the theorem is equivalent to the Boolean *prime* ideal theorem, a weakened choice principle which states that every Boolean algebra has a prime ideal.

8 Complexity

> **LEARNING OBJECTIVES**
>
> After reading this chapter you will be able to understand the following:
> - Basic ideas of algorithms, programs, and data structures
> - Analysis of algorithm in terms of resources needed, such as time and space
> - Identification of the fundamental operations used in algorithms for measuring their resource requirements (complexity)
> - Ways to express the complexity
> - Derivation of recurrence relations for the running time of recursive procedures
> - Design of efficient algorithms
> - Basic searching and sorting techniques

8.1 INTRODUCTION

Computer programming involves writing software (a collection of related programs) that allows a machine to perform various tasks. A software is just an algorithm implemented in any high level programming language. As soon as one learns to write an algorithm, it is equally necessary to learn how to analyse it. The first type of analysis is to verify the correctness of an algorithm.

But the most important role of the algorithm designer is to design efficient algorithm (which takes least possible *time* and/or *space*) for any problem. To fulfil this, we should be aware of the basics of data structure to organize data in an appropriate manner, the way of writing algorithms, techniques of measuring time and space requirements of an algorithm, and then represent them through suitable notations.

Clearly, we should be aware of the *complexity theory* which describes how difficult it is for an algorithm designer to find a solution of a problem in terms of computational resources such as time and space. The theory deals with the *optimization* of the resources used in an algorithm. However, the *computability theory* discusses whether a problem can be solved at all.

This chapter discusses the basic idea of algorithm, data structures and its primary categories, algorithms of some specific problems, and basic techniques for computing the time complexity of algorithms.

8.2 ALGORITHM

The word *algorithm* comes from the name of a Persian mathematician, Abu Ja'far Muhammad ibn Musa al-Khwarizmi (825 AD). This term refers to a *method* which can be used by a computer for the solution of a problem and it is likely to be different from words such as *process*, *technique*, or *method*. In fact, to design any software, one should possess a sound knowledge of algorithm, programming, data structures, and analysis of algorithms.

Definition An algorithm is a finite set of instructions, each of which performs a specific task. It is written by using *pseudo-code* which is generally a mixture of natural language and high level languages such as C, C++, Pascal, Java, etc. The algorithmic language followed in this book is a kind of *pseudo-code* combination, Pascal, C, and English.

In practice, a group of instructions is placed under names such as *procedure* or *function*. Further, each procedure or function must possess a unique name which generally signifies the aim of the problem. In this book, we have taken the help of procedure or function to write an algorithm.

8.2.1 Basic Criteria of Algorithm

Every algorithm written for solving any problem must satisfy the following criteria:
 (i) *Input* They must have *zero* or *more* inputs which are either fixed or supplied externally by the user.
 (ii) *Output* At least one *output* must be produced by the algorithm to check its correctness. An algorithm can also simply manipulate the value(s) of a (set of) variable(s) instead of explicitly printing an output value. In any case, each algorithm must produce output(s) in any form.
 (iii) *Definiteness* Each instruction in the algorithm must be *clear* and *unambiguous*.
 (iv) *Finiteness* Each algorithm must terminate after a *finite* number of steps.

$$\text{Program} = \text{Data structure} + \text{Algorithm}$$

Program An algorithm written for solving a problem, using the syntax of any programming language such as Pascal, C, C++, etc. is called a program.

8.3 DATA STRUCTURE

A data structure is a container (or a place-holder in computer's memory) to store data for easier and faster access, and modification. It is the *logical* representation of the data. For example, *int x*; where x is a variable (*container* to store data) of type int (integer) allowed to be declared in C, C++, and Java programming languages. Actually, it is the logical (*symbolic*) name of the *memory location* (address) allocated for x. Now, instead of accessing this variable through its allocated location (say 00011H) throughout the program, it is easier to access the same using its *symbolic* (logical) name x.

8.3.1 Operations on Data Structure

A data structure consists of a set of operations to be performed on data. The following basic operations are normally performed on any data structure:
 (i) *Insertion* This operation inserts data into a specific data structure.
 (ii) *Modification* This modifies the existing data stored in a specific data structure.

(iii) *Searching* It finds or retrieves any existing data stored in a specific data structure.
(iv) *Deletion* This operation deletes or removes any data stored in a particular data structure.

We can also perform common *arithmetic* operations such as *addition*, *multiplication*, etc., on a data structure if it supports such kinds of operations.

8.3.2 Categorizations of Data Structure

A data structure is primarily divided into two categories:

 (i) *Primitive data structure* These cannot be broken further and hence are also called *atomic data structure*. *Integer*, *real*, *character*, etc. are the examples of such data structure, as they *cannot* be split further. These primitive data types are often called in-built data type, since these are defined in the language itself (during construction).
(ii) *Non-primitive data structure* These are derived from *primitive* data structure. So, the *non-primitive* data structures can be broken further. *Array*, *structure* (record), and *union* are some examples of this type of data structure.

> **Note** We can term every *data structure* as *data type* and vice versa.

Array as a Non-Primitive Data Structure

Array

This data type is frequently arranged in arrays in which the elements are *indexed* by one or more *subscripts*. So, we can define an *array* as an ordered set of *same* data type which consists of a fixed number of *objects* (elements) accessible through subscripts. For example, *int A[10]* is a one-dimensional array *data structure* allowed to be declared in C-language. Clearly, it is a collection of 10 primitive elements of *int* (integer) data type. Each of the stored integers is represented by a particular *subscript* value of the array A.

For example, A[5], where subscript value 5 of the array A represents the 6th element out of 10 elements (since starting location of an array in C is 0). Accordingly, through the *subscript* value, *direct access* of an element (stored at a location) is possible.

However, *memory allocation* for array is *static* (i.e., at the time of *compilation*). Also, *insertion* in between two elements in an array is a very time-consuming factor, as large number of data *movements* may be needed. Furthermore, *deletion* of any element from array is very difficult and it has less practical value (as actual deletion of any element from array is *not* possible).

It is *true* that array with any number of dimensions is allowed. A one-dimensional array is called a *vector*, whereas a two-dimensional array is termed as a *matrix*. Matrices provide a means of storing huge information in such a way that each information can be identified and manipulated.

Structure as a Non-Primitive Data Structure

Structure

This data type allows to combine several types of data in a single entity. For example,

```
struct student {
            int  roll;
            char name[20];
            char dept[10];  } ;
```

is a structure (record) data type (declared in C/C++, etc.) for maintaining some information about student.

Comment Non-primitive data type is also called as *user-defined* data type, as its members/fields (the smallest pieces of information) are selected by the users depending on their requirements.

8.3.3 Abstract Data Type

Abstract data type (ADT) is a data type whose *internal representation* is not *transparent* to the user. Actually, the term *abstraction* in context of data type implies: *it is known what a data type can do but how it is done, is hidden*. Such data type consists of data as well as set of operations (like addition of data, deletion of data, etc.) performed on data.

Class (used in C++ and Java) is the basic construct for creating ADT. Of course, *structure*, *array*, *pointers*, etc. are several non-primitive data structures that are also used as the basic components in the implementation of ADT. Stack, queue, tree, graph, etc., are examples of such data type.

Such type of data can be implemented in two ways: (i) sequentially (i.e., by using array) and (ii) dynamically (i.e., by pointer/link).

8.3.4 Linear and Non-Linear Data Structure

On the basis of *relationship* of adjacency, some data structures are called as *linear* and some are *non-linear*. However, the basic components of linear as well as non-linear data structures may be primitive or non-primitive data structures. Again, both the linear and non-linear data structures can be implemented *sequentially* or *dynamically*.

- *Linear data structures* are those in which each item is attached to at most one next and at most one previous item. Examples of linear data structure are *array*, *stack*, *queue*, etc. The linear data structures like stack and queue can be implemented with the help of another linear data structure: array or pointer.

- *Non-linear data structures* are those in which each item may be attached to more than *two* items. Tree, graph, etc., are the examples of non-linear data structures. Such non-linear data structures are also represented by array or pointer.

8.4 COMPLEXITY

Algorithm and complexity of algorithm comprise the study of computational models, its inherent powers, limitations, and the precise description of the problems to be solved. Complexity is the measure of resources required by an algorithm. The most commonly measured resources are time and space. The resource considered for time is CPU, and the resource for space is the memory of computer. In fact, time complexity measures the number of steps needed by an algorithm, whereas space complexity measures the memory space occupied by the algorithm. The most commonly used term *algorithm analysis* means prediction of resources required by an algorithm.

We can express the running time of any algorithm in two ways:

(i) *Exactly (empirically)* This measures exact time of an algorithm when the algorithm is implemented into a program by setting *timer* function supported by the language. However, the time may vary from machine to machine.

(ii) *Inexactly (theoretically)* This is the most useful course in practice. In this way, we count the number of designer-specified operations like +, −, *, /, comparison, assignment, etc., in an algorithm (in terms of the input size), and express by numerical function called as *complexity function*. In particular, the designer-specified operation such as *addition* does not emphasize here on the actual amount of time taken by the machine by applying its logical unit such as ADDER.

Then, we should check how the complexity grows. In fact, the total number of operations (i.e., complexity function) is finally represented by an appropriate growth function through *asymptotic notation*.

More clearly, we need notation to classify *complexity functions* according to their approximate rate of *growth*.

Note Growth function is a simpler function such as c (constant), n, $\log_2 n$, n^2, n^3, ..., i.e., it is, in fact, the highest degree term with (+)ve coefficient in an expression. Obviously, the lower order terms are ignored, since the highest degree term only is sufficient to express the rate of growth of a complete function. For example, $f(n) = 2n^2 + 3\sqrt{n} + 7$. In this example, the highest degree term with (+)ve constant is $2n^2$. Therefore, here the simpler function is n^2.

8.4.1 Idea on Complexity Function of Any Algorithm

If the input size of an algorithm is expressed in terms of only *one* variable (say n), then the number of operations performed by that algorithm is also expressible by the same *single* variable function, say, $f(n)$. For example,

$$f(n) = a * n^2 + b * n + c$$

where $(a * n^2 + b * n + c)$ returns the total number of operations.

Further, it may be seen that the *total* number of operations in an algorithm may be expressible through some (+)ve and (−)ve *coefficient* terms, but the coefficient of the *highest* degree term should be (+)ve for an algorithm of any *real-world problem*. In fact, the *negative* coefficient terms may occur due to simplification of the computed operations. For example, suppose the total number of operations of an algorithm is expressed by the series $S = n \times 1 + n \times 2 + n \times 2^2 + n \times 2^3 + \cdots +$ up to nth term. Now, S returns $(n2^n - n)$ on simplification.

Again, more than one variable may be needed to express input size for certain problems. For example, consider *multiplication* of two matrices of order $m \times n$. In this problem, the amount of data (i.e., input) is expressed by two variables: m and n. Clearly, the number of operations performed by the *matrix-multiplication algorithm* must be expressed by two variables (m and n), and the time complexity function is then denoted by $f(m, n)$. [*Counting strategy* of operations in an algorithm (i.e., which operations in an algorithm should be counted and which are to be ignored) is explained in detail in Section 8.4.4.]

Whatsoever, complexity function of any algorithm (with any number of involved variables) is finally expressed in terms of asymptotic notations.

8.4.2 Asymptote and Its Behaviour

Before expressing *inexact* time of an algorithm, the basic idea on *asymptote* and its behaviour are essential. An asymptote is a *line* or *curve* (represented by a function, say, $g(n)$) that approaches a given curve [say $f(n)$] arbitrarily closely, but never reaches, as they tend to infinity. No matter, how far, we go into *infinity*, the line will not actually reach to the given curve, but always get *closer* and *closer*. In fact, one should not demand that the function $f(n)$ is *equal* to the function $g(n)$. However, it can be *utmost* said that $f(n)$ *grows* like $g(n)$, i.e., growth function of $f(n)$ can be considered as $g(n)$. More clearly, $f(n)$ can be asymptotically *bounded* by $g(n)$, and we need some notations called as *asymptotic notations* (which are, in fact, derived from mathematics) to do it. O (*big oh*), θ (*theta*), Ω (*omega*), etc., are the examples of the same notations, each of which has some special meaning.

8.4.3 Why Asymptotic Notations to Express Inexact Running Time?

In measuring inexact running time of an algorithm, we may *never* expect *exact* time of that algorithm, since exact amount of execution time of each considered operation in the algorithm is not known. Thus, to compute it theoretically, we should be interested in the asymptotic behaviour of complexity function, i.e., expressing running time can also be treated like asymptotic nature between two functions, say, $f(n)$ (calculated from the *given* algorithm) and $g(n)$ (suggested). Accordingly, we should not conclude that the function $f(n)$ is equal to the function $g(n)$. However, it can be said that $f(n)$ *grows* like $g(n)$. That is why we need some asymptotic notations to describe the behaviour of a complexity function.

In particular, the said notations are used to estimate (or to bound) the *growth* (i.e., rate of increasing) of a complexity function $f(n)$ in terms of another function $g(n)$ [usually *simpler* like c (constant), $\log_2 n$ (logarithmic), n (linear), n^2 (quadratic), etc., called as *growth* functions].

Further, asymptotic notations generally *ignore* the measures that grow more *slowly* because eventually the measure that grows more *quickly* will dominate the lower measures. A common notation that follows such strategy is *big oh* (O).

For example, let us consider a time complexity function $f(n) = 2 + 3n$, and express it by notation *big oh* (O). Now, we can express it as $f(n) = 2 * O(1) + 3 * O(n)$. But $O(1)$ (i.e., constant) grows slowly as compared to $O(n)$. Thus, we can simplify it as $f(n) = O(n)$, *dropping* the lower degree term 2 and *ignoring* the constant coefficient 3. Obviously, here the *simpler* growth function is $g(n) = n$. Hence, in expressing complexity in terms of this notation, there is *no point* in specifying all the lower order terms, as these are all understood to be included in the asymptotic function denoted by the higher degree term. Thus, it holds true that *big oh* (O) notation gives the upper bound of the complexity function.

Further, sometimes we are interested in *lower* bound but not in upper bound. In such a case, requirement is on the *least/minimum* amount of running time of an algorithm, and it is then expressed in terms of Ω (*omega*). Also, the notation θ (*theta*) says that both the upper and the lower bounds of a complexity function are same, i.e., asymptotically *tight bound*, and it is obviously *good*. [Details about expressing complexity in terms of different notations are explained in Section 8.4.6.]

> **Note** Time complexity function is asymptotically positive means '*if the input size increases then execution time must increase*'. But it is constant which means '*execution time is fixed*' (i.e., independent on input size). Further, time complexity function of any *practical* problem may not behave as '*if the input size increases, then execution time decreases*'. In other words, the *coefficient* of the *highest* degree term of complexity function (of any practical problem) must be *positive*. For example, in complexity function $f(n) = a * n^2 + b * n + c$, the *highest* degree term is $a * n^2$, wherein a is the coefficient. The coefficient a may *not* be negative.

8.4.4 Counting Strategy of Operations in Algorithm

To compute inexact time complexity of an algorithm, *step-count* value (*weight*) of each statement (involving some operation) is provided. Indeed, all the operations in an algorithm do not affect the time complexity. Only some basic operations play this role. A brief list (with step-count value for each) is presented below.

(a) Comment statement has step-count value *zero* (0). [In this book, the comment statements either start with // (for single-line comments) or enclosed within /* */ for multi-line comments.]
(b) An assignment statement, which does not involve any procedure or function call, is counted as *one* (1) step-count.
(c) For an iterative statement such as *for, while, repeat–until*, consider step-count only for the *control part* of the statement (taking care on starting part of that iterative statement).
(d) Simple comparison statement has step-count value *one*.
(e) Return statement has step-count value *one*.
(f) Simple procedure or function name has *no* step-count.
(g) Begin ({), end (}) have step-count value *zero*.

Let us consider the following *algorithm* that finds the *sum* of n elements.

```
1.  Procedure Sum( a, n )  ... step-count is 0.
2.     begin  ... step-count is 0.
       // a is array of size n, where n elements are to be stored.
3.     S := 0  ... step-count is 1.
       // S is the variable where the sum of n elements will be stored.
4.     for i :=1 to n do  ... total step-count is n + 1.
5.        S:= S + a[i]  ... total step-count is n*1 (one per iteration).
6.     return S  ... step-count is 1.
7.  end  ... step-count is 0.
```

Clearly, the *total* number of operations performed by this algorithm is counted in sequence (starting from steps 1 to 7) as follows:

$$f(n) = 0 + 0 + 1 + (n + 1) + (n * 1) + 1 + 0 = 2n + 3$$

where n is the *input size* and $f(n)$ is the function to express the number of operations to be performed by the algorithm. It can be expressed by asymptotic *upper bound* as $O(n)$.

In particular, step 5 of this algorithm is the body of the loop (represented through step 4) and consists of a single statement. Clearly, it will be executed n number of times. Accordingly, total step-count for this single statement is $n * 1$.

Comment If the number of operations in a program or algorithm segment can, in no way, be expressed in terms of generalized input size like n, m, etc., then time complexity of that segment is to be considered as *constant*.

For example, let us consider the following segment:

```
for i = 1 to 10 do
    a[i]=a[i] +3;
```

It is clear that the above segment will be executed only 10 times. So, it is *fixed*. Thus, its complexity is $O(c)$.

However, in this segment, there is a *scope* of varying the number of times of execution of the loop, i.e., instead of considering only 10, we may *increase* or *decrease* this fixed number. Therefore, in *general case*, the above segment can be expressed as $O(n)$ (in terms of input size n).

Note Whenever there is a loop, a factor of n is introduced in the complexity. Here, there is a *for* loop. So, it is not *safe* to assume *constant* complexity even if the number of iterations is *hard-coded*. The programmer can always change the number of iterations and this will affect the complexity. In other words, the complexity depends on the number of times the loop is executed. Thus, the *general* case of such segment should be treated as $O(n)$.

8.4.5 Discussion on Order of Complexity

The *orders* of time complexity of practical problems are ranked below in ascending order, using asymptotic notation *big oh* (O).

1. $O(c) = O(1)$ means no growth, i.e., it always takes a fixed amount of running time for any input size; here c means constant which can be represented by 1 also. For example, printing the first element in a list.
2. $O(\log_2 n)$ or $O(\log n)$ represents *logarithmic*, where n is the input size.
3. $O(n)$ means *linear* (i.e., growth is linear), where n is the input size.
4. $O(n^2)$ implies *quadratic*, where n is the input size.
5. $O(n^3)$ stands for *cubic*, where n is the input size.
6. $O(2^n)$ represents *exponential*, where n is the input size.

It is, in fact, true that we may represent time complexity function by asymptotic notation like O (big oh), θ (theta), Ω (omega), etc.

8.4.6 Mathematical Definitions of Some Useful Asymptotic Notations

Let us look at some mathematical definitions of a few useful asymptotic notations such as big oh, big omega, theta, little oh, and little omega.

Big Oh (Upper Bound)

Formally, the big oh notation is defined as follows.

A function $f(n) = O(g(n))$, if and only if there exist (+)ve constants c and n_0 such that $0 \leq f(n) \leq c * g(n)$ for all $n \geq n_0$. [The growth function $g(n)$ may be chosen from the terms of $f(n)$ itself or outside, i.e., which is not a term or part of a term in $f(n)$].

The idea is that if $f(n) = O(g(n))$, then $f(n)$ does not grow faster (and possibly slower) than $g(n)$. In fact, $g(n)$ is an *upper* bound on $f(n)$, i.e., it is *as small as* a function of n for which $f(n) = O(g(n))$. More clearly, *several* forms of $g(n)$ with different higher degrees (like $n, n^2, \ldots$) for $f(n)$ may be considered, but we must select the $g(n)$ with the smallest degree for which it satisfies $0 \leq f(n) \leq c * g(n)$ for all $n \geq n_0$.

Thus, it is necessary to determine the constants c and n_0 such that $0 \leq f(n) \leq c * g(n)$ for all $n \geq n_0$. Now, finding $g(n)$ and one value of constant c will be *easier*, if one simplifies the absolute value of $f(n)$ in the form of the *mathematical inequality* $|f(n)| \leq c|g(n)|$, where the symbol $||$ is used to return *absolute value* (i.e., consideration of absolute value of each term makes easier to obtain one constant value) and $g(n)$ represents the simplified function obtained from $|f(n)|$. For example, consider $f(n) = 5n^3 - 7n^2 + 13n - 4$, and then

$$|5n^3 - 7n^2 + 13n - 4| \leq |5n^3| + |7n^2| + |13n| + 4 \leq 5|n^3| + 7|n^3| + 13|n^3| + 4|n^3|$$
$$= 29n^3, \text{ for all } n \geq 1$$

So, $f(n) \leq 29 n^3$. Hence, primarily we may select $c = 29$, $n_0 = 1$, and $g(n) = n^3$.

Thus, $f(n) = O(n^3)$, but $c = 29$, $n_0 = 1$ is *not* the only possible solution to express $f(n)$. We may consider the *smaller* value of c, but the value of n_0 needs to be *increased*. Accordingly, there may exist a number of possible combinations of c and n_0.

Choosing Appropriate Growth Function for Big Oh

In order to choose the appropriate *growth function* $g(n)$ of a given complexity function $f(n)$, i.e., to choose $g(n)$ as small as function of n to express in terms of *big oh*, we can follow three important considerations which are presented below with an example.

Let us consider $f(n) = 3n + 2$.

(a) We can express $f(n)$ as $f(n) = 3n + 2 \leq 3n^2$ for $n \geq n_0 = 2$. Here, selected c is 3, $g(n) = n^2$, and $n_0 = 2$. Thus, $f(n) = O(g(n)) = O(n^2)$. [Different combinations of c and n_0 are allowed to satisfy $3n + 2 \leq c * n^2$.] Hence, chosen $g(n)$ is *true* to express $f(n)$ asymptotically, using *big oh*.

Clearly, $f(n) = O(n^3) = O(n^4)$ are also *true*, as these are of higher order growth functions than n^2.

(b) Let us attempt to select a *lower* order $g(n)$ than the assumed $g(n)$ in (a) in order to express $f(n)$ as $f(n) \leq c * g(n)$ for all $n \geq n_0$. To do so, consider $f(n) = 3n + 2 \leq 4n$ for all $n \geq n_0 = 2$. Here, selected c is 4, $g(n) = n$, and $n_0 = 3$. Thus, $f(n) = O(g(n)) = O(n)$. [Again, different combinations of c and n_0 are allowed to satisfy $3n + 2 \leq c * n$.] Therefore, the selected $g(n)$ here is also *true* to express $f(n)$ asymptotically in terms of big oh.

(c) Now, let us try to find any $g(n)$ that is of *lower order* than the chosen $g(n)$ in (b), and it (if any) must satisfy the definition of *big oh*.

Suppose $f(n)$ is expressed as $f(n) = 3n + 2 \leq 100 * n^0$ for $n \leq n_0 = 32$. Clearly, $c = 100$, $g(n) = n^0$. But it is *not* satisfying the necessary condition (of big oh) *for all* $n \geq n_0$. Obviously, it is true only up to the values of $n = 32$. But when value of n reaches 33 and exceeds, $f(n) = 3n + 2 \leq 100$ does *not* hold true.

So, whatever *large* value of constant c is considered, $f(n) \leq cn^0$ fails after certain value of n. Accordingly, we cannot choose such *smaller* $g(n)$ for the given $f(n)$, as some *break point* arises.

Conclusion

Comparing $g(n) = n^2$ [obtained from (a)] and $g(n) = n$ [obtained from (b)], we can infer that $g(n) = n$ is the *smaller* one, and it is the function which is as small as function of n. Hence, time complexity of $f(n) = 3n + 2$ will be $O(n)$ (when expressed in terms of big oh).

Note The following operations on *big oh* are *true*.

(a) $k * O(f(n)) = O(f(n))$, k is constant
(b) $O(O(f(n))) = O(f(n))$
(c) $O(f(n)) * O(g(n)) = O(f(n) * g(n))$

These operations are *true* for other notations also.

EXAMPLE 8.1 Use the definition of big O to prove that $f(n) = 5n^4 - 37n^3 + 13n - 4 = O(n^4)$.

Solution To find integers c and n_0 such that $5n^4 - 37n^3 + 13n - 4 \leq cn^4$ for all $n \geq n_0$ proceed as follows:

$$|5n^4 - 37n^3 + 13n - 4| \leq |5n^4| + |37n| + |13n| + 4 \leq 5|n^4| + 37|n^4| + 13|n^4| + 4|n^4|$$
$$\leq 59 n^4, \quad \text{for all } n \geq 1$$

So, $f(n) \leq 59 n^4$ for all $n \geq n_0 = 1$. Hence, $c = 59$. Therefore, $f(n) = O(n^4)$ is *correct*.

EXAMPLE 8.2 Give a big O estimate for each of these functions. Use a simple function in your big O estimate.

(a) $3n + n^3 + 4$
(b) $1 + 2 + 3 + \cdots + n + 3n^2$
(c) $\log_{10}(2^n) + 10^{10}n^2$

Solution

(a) Since $3n + n^3 + 4 \leq 3n^3 + n^3 + 4n^3 = 8n^3$ for $n \geq 1$, therefore $3n + n^3 + 4 = O(n^3)$.
(b) $1 + 2 + 3 + \cdots + n \leq n + n + n + \cdots + n = n \times n = n^2$.
   Hence, $1 + 2 + 3 + \cdots + n + 3n^2 \leq n^2 + 3n^2 = 4n^2$.
   Therefore, $1 + 2 + 3 + \cdots + n + 3n^2 = O(n^2)$.
(c) $\log_{10}(2^n) + 10^{10}n^2 = n \log_{10} 2 + 10^{10}n^2 \leq n^2 \log_{10} 2 + 10^{10}n^2 \leq (\log_{10} 2 + 10^{10})n^2$,
   if $n > 1$. But $\log_{10} 2 + 10^{10}$ is a constant. Therefore, $\log_{10}(2^n) + 10^{10}n^2 = O(n^2)$.

EXAMPLE 8.3 Use the definition of big oh to prove that $|(3n^4 - 2n)/(5n - 1)| = O(n^3)$.

Solution Now, it is necessary to find positive integers c and n_0 such that for all $n \geq n_0$, $|(3n^4 - 2n)/(5n - 1)| \leq cn^3$.

The above expression can be written as follows to make the *numerator* larger and the *denominator* smaller.

$$|(3n^4 - 2n)/(5n - 1)| \leq |3n^4/(5n - 1)| + |2n/(5n - 1)|$$
$$\leq 3|n^4/(5n - 1)| + 2|n^4/(5n - 1)|$$
$$\leq 3|n^4/(5n - n)| + 2|n^4/(5n - n)| = (5/4)||n^4/n| = (5/4)|n^3| = O(n^3)$$

or we can solve as follows:

$$f(n) = (3n^4 - 2n)/(5n - 1) = 0.6n^3 + 0.12n^2 + 0.024n - 0.3952 - 0.3952/(5n - 1)$$
$$|f(n)| \leq |0.6n^3| + |0.12n^2| + |0.024n| + (0.3952 + |0.3952/(5 * n)|$$

$$\leq |0.6n^3| + |0.12n^3| + |0.024n^3| + 0.3952n^3 + |(0.3952/5) * n^3|, \text{ for all } n \geq 1$$
$$\leq 1.21824n^3, \text{ for all } n \geq 1$$
So, $f(n) = O(n^3)$.

Big Omega (Lower Bound)

The big omega notation is defined as follows. A function $f(n) = \Omega(g(n))$ iff there exist (+)ve constants c and n_0 such that $f(n) \geq c * g(n)$ for all $n \geq n_0$. [The *growth function* $g(n)$ is only a *lower* bound on $f(n)$, i.e., it should be *as large as* function of n for which $f(n) = \Omega(g(n))$.]

It is now *necessary* to determine the constants c and n_0 such that $f(n) \geq c * g(n)$ for all $n \geq n_0$. Actually, determination of $g(n)$ and one value of constant c will be *easier* if the mathematical inequality $|f(n)| \geq c|g(n)|$ is considered, where symbol $||$ is used to return *absolute value*. But it is true that there may exist number of possible combinations of c and n_0 to satisfy $f(n) \geq c * g(n)$.

Choosing Appropriate Growth Function for Big Omega

To choose the growth function ($g(n)$) *as large as* function of n to express in terms of Ω, we can follow *three* considerations, which are discussed below with example.

Let us consider the time complexity function of an algorithm as $f(n) = 3n + 2$. We will now select the *appropriate* $g(n)$ to express $f(n)$ in terms of big omega by analysing the discussion presented below.

(a) First, express $f(n)$ as $f(n) = 3n + 2 \geq 3n^0$ for $n \geq n_0 = 1$. Clearly, selected c is 3, $g(n) = n^0$, and $n_0 = 1$.

Hence, $f(n) = \Omega(g(n)) = \Omega(n^0) = \Omega(c)$ (of course, different *combinations* of c and n_0 are allowed to satisfy $3n + 2 \geq cn^0$).

(b) Let us consider the *inequality* $f(n) = 3n + 2 \geq 3n$ for all $n \geq n_0 = 1$, expecting a higher order $g(n)$ than the $g(n)$ selected in (a).

Clearly, the suggested $g(n) = n$, value of c is taken as 3, and $n_0 = 1$ (different combinations of c and n_0 are permitted to satisfy $3n + 2 \geq c * n$). Therefore, $f(n) = \Omega(g(n)) = \Omega(n)$. Hence, the proposed $g(n)$ in (b) is also *true* to express $f(n)$ in terms of big omega.

(c) Now, let us try to search any $g(n)$ which is *larger* than the chosen $g(n)$ in (b), satisfying the definition of big omega. Let us express $f(n)$ as $f(n) = 3n + 2 \geq 2n^2$ for $n \leq n_0 = 2$.

Here, $c = 2$, $n_0 = 2$, and $g(n) = n^2$. But it is not satisfying the necessary condition *for all* $n \geq n_0$. It is *true* only *up to* values of n equals to 2. But when value of n reaches 3 and *above*, $f(n) = 3n + 2 \geq 2n^2$ does *not* hold true. So, whatever *small* (+)ve value of c is considered, $f(n) \geq c * n^2$ *fails* after certain value of n. Therefore, *such larger* $g(n)$ such as $n^2, n^3, \ldots$, cannot be chosen, as some *break point* arises.

Conclusion

Comparing $g(n) = n^0$ [obtained from (a)] and $g(n) = n$ [obtained from (b)], we select $g(n) = n$ as the *larger* one, and it is the function which is *as large as* function of n for the given $f(n)$. Hence, time complexity of $f(n) = 3n + 2$ will be $\Omega(n)$, when expressed using big omega notation.

Note Big oh gives us an UPPER BOUND. Accordingly, it says that 'your algorithm is at least *as good as g(n)*'. Big omega gives us a LOWER BOUND. So, it says that 'your algorithm is at least *as bad as g(n)*'.

Theta (Tight Bound)

The function $f(n) = \theta(g(n))$ iff $g(n)$ is both an upper and lower bound on $f(n)$. More clearly, $f(n) = \theta(g(n))$ iff there exist (+)ve constants c_1, c_2, and n_0 such that $c_1 * g(n) \leq f(n) \leq c_2 * g(n)$ for all $n \geq n_0$ [i.e., $f(n) = O(g(n)) = \Omega(g(n))$].

So, we can write $\theta(g(n)) = O(g(n)) \cap \Omega(g(n))$, if $f(n)$ can be expressed as $\Omega(g(n))$ (here, $\cap$ is intersection operation). For example, $f(n) = 3n + 2 = O(n) = \Omega(n)$. So, $f(n) = \theta(n)$.

We may express complexity function in terms of asymptotic notations through their graphical representations also. Such a representation is shown in Figure 8.1.

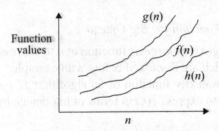

FIGURE 8.1 Graphical representation of complexity function

Here, the values of the functions with respect to the input size n (along the lower axis) are plotted along the upper axis, where $f(n)$ is the complexity function of an algorithm. Now, we can write $f(n) = O(g(n))$ and $f(n) = \Omega(h(n))$ if and only if the plotted functions follow the nature shown like Figure 8.1 for all $n \geq n_0$. However, value of n_0 may not be same in $f(n) = O(g(n))$ and $f(n) = \Omega(h(n))$.

Little Oh (o) and Little Omega (ω)

It is often observed that the complexity function $f(n)$ is structurally very much complicated, and so it is very difficult to determine any suitable growth function $g(n)$ for $f(n)$. In such a situation, we normally prefer complexity by using little oh or little omega. These are mathematically explained below.

Little Oh (o)

The function $f(n) = o(g(n))$ iff

$$\lim_{n \to \infty} \frac{f(n)}{g(n)} = 0$$

or one can say that $f(n) < c * g(n)$. Let $f(n) = 3n + 2$. Then

$$f(n) = o(n^2), \text{ since } \lim_{n \to \infty} \frac{3n + 2}{n^2} = 0 \qquad [\because g(n) = n^2]$$

Little Omega (ω)

The function $f(n) = \omega(g(n))$ iff

$$\lim_{n \to \infty} \frac{g(n)}{f(n)} = 0$$

or one can say that $f(n) > c * g(n)$. Let $f(n) = 3n + 2$. Then

$$f(n) = \omega(1), \text{ as } \lim_{n \to \infty} \frac{1}{3n+2} = 0 \qquad [\because g(n) = 1]$$

The notations such as *little oh* (o) and *little omega* (ω) are useful to represent the time complexity function asymptotically, when the function is very *complex*. [Reader should try to express all the examples discussed above in terms of big omega (Ω) and theta (θ), little oh (o), little omega (ω) notations.]

Note If $f(n)$ and $g(n)$ are, respectively, the complexity functions and the suggested growth function for $f(n)$, then on the basis of *limiting* value of

$$\lim_{n \to \infty} \frac{f(n)}{g(n)} = L$$

we can conclude the following:

Case I $0 < L < \infty$. In fact, the range gives one finite value. However, the inequality is true only if both f and g are of *same* order.
Case II $L = 0$. It is possible only if f has *smaller* order than g.
Case III $L = \infty$. This is possible when f has *larger* order than g.

8.4.7 Standard Cases

The following three *standard* cases related to time complexity of an algorithm are briefly explained.
 (i) *Best case* When minimum number of operations are executed by an algorithm.
 (ii) *Worst case* When all the operations in the algorithm are executed.
 (iii) *Average case* It occurs when the number of operations executed by an algorithm exists between the best case and the average case. So, the average case complexity of an algorithm is the function defined by the average number of steps taken on any input instance of size n. Clearly, the resource usage is average here.

In practice, the average running time is often much harder to determine than the worst case running time because finding 'average' steps frequently has no obvious meaning. Let us consider a very simple way to measure the same. Suppose an algorithm handles a typical input data of size n. Now, we can compute the *average running time* by averaging the running time of the algorithm for all possible inputs of size n. However, the *big* question arises here is, if there exists any specific rule to select the value of n as well as the n input instances since the running time may depend on the nature of the selected instances also.

Thus, the case of time complexity may not be considered a good measure as in this case we have to assume the underlying probability distribution for the input which, if in practice, is violated, then the determination of average case time complexity will be meaningful. Thus, it is useful but more difficult *to* compute.

So, we shall use worst case running time as the *principle* measure of time complexity, although we shall mention average case complexity wherever we can do so meaningfully.

8.4.8 Some Properties of Time Complexity Functions

(a) If $f_1(n)$ and $f_2(n)$ are two asymptotic time complexity functions, then

$$f_1(n) + f_2(n) = (\max(f_1(n), f_2(n))).$$

Here, '+' implies that two segments with complexities, respectively, $f_1(n)$ and $f_2(n)$ are distinctly placed in an algorithm (i.e., these are not cascaded), and so these are executed separately.

(b) If $f(n) = O(g(n))$ and $g(n) = O(h(n))$, then $f(n) = O(h(n))$ [*transitivity property*].

Proof Since $f(n) = O(g(n))$, therefore $f(n) \leq c_1 g(n)$ for all $n \geq n_1$ (as per definition of big oh), where c_1 and n_1 are positive constants. Obviously, here $g(n)$ is the least upper bound of $f(n)$. So, its degree is assumed to be the same as that of $f(n)$.

On the other hand, $g(n) = O(h(n))$, so $g(n) \leq c_2 h(n)$ for all $n \geq n_2$, where c_2 and n_2 are positive constants. Clearly, $h(n)$ is considered here the least upper bound of $f(n)$, i.e., its degree can be assumed to be the same of $g(n)$.

Now, from the above two considerations, we may conclude that all the functions: $f(n)$, $g(n)$, and $h(n)$ are of same degree. Further, combining the above mathematical inequalities, we may express as follows:

$$f(n) \leq c_1 g(n) \leq c_1 c_2 h(n), \quad \text{for all } n \geq \max(n_1, n_2)$$

Putting $c_1 \cdot c_2 = c$, we get, $f(n) \leq ch(n)$, for all $n \geq \max(n_1, n_2)$. This is indeed a mathematical inequality for *big oh*, and so $f(n) = O(h(n))$. [Check the property for other asymptotic notations as per their definitions.]

(c) If $f(n) = O(g(n))$, then $g(n) = \Omega(f(n))$.

Proof Here, $f(n) = O(g(n))$. Therefore, $f(n) \leq c_1 g(n)$ for all $n \geq n_1$ (as per the definition of big oh), where c_1 and n_1 are positive constants.

So, we may write $g(n) \geq c_2 f(n)$ for all $n \geq n_2$ (as per the definition of big omega), where c_2 and n_2 are positive constants.

Thus, $g(n) = \Omega(f(n))$.

(d) If $f_1(n)$ and $f_2(n)$ are two asymptotic time complexity functions, then

$$f_1(n) * f_2(n) = O(f_1(n) * f_2(n)),$$

i.e., the highest (+)ve degree term from $(f_1(n) * f_2(n))$ is selected. Here, '*' implies that two segments with complexities, respectively, $f_1(n)$ and $f_2(n)$ are cascaded and executed dependently.

(e) If $f_1(n)$ is the complexity function of a segment and a segment with complexity $f_2(n)$ is deducted from $f_1(n)$, then the idea can be expressed like $f_1(n) - f_2(n)$. However, it possesses *less* practical applicability. Example of the property $f_1(n) * f_2(n)$ is discussed below.

Suppose one function $DO(m)$ is called n times by another function $TOTAL()$ as follows:

```
function DO ( int m )
    begin
    for i := 1 to m do
        print("TEST");
    end,
```

The number of operations performed by the function $DO(m)$ is $f_1(m) = 2m$. So, the asymptotic time complexity of the function $DO(n)$ is $O(n)$, in terms of input size n.

```
function TOTAL(int n )
   begin
   for i := 1 to n do
      call DO(n);
   end.
```

The *number* of operations performed by this function is $f_2(n) = n * f_1(n)$. Clearly, the time complexity of the function TOTAL( ) is $O(n * n) = O(n^2)$.

EXAMPLE 8.4 Is $2^{n+1} = O(2^n)$?

Solution As per the given question, it is to be checked whether $2^{n+1} = O(2^n)$ or not. Let us assume that it is *true*. Hence, the *growth* function will be $g(n) = 2^n$.

Now, it is necessary to find the values of constants c and n_0 (if any) to satisfy the definition of *big oh*. Let us define an *inequality* as follows:

$$2^{n+1} \leq c2^n, \quad \text{for all } n \geq n_0$$
$$\Rightarrow 2 \cdot 2^n \leq c2^n, \quad \text{for all } n \geq n_0$$

The above *inequality* gives the value of $c = 2^{n+1}/2^n = 2$. Therefore, it holds *true* for $c = 2$ and $n_0 = 1$, as one of the different possible combinations of c and n_0. Thus, $2^{n+1} = O(2^n)$ is true.

EXAMPLE 8.5 Is $2^{2n} = O(2^n)$?

Solution According to the given question, it is to be checked if $2^{2n} = O(2^n)$ or not. Let us assume that it is *true*. Clearly, the *growth* function will be $g(n) = 2^n$.

Now, we need to determine the values of constants c and n_0 (if any) to satisfy the definition of *big oh*. Let us define as follows:

$$2^{2n} \leq c2^n, \quad \text{for all } n \geq n_0$$

The inequality gives the value of $c = 2^{2n}/2^n = 2^n$. But the constant c cannot depend on n, i.e., $c = 2^n$ is not a constant. Thus, $2^{2n} = O(2^n)$ is *not* true.

8.4.9 Complexity of Recursive Procedure

It is well known that *recursive procedure* calls the same procedure to compute outputs on n inputs until the stopping *criteria* is satisfied. Now, if we have an algorithm with procedures, none of which is recursive, then we can compute the running time of the various procedures one at a time, starting with those procedures that make no calls on other procedures. On the other hand, if there is at least one recursive procedure present in the algorithm, we then evaluate the running time of that algorithm, starting at the procedure which makes no call (i.e., size of input reaches to stop further call), although it is called by the other. Obviously, its running time is known. However, the unknown running time of the simultaneous earlier procedure calls are calculated from the known values of the respective procedures, and we may establish a *recurrence relation* using an unknown time function $T(n)$ to measure the complexity of the recursive calls, where n measures the size of input data. Generating *Fibonacci series* is a good example of recursion.

In context of recursive procedure, we discuss here a very well-known approach named as divide-and-conquer. In this *approach*, a problem P with size n is first divided into k ($1 < k \leq n$) distinct *smaller* sub-problems: $P_1, P_2, \ldots, P_k$ with sizes $n_1, n_2, \ldots, n_k$, respectively. Then, each of the sub-problems is further divided into distinct smaller sub-problems. Accordingly, each of the sizes: $n_1, n_2, \ldots, n_k$ is further splited into next distinct smaller sub-parts like n_1 into $n_{11}, n_{12}, \ldots, n_{1k}$, where each of these is smaller than n_1; n_2 into $n_{21}, n_{22}, \ldots, n_{2k}$, where each of these is smaller than n_2, and so on. Now, division of any sub-problem constructed at any stage stops, if its size reaches a specific value. This phase is, in fact, termed as divide phase of the approach. On the other hand, when division of any sub-problem stops, the desired solution is normally achieved from the input of the same sub-problem and combined with the *partial* solutions obtained from the respective sub-problems (if splitting on them is stopped), and so on to get the complete solution of the original problem P, moving to the stage where it starts to split. This operation is named as *conquer* (or combining).

Obviously, the aim of splitting is to reapply the same approach *divide-and-conquer* on the generated sub-problems (depending on the specified condition). Although, at each step, all the sub-problems may not necessarily reapply the approach, i.e., some can do and some cannot. Further, no work may be necessary during conquer phase. However, all these depend on the nature of the problem.

In practice, divide-and-conquer strategy is accomplished by recursive approach, and many real-world problems are solved by using divide-and-conquer strategy. Now, if a problem is solved recursively, using divide-and-conquer approach similar to the fashion adopted by problem P (discussed above), then we can describe its *computation time* by the *recurrence* relation as follows:

$$T(n) = g(n), \quad \text{when} \quad n \text{ is small}$$
$$= T(n_1) + T(n_2) + \cdots + T(n_k) + f(n) \quad \text{otherwise}$$

where $g(n)$ is the time to compute the answer directly for *small* inputs, whereas $f(n)$ is the time for *dividing* problem P into k sub-problems and *combining* the solutions of these sub-problems. So, $f(n)$ can be written as $f(n) = D(n) + C(n)$, where $D(n)$ is the function for dividing P and $C(n)$ is the function for combining the solutions of the sub-problems. Thus, at the next step, we can express $T(n_1), T(n_2), \ldots, T(n_k)$ as follows:

$$T(n_1) = T(n_{11}) + T(n_{12}) + \cdots + T(n_{1k}) + f(n_1)$$
$$T(n_2) = T(n_{21}) + T(n_{22}) + \cdots + T(n_{2k}) + f(n_2), \ldots, \text{and so on}$$

Clearly, at each step, if the *current* problem is divided into k sub-problems each with *equal size* and divide-and-conquer strategy is reapplied on all these k sub-problems (until size of each becomes very small to split), then we can express the time complexity, assuming same running time for each sub-problem to solve, by the *recurrence* as follows:

$$T(n) = g(n)$$

when n is small and splitting (division) is stopped.

$$kT(n/k) + f(n), \text{ otherwise}$$

[Obviously, the value of n of $f(n)$ changes here from *level* to *level* of the recursion, and it is indeed implicitly assumed.]

Usually, the complexity of many real-world problems, when solved using recursion as well as divide-and-conquer strategy, is expressed by the following *recurrence* relation.

$$T(n) = T(1), \quad n = 1$$
$$= aT(n/b) + f(n), \quad n > 1$$

where a and b are (+)ve constants. The above relation implies that, at every step of the *recursion*, the input size of each of the current sub-problems reduces by a factor $1/b$, where b is the total number of sub-problems as *outcome* at every step, although divide-and-conquer is reapplied only on a sub-problems *out* of b sub-problems (if $b \leq a$). However, we often find that the value of a in such a relation is greater than that of b, i.e., $a > b$. This implies that some structurally same sub-problems are called more than once at every step.

For example At every step of *binary search*, we get two *partitions*. However, depending on the result of the imposed condition, we select only *one* partition.

A simple example of *recursion* based on divide-and-conquer strategy is presented below. The designed algorithm finds recursively the sum of n elements stored in an array a.

```
1.  Procedure RSum(a, n)
2.     begin
3.        if ( n <= 0 ) then
4.           return 0.0
5.        else
6.           return( Rsum(a, n-1) +a[n])
7.     end.
```

In the above segment, the *RSum*( ) procedure calls *itself* until $n \leq 0$. Clearly, at every call, during *division* of the problem, only *one* basic operation: *condition checking* is executed. On the other hand, the following *two* basic operations are left to be executed during *combination* (conquer).

(a) *Summing* the element involved in the current call with the *partial* sum returned from the immediate next call element (i.e., updating of the *partial* sum).
(b) *Returning* the updated *partial* sum to the function from where it was called.

Since at each call (in this algorithm), the input size reduces by *one*, so the time complexity of this recursive procedure can be described by *recurrence* relation as follows:

$$T(n) = T(n-1) + f(n) = T(n-1) + (1+2), \quad \text{when } n > 1$$
$$[\text{here, } f(n) = D(n) + C(n) = 1 + 2]$$
$$= 2 \quad \text{when } n = 0$$

(due to execution of two basic operations: (i) condition checking and (ii) return)

$$T(n) = T(n-1) + 3 = [T(n-2) + 3] + 3 = [T(n-3) + 3] + 3 + 3$$
$$\text{(applying substitution method)}$$
$$= 3 + 3 + \cdots + 2, \quad \text{when } n = 0$$
$$= 3 + 3 + \cdots \text{ up to } n\text{th term}$$
$$= 3(n-1) + 2 = 3n - 1.$$

Thus, we can express the simplified solution in terms of O (big oh) as $O(n)$.

8.4.10 Solving Recurrence Relation: $T(n) = aT(n/b) + f(n)$, $a \geq 1, b > 0$

The complexity of an algorithm is expressed in recurrence as $T(n) = aT(n/b) + f(n)$, where $a \geq 1$ and $b > 0$. Let us try to solve it in simpler way. Here, the coefficient in the *recursive term* $(T(n/b))$ is a, *additive term* is $f(n)$, and input n is reduced by $(1/b)$ at each step.

Assume that $n = b^k$, $k \geq 0$. Accordingly, number of *splitting step* will be k. In order to compute the solution, we can proceed, following the steps mentioned below.

(i) The size of the input n is reduced as follows, starting from start steps onwards.

$$n/b, \; n/b^2, \; n/b^3, \ldots, \text{ up to } k\text{th term (due to } k\text{th step)}$$

(ii) The corresponding additive terms will be as follows:

$$af(n/b), \; a^2 f(n/b^2), \; a^3 f(n/b^3), \ldots, \text{ up to } k\text{th term (due to } k\text{th step)}$$

Thus, the sum of the additive terms obtained from simultaneous division of the input can be expressed by a series as follows:

$$S = f(n) + af(n/b) + a^2 f(n/b^2) + a^3 f(n/b^3) + \cdots \text{ up to } k\text{th term}$$

On the other hand, due to splitting, the recursive term $T(n/b)$ results its associate coefficient as a^k at the *final* step (i.e., at the kth step).

Hence, $T(n) = a^k + S$, and the complexity is the *maximum* between these two terms.

EXAMPLE 8.6 Consider the recurrence relation $T(n) = 2T(n/2) + n^2$, and solve it using the above approach.

Solution Here, $a = 2, b = 2$, and $f(n) = n^2$. Assume that $n = 2^k$. Therefore, the number of steps (by splitting) will be k, and the size of the input n is reduced as follows (starting from start steps):

$$n/2, \; n/2^2, \; n/2^3, \ldots, \text{ up to } k\text{th term}$$

The corresponding *additive* terms will be as follow:

$$f(n), \; 2f(n/2), \; 2^2 f(n/2^2), \; 2^3 f(n/2^3), \ldots, \text{ up to } k\text{th term}$$
$$\Rightarrow \quad n^2, \; 2(n/2)^2, \; 4(n/2^2)^2, \; 8(n/2^3)^2, \ldots, \text{ up to } k\text{th term}$$
$$\Rightarrow \quad n^2, \; 2(n^2/4), \; 4\,(n^2/16), \; 8(n^2/64), \ldots, \text{ up to } k\text{th term}$$
$$\Rightarrow \quad n^2, \; n^2/2, \; n^2/4, \; n^2/8, \ldots, \text{ up to } k\text{th term}$$
$$\Rightarrow \quad n^2, \; n^2/2, \; n^2/4, \; n^2/8, \ldots, \text{ up to } k\text{th term}$$
$$\Rightarrow \quad n^2, \; n^2/2, \; n^2/2^2, \; n^2/2^3, \ldots, \text{ up to } k\text{th term}$$

Thus,

$$\begin{aligned}
S &= n^2 + n^2/2 + n^2/2^2 + n^2/2^3 + \cdots \text{ up to } k\text{th term}\\
&= n^2(1 + 1/2 + 1/2^2 + 1/2^3 + \cdots \text{ up to } k\text{th term}\\
&= n^2 \, (1 \cdot ((1/2)^k - 1))/(1/2 - 1)\\
&= n^2 \, (1 - 2^k)/(1 - 2) \cdot (1/2^{k-1})\\
&= n^2 \, (2^k - 1)/2^{k-1}\\
&= n^2 2^k/2^{k-1} - n^2/2^{k-1}\\
&= n^2/2 - 2^{2k}/2^{k-1}\\
&= n^2/2 - 2^{k+1} = n^2/2 - n/2
\end{aligned}$$

Now, the *recursive* term $2T(n/2)$ results 2^k at the final step (i.e., at the kth step).
Hence, $T(n) = 2^k + n^2/2 - n/2$
$= n + n^2/2 - n/2$ [as $n = 2^k$]
$= n^2/2 + n/2$
$= O(n^2)$ [as $\max(n^2, n)$ is n^2]

EXAMPLE 8.7 $T(n) = T(n-1) + n$ [here, $f(n) = D(n) + C(n) = n$]

Solution Remark: $T(n)$ is not equal to $O(n)$.

To understand the remark, the reader needs to be thorough with the concept of degree of a polynomial. So, some essential mathematical background is discussed here. A *polynomial* function is expressed as

$$P(x) = a_n x^n + a_{n-1} x^{n-1} + a_{n-2} x^{n-2} + \cdots + a_2 x^2 + a_1 x + a_0$$

An important point to be noted here is that the function $P(x)$ is a polynomial function of degree n if and only if each of the following conditions hold *true*.
(i) All the coefficients $a_0, a_1, a_2, \ldots, a_n$ are *real numbers*.
(ii) n must be a positive integer.
(iii) The coefficient $a_n \neq 0$.
(iv) For any $0 \leq k \leq n$, there must be at most one term containing x^k.

However, the last condition is often ignored by students.

For example

(i) The function $P(x) = 3x^2 + 4x + 4x^2 + x + 8$ is NOT a polynomial function. $P(x)$ can be simplified as $7x^2 + 5x + 8$, which is definitely a polynomial function in x with degree 2.
(ii) $F(n) = n + n + n + \cdots + n$ (total n terms) is NOT of degree n. $F(n)$ is not a polynomial function in n. It violates condition (iv) defined above. There are more than one terms having degree of n as 1.

Since $F(n)$ written in above form is not a polynomial function, so one cannot define its *degree*.
On simplification, we get $F(n) = n^2$, which is a polynomial of degree 2. Hence, we can say $F(n) = O(n^2)$. Here, $T(n) = T(n-1) + n$. Before talking of *degree* of $T(n)$, one has to express $T(n)$ as a polynomial in n.

The above expression is not a valid polynomial expression, because $T(n-1)$ is yet another function, and not a simple algebraic term of the type ax^n, where a is a real number and n is a positive integer.

So, to express $T(n)$ as a polynomial, expand $T(n-1)$ which gives

$$T(n) = n + (n-1) + T(n-2)$$

Again, it is *not* a polynomial by similar explanation. So, expand again. Continuing this process, it can be ended up with

$$T(n) = n + (n-1) + (n-2) + \cdots + 3 + 2 + 1$$

$T(n)$ is still not in a *valid* polynomial form because there are *multiple* terms containing n^1, and hence *violating* condition (iv).

Simplifying, we obtain $T(n) = n(n+1)/2 = 0.5n^2 + 0.5n$, which is a *valid* polynomial. So, degree of $T(n)$ is 2, and NOT 1.

Clearly, for all $n \geq 1, n \leq n^2$. So, $0.5n \leq 0.5n^2$. Thus, $T(n) \leq 0.5n^2 + 0.5n^2 = n^2$. Hence, by definition of big oh notation, $T(n) = O(n^2)$.

Note The number of anonymous function in an expression is understood to be *equal* to the number of times the asymptotic function appears. For example, in the expression $\sum_{i=1}^{n} O(i)$, there is a single anonymous function of i. It can be thought as $O(n*i)$. It is not same as $O(1) + O(2) + \cdots + O(n)$, which does not give a *clean interpretation*.

In fact, on the expression: $O(1) + O(2) + \cdots + O(n)$, if one applies the property, maximum, mentioned in Section 8.4.8 (a), then he/she gets simply $O(n)$ as result which is not equal to $O(n*i)$.

Further, $O\left(\sum_{i=1}^{n}(i)\right)$ is equal to $O(1 + 2 + 3 + \cdots + n) = O(n^2)$. The following cascaded segment should be safely expressed as $O\left(\sum_{i=1}^{n}(i)\right)$

Segment ...

```
for i:= 1 to n do
    for j:= i to n do
        S:= S + i;
```

EXAMPLE 8.8 Find the asymptotic upper bound for the recurrence relation,

$T(n) = 2T(n/2) + 1$, when $n > 2$
$\quad\quad = 1$, when $n = 1$

Solution $T(n) = 2T(n/2) + 1 = 2[2T(n/4) + 1] + 1 = 2^2 T(n/4) + 2^1 + 1$
$\quad\quad = 2^2 (2T(n/8) + 1) + 2^1 + 1 = 2^3 T(n/8) + 2^2 + 2^1 + 1$
$\quad\quad = 2^k + 2^{k-1} + \cdots + 2^1 + 1$ (assuming $n = 2^k, k >= 0$)
$\quad\quad = 2^{k+1} - 1$ [using the formula for finding the sum of arithmetic progression]
$\quad\quad = 2*2^k - 1$
$\quad\quad = O(n)$ \quad\quad\quad [as $2^k = n, k \geq 0$]

EXAMPLE 8.9 Solve $T(n) = T(n-1) + \log_2 n, T(1) = 1$.

Solution Expanding the given recurrence relation, we get

$T(n) = \log_2 n + \log_2(n-1) + \log_2(n-2) + \cdots + \log_2 3 + \log_2 2 + 1$
$\quad\quad = \log_2(n*(n-1)*(n-2)*\cdots*3*2) + 1$ (using properties of logarithms)
$\quad\quad = \log_2(n!) + 1$
$\quad\quad \leq \log_2(n^n) + 1$ [i.e., $\log_2(n!)$ it can be bounded as $\log_2(n^n)$]
$\quad\quad = n\log_2 n + 1$
$\quad\quad = O(n\log_2 n)$

EXAMPLE 8.10 $T(n) = T(n/2) + \log_2 n, T(1) = 1$.

Solution Expanding the given recurrence relation, we get

$$T(n) = \log_2 n + T(n/2)$$
$$= \log_2 n + \log_2(n/2) + T(n/2^2)$$
$$= \log_2 n + \log_2(n/2) + \log(n/2^2) + T(n/2^3)$$
$$= \cdots$$
$$= \log_2 n + \log_2(n/2) + \log(n/2^2) + \cdots + \log(n/2^{k-1}) + T(n/2^k)$$

Let $n = 2^k$. This gives $k = \log_2 n$. So, we get

$$T(n) = \log_2[n*(n/2)*(n/2^2)*\cdots*(n/2^{k-1})] + 1$$

[using properties of logarithms and putting $T(1) = 1$]

$$= 1 + \log_2(n^k) - \log_2[2*2^2*2^3*\cdots*2^{k-1}]$$
$$= 1 + k\log_2 n - \log_2[2^{1+2+3+\cdots+(k-1)}]$$
$$= 1 + k\log_2 n - \log_2[2^{k(k-1)/2}]$$
$$= 1 + k\log_2 n - k(k-1)/2$$
$$= 1 + k^2 - k(k-1)/2 \qquad \text{[using } k = \log_2 n\text{]}$$
$$= 1 + k(k+1)/2$$

Hence, $T(n) = 1 + (\log_2 n)(\log_2 n + 1)/2 = O((\log_2 n)^2)$.

EXAMPLE 8.11 What will be the time complexity of the following program segment (in C)?

```
int recursive(int n)
    {
    if (n==1) return(1)
    else
    return (recursive(n-1) + recursive(n-1))
    }
```

Solution The time complexity of the above recursive segment can be expressed as

$$T(n) = 2T(n-1) + 3, T(1) = 2, \text{ which gives solution as } O(2^n).$$

EXAMPLE 8.12 Solve

$$T(n) = \sum_{i=1}^{n-1} T(i) + n \quad \text{and} \quad T(1) = 1$$

Solution Actually, here $T(n) = T(1) + T(2) + \cdots + T(n-2) + T(n-1) + n$.

Now, we can write it as

$$T(n) = n + T(n-1) + \sum_{i=1}^{n-2} T(i) \text{ (by range transformation)}$$
$$= n + T(n-1) + T(n-1) - (n-1) = 2T(n-1) + 1 = O(2^n)$$

$$[\text{as } T(n) = \sum_{i=1}^{n-1} T(i) + n, \text{ so } \sum_{i=1}^{n-1} T(i) = T(n) - n]$$

8.4.11 Comparison of Complexity

Suppose there are often several different algorithms which correctly solve the same problem. How can one choose the best among them? There are several different *criteria* to achieve it. These are pointed out below.

- Ease of implementation
- Ease of understanding
- Efficiency in *time* and *space*

The first two are somewhat subjective. However, efficiency is something a mathematical analysis, to gain insight as to which is the *fastest* algorithm for a given problem. The better the time complexity of an algorithm is, the faster the algorithm will carry out his work in practice.

However, there is often a *time–space–tradeoff* involved in a problem, i.e., it cannot be solved with low computing time along with low memory consumption. One then has to make a compromise and to exchange computing time for memory consumption or vice versa, depending on which algorithm one chooses and how one parameterizes it.

Now, based on measured *complexities* of several designed algorithms of a problem, one can *analyse* which algorithm is the *best* or the *worst* for that problem. One such analysis is presented through Example 8.13.

EXAMPLE 8.13 Three students wrote algorithms for the same problem. They tested the three algorithms with two sets of date as shown below.

Case I $n = 10$
  Runtime for student 1:1
  Runtime for student 2:1/100
  Runtime for student 3:1/1000

Case II $n = 100$
  Runtime for student 1:10
  Runtime for student 2:1
  Runtime for student 3:1

What is the *complexity* of each algorithm? Which is the best? Which one is the worst?

Solution Since runtime of each of the *three* algorithms (designed by three different students) is given only for two values of n, so it is very difficult to find *unique* growth function for each algorithm (by drawing the *plots* or some other *ways*).

In fact, to draw some curve close to the exact curve (or to get some equation close to the exact equation), *at least* three points are necessary. Therefore, several possible functions can be considered for each algorithm using only two values of n.

Let us try to find out one possible answer by designing the following *growth* functions corresponding to the given runtimes.

Algorithm 1 $f(n) = n/10 = O(n)$, it satisfies $f(n) = 10/10 = 1$, when $n = 10$, and $f(n) = 100/10 = 10$, when $n = 100$. *Algorithm 2* $f(n) = n^2/10000 = O(n^2)$, it satisfies $f(n) = 100/10000 = 1/100$, for $n = 10$, and $f(n) = 10000/10000 = 1$, when $n = 100$. *Algorithm 3* $f(n) = n^3/1000000 = O(n^3)$, since $f(n) = 1000/1000000 = 1/1000$, when $n = 10$ and when $n = 100$, $f(n) = 1000000/100000 = 1$.

As per the above *selected* growth function, it is clear that the time complexities of alg-1, alg-2, and alg-3 are, respectively, $O(n)$, $O(n^2)$, and $O(n^3)$. Hence, it can easily be concluded that among these algorithms, the alg-1 is the *best* algorithm, as it is taking *linear* time; and alg-3 is the *worst* one, as it is taking *cubic* time.

Another possible answer can be inferred from the *following illustration*.

Consideration of alg-1 For algorithm 1, analysing the runtimes for *two* separate values of n (input data), we can say that its time complexity should be $O(n)$.

Consideration of alg-2 For algorithm 2, let us try to analyse the runtimes for *two* separate input values. It is given that runtime $(t_1) = 1/100$, when $n = 10$; and runtime $(t_2) = 1$, when $n = 100$.

Clearly, from the above results, we can write $t_2 = 100 * t_1$. It implies that due to 10 times increase of initial data ($n = 10$), time is being increased by 100 times of t_1.

Consideration of alg-3 For algorithm 3, let us once again try to analyse the runtimes for *two* separate input values. It is given that runtime $(t_1) = 1/1000$, when $n = 10$; and runtime $(t_2) = 1$, when $n = 100$.

Hence, it is *true* that the above results are based on the expression: $t_2 = 1000 * t_1$, i.e., it implies that due to 10 times increase of initial data ($n = 10$), running time increases 1000 times of t_1.

Finally, from the above discussion on the three algorithms, we can conclude that *alg-1* is the *best* among the three algorithms; and in between *alg-2* and *alg-3*, alg-2 is *better* than alg-3. Hence, alg-3 is the *worst* one.

Comment In latter approach, we find that all these algorithms are of linear time complexity but algorithm 1 must show less *slop* as compared to the others while drawing *plots* using the given results. This is, in fact, easier and more reliable than the other.

Note Algorithm designer must pay attention in designing efficient algorithm. One efficient algorithm is presented in Example 8.14.

EXAMPLE 8.14 An element in an array X is called a *leader*, if it is greater than all the elements to the *right* of it in X. Design a linear time algorithm to find all the leaders in the array X.

Solution An array sorted in descending order from left to right is a special case, each element in that array (except the last element—depending on your convention) is a *leader*. The $O(n)$ algorithm for finding leaders in array is as follows:

```
          Procedure Leaders(int A[], int N)
       // assuming array indexing starts at zero
          begin
             max := A[N - 1];
             for i:= N-2 to 0 step -1 do
             begin
        if(A[i] > max) then
//strictly >, i.e., not >=, as per definition of leader
                begin
                   A[i] is a leader
                   max:= A[i]
             end
          end
       end
```
For example 8, **12**, 9, 10, 7, 2, 5, **6**, 4, 1, 3

The elements in *bold* font are all leaders. Each of these elements is *greater* than all elements to their *right* side.

EXAMPLE 8.15 Arrange the elements of an array in linear time so that all the negative elements appear before the positive elements.

Solution The efficient algorithm for arranging the elements as per the given instruction is presented below.

```
              Procedure ARRANGE(a, n)
/* a[ ] is a list of elements consisting of negative and posi-
tive elements. n is the size of the array */
  begin
  while(i < j)do
    begin
/* when element take from L.H.S. is negative and element from
R.H.S. is positive. Then no swap, only L.H.S. index is incre-
mented by 1. So, case-1 is performed.*/
    case1: if ((a[i] < 0) and(a[j] > 0))then
              i := i+1
    /* when elements taken from L.H.S. and R.H.S. are both pos-
itive. Then no swap, only R.H.S. index is incremented by 1. So,
case-2 is performed.*/
      case2: if ((a[i]> 0)and (a[j]> 0)) then
              j := j+1
    /* when elements taken from L.H.S. and R.H.S. are both neg-
ative. Then no swap, only L.H.S. index is incremented by 1. So,
case-3 is performed.*/
     Case3: if ((a[i]< 0)and (a[j])< 0)then
              j := j+1
    /* when element taken from L.H.S. is positive but element from
R.H.S. is negative. Then swap, values of both i and j are incre-
mented by 1, case-4 is performed.*/
     Case4: if ((a[i]> 0)and (a[j]< 0)) then
              j :=j+1
             end
       end
```

The above procedure, in fact, takes linear time $O(n)$.

8.5 SEARCHING AND SORTING

Let us look at some techniques for searching and sorting of data.

8.5.1 Searching

Searching means finding any data (i.e., target item) from a list of items. The *objective* is to retrieve the location of the target item in the list if it is available in the list; otherwise (if not present in the list) report that it is *not* found in the list.

The *linear* (sequential) search and the *binary* search are very straightforward and have a number of *shortcomings*. A number of search techniques involving the use of *hashing* functions increase the efficiency of searching.

Linear (Sequential) Search

The basic idea behind this search technique is that a list of items is given and a *target* item will be searched from this list. To find it from the list, target element is compared with each element of the list (starting from the first element or the last element in the list) until it is found or the processing of the list is over.

Linear searching can be accomplished by array or pointer data structures. A simple procedure, using array, is presented below. In this procedure, searching is started from the first element of the list.

```
  Procedure Linear_search( x, n, data)
/* x is a list of elements with size 'n' and data is the target
element */
  begin
  flag := 0
  for i := 1 to n do
       begin
    if ( x[i] = data) then
            begin
                flag := 1
                break
            end-if
     end-for
    if ( flag =1) then
      print( data is found at the i-th position )
       else
          print( data is not present in the list)
       end-if
  end
```

Analysis of Time Complexity

The running time of this procedure is $O(1)$ in the best case if the desired element is the first element in the array; and $O(n)$ in the worst case, when the desired element is the last or if the element being sought is not in the list. Here, the size of the list (i.e., input size) is n.

Note When searching starts from the last element of the list and the index moves *backward* by -1, then the best case time complexity $O(1)$ occurs if the last element of the list is equal to the *desired* value. But if the desired element is the first element of the list or it is not at all found in the list, then running time equals to $O(n)$ and it is the worst case complexity.

No matter, whether searching starts from the first or the last position, the *best* case is $O(1)$—'if the desired element is found at *first* search', and the *worst* case is $O(n)$—'if it is found at last search in the list or not found in the list'.

Binary Search

The basic idea behind binary search is to search a *target* element from a *sorted* list of elements. If the list is not sorted, then sort it first. Now, assuming the sorted list in ascending order, we present below the binary searching technique.

During searching, at every step, the list is divided at its *mid* point into two sub-lists: *left* and *right*. The mid is computed as

$$\left\lfloor \frac{low + high}{2} \right\rfloor$$

where *low* and *high* are, respectively, the lower and the upper limits of the current list. So, the mid here helps to split the current list.

Now, If the target data equals to the data stored at the *mid* location of the list, then it is found successfully and hence one should stop further searching. Otherwise, consider the *left half* to the *mid* if the target data is *less* than the data at the *mid* or the *right half* to the mid if the target data is greater than the data at the mid. Continue this process until the target data is found or the size of the list becomes invalid (i.e., low > high for the current list). In the implementation of binary search, only array data structure will be accomplished.

Iterative Approach for Binary Search

Here, it is assumed that the list is sorted in ascending order.

```
Procedure Binary_search( x, n, data)
/* x is a list of elements with size n and data is the target
element, low and high are respectively the lower and upper lim-
its of the current list */
    begin
        low:= 1, high:= n
        while (low <= high ) do
            begin
```
$$mid := \left\lfloor \frac{low + high}{2} \right\rfloor$$
```
            if(data=x[mid]) then print(data found at mid) and exit
                else
                    if ( data < x[mid] )then
                        high:= mid -1
                    else
                        low:= mid +1
            end
    end
```

Remark If the list is in descending order, then change the *logic* as if (data < x[mid]) then low = mid + 1 and if (data > x[mid]) then high = mid − 1.

Verification of This Procedure

Let a list of elements be $x = \{1, 2, 3, 5, 7, 9, 11, 15, 25\}$ in ascending order and the *data* to be searched is 5. Here, $n = 9$. Initially, low = 1, high = n = 9, $x_1 = 1, x_2 = 2, x_3 = 3, x_4 = 5, x_5 = 7, x_6 = 9, x_7 = 11, x_8 = 15, x_9 = 25$.

At the first pass, condition: low (= 1) ≤ high (= 9) is satisfied. Hence, *mid* is calculated as

$$\left\lfloor \frac{1+9}{2} \right\rfloor = 5$$

Now, target *data* (= 5) is not equal to x[mid] = x[5] = 7 but it is *less* than x[5]. So, the *right* part to x[5] need *not* be considered to search, i.e., the target data may be found to the left part of x[5].

Therefore, high is *shifted* here from 9 to (mid−1) i.e., (5 − 1) = 4. Hence, the current *sublist* becomes {1, 2, 3, 5} on which *searching* will now be applied. The *while* loop continues as the condition: low (= 1) ≤ high (= 4) is satisfied. Currently, *mid* is

$$\left\lfloor \frac{1+4}{2} \right\rfloor = \left\lfloor \frac{5}{2} \right\rfloor = \left\lfloor \frac{4.5}{2} \right\rfloor = 4$$

Target *data* (= 5) is now *equal* to x[mid] (i.e., x[4] = 5). Since the target *data* is found, the process stops at this pass and exact location of the list (in which data is found) is printed. Observe that the search involves here only two comparisons.

Analysis of Time Complexity

Let us consider the worst case situation of this searching. Obviously, the *worst case* situation occurs when the *target* data is present at either of the *two end* points (low or high) or it is *not* present in the list. Thus, if the size of the list is n (and it is assumed that $n = 2^k, k \geq 0$), then in worst case, *maximum* number of comparison operations will be $k + 1$ (due to k times splitting of the list) and hence the running time of binary search will be $O(\log_2 n)$ [as $k = \log_2 n$] which is also in *average* case.

The *best case* complexity occurs if the target data is found in the first splitting.

 EXAMPLE 8.16 A is an $n \times n$ matrix, where for all $i, j, 0 \leq i < n$ and $0 \leq j < n$, and the following conditions always hold *true*.

$$A[i][j] \geq A[i-1][j] \quad \text{and} \quad A[i][j] \geq A[i][j-1], \quad \text{for } i, j \geq 1$$

In other words, all the elements rowwise as well as columnwise are in *ascending* order. Consider the problem of searching for an element from this matrix.

Problem Analysis

If *linear* searching technique is applied here, then it will take $O(n^2)$ in the worst case, since number of locations in the matrix is n^2 and all the locations may be searched.

On the other hand, if *binary* searching is applied *rowwise* (since the elements are in ascending order), then $O(n \log_2 n)$ time is required in *worst case*, as number of *rows* present in the matrix is n and each row has n elements. However, we can reduce this time to $O(n)$ if searching starts from A[0][n − 1] and proceeds as follows.

If found, then processing stops. Otherwise, if the search element is *less* than the element of the current location, then decrease the value of current column in the row (since it is not available in this column, i.e., it may be found in the *earlier* column), else increase the value of row in the current column.

The above concept is implemented through a *recursive* approach (presented *below*). Assume that initial call to following algorithm is *Matrix Search* (A, 0, n − 1, x). However, we may start calling from other diagonals also such as (0,0), (n − 1, 0), etc.

```
Procedure MatrixSearch (A, i, j, x)
/* indexing "starts" at zero, both i, j indicate the location
of the matrix and 'x' is the search element */
  begin
    if x = A[i,j] then return (i; j)
    if (i= n or j=-1 )then printf("not found") and return
    if(x < A[i,j]) then MatrixSearch (A, i, j-1, x);
      else MatrixSearch(A, i+1,j,x);
  end
```
The above procedure takes only $O(n)$ time.

8.5.2 Sorting

Sorting is one of the most frequently encountered problems in the field of computer science. In fact, it is a technique to arrange the elements of a list in a certain order. The most-used orders are numerical order (may be ascending or descending) and *lexicographic* or *lexicographical order* (also known as *dictionary order* or *alphabetic order*). But if *no order* is specified, then it is assumed that ascending order is followed (or to be followed). For any sorting technique, two basic operations: *compare* and *exchange* are considered.

All algorithms presented in this section are designed for sorting the elements in ascending order. Just a little bit of observation is needed if the readers wish to modify these algorithms to arrange data in descending order.

In the present section, only two simple sorting techniques: *Merge sort* and *Bubble sort* are discussed.

Merge Sorting

It is an example of *divide-and-conquer* strategy. In this strategy, the complete problem P is divided into *smaller* sub-problems like $P_1, P_2, \ldots, P_k$, and then the solution of each sub-problem is combined to get a final solution. Here, the divide-and-conquer approach is done *recursively*. Merge sort generally uses two recursive procedures.

(a) *Mergesort( )* procedure through which the current list (at each step) is divided into two parts, finding the *mid* as

$$mid = \left\lfloor \frac{low + high}{2} \right\rfloor$$

where *low* and *high* are, respectively, the lower and upper bounds of the current list.

(b) *Merge( )* procedure is for *merging* two sorted lists. Both the lists will be in *same* order either in *ascending* or in *descending* (but not like that one is ascending and the other in descending order). Clearly, if we need the *merged* list in ascending order, then both the lists (which are to be merged) should be in ascending order.

Similarly, both the lists must be in descending order to get the resultant lists in descending order. Otherwise, merging does not work. This can be easily understood from the following examples (while merging):

(i) Suppose one is going to merge two sorted lists in ascending order but both the lists are in descending order such as $L_1 = \{7, 5, 4\}$, $L_2 = \{3, 2, 1\}$. Here, merging is not possible, unless we modify the standard logic for merging.

(ii) Suppose one is going to merge *two* sorted lists in ascending order but one list $L_1 = \{7, 5, 4\}$ in descending order and the other $L_2 = \{1, 2, 3\}$ in ascending order. Merging is not possible.

The presented *merge sorting* technique here uses two arrays: (i) one is the input array where unsorted data are available and (ii) the next is the resultant array where the final sorted list will be stored. However, during sorting, the input array is divided into smaller parts.

```
 Procedure Mergesort (low, high)
/*low and high are respectively the lower and upper limits of
the current list */
       begin
    if ( low < high) then // divide the current list
       begin
```
$$mid := \left\lfloor \frac{low + high}{2} \right\rfloor$$
```
//From every Mergesort() call,3 following calls are performed
   call Mergesort(low, mid)
   call Mergesort (mid +1, high)
   call Merge(low,mid,high) //combining the sub-solutions
       end
       end
Procedure Merge( low, mid, high)
 /* Assume that elements to be merged are stored in ascending
order in two halves of an array. Also, index h for 1st partition,
j for 2nd partition, i for final(resultant) array*/
    begin
    h:= low, i:= low, j:= mid +1
    while (( h<= mid) and (j <= high)) // condition (h<=mid) for
//1st the partition and condition (j<=high) for the 2nd partition
       begin
        if (a[h] <= a[j] ) then
/* one element from the 1st partition and the other element from
the 2nd partition are taken and then compared */
          begin
          b[i]:= a[h]
 /*if the element taken from 1st partition is smaller or equal
to the element taken from 2nd partition, then it is placed into
the resultant b[ ]*/
   h:= h + 1 //array b[] and value of 'h' is increased by 1
          end
       else
/* otherwise element taken from 2nd partition is placed into the
resultant array b[ ] and value of 'j' is increased by 1 */
          begin
          b[i]:= a[j]
          j:= j+1
          end
       i:= i + 1
/* Finally, for storing the next element in the resultant array
b[] value of 'i' is increased by 1. */
```

```
              end
   /* end of while loop, when at least one of the partitioned or
   both are copied */
         if (h > mid) then
      //means 1st partition is copied into the final list but 2nd partition
      // may have some elements which are still not copied into final list
            begin
            for k := j to high do
       // now current location of the 2nd partition is j-th element
            begin      // ( i.e., hold by index 'j' )
              b[i] := a[k]
                 i := i +1
            end
           end
         else
   //means 2nd partition is copied into the final list but 1st partition
   // may have some elements which are still not copied into final list
         begin
        for k := h to mid do
   /* now, current location of the 1st partition is h-th element,
   i.e., hold by index 'h' */
           begin
            b[i] := a[k]
              i := i +1
           end
         end
   // Now, copy from resultant array b[ ] into array a[ ]
         for k := low to high do
             a[k] := b[k]
       end //end of Merge procedure
```

Verification of the Procedure

Let us consider a list of elements $a[\ ]$ = {310, 285, 179, 659, 351, 423, 861, 254, 450, 520}. In this list, number of elements is 10. Now, apply Mergesort( ) on this list to get the data in ascending order.

An illustration of the sub-divisions of the given list [through recursive procedure call Mergesort (low, high)] at several possible levels and then merging of these sub-divisions [by procedure call Merge( )] in order to sort the list is pictured through a tree structure (shown in Figure 8.2). Every node of the tree denoted by *rectangle* (box) represents here a list. The values inside each box represent, respectively, for the *lower* limit and the *upper* limit of the current list. On the other hand, the middle location of the list is denoted by *mid* (placed outside the box). The number within parentheses shows the *level* value of the tree, whereas the values within { } on top of each node represent the *sorted* sub-lists obtained from the lower *sub-trees* of that node [by applying simultaneously the Mergesort( ) and Merge( ) procedures]. The process is explained in detail below.

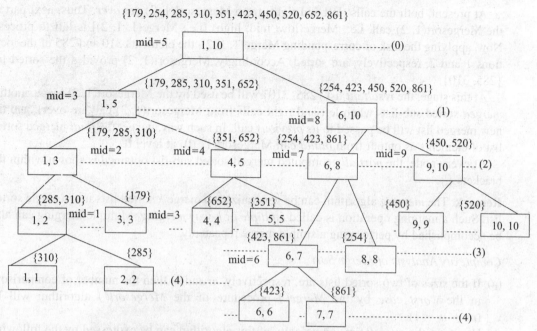

FIGURE 8.2 Tree structure for merge sort

In this example, the number of elements is 10 and initially the values of *low* and *high* are 1 and 10, respectively. So, here the first call is Mergesort(1, 10). Since (*low* = 1) < (*high* = 10), so *mid* is computed and its value is 5 at this stage.

Now, Mergesort(low, mid) is called (here, mid replaces high). Clearly, it is Mergesort(1, 5). Currently (*low* = 1) < (*high* = 5). Accordingly, *mid* is computed as 3 at this stage.

Mergesort(low, mid) is again called (here, mid replaces high). It is indeed Mergesort(1, 3). Here, (*low* = 1) < (*high* = 3). So the computed *mid* is 2 at this stage.

Accordingly, Mergesort(low, mid) is called (here, mid replaces high). Therefore, the current call is Mergesort(1, 2). Since (*low* = 1) < (*high* = 2), so *mid* is now 1.

Clearly, Mergesort(low, mid) is further called from this point. Present call is Mergesort(1, 1). Now, (*low* = 1) is not less than (*high* = 1). So, call like Mergesort(low, mid) and Mergesort(mid + 1, high) are not possible from the procedure call Mergesort(1, 1). Hence, control moves *back* to the call *from where* Mergesort(1, 1) was called, and it is Mergesort(1, 2).

Now, processing of the *rest* part of the Mergesort(1, 2) starts. However, the value of *mid* computed in this Mergesort(1, 2) call is 1 and Mergesort(low, mid) call from Mergesort(1, 2) is already over. But Mergesort (mid + 1, high) and Merge(low, mid, high) calls [from Mergesort(1, 2)] are not finished.

Hence, Mergesort(mid + 1, high) will now start from Mergesort(1, 2). Value of *mid* is 1 [already computed in the call Mergesort(1, 2)]. Thus, current call is Mergesort(2, 2) as mid is 1 and high is 2 [obtained from the call Mergesort(1, 2)].

Now, in call Mergesort(2, 2), *low* is 2 and *high* is 2. Currently, *low* is not less than *high*. So, call like Mergesort(low, mid) and Mergesort(mid + 1, high) are not possible from the procedure call [Mergesort(2, 2)]. Hence, control again moves *back* to the Mergesort(1, 2) call *where from* Mergesort(2, 2) was called, and it is Mergesort(1, 2).

At present, both the calls (left and right) from Mergesort(1, 2) are *over*. Thus, next part of the Mergesort(1, 2) call, i.e., Merge(low, mid, high) [i.e., Merge(1, 1, 2)] is left to process. Now, applying the task of corresponding Merge(1, 1, 2), the elements 310 and 285 of the locations 1 and 2, respectively, are sorted. Accordingly, Mergesort(1, 2) provides the sorted list {285, 310}.

At this stage, the *resultant* list {285, 310} will be used by the Mergesort(1, 3) to get another *merged* sorted sub-list [when left and right calls from Mergesort(1, 3) call are over], and the new merged list will be passed to its *previous* call. In such way, we get the *final* merged sorted list of elements as output from *initial* call Mergesort(1, 10) at level 0.

At every node, the *partially* computed merged sorted sub-list obtained is shown within the brackets { }.

Remark The *merging* algorithm can be generalized to merge k sorted lists into a single sorted list. Such a merging operation is called *multiple* or *k-way merging*. Multiple merging can also be accomplished by performing a simple merge repeatedly.

Complexity Analysis of Merge Sort

(a) If the sizes of two sorted lists are, respectively, m and n, then the number of comparisons in the *worst case* by the *Merge( )* procedure of the *Mergesort( )* algorithm will be $(m + n - 1)$.

(b) However, the *overall* time complexity of this algorithm can be expressed by the following *recurrence* relation:

$$T(n) = 2T(n/2) + c*n, \quad \text{when } n > 1, c \text{ is constant}$$
$$= a, \quad \text{when } n = 1, a \text{ is constant}$$

$c*n$ is the number of operations computed for merging two sub-lists at a node of a *level*. But input size varies from *level* to *level* in the constructed tree structure.

Hence, it is *true* that the number of merging operations *increases*, as the value of *level decreases*. For example, during merging, number elements in each sub-list at *level* 1 is $n/2$, it is $n/4$ at *level* 2, and so on, where n is the input size.

Now, suppose $n = 2^k$ for $k > 0$, then we can solve the above equation by successive substitutions as follows:

$$T(n) = 2(2(T(n/4) + c*n/2) + cn$$
$$= 4T(n/4) + 2cn$$
$$= 4(2T(n/8) + cn/4) + 2cn$$
$$= 8T(n/8) + 4cn/4 + 2cn$$
$$= 8T(n/8) + 3cn = 2^3 T(n/8) + 3cn$$
$$\ldots$$
$$\ldots$$
$$= 2^k T(1) + kcn$$
$$= n*a + kcn \quad (\text{as } 2^k = n)$$
$$= n*a + c*(n\log_2 n) \quad (\text{as } k = \log_2 n)$$

In this expression, the highest (+)ve degree term is $n\log_2 n$. Thus, the overall time complexity of Mergesort( ) algorithm is always $O(n\log_2 n)$ (i.e., in *best*, *average*, and *worst* cases).

Bubble Sorting

The basic idea behind bubble sorting is that the largest element in a list will be placed in the largest position in the list, the second largest element in the second largest position in the list, and so on. For example, if n elements are present in a list, then the first largest element is placed at the nth position, the next largest (i.e., the second largest element found from the list) is placed at the $(n-1)$th position in the list, and so on.

```
Procedure Bubble_sort ( a, n)
    /* a[ ] is an array of elements with size n. These n
elements are to be stored in ascending order*/
    begin
        for i := n down to 2 do
            for j := (i-1) to 1 do
                begin
                    if (a[i] < a[j] ) then
                        begin
                            temp := a[i]
                            a[i] := a[j]
                            a[j] := temp
                        end
                end
    end
```

Verification of Bubble Sorting Procedure

Let us consider a list of elements $a[\ ] = \{5, 3, 7, 2, 1\}$. Here, n (number of elements) = 5. Bubble sorting is applied on this example and each pass is discussed below separately.

$i = 5$

    $j = 4$:   5, 3, 7, 1, 2 (after interchanging between ($a[4] = 2$) and ($a[5] = 1$))
    $j = 3$:   5, 3, 2, 1, 7 (after interchanging between ($a[3] = 7$) and ($a[5] = 2$))
    $j = 2$:   5, 3, 2, 1, 7 (interchanging between ($a[2] = 3$) and ($a[5] = 7$) is not required)
    $j = 1$:   5, 3, 2, 1, 7 (interchanging between ($a[1] = 5$) and ($a[5] = 7$) is not required)
    [So, after 1st pass the largest element 7 (in the list) is placed at the largest (i.e., 5th) position in the list.]

$i = 4$

    $j = 3$:   5, 3, 1, 2, 7 (after interchanging between ($a[4] = 1$) and ($a[3] = 2$))
    $j = 2$:   5, 2, 1, 3, 7 (after interchanging between ($a[4] = 2$) and ($a[2] = 3$))
    $j = 1$:   3, 2, 1, 5, 7 (after interchanging between ($a[4] = 3$) and ($a[1] = 5$))
    [Thus, after 2nd pass the 2nd largest element 5 is placed at the 2nd largest (i.e., 4th) position in the list.]

$i = 3$

    $j = 2$:   3, 1, 2, 5, 7 (after interchanging between ($a[3] = 1$) and ($a[2] = 2$))
    $j = 1$:   2, 1, 3, 5, 7 (after interchanging between ($a[3] = 2$) and ($a[1] = 3$))
    [So, after 3rd pass the 3rd largest element 3 is placed at the 3rd largest (i.e., 3rd) position in the list.]

$i = 2$

    $j = 1$:   1, 2, 3, 5, 7 (after interchanging between ($a[2] = 1$) and ($a[1] = 2$))

[So, after 4th pass the 4th largest element 2 is placed at the 4th largest (i.e., here, 2nd) position in the list, and the smallest element is placed at the 1st position in the list.]
Finally, we get the sorted list of elements.

Time Complexity Analysis of Bubble Sorting

The required running time of this algorithm for input size n will be $O(n^2)$ in *best*, *average*, and *worst* cases. We explain below how this time is computed. In the 1st iteration of the *for* loop i, number of iterations for j is $(n-1)$. In the 2nd iteration of the *for* loop number of iterations for j is $(n-2)$, and so on.

This can be written as
$$(n-1) + (n-2) + \cdots + 1 = \frac{n(n-1)}{2} = \frac{(n^2 - n)}{2} = O(n^2)$$

Tree Sort

The sorting techniques which are based on tree representation are called *tree sorting*, for example, merge sort.

EXAMPLE 8.17 Show that any technique that sorts an array can be expanded to find all duplicates in that array.

Solution Every sorting algorithm uses comparison of elements. In other words, any sorting algorithm will compare some element $A[i]$ with some other element $A[j]$ where $i \neq j$. Now, the following three cases arise:

Case 1 The algorithm is iterative. No issues here—just include an extra check that if equality holds for $A[i]$ and $A[j]$, then $A[i]$ is a duplicate.

Case 2 The algorithm is recursive—it uses divide-and-conquer strategy where the array $A[i \cdots j]$ is partitioned into two sub-arrays $A[i \cdots k]$ and $A[k+1 \cdots j]$. In such cases, include an equality check for $A[k]$ and $A[k+1]$, for example, merge sort.

Case 3 The algorithm is recursive—it uses divide-and-conquer strategy where the array $A[i \cdots j]$ is partitioned into two sub-arrays $A[i \cdots k-1]$ and $A[k+1 \cdots j]$ with the element $A[k]$ being placed at its correct position. In such cases, include equality checks for $A[k-1]$ with $A[k]$, and for $A[k]$ with $A[k+1]$, for example, quick sort.

However, one has to implement carefully some sort of hashing technique to ensure that the same element is not reported multiple times as duplicate.

An alternative approach may be—just sort the array, and then run the following algorithm on sorted array.

```
Report_Duplicates(int A[ ], int N)
  begin
    i = 0;
    while(i < N)
    begin
      index = i;
      while((i<N - 1)&&(A[i+1]==A[index]))
        i=i+1;
      if(i > index) then
        begin
          print "A[index] is a duplicate"
        end
      i=i+1;
    end
  end
```

SUMMARY

Complexity is the study of analysis of resource usage of a given algorithm. In fact, analysis of algorithm means analysis of resource usage of a given algorithm. This chapter provides a comprehensive idea on algorithm, program, pseudo-code, data structure, and its types. We then discuss the need to measure complexity of algorithm and the ways to express it. The chapter emphasizes on designing efficient algorithms, since efficient algorithms lead to efficient program that makes better use of hardware. Further, the programmers who write efficient algorithm are more marketable than those who do not. Of greater interest, discussion on recursive procedure as an algorithm implementation tool is also presented in this chapter.

EXERCISES

1. Write short notes on the following:
   (a) Algorithm
   (b) Data structure
   (c) Recursion
   (d) Time complexity
2. What are the basic steps for writing a good algorithm/program? Explain.
3. What criteria should be fulfilled by an algorithm? Explain.
4. Define data structure and explain its need. Classify the data structures. Is data type also data structure? Justify your answer.
5. On what basis will you classify linear and non-linear data structure? Explain with example.
6. What is ADT? Give some examples.
7. Mention the parameters for considering the complexity of an algorithm. How do you measure the time complexity of an algorithm? Why the concept of asymptote is taken in measuring inexact time complexity of an algorithm?
8. (a) Define the common asymptotic notations used to analyse complexity of algorithm. Explain their behaviour also.
   (b) For $f(x) = 3x^3 + 2x^2 + 9$, show that
      (i) $f(x) = O(x^3)$
      (ii) $f(x) \neq \theta(x^2)$
      (iii) $f(x) = \omega(x)$
9. Explain how to analyse the complexity of recursive procedure. Give an example.
10. What are the different algorithm design techniques? Explain briefly each of them.
11. Give big oh estimation for the following:
    (a) 10, (b) $n + 5$, (c) $n^2 + n$, (d) 10^{10},
    (e) $n!$, (f) $(n^3 + 1)/(n + 1)$,
    (g) $n \log(n^8) + n^{3/2} = O(n^{3/2})$, and
    (h) $5n \log(n) + n^{3/2} = O(n^{3/2})$.
    Verify true or false.
12. Solve the following recurrence relations and express $T(n)$ by using big oh notation:
    (a) $T(n) = T(n/3) + n$ and $T(1) = 2$
    (b) $T(n) = \log_2 n + T(n - 1)$ and $T(1) = 1$
    (c) $T(n) = T(n/3) + \sqrt{n}$ and $T(1) = 2$
    (d) $T(n) = T(n - 1) + 2^n$ and $T(1) = 3$
    (e) $T(n) = T(n - 1) + 1$ and $T(1) = 1$
    (f) $T(n) = 2T(n - 1) + 1$ and $T(1) = 1$
    (g) $T(n) = 2T(n - 1) + n$ and $T(1) = c$ (constant)
    (h) $T(n) = T(n - 1) + \sqrt{n}$, $T(1) = 1$
    (i) $T(n) = \sqrt{n}T(\sqrt{n}) + n$ for $n > 5$, $T(n) = 2$ for $n \leq 5$
    (j) $T(n) = T(n/2) + \log_2 n$, $T(1) = 1$
    (k) $T(n) = T(n - 1) + 1/n$, $T(1) = 1$
13. Given the polynomial $f(x) = \sum a_i x_i$, where $0 \leq i < n$, all a_i are real numbers. Write an $O(n)$ time algorithm to evaluate $f(x)$ for any real number x. [Hint: You can express the polynomial $2x^3 + 4x^2 + 5x + 6$ as $((2x + 4)x + 5)x + 6$, requiring just 3 multiplications and 3 additions.]
14. Consider the following complexity functions:
    $g_1(n) = n^3$ for $0 \leq n < 10,000$, and n^2 for $n \geq 10,000$
    $g_2(n) = n$ for $0 \leq n \leq 100$, and n^3 for $n > 100$
    Can we express like $g_2(n) = O(g_1(n))$? Justify. [Hint: For solving this question, one may plot the function graphically.]

15. For a given problem with inputs of size n, Algorithms A and B are executed. In terms of running time, one of the algorithms is $O(n)$ and another is $O(n \log n)$. Some measured running times of these algorithms with respect to input n are given below.

	$n = 512$	$n = 1024$	$n = 2048$
A	70	134	262
B	42	86	182

    Identify which algorithm has what time complexity and also find the running times. Which algorithm would you select for different values of n?

16. Suppose that each row of an $n \times n$ array A consists of 1s and 0s such that in any row of A, all the 1s come before 0s in that row. Present an algorithm to count the number of 1s in A in $O(n \log_2 n)$.

17. Given a sorted array containing all unique elements (i.e., no repetitions of elements in the array). Define an operation 'RotateRight' as follows:

    void RotateRight(int A[ ], int N)
    begin
        x = A[N − 1]; //where A is array and N is number of elements in array
        for ($i = 1; i < N; i++$)
            A[i] = A[i − 1];
        A[0] = x;
    end

    The operation RotateRight( ) is performed on the array k times where $0 \leq k < N$, and k is not known to you. Develop an algorithm to find k in $O(\log_2 N)$ worst case time.
    [*Hint*: Think on the lines of binary search. Watch out for the 'interval' in which A[low] > A[mid] or A[mid] > A[high]. The value of k is same as index of minimum element in the rotated array.]

18. Write a routine BUBBLE( ) so that successive passes go in opposite direction (i.e., if 1st is from backward then 2nd from forward, and so on).

19. Design an *efficient* algorithm to find the *repeated* elements in an array.
    Examples: 7, 8, 3, 7, 5, 8, 3, 8, 9

20. Show that any technique that sorts an array can be expanded to find all the *duplicates* in that array.

21. Given an array A of length N, sorted in ascending order. Also, the value of $A[N/2]$ is known. Explain how to use linear search efficiently for searching an element in this array.

22. Given an array of n integers. Write pseudo-code for reversing the contents of the array without using another array. You can take help of an extra variable.

9 Graph Theory

LEARNING OBJECTIVES

After reading this chapter you will be able to understand the following:
- Different types of graphs and their basic terminologies
- Representations of graphs in a computer's memory
- Operations on graph
- Dijkstra's algorithm for single-source shortest path problem
- Planar, Hamiltonian, and Eulerian graphs
- Graph colouring and the maximum flow of the network

9.1 INTRODUCTION

Many situations observed in the fields of computer science, physical science, economics, and many other areas can be analysed by using techniques from a relatively new area of mathematics called *graph theory*. The idea of this theory was first introduced in 1736 from the Konigsberg bridge problem developed by a Swiss mathematician, Leonard Euler. The Konigsberg bridge problem states that the river Pregel flows through the town Konigsberg thereby dividing it into four land areas labelled A, B, C, and D on the border of the river. These land areas were connected by seven bridges. The goal is to determine whether, starting at one land area, it is possible to walk across all the bridges exactly once and return to the starting land area.

In fact, a graph can be used to represent any problem involving discrete arrangements of objects, where we are interested only in the relationship between those objects, rather than their properties. The concept of the minimum spanning tree helps in all network design problems such as telephone company, computer network, travelling agency, etc., to choose the cheapest subset of edges that keeps the graph in one connected component. We shall discuss this in Chapter 10.

In this chapter, we begin with some basic *graph terminologies* and then discuss some important concepts in graph theory such as types of graphs, storage representation, and operations. We have provided a comprehensive study on special types of graphs like *Eulerian*, *Hamiltonian*, and *Planar*. The promising topic *graph colouring* is also introduced in this chapter. Further, a summary on *applications of graphs*, following the section *flow network*, is presented at the end of this chapter.

9.2 GRAPHS AND BASIC TERMINOLOGIES

A graph is a mathematical concept which can be used to model many concepts from the real world. A graph consists of a pair of sets, represented as $G = (V, E)$, where V is a non-empty set of vertices (also called nodes) and E is a set of edges (sometimes called arcs). A *node* is a structure that normally contains some information or condition while an *edge* is the means to provide connectivity between pair of vertices/nodes. An edge can be represented as a pair of nodes (u, v) indicating an edge from node u to node v.

Two vertices/nodes x and y of G are *connected* if there is an edge xy between them, and these vertices are then called *adjacent* or *neighbour* vertices/nodes. Here, the nodes x and y are called the *endpoints* of the edge.

Further, in a graph G, a node which is not adjacent to any other node is called an *isolated node*.

A graph is *finite* if it has a finite number of vertices and a finite number of edges, otherwise it is *infinite*. If G is finite, $G(V)$ denotes the number of vertices in G and it is called the *order* of G. Similarly, $E(G)$ denotes the number of edges in G and it is called the *size* of G. The graph shown in Figure 9.1 has four vertices: a, b, c, and d. Here, (a, b) is a pair of vertices which are connected, and this connectivity represents an edge between them. Now, a and b are the *end points* of the edge (a, b). Clearly, *neighbours* of vertex a in this graph are b and c, as there are edges from a to b and a to c. Similarly, (a, c) and (b, c) are the other edges of this graph. Further, vertex d is the *isolated* vertex, as it is not adjacent to any other vertices. Obviously, it is an example of finite graph and its *order* of $G(V)$ is 4 [as $V = \{a, b, c, d\}$].

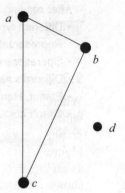

FIGURE 9.1 A graph with isolated vertex

9.2.1 Undirected and Directed Graphs

Graphs may be *directed* or *undirected*. A graph is directed (or *digraph*) when direction of edge from one vertex to another is defined, otherwise it is an undirected graph. Directed edge between vertices u and v is normally expressed as $<u, v>$, whereas undirected edge between u and v is expressed as (u, v). Thus, $(u, v) = (v, u)$ and it implies that it is possible to move from u to v and vice versa, i.e., the edge (u, v) can be used both for going from u to v and for going from v to u. On the other hand, $<u, v> \neq <v, u>$ and it means that if there is a path from u to v, then that path cannot be chosen to traverse from v to u.

A *roadmap* (without the background colouring) is an *example* of a graph in which the cities and the villages are the nodes and the roads are the edges. In fact, this is an example of an *undirected* graph: typically a road on a map can be used in both directions. But we can also make a sociogram: each person is a node and there is an edge from person A to person B if A likes B. These edges are *directed*: liking someone is not always *symmetric*. However, in both cases, if there are n nodes, then we generally number them from 0 to $n - 1$.

Graph as shown in Figure 9.1 is an example of undirected graph, whereas some directed graphs are pictured in Figure 9.2 (since its edges are directed). Clearly, the directed graph in Figure 9.2(a) has an edge from vertex a to b. Therefore, it is possible to move from a to b only, but not in the *reverse* direction since no direct edge exists from b to a. On the other hand, the edge between a and b in the graph given in Figure 9.1 is undirected. So, it is possible to move in both the directions between a and b.

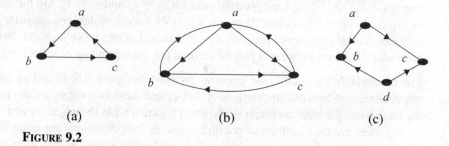

FIGURE 9.2

Note Any undirected graph G can be replaced by an equivalent directed graph, replacing each edge of G by a pair of edges, each of which has opposite direction to the other.

9.2.2 Weighted Graph

When a weight (may be cost, distance, etc.) is associated with each edge of a graph, then it is called as *weighted* graph, otherwise *unweighted* graph. The graph given in Figure 9.1 is an example of unweighted undirected graph, as no weight is assigned to the edges, whereas Figure 9.3 shows an example of weighted undirected graph (since, weight to each edge of the graph is provided).

For example In this weighted graph, weight 2 is assigned to the edge (a, b) and weight 6 to (a, c), and so on.

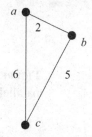

FIGURE 9.3 A weighted undirected graph

9.2.3 Self-Edge or Self-Loop

An edge which starts from a vertex and moves back to it is called a *self-edge* or *self-loop*, i.e., a loop is an edge (v_i, v_f), where $v_i = v_f$. The graph as shown in Figure 9.4 possesses a self-loop from vertex b to b.

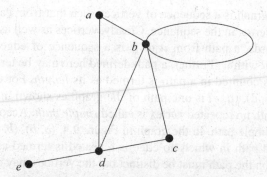

FIGURE 9.4 A graph with self-loop

9.2.4 Multiple or Parallel Edges

In some directed as well as undirected graphs, there may exist a pair of nodes joined by more than one edge and such edges are called *multiple* or *parallel* edges. In *undirected* graph, two edges: (v_i, v_j) and (v_p, v_q) are *parallel edges* if $v_i = v_p$ and $v_j = v_q$. On the other hand, in case of directed edges, two edges (between a pair of nodes) which are *opposite* in direction, are considered *distinct* (i.e., not parallel). So, in directed graph, more than one directed edges (in a particular direction) between a pair of vertices are considered as *parallel edges*.

For example In the *undirected* graph as pictured in Figure 9.5, e_1 and e_2 are two undirected edges connected between the vertices v_1 and v_2 and these two edges are the *parallel* edges. On the other hand, the *directed* graph as shown in Figure 9.6 has the edges: e_1 and e_3 between v_1 and v_2, and these are the examples of parallel edges, as their directions are same. But the edge pair: (e_1, e_2) in this graph is not an example of *parallel edges*, since the directions of the edges in the edge pair are *opposite*. Similarly, the edge pair: (e_2, e_3) in the same graph is also not parallel.

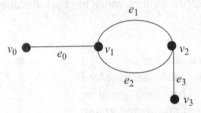

FIGURE 9.5 Undirected graph with parallel edge

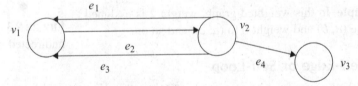

FIGURE 9.6 Directed graph with parallel edge

Path in a Graph

A *path* in a graph is a sequence of vertices such that from each of its vertices there is an edge to the next vertex in the sequence. Clearly, vertices as well as edges may be *repeated* in a path. In other words, a path from u to w is a sequence of edges $(u, v_1), (v_1, v_2), \ldots, (v_{k-1}, w)$ connecting u with w. Further, a path defined here may be termed as *walk* also. Now, the number of edges counted in a path is termed as its *length*. For example, $(a, b), (b, b), (b, d), (d, c), (c, b), (b, d), (d, e)$ is one path of the graph as shown in Figure 9.4 and its length is 7.

A path with no repeated vertex is called *simple path*. Accordingly, no duplicate edge will be present in simple path. In the graph in Figure 9.4, $(a, b), (b, d), (d, e)$ is a simple path.

Further, a path in which no edge is repeated is termed as *trail*. Thus, by the definition of trail, edges on the path must be distinct but the vertices may repeat. Again, in a closed trail, the first and the last vertices are same.

A *closed path* is a path that starts and ends at the same point, otherwise the path is *open*. Clearly, edge repetition is allowed on the closed path. For example, $(a, b), (b, d), (d, c), (c, b),$

(b, d), (d, a) is a closed path of the graph given in Figure 9.4, since the start and endpoint is a (i.e., same). Also, the edge (b, d) is repeated on this path.

A *cycle* (circuit/tour) is a closed path of non-zero length that does not contain any repeated edge(s). However, vertices other than the *end* (i.e., *start*) vertex may also be repeated. In the graph in Figure 9.4, (a, b), (b, b), (b, d), (d, a) is a cycle. Here, all the edges are *distinct* but b (which is not start/end vertex) is repeated, which, in turn, is forming one more circuit within another. In other words, a cycle may contain another cycle within itself. On the other hand, a *simple cycle* is a cycle that does not have any repeated vertex except the first and the last vertex. (a, b), (b, d), (d, a) is an example of simple cycle of the same graph.

A graph without cycles is called *acyclic*. A *tree* is an acyclic and connected graph. A *forest* is a set of trees.

Connected Graph

A graph is called *connected* if and only if for any pair of nodes u, v, there is at least one path between u and v. Otherwise, it is disconnected. Clearly, the graph in Figure 9.4 is a connected undirected graph, whereas the graph given in Figure 9.1 is disconnected.

Comment Some authors may use 'path' and 'simple path' synonymously. However, in this book, these two types of paths are considered distinctly.

Corollary 9.1 Repetition of vertex on a path may not ensure the repetition of edge, whereas repetition of edge ensures the repetition of vertices

Proof It is obvious that a vertex may be repeated without repetition of any edge on a path. For example, when a cycle is formed, only start vertex repeats, i.e., both the start vertex and the end vertex coincide. However, no edge is repeated here. On the other hand, in case of repetition of edge, the vertices representing the edge must be repeated because when an edge is traversed twice (i.e., repeated), the vertices representing the edge must also be traversed *twice*.

Thus, edge repetition on a path guarantees the repetition of vertices on the edge.

Types of Connectivity in Graphs

By the definition of connectivity, a connected graph must have at least two vertices. Thus, a graph with a single vertex is not considered as a connected graph.

For directed graphs, we mostly speak of *strongly* connected as we take the direction of the edges into account for the paths. A graph is *strongly connected* if and only if every pair of vertices in the graph are reachable from each other, i.e., if there are paths in both directions between any two vertices. Otherwise, we speak of *weakly* or *unilaterally* connected. The graph in Figure 9.2(b) is an example of *strongly* connected graph, whereas the graph in Figure 9.2(a) is an unilaterally connected graph, as it has a path from a to c but no path exists from c to a, and so on.

Also, we say that a graph is *strictly weakly* connected if it is not unilaterally connected. Thus, a strictly weakly connected graph may have many *sources* and *sinks* (destinations). The graph given in Figure 9.2(c) is an example of *strictly* weakly connected graph.

Simple Graph, Multi-Graph, and Pseudo-Graph

A directed or undirected graph which has neither self-loops nor parallel edges is called *simple graph*. However, cycle(s) is (are) allowed in a simple graph. Further, a simple graph may contain *isolated* vertex also.

The graph as shown in Figure 9.7(a) is a simple connected graph, since it has no self-loop and parallel edges. But the graphs (b) and (c) in the same figure are not simple, since they contain, respectively, parallel edges and self-loop. Further, the graph presented in Figure 9.2(a) is a *simple directed* graph, as it has no self-loop or parallel edges. On the other hand, the graph in Figure 9.1 is a *simple disconnected* graph.

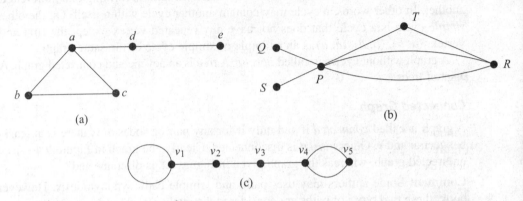

FIGURE 9.7

Any graph (directed or undirected) which contains some parallel edges is called a *multi-graph*. In multi-graph, no self-loop is allowed but cycle may be present. The undirected graph given in Figure 9.5 is an example of multi-graph, since it has parallel edges but no self-loop, whereas the graph in Figure 9.6 is a directed multi-graph.

A directed or undirected graph in which self-loop(s) and parallel edge(s) are allowed is called a *pseudo-graph*.

Degree of Vertex

The degree of a vertex of an undirected graph is the number of edges incident on it, except the fact that a loop contributes *twice* to the degree of that vertex. The degree of a vertex v in a graph G may be denoted by $\deg_G(v)$.

For example In the undirected graph in Figure 9.4, the degree of a [i.e., $\deg_G(a)$] is 2, degree of b [i.e., $\deg_G(b)$] is 5 (since there is a self-loop at b), degree of c [i.e., $\deg_G(c)$] is 2, and so on.

In directed graph G, we consider two types of degrees of vertices: (a) *in-degree* and (b) *out-degree*. The in-degree of a vertex v of G, denoted by $\deg_G^-(v)$, is computed by the number of edges moving into that vertex, whereas the *out-degree* of v, denoted by $\deg_G^+(v)$, is computed by the number of edges moving out from that vertex. The sum of the *in-degree* and the *out-degree* of a vertex is called the *total degree* of that vertex.

For example In the directed graph in Figure 9.2(c), the in-degree of a, $\deg_G^-(a)$ is 0 (zero) (since it has no incoming edge) but its out-degree, $\deg_G^+(a)$ is 2 (since it has two outgoing edges from itself). Hence, the total degree of a is $\deg_G(a) = \deg_G^-(a) + \deg_G^+(a) = 0 + 2 = 2$. Similarly, we can find the degrees of the other vertices: b, c, and d of that graph.

Further, a vertex with zero in-degree is called a *source* vertex and a vertex with zero out-degree is called a *sink* vertex. For example, in the directed graph in Figure 9.2(c) a is a source vertex, since its in-degree is zero, whereas vertex c is the sink vertex (since its out-degree is zero). Usually, source and sink vertices are considered only in directed graphs.

A vertex of degree 0 (zero) is called *isolated* vertex. A vertex is *pendant* vertex if and only if its degree is 1. The vertex d in the graph presented in Figure 9.1 is an isolated vertex, as its degree is zero, whereas the vertex e of the graph in Figure 9.4 is a pendant vertex (because of its degree 1).

> **Note** We may consider vertices such as isolated, pendant, etc., for both undirected and directed graphs.

Degree Sequence of Graph

If $v_1, v_2, v_3, \ldots, v_n$ are the vertices of G, then the sequence $(d_1, d_2, d_3, \ldots, d_n)$, where $d_i = \deg_G(v_i)$ is the degree sequence of G. In general, the *degree sequence* is expressed as a monotonically increasing sequence. For example, the degree sequence of the graph in Figure 9.4 is (1, 2, 2, 4, 5). More clearly, $\deg_G(e) = 1$, $\deg_G(a) = 2$, $\deg_G(c) = 2$, $\deg_G(d) = 4$, $\deg_G(b) = 5$.

> **Note** The *degree* of a node u is the number of edges with endpoint u. The degree of a graph G is the *maximum* of the degrees of all nodes in G. If the number of edges $m = O(n)$ (where n is the number nodes in the graph), then the graph is said to be *sparse*. If m is larger than linear order of n, i.e., $m = O(n^2)$ (but as long as there are no multiple edges), then the graph is called *dense*.

Theorem 9.1 A simple graph with $n \geq 2$ vertices contains at least two vertices of same degree.

Proof Let G be a simple graph with $n \geq 2$ vertices. Here, the graph G is a simple graph. Hence, it has no loop and parallel edge. So, the degree of each vertex is $\leq n - 1$. Again, it is assumed that all the vertices of G are of distinct degrees.

Thus, the degrees, 0, 1, 2, 3, ..., $n - 1$, are possible for n vertices of G. Let u be the vertex with degree 0. Clearly, u is the isolated vertex. Let v be the vertex with degree $n - 1$, then v must have $n - 1$ adjacent vertices. In fact, it is possible if the vertex v is adjacent to each vertex of the graph G (even with u too), since G has no loop or parallel edges. But it is assumed that u is an isolated vertex, i.e., it is not adjacent with any vertex of G.

Hence, either u is not an isolated vertex or the degree of v is not $n - 1$. So, *contradiction* occurs on the *assumption* of different distinct degrees of vertices of G. Thus, the contradiction proves that a simple graph contains at least two vertices of same degree.

The above discussion can be clearly understood by taking some examples. Suppose, the simple graph G has two vertices v_1 and v_2. Since it is a *simple* graph, so it has no loop or parallel edges. First, consider that both the vertices are isolated. Hence, $\deg_G(v_1) = \deg_G(v_2) = 0$. So, both the vertices have same degree 0. Now, suppose that they are adjacent to each other (but no parallel edges); hence, $\deg_G(v_1) = \deg_G(v_2) = 1$. So, both the vertices have same degree 1.

Similarly, consider three vertices v_1, v_2, and v_3. The graph is a simple graph, so it has no loop or parallel edges. First, consider that all the vertices are isolated. Hence, $\deg_G(v_1) = \deg_G(v_2) = \deg_G(v_3) = 0$. So, at least two vertices (however, here all the vertices) have same degree 0. Now, suppose, one vertex is isolated and the remaining two are adjacent to each other. Then, degree sequence is (0, 1, 1) and it means at least two vertices out of three have same degree. Now, it can be verified by considering no isolated vertices out of three. Further, it is true for a simple graph with any number of vertices.

The theorem is true for both directed as well as undirected graphs.

Theorem 9.2 (The Handshaking theorem) If $G = (V, E)$ is a graph with e number of edges, then

$$\sum_{v \in V} \deg_G(v) = 2e$$

i.e., the sum of degrees of the vertices of G is always even. For directed graph,

$$\sum_{v \in V} \deg_G(v) = \sum_{v \in V} \deg^-(v) + \sum_{v \in V} \deg^+(v)$$

i.e., the sum of degrees of the vertices is the sum of the in-degrees and the out-degrees of the vertices.

Proof Let us first consider that the graph G is an *undirected* graph. The degree of a vertex of G is the number of edges incident with that vertex. Now, every edge is incident with exactly two vertices. Hence, each edge gets counted *twice*, once at each end. Thus, the sum of the degrees equals twice the number of edges.

If we consider the graph G to be *directed*, then in-degree and out-degree of each vertex of G are considered. However, the sum of the in-degree and out-degree of a vertex is the total degree of that vertex. Further, every edge is incident with exactly two vertices. So, here also, each edge gets counted *twice*: one as in-degree and the other as out-degree. Thus, the sum of the degrees (in-degrees and out-degrees) of all the vertices equals twice the number of edges.

Note The name of this theorem is *handshaking* because if several people shake hands, the total number of hands involved must be even (since for every handshaking, two hands are required). This theorem applies even if multiple edges and self-loops are present in a graph, and it is true for both connected and disconnected graphs. If *sum of degrees of the vertices of a graph is given, then the number of edges present in that graph can be computed. But the *reverse* is not possible.

Corollary 9.2 In a graph, total number of odd-degree vertices is even.

Proof Let $G = (V, E)$ be a graph, where K_1 and K_2 are the number of vertices with odd degree and even degree, respectively.

Now,

$$\sum_{v_i \in V} \deg_G(v_i) = \sum_{v_i \in K_1} \deg_G(v_i) + \sum_{v_i \in K_2} \deg_G(v_i)$$

$$\Rightarrow \quad 2e - \sum_{v_i \in K_1} \deg_G(v_i) = \sum_{v_i \in K_2} \deg_G(v_i)$$

[Since sum of the degree of vertices is twice the number of edges (e), and it is always *even*.] Further, sum of the even-degree vertices is even.

Clearly,

$$\sum_{v_i \in K_1} \deg_G(v_i)$$

i.e., the sum of the odd-degree vertices is also even, since $2e$ is even. Again,

$$\sum_{v_i \in K_1} \deg_G(v_i)$$

is even only if K_1 is even.
[For example, suppose three vertices contain odd degrees 1, 3, 5, respectively. Clearly, there sum may not be even, since number of vertices is 3 which is an odd number.]

Note The sum (S) of two numbers (say, n_1 and n_2) gives *even* if both of n_1 and n_2 are either odd or even, i.e., odd + odd = even, even + even = even.

Theorem 9.3 If $G = (V, E)$ be a directed graph with e number of edges, then

$$\sum_{v \in V} \deg^-(v) = \sum_{v \in V} \deg^+(v)$$

i.e., the sum of the out-degrees of the vertices of G equals the sum of the in-degrees of the vertices, which equals the number of edges in G.

Proof Any directed edge of G contributes 1 out-degree and 1 in-degree. Also, a self-loop contributes two (1 out-degree and 1 in-degree). Hence, the theorem is proved.

For example In the directed graph in Figure 9.2(a), the in-degree of a, $\deg_G^-(a)$ is 1 and its out-degree $\deg_G^+(a)$ is 1. Hence, the total degree of a is $\deg_G(a) = \deg_G^-(a) + \deg_G^+(a) = 1 + 1 = 2$, $\deg_G(b) = \deg_G^-(b) + \deg_{G^+}(b) = 1 + 1 = 2$, $\deg_G(c) = \deg_G^-(c) + \deg_G^+(c) = 1 + 1 = 2$.

Hence, $\deg_G^-(a) + \deg_G^-(b) + \deg_G^-(c) = 1 + 1 + 1 = 3 = e$ (number of edges) and $\deg_G^+(a) + \deg_G^+(b) + \deg_G^+(c) = 1 + 1 + 1 = 3 = e$ (number of edges).

EXAMPLE 9.1 Show that the degree of a vertex of a simple graph G on n vertices cannot exceed $n - 1$.

Solution Let v be a vertex of G. Since G is simple, no multiple edges or self-loops are allowed in G. Thus, v can be adjacent to at most all the remaining $n - 1$ vertices of G. Hence, v may have maximum degree $n - 1$ in G. If the degree of v becomes more than $(n - 1)$, then there must have self-loops or parallel edges in the graph, i.e., there must be self-loops or paralled edges in the graph, which is not allowed in *simple graph*. So, the degree of a vertex $v \in V(G)$ in a simple graph lies in the range: $0 \leq \deg_G(v) \leq n - 1$. In particular, it is 0 if the vertex is isolated.

The above inequality is true for both *directed* and *undirected* simple graphs.

EXAMPLE 9.2 Show that the maximum number of edges in a simple undirected graph with n vertices is $n(n - 1)/2$.

Solution By the handshaking theorem, we know

$$\sum_{v \in V} \deg_G(v) = 2e$$

where e is the number of edges with n vertices in the graph G. This implies

$$d(v_1) + d(v_2) + d(v_3) + \cdots + d(v_n) = 2e \qquad (9.1)$$

Since maximum degree of each vertex in a simple graph can be $(n - 1)$. Therefore, Eq. (9.1) can be written as follows:

$$(n - 1) + (n - 1) + \cdots + \text{up to } n \text{ terms (considering maximum degree for each vertex)}$$
$$= n(n - 1) = 2e$$

Hence, e (maximum number of edges in a simple graph with n vertices) $= n(n - 1)/2$.

[Note that maximum number of edges in a *simple directed graph* G is $2n(n - 1)/2 = n(n - 1)$, since in a simple directed graph, edges with opposite direction between any pair of vertices are allowed.]

EXAMPLE 9.3 For a simple graph with n vertices, what is the minimum number of edges required to ensure that the graph is connected?

Solution Let $S \in V$ be a set of vertices for which each vertex in S has degree 0. If S has just one vertex (the minimum case of a disconnected graph), then $(n - 1)$ edges are possible between S and $(V - S)$.

Therefore, the maximum possible number of edges in a disconnected graph is

$$n(n - 1)/2 - (n - 1) = (n - 1)(n - 2)/2$$

Clearly, the minimum number of edges in a connected graph is $1 + (n - 1)(n - 2)/2 = (n^2 - 3n + 4)/2$. Some useful tips to check the existence of a graph when its degree sequence is given.

Tips

To solve such problem on any graph (directed or undirected), we can primarily concentrate on the following points:

1. If the sum of the degrees of the vertices of the graph is not even, then graph corresponding to the given degree sequence cannot be drawn (application of *handshaking theorem*).
2. If the total number of odd degree vertices (counted from the given degree sequence) is odd, then graph corresponding to the given degree sequence cannot be drawn.

In fact, the second point depends on the first point because if the sum of the degree of vertices is even, then the number of odd-degree vertices must be even, otherwise, it is odd. It is mathematically shown in Corollary 9.2.

Hence, for the existence of any graph G, the number of odd-degree vertices must be even, and this point can be applied only for confirmatory checking, i.e., it is not compulsory to consider.

Now, if both the above-mentioned conditions are *false* (i.e., when the sum of the degrees is even and the number of odd-degree vertices is also even), then it is certainly possible to draw one graph, but it may not be possible to draw a simple graph following the given degree sequence. For checking the existence of a simple graph, we must concentrate on its properties. Some examples on degree sequence are given below.

EXAMPLE 9.4 Is there a simple graph corresponding to the following degree sequences?
(a) (1, 1, 2, 3)
(b) (2, 2, 4, 4)

Solution

(a) It is known that the total number of odd-degree vertices in a graph is even. Hence, it is true for a simple graph too. In the graph with degree sequence, (1, 1, 2, 3), the number of odd-degree vertices is 3, and these vertices are of degrees 1, 1, 3, respectively. Hence, no graph corresponding to this degree sequence can be drawn. [*Second method*: The sum of degrees = 1 + 1 + 2 + 3 = 7, which is odd. By handshaking theorem, the sum of degrees of any simple graph must be even. Hence, no graph exists for this case.]

(b) In the given graph, the sum of the degree of the vertices is 12 which is even. Also, the number of the odd-degree vertices is 0. It is even too. So, a graph is possible to draw, using the given degree sequence. Now, let us check if any simple graph is possible to draw or not. Clearly, the number of vertices is 4 and the maximum degree of two vertices is 4. However, the degree of any vertex in a simple graph G on n vertices cannot exceed $n-1$. Hence, no simple graph corresponding to the given degree sequence can be drawn.

EXAMPLE 9.5 Does there exist a simple graph with seven vertices having degrees (1, 3, 3, 4, 5, 6, 6)?

Solution Here, the sum of the degrees of the vertices is $1 + 3 + 3 + 4 + 5 + 6 + 6 = 28$ and it is an even number. Also, the number of odd-degree vertices is even. So, certainly, the graph corresponding to the given degree sequence exists.

Now, let us check whether any simple graph exists or not. Assume that it exists. Here, two vertices out of seven have degree 6. So, each of these two vertices is adjacent to the rest six vertices of the graph. Accordingly, the degree of each vertex should be at least 2, i.e., it may not be 1. But in the degree sequence, no vertex with degree 2 is provided. Moreover, a vertex with degree 1 is given.

Therefore, we arrive at a *contradiction* in our *assumption*. Thus, no simple graph, following the given degree sequence, can be drawn.

EXAMPLE 9.6 For the graph G as shown in Figure 9.8, write the degree sequence of G.

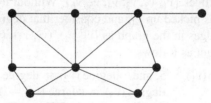

FIGURE 9.8

Hence, find the number of odd-degree vertices and the number of edges in the graph G.

Solution The degree sequence is given as {7, 4, 3, 3, 3, 3, 3, 3, 1}. Hence, the number of the odd-degree vertices is 8, which is even as per the corollary of handshaking theorem.

Now, the sum of degrees of all vertices is $2e$, where e is the number of edges. So, we get

$$7 + 4 + 3 + 3 + 3 + 3 + 3 + 3 + 1 = 2e$$
$$\Rightarrow \quad 2e = 30$$
$$\Rightarrow \quad e = 15$$

Hence, the number of edges in the given graph is 15 and can be verified by counting.

EXAMPLE 9.7 For each of the following degree sequences, determine if there exists a graph whose degree sequence is given. If possible, draw the graph or explain why such a graph does not exist.
  (i) (5, 4, 3, 2, 1, 1)
  (ii) (3, 3, 3, 1)

Solution

(i) The given sequence is (5, 4, 3, 2, 1, 1). Thus, the sum of degrees = 5 + 4 + 3 + 2 + 1 + 1 = 16 = even number. Also, the number of odd-degree vertices is 4 (even), and the degrees of these are, respectively, 5, 3, 1, 1. Hence, it is certainly possible to draw graph, considering the given sequence. One possible graph, following the given sequence is shown in Figure 9.9.

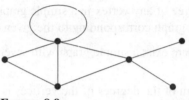

FIGURE 9.9

However, this is not a simple graph because a self-loop is present. Now, let us try to check whether it is possible to draw a simple graph by this degree sequence or not.

Sum of degrees = 5 + 4 + 3 + 2 + 1 + 1 = 16 = even number

Therefore, e = number of edges can be computed as 16/2 = 8; n (number of vertices in graph) = 6.

We know that the maximum degree of any vertex of a simple graph = $n - 1$ = 5. Further, the maximum number of edges for the graph to be connected = $n(n - 1)/2$ = 15. So, simple graph may or may not exist.

Let us take 6 vertices $\{v_1, v_2, v_3, v_4, v_5, v_6\}$. Without loss of generality, we can pick up any vertex, say, it is picked up v_1, and consider that degree (v_1) = 5. So, this means that there are these 5 edges in the graph as $(v_1, v_2), (v_1, v_3), (v_1, v_4), (v_1, v_5)$, and (v_1, v_6). So, after this step, we get as follows:

$$\text{degree}(v_1) = 5, \text{ and } \text{degree}(v_2) = \text{degree}(v_3) = \text{degree}(v_4)$$
$$= \text{degree}(v_5) = \text{degree}(v_6) = 1$$

Now, it needs to introduce more edges so that the degree of one vertex becomes 4. It is already shown that degree (v_1) = 5, and the degree of each vertex in the set $\{v_2, v_3, v_4, v_5, v_6\}$ is 1. Again, without loss of generality, we can pick up any arbitrary vertex, say, v_2, from the above set, and try to make its degree as 4. Current value of degree (v_2) = 1, and it is already connected to v_1. Let us introduce 3 more edges, taking 3 vertices out of the remaining set $\{v_3, v_4, v_5, v_6\}$. Let us introduce the edges $(v_2, v_3), (v_2, v_4), (v_2, v_5)$. Now, the degrees are given by degree (v_1) = 5, degree (v_2) = 4, degree (v_3) = 2, degree (v_4) = 2, degree (v_5) = 2, degree (v_6) = 1.

Everything is fine except the fact that, at this stage, we cannot have 2 vertices with degree 1, as defined by the degree sequence. Hence, we can arrive at the above-mentioned conclusion by doing this graph construction in any sequence. So, simple graph does not exist for the degree sequence (5, 4, 3, 2, 1, 1).

(ii) The given degree sequence is (3, 3, 3, 1). Clearly, the sum of the degrees of the vertices = 3 + 3 + 3 + 1 = 10 = even number. So, the number of edges can be computed as 10/2 = 5, i.e., $e = 5$. Also, the number of vertices in the graph is 4, i.e., $n = 4$. We may now go for checking the number of odd-degree vertices. It is 4 (an even number), and these are 3, 3, 3, 1. Hence, the graph corresponding to the degree sequence (3, 3, 3, 1) can be certainly drawn.

EXAMPLE 9.8 For each of the following degree sequences, determine if there exists a graph whose degree sequence is given. If possible draw the graph or explain why such a graph does not exist.

(i) (1, 1, 1, 1, 1)
(ii) (1, 1, 1, 1, 1, 1)

Solution

(i) The given degree sequence is (1, 1, 1, 1, 1).
   Sum of the degrees of the vertices = 1 + 1 + 1 + 1 + 1 = 5 = odd number
   Hence, it is not possible to draw any graph corresponding to the degree sequence (1, 1, 1, 1, 1).

(ii) The given degree sequence is (1, 1, 1, 1, 1, 1).
   Sum of degrees = 1 + 1 + 1 + 1 + 1 + 1 = 6 = even number

Therefore, e = number of edges can be computed as 6/2 = 3. Here, n (number of vertices in the graph) = 6. Also, number of odd-degree vertices is 6 and it is an even number. Hence, the graph corresponding to the given sequence (1, 1, 1, 1, 1, 1) can be drawn.

It is obvious to see that a disconnected graph exists as shown in Figure 9.10, corresponding to the given sequence.

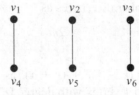

FIGURE 9.10

EXAMPLE 9.9 Let G be a simple graph with 12 edges. If G has 6 vertices of degree 3 and the rest of the vertices have degree less than 3, then find the (a) minimum number of vertices and (b) maximum number of vertices.

Solution

(a) Suppose the total number of vertices in G is p. However, it is given that 6 vertices have degree 3. Hence, the total degree computed from these 6 vertices is $3 \times 6 = 18$. Clearly, the rest $(p - 6)$ vertices have degree less than 3, i.e., their degree lies inclusively between 0 and 2.

Here, as we need to find the *minimum* number of vertices, so the rest $(p - 6)$ vertices must have *maximum* degree [here, it is 2 taken as maximum value of the range (0, 2)].
Therefore, applying the handshaking theorem, we get
$$18 + 2(p - 6) = 2e = 24 \Rightarrow p = 9.$$

(b) In such case, we need to calculate the *maximum* number of vertices. In particular, number of vertices becomes more (satisfying the constraint of fixed number of edges) if degree of vertices reduces. Thus, we can write here:

$$18 + (p - 6) = 2e = 24 \text{ (assuming that the graph is a simple connected graph)}$$
$$\Rightarrow p = 12$$

Note For a given number of edges in a graph G, the number of vertices and the degree of vertices are *inversely* proportional to each other. Some examples are discussed below.

Theorem 9.4 The largest possible number of vertices in a graph with m edges, all vertices having degree at least r is

$$\left\lfloor \frac{2m}{r} \right\rfloor$$

if $2m$ is not a multiple of r, else

$$\frac{2m}{r}$$

if $2m$ is an exact multiple of r.

Proof Let $2m = kr + x$, where k and x are positive integers and $0 \leq x < r$. Consider the case $x > 0$. Clearly,

$$\left\lfloor \frac{2m}{r} \right\rfloor = k \text{ and } \left\lceil \frac{2m}{r} \right\rceil = k + 1$$

[We know that $\lceil x \rceil$ (i.e., *ceil(x)*) returns the *largest* integer which is *less* than or *equal* to x. For example, $\lceil 1.2 \rceil = 2$. Also, $\lfloor x \rfloor$ (i.e., *floor(x)*) returns the *smallest* integer which is *greater* than or *equal* to x. For example, $\lfloor 1.2 \rfloor = 1$. Further, $\lceil -x \rceil = -\lfloor x \rfloor$ and $\lfloor -x \rfloor = -\lceil x \rceil$]

If we consider the number of vertices as

$$\left\lfloor \frac{2m}{r} \right\rfloor = k$$

then we can write $2m = (k - 1)r + (r + x)$, as $2m = kr + x$. This gives $(k - 1)$ vertices with degree exactly r, and 1 vertex with degree $(r + x) \geq r$, as $0 \leq x < r$.

On the other hand, if we take the number of vertices as

$$\left\lceil \frac{2m}{r} \right\rceil = (k + 1)$$

then as $2m = kr + x$, we get k vertices of degree exactly r, and 1 vertex with degree x, where $0 \leq x < r$.

Thus, it gives 1 vertex with degree less than r, leading to violation. Also, if $x = 0$, i.e., if $2m$ is an exact multiple of r, then it is trivially seen that the number of vertices $= 2m/r$. This proves the theorem.

EXAMPLE 9.10 Find the largest possible number of vertices in a graph with 31 edges, all vertices having degree at least 3.

Solution

Here, e = number of edges = 31. Obviously, S = sum of degrees of the vertices = $2e$ = 62.

Now, *each vertex has degree* ≥ 3.

Case I When
$$\left\lfloor \frac{s}{3} \right\rfloor = \left\lfloor \frac{62}{3} \right\rfloor = 20$$

is considered to find the number of vertices. This gives us 19 vertices of degree exactly 3 each, plus 1 vertex of degree 5; or 18 vertices of degree exactly 3 each, and 2 vertices of degree 4 each (both of the above cases satisfy the rule degree ≥ 3).

Case II When
$$\left\lceil \frac{s}{3} \right\rceil = \left\lceil \frac{62}{3} \right\rceil = 21$$

is considered to find the number of vertices. This gives us 20 vertices of degree 3 each, plus 1 vertex of degree 2, and this *violates* the condition that each vertex must have degree ≥ 3. Therefore, the maximum possible number of vertices in the graph = 20.

Note If the number of vertices in a simple graph is n and m ($m < n$) number of vertices have degree ($n - 1$), then the degree of each vertex should be *at least m* but at most ($n - 1$). When $m = n$, i.e., each vertex of the simple graph has degree ($n - 1$), then that simple graph is said to be a complete graph. (Proof has been left as an exercise for the readers.)

9.3 TYPES OF GRAPHS

Some important types of graphs are introduced here. These are often used in many applications.

9.3.1 Null Graph

A graph which contains only isolated nodes is called a null graph, i.e., the set of edges in a null graph is empty. Such graph on n vertices is denoted by N_n. For example, null graph (N_4) with 4 vertices is shown in Figure 9.11.

FIGURE 9.11 A null graph with 4 vertices

Note Some authors consider a graph with zero vertices and zero edges as a Null graph.

9.3.2 Complete Graph

A simple graph G is said to be complete if every vertex of G is connected with every other vertex of G, i.e., every pair of distinct vertices contains exactly one edge. It is usually denoted by K_n, where n is the total number of vertices in G. Some complete graphs, K_1, K_2, K_3, K_4, K_5, are shown in Figure 9.12.

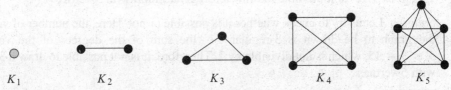

FIGURE 9.12 Some undirected complete graphs

Note A complete graph G is a simple graph and it may be *directed* as well as *undirected*. Whatsoever, any complete graph K_n with n vertices has exactly $n(n-1)/2$ edges. A *directed* graph K_3 is shown in Figure 9.13. It must be noted that the directions of the edges among a, b, and c are not necessary to be exactly like the directed graph K_3 as shown below. But it is true that the number of edges will be exactly 3 for K_3 (since exactly $n(n-1)/2$ edges for K_n).

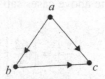

FIGURE 9.13 A directed K_3 graph

9.3.3 Regular Graph

A graph in which all the vertices are of same degree is called a *regular graph*. If the degree of each vertex is r, then the graph is called a regular graph of degree r, and it is denoted by K_r.

It should be noted that a regular graph may be *directed* or *undirected*. When it is directed, then the degree of each vertex is computed as the sum of its *in-degree* and *out-degree*. Further, it is not necessary that a regular graph needs to be a simple graph, i.e., self-loop and parallel edges are allowed. But the only important point is that the degree of each vertex is same.

FIGURE 9.14 A 2-regular graph with 4 vertices

A complete graph K_n is a regular graph of degree $n-1$ or it is called $(n-1)$-regular graph. Obviously, if a graph is *null* graph, then it is 0-regular (as degree of each vertex is 0).

A 2-regular graph with 4 vertices is given in Figure 9.14.

Note If a graph G with n vertices is regular of the degree of r, then G has $r \times n/2$ edges. Since the graph has n vertices and it is r-regular, so the sum of the degree of the vertices is $n \times r$. Again, it is known that the sum of the degrees of the vertices of a graph equals to *twice* the number of edges. Hence, the number of the edges of the regular graph with n vertices is $r \times n/2$.

EXAMPLE 9.11 Find the number of edges of a 4-regular graph with 6 vertices.

Solution Number of edges $(e) = r \times n/2 = 4 \times 6/2 = 12$.

EXAMPLE 9.12 Is it possible to draw a 3-regular graph with 5 vertices?

Solution Let us try to check whether it is possible or not. Here, the number of vertices is 5 and the graph to be drawn is 3-regular. So, the sum of the degrees of the vertices will be $5 \times 3 = 15$, which is not divisible by 2. Therefore, it is not possible to draw a 3-regular graph with 5 vertices.

Note A graph with n vertices is r-regular if either r or n or both are even.

Cycles

The cycle C_n, $n \geq 3$, consists of n vertices and n edges so that the second endpoint of the last edge coincides with the starting vertex. One example of cycle is shown in Figure 9.15. Here, n is 4.

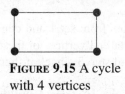

FIGURE 9.15 A cycle with 4 vertices

9.3.4 Bipartite Graph

A graph $G = (V, E)$ is a bipartite graph if the vertex set V can be partitioned into two disjoint subsets, say, V_1 and V_2 such that every edge in E connects a vertex in V_1 and a vertex V_2 (but no edge in G connects either of the two vertices in V_1 or two vertices in V_2). Also, (V_1, V_2) is called a *bipartition* of G. Some examples of bipartite graph are shown below.

$V_1 = \{v_2, v_4, v_3\}$ $V_1 = \{v_1, v_2\}$ $V_1 = \{v_1, v_4, v_6, v_7\}$
$V_2 = \{v_1, v_5\}$ $\{V_2 = v_3, v_4\}$ $V_2 = \{v_2, v_3, v_5, v_8\}$

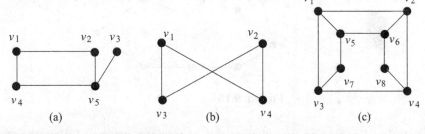

FIGURE 9.16 Some bipartite graphs

The graph in Figure 9.16(a) is a bipartite graph. As an example, if we select the distinct vertex sets as $\{v_2, v_4, v_3\}$ and $\{v_1, v_5\}$, then there do not exist direct edges connecting members within a set. Same situation holds true for whatever vertices are selected as the two distinct sets of vertices. Again, the two distinct sets of vertices of the graph as shown in Figure 9.16(b) are $\{v_1, v_2\}$ and $\{v_3, v_4\}$. Further, $\{v_1, v_4, v_6, v_7\}$ and $\{v_2, v_3, v_5, v_8\}$ are the two distinct sets of vertices of the graph in Figure 9.16(c).

EXAMPLE 9.13 Show that the graph C_6 as shown in Figure 9.17 is bipartite.

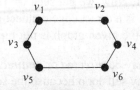

FIGURE 9.17

Solution In this graph, the two distinct sets of vertices are $\{v_1, v_5, v_4\}$ and $\{v_2, v_3, v_6\}$. Hence, C_6 is bipartite.

A simple procedure to check if a graph G is bipartite or not: Some basic outlines are pointed below to check whether a graph G is bipartite.

Step 1 Arbitrarily select a vertex from G and include it into set 1.
Step 2 Consider the edges directly connected to that vertex and put the other end vertices of these edges into set 2.
Step 3 Now, pick up one vertex from set 1 and consider the edges directly connected to that vertex, and put the other end vertices of these edges into set 1.
Step 4 At each step, step 2 and step 3, check if there is any edge among the vertices of set 1 or set 2.
   If so, construction of sets is stopped and the given graph is not bipartite graph, and then return.
   Else continue step 2 and step 3 alternately until all the vertices are included in the union of set 1 and set 2.
Step 5 If two computed sets following the above steps are distinct, then it is bipartite.

EXAMPLE 9.14 Is G a bipartite graph? If it is bipartite, find the two distinct sets V' and V'' in partition of vertices of G.

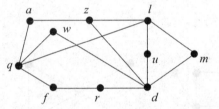

FIGURE 9.18

Solution For finding two distinct sets of vertices, we may follow the steps mentioned in Figure 9.18 to check whether a graph G is bipartite or not.

Now, considering the procedure, put vertex d in set 1. So, the vertices joined to d through direct edges will be in set 2. Accordingly, we get set $1 = \{d\}$ and set $2 = \{r, w, z, u\}$. Vertices in set 2 are not connected among themselves by direct edges.

Next, consider the vertices which are connected to z through direct edges. These vertices must be put in set 1 because $z \in$ set 2.

Clearly, at this stage, we have set $1 = \{a, l, d\}$ and set $2 = \{r, w, z, u\}$. Again, vertices in set 1 are not connected among themselves by direct edges.

Consider the vertices which are connected to a through direct edges. These vertices must be put in set 2 because $a \in$ set 1. Now, we get set $1 = \{a, l, d\}$ and set $2 = \{r, w, z, u, q\}$.

In set 2, we find that there is a direct edge connecting w and q, i.e., it violates the condition for a bipartite graph. Hence, the given graph is not bipartite.

Note A bipartite graph may be *directed* or *undirected* but it must be connected. The graph must not possess any self-loop because the self-loop connects the same vertex and it is not allowed here. However, a bipartite graph may have parallel edges, since the presence of parallel edge does not violate the definition of bipartite graph.

9.3.5 Complete Bipartite Graph

A bipartite graph G is a complete bipartite graph if there is an edge between every pair of vertices taken from two disjoint sets of vertices (one vertex from one set, say, V_1 and the other from another set, say, V_2). Complete bipartite graph G is denoted by $K_{m,n}$, where m and n are the number of vertices in two distinct subsets: V_1 and V_2. Some examples of complete bipartite graphs are shown in Figure 9.19.

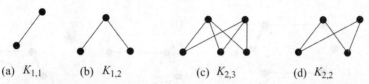

(a) $K_{1,1}$ (b) $K_{1,2}$ (c) $K_{2,3}$ (d) $K_{2,2}$

FIGURE 9.19 Some complete bipartite graphs

EXAMPLE 9.15 How many edges do the complete bipartite graph, $K_{m,n}$, have?

Solution The vertex set of $K_{m,n}$ consists of two disjoint sets, say, A and B such that A contains m vertices and B contains n vertices; each vertex in A is adjacent to each vertex in B; no two vertices either in A or in B are adjacent. Hence, the degree of each vertex in A is n, and the degree of each vertex in B is m. Therefore, the sum of the degrees is $2 \times m \times n$, and so there are $m \times n$ edges (as per the *handshaking* theorem).

Note A complete bipartite graph $K_{m,n}$ has $m + n$ vertices and $m \times n$ edges. Further, $K_{m,n}$ is regular if $m = n$. From Figure 9.19, it is obvious that the complete bipartite graphs: $K_{1,1}$ and $K_{2,2}$ are regular.

EXAMPLE 9.16 Prove that a graph which contains a triangle cannot be bipartite.

Solution As per the basic idea of bipartite graph, the vertices should be divided into two distinct set of vertices. Certainly, the number of vertices of the given graph is 3, as it is triangle. So, it is not possible to divide the vertices into two disjoint set of vertices (since each edge is joined by the rest two edges). Hence, this graph may not be a bipartite graph.

9.4 SUBGRAPH AND ISOMORPHIC GRAPH

9.4.1 Subgraph

In some problems, instead of considering the entire graph, only part of the graph or the smaller graph needs to be considered. Such graph is called a *subgraph* of the original graph. *For instance*, we may have interest about the part of a large computer network that involves two specific computer centres.

If G and H are two graphs with vertex sets $V(G)$ and $V(H)$ and edge sets $E(G)$ and $E(H)$, respectively, such that $V(H) \subseteq (G)$ and $E(H) \subseteq E(G)$, then we say that H is a subgraph of G (or G is a super-graph of H). In other words, if H is a subgraph of G, then all the vertices and the edges of H are in G and each edge of H has the same endpoints as in G.

Now, if $V(H) = V(G)$ and $E(H) \subseteq E(G)$, then we say that H is a *spanning subgraph* of G. Hence, a spanning subgraph of a graph G contains all the vertices of G but not necessarily all the edges of G. Concisely, if H is a subgraph of G, then

(a) All the vertices of H are in G.
(b) All the edges of H are in G.
(c) Each edge of H has the same endpoints in H as in G.

For example One original graph G is shown in Figure 9.20(a) and its one subgraph is shown in Figure 9.20(b), but the graph as shown in Figure 9.20(c) is not a subgraph of G, as no edge between v_3 and v_2 is present in the original graph G.

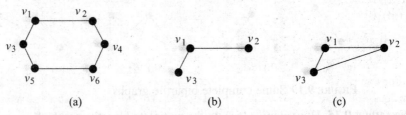

FIGURE 9.20

Note Suppose a graph G has n number of vertices (i.e., $|V| = n$) and m number of edges (i.e., $|E| = n$). Then, we find the total non-empty *subsets* of V as $2^n - 1$ and the total number of subsets of E as 2^m. Thus, the total number of non-empty subgraphs that can be obtained from the original graph G is $(2^n - 1) \times 2^m$. In particular, the number of spanning subgraphs which can be constructed from the graph G is 2^m because all the vertices of G must be included in a spanning subgraph. Also, in a spanning subgraph of G, the number of vertices will be n but number of edges may vary from 0 to m. Accordingly, a spanning subgraph may also be a disconnected graph, consisting strictly of n vertices. Let us try to understand how the number of spanning subgraphs is 2^m.

Number of spanning subgraphs with 0 (zero) edge and m vertices is $^mC_0$.
Number of spanning subgraphs with 1 edge and m vertices is $^mC_1$.
Number of spanning subgraphs with 2 edges and m vertices is $^mC_2$.
... ...
... ...

Proceeding in this way, the total number of spanning subgraphs with m vertices from the original graph G consisting of m vertices and n edges will be found as

$$^mC_0 + {}^mC_1 + {}^mC_2 + \cdots + {}^mC_m = 2^m \text{ (by binomial theorem)}$$

 EXAMPLE 9.17 For the graph G, as shown in Figure 9.21, draw the subgraphs.

(a) $G - e$ (here, e is one edge)
(b) $G - a$ (here, a is one vertex)

FIGURE 9.21

Solution The subgraphs are shown in Figure 9.22.

(a) Deleting edge e (b) Deleting vertex a

FIGURE 9.22

9.4.2 Isomorphic Graph

Two graphs $G = (V, E)$ and $G' = (V', E')$ are said to be isomorphic, if there exists a bijection $f: V \to V'$ such that $(u, v) \in E$ if and only if $[f(u), f(v)] \in E'$. In other words, we can relabel vertices of G to be vertices of G', maintaining the corresponding edges in G and G'.

The necessary conditions for two graphs G and G' to be isomorphic are as follows:
  (i) Both G and G' have same number of vertices.
 (ii) Both G and G' have same number of edges.
(iii) Both G and G' have same degree sequences.

If any of the above three differs in two graphs, then they cannot be isomorphic. However, these conditions are by no means sufficient.

For example Two graphs as shown in Figure 9.23 have same number of vertices, same number of edges, and same degree sequences, yet they are not isomorphic. Here, we may appropriately choose a vertex y in Figure 9.23(b) corresponding to the vertex x in Figure 9.23(a). Obviously, x has adjacent vertices u and v, respectively but each of u and v is not adjacent to any vertex except x. On the other hand, y in Figure 9.23(b) has adjacent vertices and each of which, in turns, is adjacent to vertex other than y. Thus, these two graphs are not isomorphic as per the properties of isomorphism.

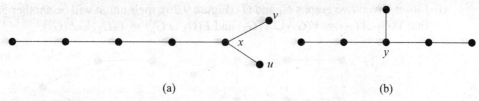

FIGURE 9.23

 EXAMPLE 9.18 Show the two graphs as shown in Figure 9.24 are isomorphic.

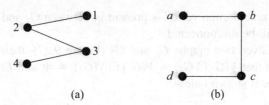

FIGURE 9.24

Solution The above two graphs are isomorphic, since all the necessary conditions to be isomorphic for two graphs are satisfied and there is 1-1 mapping of G into G', i.e., $f(1) = f(a)$, $f(2) = f(b)$, $f(3) = f(c)$, and $f(4) = f(d)$. Further, $\{1, 2\} \in E$ and $\{f(1), f(2)\} = \{a, b\} \in E'$, $\{2, 3\} \in E$ and $\{f(2), f(3)\} = \{b, d\} \in E'$, $\{3, 4\} \in E$ and $\{f(3), f(4)\} = \{c, d\} \in E'$. The other edges are not present in both the graphs. Hence, f preserves adjacency as well as non-adjacency of the vertices.

EXAMPLE 9.19 Are the graphs G' and G'' isomorphic? Justify.

Solution No, the two graphs, as shown in Figure 9.25, are not isomorphic. Observe the vertices q and r of G; degree $(q) = 2$ and degree $(r) = 4$. All other vertices of G have degree $= 3$. So, the one and only corresponding mapping of these two vertices in G' is defined by $q \to C$ and $r \to F$. Also, note that there is an edge (q, r) in G, but there is no edge (C, F) in G'.

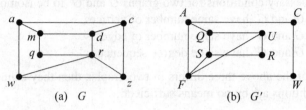

FIGURE 9.25

So, there is no possible 1-1 mapping of G into G'. Hence, the given graphs are not isomorphic.

9.5 OPERATIONS ON GRAPHS

In this section, some basic operations on graph are discussed.

(i) *Union* Given two graphs G_1 and G_2 (Figure 9.26), their union will be another graph G such that $V(G_1 \cup G_2) = V(G_1) \cup V(G_2)$ and $E(G_1 \cup G_2) = E(G_1) \cup E(G_2)$.

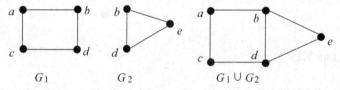

FIGURE 9.26 Union of two graphs

Note If no common vertex is present in between G_1 and G_2, then the resultant graph will be disconnected.

(ii) *Intersection* Given two graphs G_1 and G_2 (Figure 9.27), their intersection will be another graph G such that $V(G_1 \cap G_2) = V(G_1) \cap V(G_2) \neq \Phi$ and $E(G_1 \cap G_2) = E(G_1) \cap E(G_2)$. The operation is shown below.

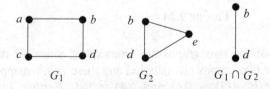

FIGURE 9.27 Intersection on two graphs

(iii) *Sum of two graphs* If the graphs $G_1 = (V_1, E_1)$ and $G_2 = (V_2, E_1)$ are such that $V_1 \cap V_2 = \Phi$, then their sum $G_1 + G_2$ is defined as the graph G in which vertex set is $V_1 + V_2$ and the edge set consists of the edges in E_1 and E_2, and the edges joining each vertex of V_1 with each vertex of V_2. One example of sum operation is shown in Figure 9.28.

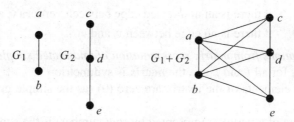

FIGURE 9.28 Sum of two graphs

(iv) *Complement* The complement G' of G is defined as a simple graph (parallel edge and self-loop are ignored) with the same vertex set as G, and where two vertices u and v are adjacent only when they are not adjacent in G. One example of complement graph G' of a graph G is shown in Figure 9.29.

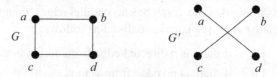

FIGURE 9.29 Complement of graph

It is observed here that G' of G is disconnected.

(v) *Product of graphs* Product of two graphs G_1 and G_2 is defined as $G = (V_1 \cup V_2, V_1 \times V_2)$, where $V_1 \cup V_2$ is the union of the vertex sets V_1 of G_1 and V_2 of G_2, and $V_1 \times V_2$ is the *cross product* to compute the edge set of the resultant graph G.

9.6 REPRESENTATION OF GRAPH

Diagrammatic (graphical) representation of a graph is very convenient for visual study, but it is practically feasible only when the number of vertices and edges of the graph is reasonably small. So, we need some other reasonable ways to represent graphs with large number of vertices and edges. These representations are also expected to be useful in computer programming. Some representations for undirected as well as directed graphs are discussed below.

9.6.1 Matrix (Adjacency Matrix) Representation

The adjacency matrix is commonly used to represent graphs for computer processing. In fact, it is more suitable for dense graphs (in which each vertex of a graph has almost all the possible edges). In such representation, an $n \times n$ *Boolean* (1, 0) matrix is used where a 1 at position (u, v) indicates that there exists an edge from vertex u to v, and a 0 at position (u, v) indicates that there is no edge reachable directly from u to v. Obviously, if the graph is undirected, then its corresponding adjacency matrix will be *symmetric*.

(i) *Matrix representation of undirected graph* If an undirected graph G consists of n vertices (assuming that the graph has no parallel edge), then the adjacency matrix of G is an $n \times n$ matrix $A = [a_{i,j}]$ and defined as follows:

$$a_{i,j} = \begin{cases} 1, & \text{if there is an undirected edge between vertices } v_i \text{ and } v_j \\ 0, & \text{if there is no edge between } v_i \text{ and } v_j \end{cases}$$

Some observations from matrix representation of undirected simple graph:
(a) $a_{i,j} = a_{j,i}$, for all i and j, i.e., the matrix is symmetric.
(b) Diagonal elements of the matrix are zero (0) (as the simple graph possesses no self-loop).
(c) The degree of a vertex (represented by row number) is the sum of the 1s in that row.
(d) Let G be a graph with n vertices: $v_1, v_2, v_3, \ldots, v_n$ and A be the adjacency matrix of G. Let B be the matrix computed as follows:

$$B = A + A^2 + A^3 + \cdots + A^n \ (n > 1)$$

Now, B is connected if and only if B has no zero entry.

(ii) *Matrix representation of directed graph* If a directed graph (digraph) G consists of n vertices (assuming that the graph has no parallel edge), then the adjacency matrix of G is an $n \times n$ matrix $A = [a_{i,j}]$, and is defined as follows:

$$a_{i,j} = \begin{cases} 1, & \text{if there is a directed edge from vertex } v_i \text{ to vertex } v_j \\ 0, & \text{if there is no edge from } v_i \text{ to } v_j \end{cases}$$

Some observations from the matrix representation of directed simple graph

(a) $a_{i,j} \neq a_{j,i}$, for all i and j, i.e., if there is an edge from vertex v_i to v_j, it means that there will not necessarily exist an edge from v_j to v_i. Hence, the represented matrix A is not necessary to be symmetric.
(b) Diagonal elements of the matrix A are zero (0) (as the simple graph has no self-loop).
(c) The sum of 1 in any column j of A is equal to the number of edges directed towards vertex v_j, i.e., it is the *in-degree* of vertex v_j.
(d) The sum of 1 in any row i of A is equal to the number of edges directed away from vertex v_i, i.e., it is the *out-degree* of vertex v_i.

Note To check whether two graphs are isomorphic or not, we may take adjacency matrix representation of the given two graphs. If both the matrices (say A and B) have the same size, then match a row of the first matrix A with any row of the second matrix B. If matching is found, then cross out those rows from A and B, and go for matching the next row of A, and so on until all the rows of A and B are crossed out. If all the rows are crossed out, then we conclude that both the graphs are isomorphic, otherwise not.

9.6.2 Linked List (Adjacency List) Representation

In such representation, each node u in the graph (assuming that the graph has no parallel edges) contains a linear linked list, consisting of the vertices directly reachable from u. Thus, if n vertices are present in a graph G, then the number of lists will also be n, i.e., the n rows of the adjacency matrix are represented by n lists. In particular, linked representation is a particularly good idea for *sparse* graphs.

The adjacency matrix and the linked list representations of the following graphs are given in Figure 9.30.

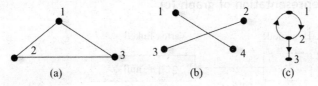

FIGURE 9.30

First, we present the *adjacency matrix* representations of the graphs in Figures 9.30(a), (b), and (c), respectively.

$$\begin{array}{c} \begin{array}{ccc}1 & 2 & 3\end{array}\\ \begin{array}{c}1\\2\\3\end{array}\left[\begin{array}{ccc}0 & 1 & 1\\ 1 & 0 & 1\\ 1 & 1 & 0\end{array}\right]\end{array} \qquad \begin{array}{c} \begin{array}{cccc}1 & 2 & 3 & 4\end{array}\\ \begin{array}{c}1\\2\\3\\4\end{array}\left[\begin{array}{cccc}0 & 0 & 0 & 1\\ 0 & 0 & 1 & 0\\ 0 & 1 & 0 & 0\\ 1 & 0 & 0 & 0\end{array}\right]\end{array} \qquad \begin{array}{c} \begin{array}{ccc}1 & 2 & 3\end{array}\\ \begin{array}{c}1\\2\\3\end{array}\left[\begin{array}{ccc}0 & 1 & 0\\ 1 & 0 & 1\\ 0 & 0 & 0\end{array}\right]\end{array}$$

(a) Matrix representation of (b) Matrix representation of (c) Matrix representation of

Now, the list representations of the above graphs (shown in Figure 9.30) are cited below.

Linked representation of graph (a)

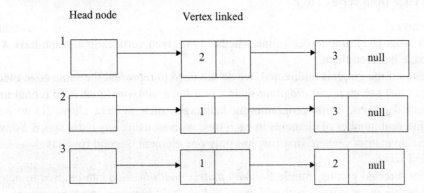

Linked representation of graph (b)

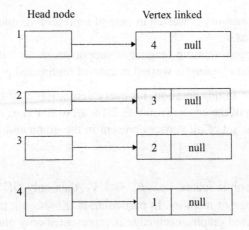

Linked representation of graph (c)

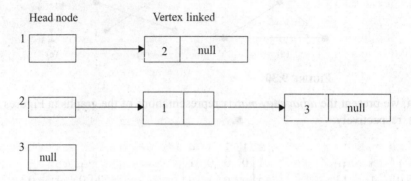

9.6.3 Advantages and Disadvantages of Matrix and Linked List Representations

(a) *Matrix representation* In this scheme, we use a matrix M of order $V \times V$ to represent the graph, where V is the number of vertices in the graph. Obviously, $M[i][j] = 1$ implies that there is an edge from vertex i to j. On the other hand, $M[i][j] = 0$ means that there is no direct edge from vertex i to j.

Advantages
 (i) It takes $O(1)$ time to determine whether two given vertices in a graph have a direct edge between them.
 (ii) Even if the graph is undirected, we do not need to represent the same edge twice. We can just use an *upper diagonal* matrix or a *lower diagonal* matrix to represent undirected graphs. Some programming languages such as Java allow 2D arrays with different number of elements in each row, without using any extra space. So, we can create matrices where first row has only one element, second row has two elements, and so on.
 (iii) For directed graphs, simple *Boolean matrix multiplication* can be used to determine whether there exists a path of length $= k$ ($k \geq 2$) between any two vertices.

Disadvantages
 (i) A lot of memory is wasted in case of representing sparse graphs—directed as well as undirected.
 (ii) Not all programming languages support matrices with variable sized rows. Hence, half of the memory is wasted in case of undirected graphs.

(b) *Adjacency list* In this notation, for every vertex $v \in V$ of the graph, we maintain a *linked list* consisting of the edges: $(v, w) \in E$ for all $w \in V$ (i.e., reachable directly from v to w), where V is the set of all vertices present in the graph and E is the set of all edges of that graph.

Advantages
 (i) If the graph is sparse, i.e., $|E| \ll |V|$, then only $O(|V|)$ space is needed, hence *no wastage* of memory. In particular, if $|E| \ll |V|$, then we can say $|E| = O(|V|)$.
 (ii) For directed graphs, each edge is represented only once.

Disadvantages
  (i) It may take a worst case time of $O(|E|)$ to determine whether there exists a direct edge between any two given vertices in the graph, since there is no provision of checking for the existence of the direct connectivity between any two vertices in a graph.
  (ii) For undirected graphs, each edge needs to be represented twice, for example for an edge: (v, w), we need to put v in the adjacency list of w, and we also need to put w in the adjacency list of v.
  (iii) For directed graphs, it is practically impossible to use adjacency list representations to determine if there exists a path of length k ($k \geq 2$) between any two vertices.

Note If a graph G possesses parallel edges, then we need a bit modification of the above representations. For example, in matrix representation, instead of providing value 1 at location, say, $A[i][j]$ if edge between vertices i and j exists, we may assign a value k at the same location $A[i][j]$ if k edges between vertices i and j exist. Otherwise, we may prefer the *incidence matrix representation* (discussed below) for such type of graph.

9.6.4 Incidence Matrix Representation of Graph

In such representation of graph, the matrix is defined as follows:

$$A_{i,j} = 1, \quad \text{if } j\text{th edge } e_j \text{ is incident to } i\text{th vertex}$$
$$= 0, \quad \text{otherwise}$$

Let us consider a graph G as shown in Figure 9.31.

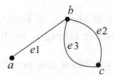

FIGURE 9.31

The corresponding *incidence matrix* is given in Figure 9.32.

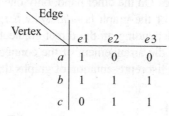

FIGURE 9.32

9.7 GRAPH ALGORITHMS

There are many graph algorithms (dealing with problems on graphs). The most elementary ones are *graph traversal*: visiting all nodes of a graph in a systematic way and finding shortest paths. These operations are discussed below.

Two basic traversal techniques are applied on a graph: (a) BFS (breadth first search) and (b) DFS (depth first search).

9.7.1 Breadth First Search

The basic idea behind BFS is that searching starts from a vertex, say, u and traverses all the neighbours of u, and then neighbours of all the neighbours of u, and so on, i.e., it traverses level by level. So, it is also called as *level-order searching*. Also, within a level, vertices are visited from left to right, and no vertex is visited more than once. The process continues until all the vertices are visited.

We may apply BFS both on directed and undirected graph. If it is applied on unweighted connected graph, we are then able to find the shortest paths from a source vertex (from which traversal starts) to the other reachable destination vertices. However, on measuring shortest paths from such unweighted connected graph, we focus only on the *number of intermediate* node traversals from source to destination.

Algorithm BFS (G)

/* $G = (V, E)$ is a connected unweighted graph, and *labelling* a vertex means assigning value such as 1 (true) to mark it as already visited */

Step 1 Start with a vertex, say, s, and label it.
Step 2 Find all the unlabelled vertices in G which are adjacent to a labelled vertex, say, v, and then label those.
Step 3 Step 2 continues until all the vertices in G are labelled.

Time complexity of BFS is $O(|V| + |E|)$.

EXAMPLE 9.20 Consider the graph as shown in Figure 9.33 and apply BFS on it to find the traversal sequence taking w as the starting vertex.

Solution Here, the starting vertex is w. Now, one possible BFS sequence is $wxyuv$. The other sequences may be like: $wyxuv$, $wuvxy$, etc., since all the vertices x, y, u, v are at the same level and adjacent to w.

However, we find such possible sequences only when the graph is represented by *adjacency matrix*. On the other hand, only one sequence is expected if the representation of the graph is *adjacency list*, since each vertex possesses a linked list depending on the direct connectivity of the vertex with other vertices, and the arrangement of the connected vertices is initially fixed. For more clarity, refer to the linked list representation of graphs discussed in the previous section.

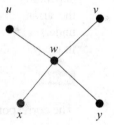

FIGURE 9.33

9.7.2 Depth First Search

The basic idea behind DFS is that it starts from a vertex, say, u, and a complete path from u is traversed, i.e., a path is traversed until a *dead* vertex (already traversed) is found or further expansion is not possible (i.e., no more link is present from the vertex currently traversed). Now, the searching comes back (i.e., it backtracks) to the earlier node to take an alternate path (if exists and not already traversed). Clearly, it is similar to *pre-order traversal* of tree. The process continues until all the vertices are visited. However, during traversal, no vertex is visited more than once. We may apply DFS both on directed and undirected graph.

Algorithm DFS (G)

/* $G = (V, E)$ is a connected unweighted graph, and *labelling* a vertex means assigning value such as 1 (true) to mark it as already visited */

Step 1 Start with a vertex, say, s and label it.
Step 2 Find an unlabelled vertex in G which is adjacent to a labelled vertex, say, v, and then label it. If no such unlabelled vertex is found from the current labelled vertex v, then *back track* to the parent node of v to select the next alternate path if exists.
Step 3 Step 2 continues until all the vertices in G are labelled.

Time complexity of DFS is $O(|V| + |E|)$

Note Applying DFS procedure, we can easily check in linear time whether a connected graph G has a *cycle* or not.

EXAMPLE 9.21 Consider the graph as shown in Figure 9.34 and apply DFS on it to find the traversal sequence taking a as the starting vertex.

Solution Here, starting vertex is a. Now, one possible DFS sequence is *abdec*. The other sequences may be like: *abedc*, *acbde*, etc., since the vertices b and c are at the same level and adjacent to a. It is obvious that we find such possible sequences only when the graph is represented by *adjacency matrix*. But only one sequence is expected if the representation of the graph is *adjacency list*, since each node possesses a linked list depending on the direct connectivity of the node with other nodes and the arrangement of the connected nodes is initially fixed. For better understanding, refer to the linked list representation of graphs discussed in the previous section.

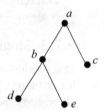

FIGURE 9.34

Note If DFS is applied on unweighted connected graph, then we may not ensure to find the shortest paths from a *source* vertex (from which traversal starts) to the other reachable destination vertices. The reason is that despite existence of a direct path from the source to any destination, multiple intermediate nodes may be visited between these nodes. But we focus only on the number of intermediate node traversals from *source* to *destination*, as no *weight* is provided to the edges.

EXAMPLE 9.22 Give a counter example to prove that if there is a path from u to v in a directed graph G, and if $d[u] < d[v]$ in a DFS of G, then v is a descendant (child) of u in the DFS forest produced, where $d[x]$ means the distance of the vertex x from the root (i.e., starting vertex).

Solution Consider the following directed graph (Figure 9.35).

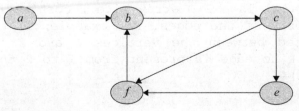

FIGURE 9.35

Assume that DFS starts at the vertex a of the presented graph. Now, we denote parent of a vertex v as $\pi[v]$. Let us start from the vertex a.

Therefore, $d[a] = 1$ and $\pi[a] = \Phi$ (i.e., vertex a has no parent).

Clearly, next node will be b, and so $d[b] = 2$ and $\pi[b] = a$.

At this stage, next node is c. Thus, $d[c] = 3$ and $\pi[c] = b$.

Now, both f and e are present in the adjacency list of c, and assume that vertex f is appearing there first. So, $d[f] = 4$ and $\pi[f] = c$

Further, from f, b is the only reachable node to be found in the adjacency list of f. But b is already traversed. So, next node to be traversed is e which is to be obtained from the adjacency list of c. Accordingly, $d[e] = 5$ and $\pi[e] = c$.

Thus, the DFS forest is generated as shown in Figure 9.36.

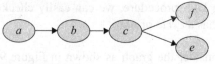

FIGURE 9.36

Certainly, $d[f] = 4$ and $d[e] = 5$, so $d[f] < d[e]$. Also, there is a path from f to e as: $f \to b \to c \to e$, but e is not a descendant of f in the DFS forest.

9.7.3 Single-Source Shortest Path Problem

In a single-source shortest path problem, we must find the shortest path from a specified vertex s (source) to all other vertices (destinations) in a weighted connected graph. Several algorithms such as Moore's, Belman-Frod's, Dijkstra's, etc., exist on this problem. Dijkstra's algorithm is presented below.

Dijkstra's Algorithm

This algorithm is used to find the shortest paths from any source vertex to all other vertices. We can apply this procedure both on directed and undirected graphs. However, cost of the edge must be non-negative.

```
                Algorithm Dijkstra

    /* Let G=(V,E) be a connected weighted graph, |V| = n.
    D[1..n] is an 1-D array to store the content of shortest path
    from source vertex to every other vertices. For example, d[2]=k
    means currently the value of the shortest path from source
    vertex to 2 is k. However, when the algorithm terminates, each
    location of D[] contains the value of the shortest path from
    source vertex to the vertex specified in d[ ]. */
    /* W[1..n][1..n] is a 2-D array (i.e., cost matrix) which con-
    tains the costs of the edges. For example, if the cost of the
    edge connected between the vertices 1 and 2 is say m, then
    W[1][2]=m, if no edge is present from 1 to 2 then W[1]2]= +∞
    (large value)*/
       begin
         S = v₀
       /* S is the selected set of vertices, it initially consists of
```

```
  v₀ (source) */
    D[v₀] = 0
    for each v in V - {v₀} do
        D[v] = W(v₀, v)
    //If (v₀, v) is not an edge, take the value as +∞
    endfor
      while (S ≠ V) do
         begin
   Choose a vertex u in (V-S) such that D[u] is minimum.
     Add u to S.
        for each v in (V-S) do
            D[v] = min(D[v], D[u] + W(u, v))
        endfor
     endwhile
  end
```

Overall the complexity of this algorithm is $O(n^2)$ if n vertices are present in G.

EXAMPLE 9.23 Find the length of shortest paths from vertex a to every other vertices by Dijkstra's algorithm for the graph as shown in Figure 9.37.

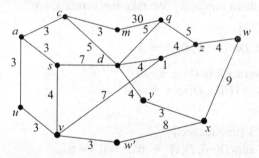

FIGURE 9.37

Solution Going by Dijkstra's algorithm, we shall end up with

$D[a] = 0, D[c] = 3, D[s] = 3, D[u] = 3, D[m] = 6, D[v] = 6, D[d] = 8,$
$D[w'] = 9, D[y] = 12, D[i] = 12, D[q] = 13, D[x] = 15, D[z] = 16, D[w] = 20$

The above values obtained are shown below step by step.

Step 1 Initialization

$D[a] = 0, D[c] = W(a, c) = 3, D[s] = W(a, s) = 3, D[u] = W(a, u) = 3$

There are no direct edges from a to any other vertex. So, D values of all other vertices will be set to $+\infty$ (large value).

$D[m] = D[d] = D[v] = D[q] = D[i] = D[y] = D[w'] = D[z] = D[x] = D[w] = +\infty$

$S = \{a\}$ initially

Step 2 Choose a vertex u in $(V - S)$ such that $D[u]$ is minimum. There are multiple possibilities for this statement: $D[c] = 3, D[s] = 3, D[u] = 3$. All these have minimum D values. Choose anyone of them randomly. We take the vertex c, so

$S = \{a, c\}, D[a] = 0, D[c] = 3$

for each vertex k in $(V - S)$ *do*
$D[k] = \min(D[k], D[c] + W(c, k))$
endfor

This will, in fact, give $D[s] = 3, D[u] = 3$.

$D[m] = \min(D[m], D[c] + W(c, m)) = \min(+\infty, 3 + 3) = 6$
$D[d] = \min(D[d], D[c] + W(c, d)) = \min(+\infty, 3 + 5) = 8$
$D[v] = \min(D[v], D[c] + W(c, v)) = +\infty$
$D[q] = \min(D[q], D[c] + W(c, q)) = +\infty$
$D[i] = \min(D[i], D[c] + W(c, i)) = +\infty$
$D[y] = \min(D[y], D[c] + W(c, y)) = +\infty$
$D[w'] = \min(D[w'], D[c] + W(c, w')) = +\infty$
$D[z] = \min(D[z], D[c] + W(c, z)) = +\infty$
$D[x] = \min(D[x], D[c] + W(c, x)) = +\infty$
$D[w] = \min(D[w], D[c] + W(c, w)) = +\infty$

Step 3 Choose a vertex u in $(V - S)$ such that $D[u]$ is minimum. There are multiple possibilities for this statement: $D[s] = 3, D[u] = 3$. All these have minimum D values. Choose anyone of them randomly. We take the vertex s, so

$S = \{a, c, s\}$
$D[a] = 0, D[c] = 3, D[s] = 3$

for each vertex k in $(V - S)$ *do*
$D[k] = \min(D[k], D[s] + W(s, k))$
endfor

$D[u] = 3$ (unchanged)
$D[m] = \min(D[m], D[s] + W(s, m)) = 6$
$D[d] = \min(D[d], D[s] + W(s, d)) = \min(8, 3 + 7) = 8$
$D[v] = \min(D[v], D[s] + W(s, v)) = \min(+\infty, 3 + 4) = 7$
$D[q] = \min(D[q], D[s] + W(s, q)) = +\infty$
$D[i] = \min(D[i], D[s] + W(s, i)) = +\infty$
$D[y] = \min(D[y], D[s] + W(s, y)) = +\infty$
$D[w'] = \min(D[w'], D[s] + W(s, w')) = +\infty$
$D[z] = \min(D[z], D[s] + W(s, z)) = +\infty$
$D[x] = \min(D[x], D[s] + W(s, x)) = +\infty$
$D[w] = \min(D[w], D[s] + W(s, w)) = +\infty$

Step 4 Choose a vertex u in $(V - S)$ such that $D[u]$ is minimum. The only possibility is $D[u] = 3$.

$S = \{a, c, s, u\}$
$D[a] = 0, D[c] = 3, D[s] = 3, D[u] = 3$

for each vertex k in $(V - S)$ *do*
$D[k] = \min(D[k], D[u] + W(u, k))$

endfor

$D[m] = \min(D[m], D[u] + W(u, m)) = 6$
$D[d] = \min(D[d], D[u] + W(u, d)) = 8$
$D[v] = \min(D[v], D[u] + W(u, v)) = \min(7, 3 + 3) = 6$
$D[q] = \min(D[q], D[s] + W(u, q)) = +\infty$
$D[i] = \min(D[i], D[u] + W(u, i)) = +\infty$
$D[y] = \min(D[y], D[u] + W(u, y)) = +\infty$
$D[w'] = \min(D[w'], D[u] + W(u, w')) = +\infty$
$D[z] = \min(D[z], D[u] + W(u, z)) = +\infty$
$D[x] = \min(D[x], D[u] + W(u, x)) = +\infty$
$D[w] = \min(D[w], D[u] + W(u, w)) = +\infty$

Step 5 Choose a vertex u in $(V - S)$ such that $D[u]$ is minimum. Two possibilities exist $D[m] = D[v] = 6$, choose anyone of them randomly. We take the vertex m, so

$S = \{a, c, s, u, m\}$
$D[a] = 0, D[c] = 3, D[s] = 3, D[u] = 3, D[m] = 6$

for each vertex k in $(V - S)$ *do*
$D[k] = \min(D[k], D[m] + W(m, k))$
endfor

$D[d] = \min(D[d], D[m] + W(m, d)) = 8$
$D[v] = \min(D[v], D[m] + W(m, v)) = 6$
$D[q] = \min(D[q], D[m] + W(m, q)) = \min(+\infty, 6 + 30) = 36$
$D[i] = \min(D[i], D[m] + W(m, i)) = +\infty$
$D[y] = \min(D[y], D[m] + W(m, y)) = +\infty$
$D[w'] = \min(D[w'], D[m] + W(m, w')) = +\infty$
$D[z] = \min(D[z], D[m] + W(m, z)) = +\infty$
$D[x] = \min(D[x], D[m] + W(m, x)) = +\infty$
$D[w] = \min(D[w], D[m] + W(m, w)) = +\infty$

Step 6 Choose a vertex u in $(V - S)$ such that $D[u]$ is minimum. The only possibility is $D[v] = 6$.

$S = \{a, c, s, u, m, v\}$
$D[a] = 0, D[c] = 3, D[s] = 3, D[u] = 3, D[m] = 6, D[v] = 6$

for each vertex k in $(V - S)$ *do*

$D[k] = \min(D[k], D[v] + W(v, k))$

endfor

$D[d] = \min(D[d], D[v] + W(v, d)) = 8$
$D[q] = \min(D[q], D[v] + W(v, q)) = 36$
$D[i] = \min(D[i], D[v] + W(v, i)) = \min(+\infty, 6 + 7) = 13$
$D[y] = \min(D[y], D[v] + W(v, y)) = +\infty$
$D[w'] = \min(D[w'], D[v] + W(v, w')) = \min(+\infty, 6 + 3) = 9$
$D[z] = \min(D[z], D[v] + W(v, z)) = +\infty$
$D[x] = \min(D[x], D[v] + W(v, x)) = +\infty$
$D[w] = \min(D[w], D[v] + W(v, w)) = +\infty$

Step 7 Choose a vertex u in $(V - S)$ such that $D[u]$ is minimum. The only possibility is $D[d] = 8$.

$S = \{a, c, s, u, m, v, d\}$
$D[a] = 0, D[c] = 3, D[s] = 3, D[u] = 3, D[m] = 6, D[v] = 6, D[d] = 8$

for each vertex k in $(V - S)$ *do*
$\quad D[k] = \min(D[k], D[d] + W(d, k))$
endfor

$D[q] = \min(D[q], D[d] + W(d, q)) = \min(36, 8 + 5) = 13$
$D[i] = \min(D[i], D[d] + W(d, i)) = \min(13, 8 + 4) = 12$
$D[y] = \min(D[y], D[d] + W(d, y)) = \min(+\infty, 8 + 4) = 12$
$D[w'] = \min(D[w'], D[d] + W(d, w')) = 9$
$D[z] = \min(D[z], D[d] + W(d, z)) = +\infty$
$D[x] = \min(D[x], D[d] + W(d, x)) = +\infty$
$D[w] = \min(D[w], D[d] + W(d, w)) = +\infty$

Step 8 Choose a vertex u in $(V - S)$ such that $D[u]$ is minimum. The only possibility is $D[w'] = 9$.

$S = \{a, c, s, u, m, v, d, w'\}$
$D[a] = 0, D[c] = 3, D[s] = 3, D[u] = 3, D[m] = 6, D[v] = 6, D[d] = 8, D[w'] = 9$

for each vertex k in $(V - S)$ *do*
$\quad D[k] = \min(D[k], D[w'] + W(w', k))$
endfor

$D[q] = \min(D[q], D[w'] + W(w', q)) = 13$
$D[i] = \min(D[i], D[w'] + W(w', i)) = 12$
$D[y] = \min(D[y], D[w'] + W(w', y)) = 12$
$D[z] = \min(D[z], D[w'] + W(w', d)) = +\infty$
$D[x] = \min(D[x], D[w'] + W(w', x)) = \min(+\infty, 9 + 8) = 17$
$D[w] = \min(D[w], D[w'] + W(w', w)) = +\infty$

Step 9 Choose a vertex u in $(V - S)$ such that $D[u]$ is minimum. The only possibility is $D[i] = D[y] = 12$. We randomly choose y.

$S = \{a, c, s, u, m, v, d, w', y\}$
$D[a] = 0, D[c] = 3, D[s] = 3, D[u] = 3, D[m] = 6, D[v] = 6, D[d] = 8, D[w'] = 9, D[y] = 12$

for each vertex k in $(V - S)$ *do*
$\quad D[k] = \min(D[k], D[y] + W(y, k))$
endfor

$D[q] = \min(D[q], D[y] + W(y, q)) = 13$
$D[i] = \min(D[i], D[y] + W(y, i)) = 12$
$D[z] = \min(D[z], D[y] + W(y, z)) = +\infty$
$D[x] = \min(D[x], D[y] + W(y, x)) = \min(17, 12 + 3) = 15$
$D[w] = \min(D[w], D[y] + W(y, w)) = +\infty$

Step 10 Choose a vertex u in $(V - S)$ such that $D[u]$ is minimum. The only possibility is $D[i] = 12$.

$S = \{a, c, s, u, m, v, d, w', y, i\}$
$D[a] = 0, D[c] = 3, D[s] = 3, D[u] = 3, D[m] = 6, D[v] = 6, D[d] = 8, D[w'] = 9,$
$D[y] = 12, D[i] = 12$

for each vertex k in $(V - S)$ do
$\quad D[k] = \min(D[k], D[i] + W(i, k))$
endfor

$D[q] = \min(D[q], D[i] + W(i, q)) = 13$
$D[z] = \min(D[z], D[i] + W(i, z)) = \min(+\infty, 12 + 4) = 16$
$D[x] = \min(D[x], D[i] + W(i, x)) = 15$
$D[w] = \min(D[w], D[i] + W(i, w)) = +\infty$

Step 11 Choose a vertex u in $(V - S)$ such that $D[u]$ is minimum. The only possibility is $D[q] = 13$.

$S = \{a, c, s, u, m, v, d, w', y, i, q\}$
$D[a] = 0, D[c] = 3, D[s] = 3, D[u] = 3, D[m] = 6, D[v] = 6, D[d] = 8, D[w'] = 9,$
$D[y] = 12, D[i] = 12, D[q] = 13$

for each vertex k in $(V - S)$ do
$\quad D[k] = \min(D[k], D[q] + W(q, k))$
endfor

$D[z] = \min(D[z], D[q] + W(q, z)) = \min(16, 13 + 5) = 16$
$D[x] = \min(D[x], D[q] + W(q, x)) = 15$
$D[w] = \min(D[w], D[q] + W(q, w)) = +\infty$

Step 12 Choose a vertex u in $(V - S)$ such that $D[u]$ is minimum. The only possibility is $D[x] = 15$.

$S = \{a, c, s, u, m, v, d, w', y, i, q, x\}$
$D[a] = 0, D[c] = 3, D[s] = 3, D[u] = 3, D[m] = 6, D[v] = 6, D[d] = 8, D[w'] = 9,$
$D[y] = 12, D[i] = 12, D[q] = 13, D[x] = 15$

for each vertex k in $(V - S)$ do
$\quad D[k] = \min(D[k], D[x] + W(x, k))$
endfor

$D[z] = \min(D[z], D[x] + W(x, z)) = 16$
$D[w] = \min(D[w], D[x] + W(x, w)) = \min(+\infty, 15 + 9) = 24$

Step 13 Choose a vertex u in $(V - S)$ such that $D[u]$ is minimum. The only possibility is $D[z] = 16$.

$S = \{a, c, s, u, m, v, d, w', y, i, q, x, z\}$
$D[a] = 0, D[c] = 3, D[s] = 3, D[u] = 3, D[m] = 6, D[v] = 6, D[d] = 8, D[w'] = 9,$
$D[y] = 12, D[i] = 12, D[q] = 13, D[x] = 15, D[z] = 16$

for each vertex k in $(V - S)$ do
$\quad D[k] = \min(D[k], D[z] + W(z, k))$
endfor

$D[w] = \min(D[w], D[z] + W(z, w)) = \min(24, 16 + 4) = 20$

Step 14 Choose a vertex u in $(V - S)$ such that $D[u]$ is minimum. The only possibility is $D[w] = 20$.

$S = \{a, c, s, u, m, v, d, w', y, i, q, x, z, w\}$
$D[a] = 0, D[c] = 3, D[s] = 3, D[u] = 3, D[m] = 6, D[v] = 6, D[d] = 8, D[w'] = 9,$
$D[y] = 12, D[i] = 12, D[q] = 13, D[x] = 15, D[z] = 16, D[w] = 20$
$S = V$, hence the process stops here and we have got the final D values.

9.8 EULER GRAPH

We can now consider another kind of graph problem in which we are interested to traverse each edge of the graph exactly once, starting and ending at the same point. A simple example of this is the *common puzzle* problem that asks the solver to trace a geometric figure without lifting the pencil from paper or tracing an edge more than once. Swiss mathematician Leonhard Euler first introduced such a problem.

Let $G = (V, E)$ be a directed or undirected graph. It is Euler graph if it has at least one Euler circuit.

An *Euler path* consists of every edge of G exactly once. To find such a path, vertices may be repeated. In particular, if the first and the last traversed vertices are same on an Euler path, then that path is called as *Euler circuit*. Clearly, a graph may contain an Euler path but not necessarily an Euler circuit.

Consider the following graphs to check the existence of Euler path and Euler circuit. The graph as shown in Figure 9.38(a) consists of Euler path like: A, B, C, E, D, B, as the path contains all the edges of this graph exactly once. However, it is not an Euler circuit, since the starting and the ending vertex is not same (although the vertex B other than the beginning vertex A is repeated). Further, this graph possesses many Euler paths but no Euler circuit. Similarly, the graph in Figure 9.38(d) possesses Euler path like: $G, E, D, G, F, D, B, A, C, D, A$, but no Euler circuit. Therefore, these two graphs are not Euler graphs.

Let us consider the graph as shown in Figure 9.38(b). It has both Euler path and Euler circuit. One Euler circuit is a, d, b, c, a. Accordingly, the graph is Euler graph.

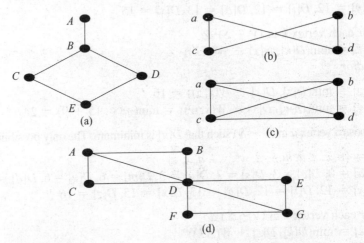

FIGURE 9.38

Again, the graph shown in Figure 9.38(c) contain neither Euler path nor Euler circuit. Hence, they are not Euler graphs.

Note As per Euler, a non-empty connected graph is Eulerian if and only if it has no vertices of odd degree. It is, indeed, called as *Euler's theorem*. Although Euler's theorem tells us whether an Eulerian circuit is possible but it does not find any such circuit for us. To do so, we can use the following algorithm.

Fleury's Algorithm If G is an Eulerian graph, then the following steps produce an Eulerian circuit (tour):

Step 1 Choose a starting vertex u.

Step 2 Traverse any available edge, choosing an edge that will disconnect the remaining graph only if there is no alternative.

Step 3 After traversing each edge, remove it (together with any vertex of degree 0 which results).

Step 4 If no edge remains, stop. Otherwise, choose another available edge and go back to step 2.

EXAMPLE 9.24 Use Fleury's algorithm on the graph as shown in Figure 9.39 to find an Eulerian circuit.

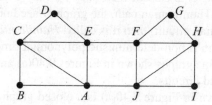

FIGURE 9.39

Solution We can start from anywhere of the graph. Let us start at vertex A. Now, we may choose AJ, then JF, to define the start of our path as AJF. At this stage, we cannot select the edge FE, since this would disconnect the graph (right part of the graph to the edge FG, i.e., in such case to possess all the edges exactly once, some edges must be repeated) and there are other alternatives. Proceeding in this way, we produce the partial trail: $AJFGHIJHF$. At this stage, we can choose edge FE, as there are no alternative. Thus, an Eulerian trail is eventually given by $AJFGHIJHFEDCBACEA$.

9.8.1 Some Useful Results on Euler Graph

(i) If every vertex of a connected graph G has even degree, then the graph is an Euler graph, i.e., no G has any vertices of odd degree.

(ii) If a connected graph G has at least one vertex of odd degree, then the graph may not be Euler graph (since no Euler circuit is present).

(iii) If a connected graph G has more than two vertices of odd degree, then there can be no Euler path. But if there are exactly two vertices of odd degree, then there exists an Euler path.

9.9 HAMILTONIAN GRAPH

We now consider a special kind of graph problem in which the task is to visit each vertex exactly once, with the exception of the starting and the ending vertex. Such type of problem was first introduced by an Irish mathematician, Sir William Rowan Hamilton. So, it is also called as *Hamiltonian cycle problem*. The prime goal of such problem is to identify a route in order to provide services to all the vertices (each of which can represent an object) on a regular basis. Many real-life problems may have same goal.

For example Serving a set of *vending machines* on a regular basis by someone through a path. Further, we may consider here another interesting real-life problem named as *travelling salesman problem* (TSP). In this problem, the vertices in the graph G might represent cities; the edges, lines of transportation; and the weight of an edge, the cost of travelling along that edge. The primary goal is to visit all the cities represented by the graph G, spending minimum cost. So, TSP is equivalent to finding a minimum Hamiltonian cycle in a connected weighted graph.

Let us consider the necessary terminologies to determine whether a connected graph $G = (V, E)$ is Hamiltonian graph.

Hamiltonian path is a path in G that contains each vertex of G exactly once.

Hamiltonian circuit is a circuit drawn from G that contains each vertex of G exactly once except the *beginning* and the *ending* vertex. Since no vertex except the start vertex is repeated, no edge will repeat.

Hence, if the graph G is *closed*, then only we may expect a Hamiltonian circuit from G. However, to obtain a Hamiltonian path, the graph G need not be closed. Further, if G contains at least one Hamiltonian circuit, then it is called *Hamiltonian graph* and determining whether a graph is Hamiltonian is non-deterministic polynomial time (NP)-complete problem.

Let us consider some graphs shown in Figures 9.40(a) and (b), to determine the presence of Hamiltonian paths and circuits.

The graph as shown in Figure 9.40(a) is a closed graph but it has no Hamiltonian circuit. However, it has Hamiltonian path such as A, E, C, B, D. On the other hand, the graph as shown in Figure 9.40(b) has neither Hamiltonian path nor Hamiltonian circuit.

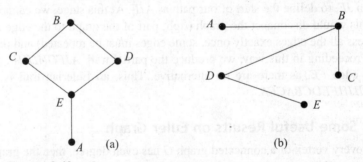

FIGURE 9.40

EXAMPLE 9.25 Prove that every complete graph K_n has $[(n - 1)!]/2$ possible Hamiltonian circuits.

Solution We know that n objects can be arranged in a line by $n!$ ways. But when we consider *circular* permutation of that n objects, then it is $(n - 1)!$; since n number of *linear* permutations give a unique circular permutation.

Now, in case of Hamiltonian cycle for K_n, there are $n!$ arrangements. But each Hamiltonian cycle gives rise to n arrangements in the same cyclic order and further n arrangements are obtained by *reversing* them. For example, both $v_1 v_2 v_3 v_4 v_5 v_1$ and $v_2 v_3 v_4 v_5 v_1 v_2$ represent the same cycle, when $v_1, \ldots, v_5$ are the vertices of K_5. So, there are $n!/(2n) = ((n - 1)!)/2$ possible Hamiltonian circuits.

9.9.1 Useful Hints on Hamiltonian Circuit

A few helpful hints are listed below to find Hamiltonian circuit from a graph $G = (V, E)$.

(i) If G has a Hamiltonian circuit, then for all $u \in V$, $\deg(u) \geq 2$.
(ii) If G has no loops and parallel edges, and if $|V| = n \geq 3$ and $\deg(u) \geq n/2$ for all $u \in V$, then G is Hamiltonian.
(iii) If a graph G with n vertices has a Hamiltonian circuit, then G must have at least n edges.
(iv) If a graph G has n vertices and m edges, then G has a Hamiltonian circuit if $m \geq (n^2 - 3n + 6)/2$.
(v) The complete bipartite graph $K_{m,n}$ is Hamiltonian if and only if $m = n$ and $n > 1$.
(vi) $K_{n,n}$ is Hamiltonian for all $n > 1$, but Eulerian only when n is even.
(vii) $K_{m,n}$, $m \neq n$ is not Hamiltonian, but Eulerian only when m and n are even integers.
(viii) K_n is Hamiltonian for all $n \geq 3$, but Eulerian only when $n \geq 3$ is odd.

[Readers are suggested to prove the above hints.]

 EXAMPLE 9.26 Is graph shown in Figure 9.41 a Hamiltonian graph?

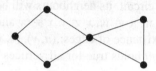

FIGURE 9.41

Solution The graph G as shown in the example has six vertices. The graph has no self-loop and parallel edges. So, it may have a Hamiltonian circuit.

Further, it is known to us that 'if G has no loops and parallel edges, and if $|V| = n \geq 3$ and $\deg(u) \geq n/2$' for all $u \in V$, then G is Hamiltonian.

Clearly, in this example, only one vertex has degree $\geq n/2 = 6/2 = 3$, and the rest has degree less than 3. So, the given graph is not Hamiltonian.

 EXAMPLE 9.27 Prove or disprove: 'A graph containing a Hamiltonian path must be cyclic'.

Solution No. Counter example exists. Let us consider a graph as $A \rightarrow B \rightarrow C$. Clearly, one of its Hamiltonian paths is $A \rightarrow B \rightarrow C$. It covers all the vertices exactly once but the graph has no cycle. Also, we may select start vertex as C.

 EXAMPLE 9.28 Prove or disprove: 'A graph containing an Euler path must be cyclic'.

Solution No. Counter example exists. Let us consider a graph as $A \rightarrow B \rightarrow C$. Clearly, one of its Euler paths is $A \rightarrow B \rightarrow C$, since it covers all the edges exactly once but the graph has no cycle. Although we may also select start vertex as C.

 EXAMPLE 9.29 Prove or disprove: 'A graph containing an Euler path as well as a Hamiltonian path must be cyclic'.

Solution No. Counter example exists. Let us consider this graph as $A \rightarrow B \rightarrow C$. Clearly, it has a Hamiltonian path: $A \rightarrow B \rightarrow C$, covering all the vertices exactly once, and with different starting and ending vertices, but the graph has no cycle.

This graph also has an Euler path: $A \to B \to C$, covering all the edges exactly once, and with different starting and ending vertices, but the graph has no cycle.

EXAMPLE 9.30 State true or false: In a graph, an Euler path can also be a Hamiltonian path in some cases. If true, then give an example of such case, else disprove this statement.

Solution True. Let us consider this graph as $A \to B \to C$. Clearly, one of its Hamiltonian paths is $A \to B \to C$, since it covers all the vertices exactly once but the graph has no cycle. Obviously, we may choose start vertex as C also.

Further, this graph also has Euler path such as $A \to B \to C$, covering all the edges exactly once but the graph has no cycle.

EXAMPLE 9.31 A graph contains two disjoint Hamiltonian circuits. What is the minimum degree of any vertices of this graph?

Solution The minimum degree of any vertices of this graph is 4. The two Hamiltonian circuits are disjoint means that they do not have any common edges.

So, for any vertex v, if its *neighbours* are u and w in the first Hamiltonian circuit, then in the second Hamiltonian circuit, its neighbours will be two entirely different vertices, say, x and y, such that $x \neq u, y \neq w, u \neq w, x \neq y, x \neq w$ and $y \neq v$.

This clearly shows existence of edges: (u, v), (v, w), (x, v), and (v, y). So, the degree of v is at least 4. *Same argument* holds true for all vertices in G.

EXAMPLE 9.32 Is it possible for a simple connected graph containing a bridge to be Hamiltonian?

Solution A bridge is an edge whose removal would disconnect the graph. Clearly, any Hamiltonian cycle in a graph with a bridge would have to include that bridge. But a bridge is not part of any cycle (since if it were, its removal would not disconnect the graph). So, any graph with a bridge is not Hamiltonian.

9.10 PLANAR GRAPH

A graph is said to be planar if it can be drawn on a plane so that no two edges *intersect*. A planar graph may be *directed* or *undirected*. Self-loop and parallel edges are allowed here. If a planar graph has no self-loop or parallel edges, then it is a simple planar graph. The basic result about planar graph known as Euler's formula is the basic computational tool for planar graph. The formula is presented below.

If a connected graph G with n vertices, e edges, and r regions is planar, then $n - e + r = 2$.

The graphs as shown in Figures 9.42(a) and (b) are planar graphs. On the other hand, the graph in Figure 9.42(c) is not a planar graph.

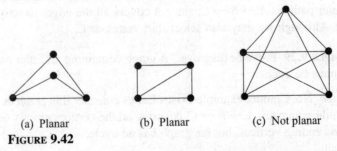

(a) Planar   (b) Planar   (c) Not planar

FIGURE 9.42

9.10.1 Properties of Planar Graph

(a) If a connected simple planar graph G has e edges and r regions, then $r \leq 2e/3$.
(b) If a connected planar graph G has e edges, n vertices, and r regions, then $n - e + r = 2$. [Euler's rule]
(c) If a connected simple planar graph G has e edges and n (≥ 3) vertices, then $e \leq 3n - 6$.
(d) If a connected simple planar graph G has e edges and n (≥ 3) vertices and no circuits of length 3, then $e \leq 2n - 4$. [One will check this inequality if the simple graph has no circuit of length 3. Otherwise, the inequality mentioned in (c) is sufficient for checking the planarity of a simple graph.]
(e) A complete graph K_n is planar if $n < 5$.
(f) A complete bipartite graph $K_{m,n}$ is planar if and only if $m < 3$ or $n > 3$.

EXAMPLE 9.33 Prove that complete graph K_4 is planar.

Solution The complete graph K_4 contains 4 vertices and 6 edges. But for a connected planar graph, it is true that $e \leq 3n - 6$. Hence, for K_4, we have $n = 4, e = 6$, and so, $3n - 6 = 12 - 6 = 6$ which satisfies $e \leq 3n - 6$. Clearly, this graph has circuit of length 3. So, we must not check the inequality $e \leq 2n - 4$. Thus, K_4 is a planar graph.

EXAMPLE 9.34 Prove that complete graph K_5 is not planar.

Solution The complete graph K_5 contains 5 vertices and exactly $5(5 - 1)/2 = 10$ edges.
For a connected graph, it is true that $e \leq 3n - 6$. Hence, for K_5, we have $n = 5, e = 10$, and so, $3n - 6 = 3 \times 5 - 6 = 9$ which does not satisfy $e \leq 3n - 6$. Since this inequality is not true for this graph, so we need not check the inequality $e \leq 2n - 4$.

EXAMPLE 9.35 Show that the graph $K_{3,3}$ is not planar.

Solution We know that a complete bipartite graph $K_{m,n}$ has $m + n$ vertices and $m \times n$ edges. Hence, the graph $K_{3,3}$ has 6 ($= 3 + 3$) vertices and 9 ($= 3 \times 3$) edges, i.e., here, $n = 6, e = 9$.
Now, first apply the inequality: $e \leq 3n - 6$. It is true here, since $e = 9 \leq 3n - 6 = 3 \times 6 - 6 = 12$. Clearly, this graph has no circuit of length 3.
So, we must apply the inequality: $e \leq 2n - 4$. But it is not true here, since $e = 9 \leq 2n - 4 = 12 - 4 = 8$ is NOT greater than or equal to 9. Hence, the graph is not planar.
[So, it may be noted that the inequality $e \leq 3n - 6$ is only a necessary condition but not sufficient for checking the planarity of a simple graph.]

EXAMPLE 9.36 A connected planar graph has 10 vertices each of degree 3. Into how many regions, does a representation of this planar graph split the plan?

Solution Here, $n = 10$ and degree of each vertex is 3.
So, $\sum d(v) = 3 \times 10 = 30 = 2e$ [by handshaking rule]. So, e (number of edges) $= 15$.
Again, by Euler's formula, we have $n - e - r = 2$. Hence, $10 - 15 + r = 2$.
Therefore, r (number of regions) $= 7$.

EXAMPLE 9.37 A connected planar graph G has 25 vertices, 13 of which have degree 1. Prove that G has at most 43 edges.

Solution G is a planar graph and it has 25 vertices in total. Out of 25, 13 vertices have degree 1.

So, these 13 vertices are connected to the remaining set of 12 vertices by exactly 1 edge each. This accounts for 13 edges for those vertices of degree 1.

Now, we are left with a cluster of 12 vertices. These 12 vertices should also form a planar graph. A planar graph means that it should be possible to draw the graph in such a way that no 2 edges *cross* each other.

Consider the *star arrangement* (see the red lines and the blue-coloured dots) of the 12 vertices.

It is a known results from geometry and symmetry that a dodecagon (12-sided polygon) of the shape, as shown in Figure 9.43 (see Plate 1), has maximum possible number of diagonals. [*Mathematically inclined readers can easily prove the result.*]

The non-criss-crossing diagonals (because we want planar graphs only) are shown by green lines. The total number of green-coloured non-criss-crossing diagonals possible for the star arrangement = 6 + 6 + 3 = 15. The 12 edges of the dodecagon account for 12 edges too.

So, till now we have accounted for 13 + 15 + 12 = 40 edges. We can have 3 more edges. Look at the pink-coloured edges as shown in Figure 9.43.

Therefore, we have accounted for 40 + 3 = 43 edges, and it is obvious that no more edges are possible. [The 13 vertices of degree 1 each can be connected by 1 edge each to any of the 3 vertices: a, b, and c marked in the figure.]

Hence, the maximum number of edges = 43.

 EXAMPLE 9.38 What is the minimum number of vertices in a 3-regular bipartite planar graph?

Solution A bipartite graph cannot contain any triangle, i.e., no cycle of length 3.

Also, it is a planar graph, so $m \leq 2n - 4$, where m and n are, respectively, the number of edges and vertices. Further, as the graph is 3-regular, so $3n = 2m$.

Applying the above inequality, we get

$$3n/2 \leq 2n - 4$$
$$\Rightarrow n/2 \geq 4$$
$$\Rightarrow n \geq 8$$

Hence, the minimum number of vertices in a 3-regular bipartite planar graph is 8.

9.11 COLOURING OF GRAPH

Colouring of graph G means colouring of vertices with different colours. A colouring is *proper* if any two adjacent vertices u and v have two different colours. The minimum number of colours needed to produce a proper colouring of a graph G is called the *chromatic* number of G. Chromatic numbers of some kinds of graphs are listed below.

(i) The chromatic number of any tree is 2.
(ii) The chromatic number of any complete graph $K_n (n \geq 3)$ is n. [Any K_n graph is $(n - 1)$-regular.]
(iii) The chromatic number of any bipartite graph $K_{m,n}$ is 2.
(iv) The chromatic number of any cycle is 2.
(v) If G is r-regular but neither complete nor complete bipartite, then the chromatic number is r.

Note The maximum number of colours to colour a connected graph G with n vertices is n.

 EXAMPLE 9.39 How many colours are necessary to colour K_5?

Solution It is obvious that the chromatic number of any complete graph $K_n (n \geq 3)$ is n. Since it is a complete graph with 5 vertices, so 5 colours are needed to colour that graph.

PLATE 1

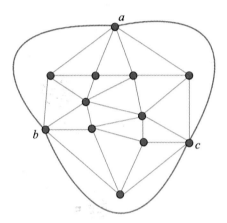

FIGURE 9.43

PLATE 2

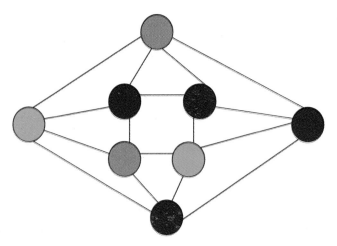

FIGURE 9.44

EXAMPLE 9.40 How many colours are necessary to colour $K_{5,2}$

Solution It is obvious that the chromatic number of any bipartite graph $K_{m,n}$ is 2. Since K_5 is a bipartite graph, so only 2 colours are sufficient to colour it.

EXAMPLE 9.41 Find the chromatic number of the following graph as shown in Figure 9.44 (see Plate 2).

Solution To colour any complete graph K_n, it requires n colours. If it is r-regular but neither complete nor complete bipartite, then r colours are needed. Since the given graph is 4-regular but neither K_4 nor complete bipartite, so 4 colours are necessary.

EXAMPLE 9.42 Let G be a connected graph, every vertex of which has degree 2. If G has even number of vertices, prove that G has perfect matching.

Solution A *matching* is defined as a subset $M \subseteq E$ for the graph $G(V; E)$, such that no 2 edges in M are incident on the same vertex.

A *prefect matching* is a matching in which every vertex is matched, i.e., each vertex has an edge incident on it.

G is given to be a connected graph with every vertex having degree 2. Also, G has even number of vertices.

Every vertex has degree 2, it means that the graph is a 2-regular graph. So, the chromatic number of this graph is 2.

Suppose that we are using *red* and *green* colours to colour the vertices of this graph. Then, we can easily arrange this graph into a set of 2 vertices such that all vertices in *set 1* are red-coloured, and all vertices in *set 2* are green-coloured.

Also, in a k-regular graph, there are equal number of vertices of each colour. [*known result*]

So, *set 1* and *set 2* have equal number of vertices. Also, there are no edges between any 2 vertices of same colour. This means that we have rearranged this graph as a bipartite graph.

It is obvious from the arrangement discussed above that for every red-coloured vertex, there exists a green-coloured vertex connected to it through a direct edge.

Since the degree of each vertex is 2, so to be more accurate, we can say that for every red-coloured vertex, there exist exactly 2 green-coloured vertices connected to it through direct edges.

Suppose that a perfect matching does not exist. This means that whatever subset M of the edge set E is taken, we shall be left with at least one unmatched vertex. Suppose that there are n vertices of colours red and green, respectively.

Case I We have taken an edge set M such that one red vertex is left unmatched, but all green-coloured vertices have been matched.

So, this means that we have picked up a set of n edges, such that these edges are incident on all n vertices of green colour, but are incident on only $(n - 1)$ vertices of red colour. This can happen if and only if 2 edges originating from the same red vertex are incident on 2 different green vertices. If that were to be true, then by definition of matching, M is not a matching at all.

This gives a *contradiction*. The same argument can also be extended to all such cases where we say that the number of unmatched red vertices is more than the number of unmatched green vertices.

Case II We have picked up a set of vertices such that k number of red vertices as well as k number of green vertices are left unmatched, where $k < n$. Since a prefect matching does not exist, so this means that if we consider the set of k unmatched red vertices, then all edges originating from those k red vertices will lead to only those green-coloured vertices which have been already matched. This would mean that the graph is disconnected, and again it leads to a *contradiction*.

So, both the cases lead to contradictions, hence we must have a perfect matching for the given graph.

9.12 COMPONENT

A connected component (or simply a component) H of a graph is a *maximal* connected subgraph. By 'maximal', we mean that G contains no other subgraph, i.e., both connected and properly contains H. Further, as it is connected, so it must have at least two vertices. An example through Figure 9.45 is presented below to make a clear idea.

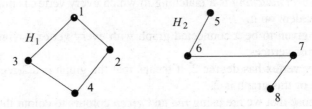

FIGURE 9.45 A graph G

The graph G has two components: H_1 and H_2. Here, neither of H_1 and H_2 contain the other component.

Note A connected component is a maximum subset of the nodes that is connected.

EXAMPLE 9.43 For the given digraph (directed graph as shown in Figure 9.46), find in-degrees and out-degrees of the vertices and prove that it is a strongly connected graph.

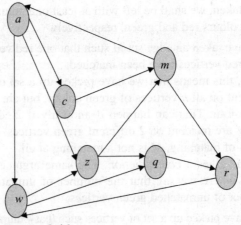

FIGURE 9.46

Solution A digraph G is strongly connected if and only if every pair of vertices in the graph is reachable from each other. In other words, for any two distinct vertices $u \in V, v \in V$ in the digraph $G(V, E)$, there exists a path from u to v as well as from v to u.

The in-degrees and out-degrees of all vertices are given in Table 9.1.

TABLE 9.1

Vertex	In-degree	Out-degree
a	1	3
c	1	1
m	2	2
z	2	2
w	2	2
q	2	1
r	2	1

In-degree sequence of the graph is given by (2, 2, 2, 2, 2, 1, 1). Out-degree sequence of the graph is given by (3, 2, 2, 2, 1, 1, 1). Further, no vertex is disconnected from the graph, since degree of any vertex of this graph is not 0 (zero). Again, from the graph, it is clear that for any two distinct vertices $u \in V, v \in V$ in this digraph $G(V, E)$, there exists a path from u to v as well as from v to u. We can check its strong connectivity by running DFS.

There are no self-loops or multiple edges in the given graph. Also, in-degree as well as *out-degree* of each vertex is at least 1. Further, no vertex is disconnected from the graph.

Let us try to find the *connected components* of the graph by running DFS.

Start DFS from vertex a. One possible DFS sequence is $a \to c \to m \to r \to w \to z \to q$

The DFS numbers of the nodes are $a = 1, c = 2, m = 3, r = 4, w = 5, z = 6$, and $q = 7$.

We start from the vertex with maximum DFS number, i.e., q.

Now, run the DFS on the transpose of the given graph, i.e., on the graph obtained by reversing all edge directions. Starting from q, we get the DFS sequences as $\{q, w, a, m, c, r, z\}$.

We get one connected component. Hence, the graph is strongly connected.

9.13 CUT VERTEX

We sometimes see that the removal of a vertex and all the edges incident with it produces a subgraph with more connected components. A cut vertex of a connected graph G (with at least 3 vertices) is a vertex whose removal increases the number of components. Clearly, if v is a cut vertex of connected graph G, then $G - \{v\}$ is a disconnected graph. A cut vertex called as a *cut point*, is also known as *articulation point*. An example through Figure 9.47 is illustrated below.

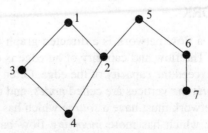

FIGURE 9.47 G

The graph G as shown in Figure 9.47 has a cut vertex at 2 because its deletion creates two components: H_1 and H_2 as shown in Figure 9.48.

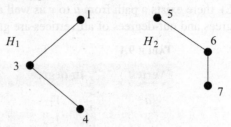

FIGURE 9.48 G_1

Block A connected graph with no cut vertex is called as *block*. An example of block is shown in Figure 9.49.

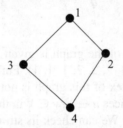

FIGURE 9.49 G

Cut edge An edge $e = (x, y)$ of a connected graph G is a *cut edge* or *bridge* if and only if its deletion increases the number of components. In a tree, every edge is a cut edge. In the graph G in Figure 9.50, the edge $(3, 4)$ is an example of cut edge because its removal creates two *components* as shown in Figure 9.51.

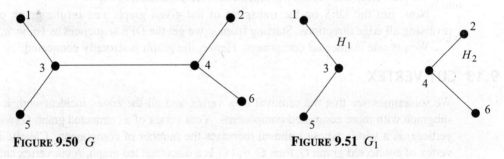

FIGURE 9.50 G **FIGURE 9.51** G_1

9.14 FLOW NETWORK

In graph theory, a flow network is a directed graph where each edge has a *capacity* and it receives a *flow*. The flow and capacity of an edge is denoted by *f/c*. The amount of flow on an edge cannot exceed the capacity of the edge. Often in operations research, a directed graph is called a *network*, the vertices are called *nodes*, and the edges are called *arcs*.

Every flow network must have a *source* which has more outgoing flow but zero incoming flow, and a *sink* which has more *incoming* flow but zero outgoing edge. Further, a flow network fulfils the *conservation condition* presented below. As per this condition, for every vertex j in V, where j is not the *source s* or the *sink t*, the sum of the flow into j equals the sum of the flow out of j. A flow that satisfies the conservation condition is called a *feasible flow*.

Applications Picture a series of water pipes, fitting into a network. Each pipe is of a certain diameter, so it can only maintain a flow of a certain amount of water. Wherever two or more pipes meet, the total amount of water coming into that junction must be equal to the amount going out, otherwise we would quickly run out of water, or we would have a build up of water. We have a water *inlet*, which is the source, and an *outlet*, the sink. A flow would then be one possible way for water to get from source to sink so that the total amount of water coming out of the outlet is consistent. Intuitively, the total flow of a network is the rate at which water comes out of the outlet.

Flows can pertain to people or material over transportation networks, or to electricity over electrical distribution systems. For any such physical network, the flow coming into any intermediate node needs to equal the flow going out of that node. Bollobás characterizes this constraint in terms of Kirchhoff's current law, while later authors (i.e., Chartrand) mention its generalization to some conservation equation.

Flow networks also find applications in ecology: flow networks arise naturally when considering the flow of nutrients and energy between different organizations in a *food web*. The mathematical problems associated with such networks are quite different from those that arise in networks of fluid or traffic flow.

Edges—saturated, free, and positive Let f be a feasible flow in G. Edge (i, j) is said to be

(a) Saturated if $f(i, j) = c(i, j)$
(b) Free if $f(i, j) = 0$
(c) Positive if $0 < f(i, j) < c(i, j)$

Augmenting path An augmenting path is an alternating sequence of vertices and edges of the form $s, e_1, v_1, e_2, v_2, \ldots, e_k, t$ in which no vertex is repeated, no forward edge is saturated, and no backward edge is free. Clearly, augmenting path contains more flow.

The *residual capacity* (rc) of an edge (i, j) equals $c(i, j) - f(i, j)$ when (i, j) is a *forward* edge, and equals $f(i, j)$ when (i, j) is a *backward* edge, as shown in Figure 9.52.

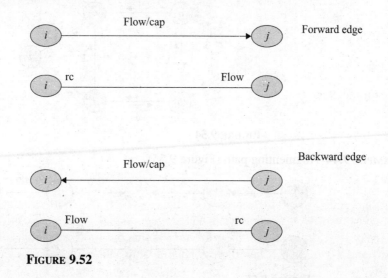

FIGURE 9.52

For better understanding, we may see a flow network (shown in Figure 9.53) with source labelled S, sink t, and four additional nodes. The flow and capacity is denoted by f/c. The total amount of flow from S to t is 5, which can be easily seen from the fact that the total outgoing flow from S is 5, which is also the incoming flow to t. We know that no flow appears or disappears in any of the other nodes.

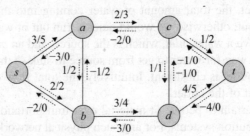

FIGURE 9.53

In Figure 9.54, we see the *residual network* for the given flow. Notice how there is positive residual capacity on some edges where the original capacity is *zero*, for example for the edge (d, c). This flow is not a maximum flow.

Actually, the maximum flow problem is to find a feasible flow through a single-source, single-sink flow network that is maximum. Sometimes it is defined as finding the value of such a flow. There is available capacity along the paths (s, a, c, t), (s, a, b, d, t), and (s, a, b, d, c, t), which are then the augmenting paths.

The residual capacity of the first path is

$$\min(c(s, a) - f(s, a), c(a, c) - f(a, c), c(c, t) - f(c, t)) = \min(5 - 3, 3 - 2, 2 - 1)$$
$$= \min(2, 1, 1) = 1$$

Notice that augmenting path (s, a, b, d, c, t) does not exist in the original network, but we can send flow along it, and still get a legal flow.

If this is a real network, there might actually be a flow of 2 from a to b, and a flow of 1 from b to a, but we only maintain the *net* flow.

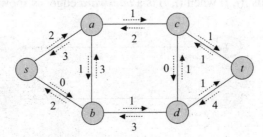

FIGURE 9.54

 EXAMPLE 9.44 Augmenting path (Figure 9.55)

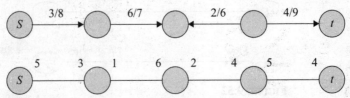

FIGURE 9.55

The excess flow capacity of an augmenting path equals the minimum of the residual capacities of each edge in the path. We can increase the flow in the path from S to T in the above diagram by determining the excess flow capacity of this path. From left to right the residual capacities (the amount of flow can be increased on the edge) are the first number on each edge (see Figure 9.56).

$$\text{minimum}(5, 1, 2, 5) = 1$$

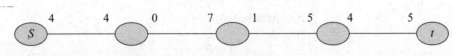

FIGURE 9.56

9.14.1 Ford–Fulkerson Algorithm

The Ford–Fulkerson algorithm determines the maximum flow of the network. The high-level description of the approach is presented below.

```
Algorithm FORD_FULKERSON(G, s, t)
    /* Here, G is the connected graph, s and t are source and
sink vertices */
    /* f(u,v) represents flow through the edge(u,v), whereas
C_f(u,v) its residual capacity. */
    /* Residual capacity of an edge(u,v), C_f(u,v)=current capacity
- current flow. C_f(P) represents  residual capacity of path P.
Initialize the edges in E(G)  which  exist in G */
            begin
                for each edge (u,v) ∈ E(G) do
                    f(u, v) ← 0
while(there exists a path P from s to t in resudual network G_f)
            begin
                C_f(P) ← min{C_f(u,v) | (u,v) is in P}
                    for each edge (u,v) ∈ P do
                        begin
                            f(u,v) ← f(u,v)+C_fP)
                            C_f(u,v) ← C_f(u,v)-C_fP)
                        end
            end
        end //of the  algorithm
```

Clearly, the algorithm terminates when no augmenting (residual) path exists.

Theorem 9.5 A flow in a capacitated network is a maximum flow if and only if there is no augmenting path in the network (see Figures 9.57–9.61).

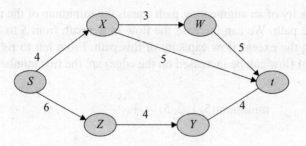

Saturated edge

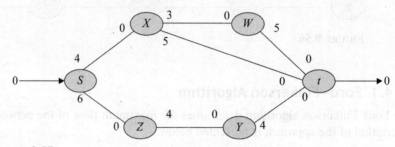

Figure 9.57

Augmenting path: $s \to X \to W \to t$
Excess capacity of $s \to X \to W \to t = \min(4, 3, 5) = 3$

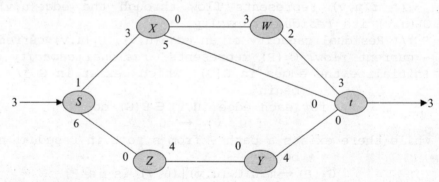

Figure 9.58

Augmenting path: $s \to X \to t$

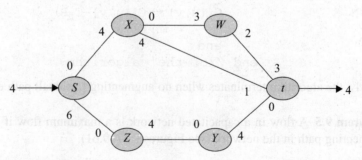

Figure 9.59

Excess capacity of $s \to X \to t = \min(1, 5) = 1$
Augmenting path: $s \to Z \to Y \to t$
Excess capacity of $s \to Z \to Y \to t = \min(6, 4, 4) = 4$

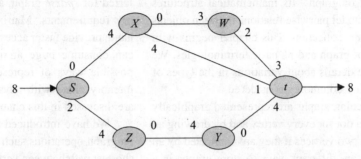

Figure 9.60

At this point, there are no remaining augmenting paths. Therefore, the maximum flow is 8.

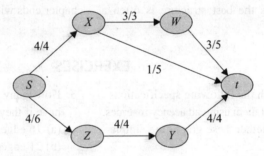

Figure 9.61

Some Important Applications of a Graph

(i) *Creating optimal public transport routes in a big city* Various prominent locations in a city can be treated as nodes (vertices) of a graph and all available roads connecting those points can be treated as edges. This design can be used in designing an optimal route scheme for city buses or local trains running in the city, so as to cover every prominent location with minimum possible number of buses/trains in minimum possible time, with suitable frequency for each location.

(ii) *Designing of municipal water supply system for a city* For a big city, when providing municipal water supply connections to various houses, the most important factor to be considered is the usage of minimum possible length of pipes. This problem can be modelled as a variant of *minimum spanning tree* problem, where edge weights are the lengths of the corresponding pipes. One node is the source node, from where water is supplied to entire city, and all other nodes are either the houses or intermediate water storage tanks.

(iii) *Laying out telephone cables in a city* Similar to above, the constraint is to use the minimum possible length of cable wires.

SUMMARY

In mathematics and computer science, graph theory is the study of graphs. Its mathematical structures are used to model pairwise relations between objects from a certain collection. This chapter begins with definitions of graph and its basic terminologies. We then provide details about variations in the types of graphs that are commonly considered.

In particular, graphs are represented graphically by drawing a dot for every vertex and by drawing an *arc* between two vertices if they are connected by an edge. There are different ways to store graphs in a *computer memory*. The data structure used depends on both the graph structure and the algorithm used for manipulating the graph. Theoretically, one can distinguish between *list* and *matrix structures* but in concrete applications the best structure is often a combination of both. List structures are often preferred for *sparse* graph, as they have smaller memory requirements. Matrix structures, on the other hand, provide faster access for some applications but can consume huge amounts of memory. All the possible ways of representing graph in computer memory along with advantages and disadvantages are discussed in this chapter.

We have introduced here a concrete discussion on graph operations such as DFS, BFS, and finding shortest path between vertices. Further, some special kinds of graphs such as *Euler* and *Hamiltonian* are discussed in this chapter. The promising topics like *graph colouring* and *the maximum flow of the network* are also dealt with examples. Finally, this chapter ends with an application section.

EXERCISES

1. Draw graphs with the following specifications and then represent them using adjacency matrices. Also, explain whether these graphs are simple graphs or not.
   (a) Graph with four vertices of degree 1, 1, 2, and 3
   (b) Graph with four vertices of degree 1, 1, 3, and 3
   (c) Graph with five vertices of degree 0, 1, 2, 2, and 3
   (d) Graph with five vertices of degree 3, 3, 3, 3, and 2
2. For each of the following degree sequences, determine if there exists a simple graph whose degree sequence is given. If possible, draw the graph or explain why such a graph does not exist.
   (a) (2, 3, 3, 4, 4, 5)   (b) (2, 3, 4, 4, 5)
   (c) (1, 1, 1, 1, 4)   (d) (1, 3, 3, 3)
   (e) (1, 2, 2, 3, 4)   (f) (1, 3, 3, 4, 5, 6, 6)
3. How many vertices do the following graphs have if they contain
   (a) 16 edges and all vertices of degree 2
   (b) 21 edges, 3 vertices of degree 4, and others each of degree 3
4. Show that there is no graph G with n (number of vertices) = 12 and e (number of edges) = 28 in which each vertex is of degree either 3 or 6.
5. Prove that, in a graph, total number of odd degree vertices is *even* but the number of even degree vertices may be odd.
6. Every set with n elements is known to have 2^n subsets, including the null set. Suppose each of these subsets represents a vertex of a graph. We define that 2 vertices of this graph have an edge between them if and only if the corresponding subsets represented by those vertices intersect in exactly 3 elements. How many isolated vertices exist in this graph?

7. (a) Let A be the adjacency matrix of a graph G with n vertices. Prove that G is a connected graph if and only if the matrix $(I^n + A)^{n-1}$ contains no zeros.
   (b) Let G be some graph and A be its adjacency matrix. Find the necessary and sufficient conditions for A^3 to be adjacency matrix of G.
8. A graph has the following adjacency matrix. Show that it is connected.

$$\begin{bmatrix} 0 & 1 & 0 & 0 & 0 \\ 1 & 0 & 1 & 1 & 1 \\ 0 & 1 & 0 & 1 & 1 \\ 0 & 1 & 1 & 0 & 0 \\ 0 & 1 & 1 & 0 & 0 \end{bmatrix}$$

9. Determine whether the following graphs are isomorphic.

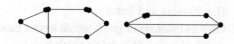

10. Find the complements of the following graphs.

11. (a) Determine whether the graphs are isomorphic or not.

(b) Find the union and complement of the above graphs.

12. Any graph which is isomorphic to its complement graph is called a self-complementary graph. Let n be the number of vertices in a self-complementary graph. Prove that either n or $(n - 1)$ is a multiple of 4.

13. Determine whether the graph as shown below is a simple graph, multi-graph, or pseudo-graph. Find the degree of each vertex. Also, find a simple path.

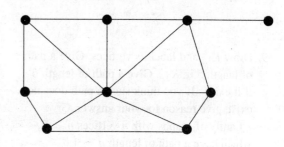

14. Suppose that a graph G is regular of degree r, where r is odd.
    (i) Prove that G has an even number of vertices.
    (ii) Prove that the number of edges in G is a multiple of r.

15. Given an undirected graph containing m edges. The degree of every vertex v in this graph is strictly defined by $k_1 \leq d(v) \leq k_2$. Also, $k_1 \leq k_2/2$ and $k_2 < \sqrt{m}$. What is the minimum and maximum possible number of vertices in this graph?

16. Prove that a graph is bipartite if and only if it does not have 2 adjacent vertices that have the same finite distance to a third vertex.

17. (i) How many complete bipartite graphs have k vertices?
    (ii) What is the maximum number of edges in a simple bipartite graph with k vertices?
    (iii) Check whether the graph is bipartite or not.

18. Show that the graph is bipartite.

19. Draw $K_{3,3}$ and label its vertices. Give a path of length 5 in $K_{3,3}$. Give a path of length 6 if it exists. If you think such a path does not exist, give reason for your answer. Give a family of graphs with n vertices $n \geq 2$, which have a path of length $n - 1$.
20. Draw a 4-regular graph on 6 vertices.
21. Draw K_5 and $K_{3,3}$. What is so special about them?
22. Let $G_1 = (V, E_1)$ and $G_2 = (V, E_2)$ be connected graphs on the same vertex set V with more than 2 vertices. If $G_1 \cap G_2 = (V, E_1 \cap E_2)$ is not a connected graph, then prove that the graph $G_1 \cup G_2 = (V, E_1 \cup E_2)$ must have at least one cycle.
23. Let G be a connected bipartite graph. Show that if G is d-regular with $d \geq 2$, then G does not contain any cut-vertex.
24. Let T be a DFS tree of an undirected graph G. Vertices u and v are leaves of this DFS tree. In the graph G, the degrees of both u and v are at least 2. Prove that G must have a cycle containing both u and v.
25. Consider the following greedy algorithm for finding the shortest path from a start vertex to a goal vertex in any graph.
    (i) Initialize path to start
    (ii) Initialize visited vertices = {start}
    (iii) If (start = goal) then return path and exit, else continue
    (iv) Find the edge (start, v) of minimum weight such that v is adjacent to start, and v is not in visited vertices
    (v) Add v to path
    (vi) start = v, and then go to step (iii)
    Prove or disprove – this algorithm always generates a shortest path correctly.
26. Find the minimum distance between two vertices K and L of the graph, using Dijkstra's algorithm.

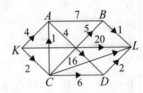

27. Find all the bridges of the graph as shown below

28. Give an example of graph which is both Hamiltonian and Euler.
29. Give an example of graph which has an Eulerian circuit but not a Hamiltonian circuit.
30. Give an example of graph which has a Hamiltonian circuit but not an Eulerian circuit.
31. Give an example of graph which has neither a Hamiltonian circuit nor an Eulerian circuit.
32. Prove that any Hamiltonian path of any graph cannot contain the same edge more than once.
33. Prove that if a graph has no cycles then any Euler path (if existing) in this graph cannot contain the same vertex multiple times.
34. A graph G has k number of mutually disjoint Hamiltonian circuits, i.e., no two Hamiltonian circuits have any common edges. What is the maximum value of k for a fully connected graph containing n vertices?

35. State Euler's formula for a planar graph. Give an example of a planar graph with 5 vertices and 5 regions, and verify Euler's formula for your example.
36. Consider the graph as shown below. Is there any Euler circuit in this graph?

37. Give a planar drawing of $K_{2,4}$. Verify Euler's formula for $K_{2,4}$.
38. Apply Fleury's algorithm, beginning with vertex A, to find an Eulerian trail in the following graph. In applying the algorithm at each stage choose the edge (from those available) which visits the vertex that comes first in alphabetical order.

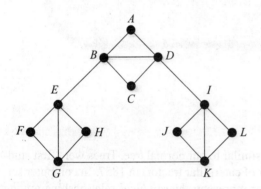

39. The *skewness* of a graph is the minimal number of edges which need to be removed from the graph to make it planar. For any graph G with $n \geq 3$ vertices and m edges, prove that skewness $(G) \geq m - 3n + 6$.
40. (i) Show that a connected simple planar graph all of whose vertices have degree at least 5 must have at least 12 vertices.
    (ii) Show that a connected simple planar graph with fewer than 30 edges has at least one vertex of degree at most 4.
41. An airline operates flights between 9 cities. What is the minimum number of flights the airline should operate so that it is possible to travel by its flights between any two cities, changing flights if necessary. Solve the problem using graph theory.
42. Give a spanning tree of the graph as shown below.

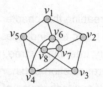

43. Let G be a graph on at least 3 vertices with the property that there is a Hamiltonian path between any pair of vertices u and v, i.e., starting at u and ending at v. Prove that G has a Hamiltonian cycle.
44. In an undirected graph, a Hamiltonian circuit is also the Euler circuit for the graph. Prove that the graph is 2-regular.
45. What is the minimum number of vertices in a simple graph G so that either G or its complement graph G' must have a cycle?
46. Distinguish between k-colouring of a graph and chromatic number of a graph.
47. An *edge colouring* of a graph is an assignment of colours to edges of a graph such that all edges incident on any vertex have distinct colours. Prove that a d-regular bipartite graph can be edge-coloured using exactly d number of colours.

Tree

> **LEARNING OBJECTIVES**
>
> After reading this chapter you will be able to understand the following:
> - The basic terminologies related to tree
> - m-ary tree, binary tree, and its variations
> - Representation of binary tree in computer memory
> - Traversal techniques of tree
> - Creation of a unique binary tree
> - Binary search tree, AVL tree, and expression tree
> - Minimum spanning tree
> - Application of tree structure

10.1 INTRODUCTION

A tree is a type of data structure that looks very similar to our normal tree. Trees were first studied by the British scientist, Cayley (the inventor of caterpillar tractor) in 1857. In computer science, 'a tree' is a non-linear data structure that represents *hierarchical* relationships among individual data items, i.e., on the basis of the hierarchies of the objects, branches are constructed from each object (item) and it takes a little imagination to see a natural tree. For example, we may consider the hierarchy of a *family* such as a grandfather (at the root of the tree) and then its next generations in the subsequent levels. However, a *computer data tree* is different from a *plant tree* in that the root is at the top and the leaves are at the bottom. In fact, the tree, as represented in computer science, is a linked list, where the link between two objects (nodes) is a branch. Many real-world examples such as business organization, educational organization, and so on, can be easily represented and operated by a tree structure.

Trees are used to organize information in database systems and to represent the syntactic structure of source programs in compilers. Directory structure of operating systems and the symbol table part of compiler are also organized as trees. They are also helpful in analysing electrical circuits. Some variations of trees such as binary search tree and AVL tree allow rapid access of information stored in their nodes. Further, the concept of tree is used to express the solution of some complicated problems in easier way. Moreover, the concept of the minimum spanning tree (MST) helps in all network design problems such as telephone company, computer network, travelling agency, and so on, to choose the cheapest subset of edges that keeps the graph in one-connected component.

This chapter presents a detailed description of trees from basic concepts to their representations, manipulations, discussion on several of its types, and important applications.

10.2 TREE

A tree (or *free tree*) can be described as any connected, acyclic graph. Again, a *rooted* tree is a free tree in which one vertex is specifically identified as a *root*. The term *tree* commonly means a *rooted* tree. In particular, a tree with single vertex is called a *trivial* tree. On the other hand, a non-trivial tree has *more* than one vertex.

Mathematically, a tree (T) can be defined as $T = (V, E)$, where V is a finite set of vertices (nodes): $V = \{v_1, v_2, \ldots, v_n\}$ and E is a finite set of edges: $E = \{e_1, e_2, \ldots, e_m\}$. Here, v_i represents the ith vertex $(1 \leq i \leq n)$, and e_j, the jth edge $(1 \leq j \leq m)$ in the tree.

Concisely, we can say that 'a tree' is a finite set of one or more connected nodes. All the vertices of a tree are *connected* (linked) through edges in such a way that there is only *one* path in between any *two* vertices, i.e., there must not be more than one path in between any two vertices in a tree. If more than one path exists between two vertices, then a *cycle* (loop) will be formed, which is not permissible for a tree. Clearly, if n number of vertices are present in a tree, then the number of edges must be $n - 1$.

Further, a tree may be *directed* or *undirected*, and can be implemented in two ways: (a) sequentially and (b) dynamically. Actually, in sequential representation, memory is allocated at the time of compilation (i.e., static allocation of memory) by using array, whereas dynamic allocation means provision of memory allocation at the time of execution by the use of pointer.

Note It is very obvious from the definition of a tree that the maximum degree of each vertex in a tree with n vertices is $n - 1$.

10.2.1 Common Terminologies of Tree

Since a tree is a graph, so some terminologies such as edge (directed and undirected), degree of vertex (both in-degree and out-degree), etc., are already introduced in Chapter 9. However, some more important terminologies on trees are explained below. Also, the definitions are examined through some examples with diagrams presented in Section 10.2.3.

Child node The child nodes of a tree are continued to consider from its specified *root* node placed at the starting level (normally taken as zero). In particular, the child nodes of a *parent* are the nodes of its just lower level. Each node has zero or more child nodes.
Leaf node Nodes with no children are called leaf nodes.
Internal node Nodes with at least one child are called internal nodes.
Siblings Children with same parent are called siblings, i.e., siblings share the same parent.

Path The path between any two nodes (source and destination) of a tree is found by traversing the edges to move to the destination starting from the source, and its *path distance* is computed by the *number* of edges required to reach the destination.

Level The level of a node is the *distance* from the root to that node. The distance is measured by counting the number of edges through that path. Normally, the level of the root node is set at level 0. Now, if a node is at level p, then its *children* will be at level $(p + 1)$.

Depth The depth of a vertex v in a tree is the level in which it is placed.

Height It is the *maximum* level of a tree. In fact, it is measured by counting the number of edges in the path (from root) with maximum distance.

Average height If n nodes are present in a tree T, then the level at which approximately $n/2$ nodes (average number of nodes) are present is considered as the average height of the tree T.

Sub-tree Any node in a tree T, together with all of its descendant nodes (i.e., its child nodes and their children, grandchildren, and so-on), comprises a sub-tree of T.

Note In directed tree, the degree of a vertex (node) is the *sum* of its in-degree and out-degree. Further, the in-degree of the root node and the out-degrees of the leaf nodes are *zero*. However, in any tree (directed or undirected) the total degree of any leaf node is exactly 1, whereas it is greater than 1 for any intermediate node.

10.2.2 Labelled Tree

A tree T is labelled when its n vertices are distinguished from one another by names such as $v_1, v_2, \ldots, v_n$. The vertices of a labelled tree on n vertices are typically given the labels $1, 2, \ldots, n$. Two-labelled trees are considered to be *distinct* if they have different vertex labels even though they might be isomorphic.

According to Cayley's tree formula, there are n^{n-2}-labelled trees on n vertices. For deducing the formula, let $v \in V(T)$ be a node of tree T. Obviously, $1 \leq \deg(v) \leq k = (n - 1)$, since we know that maximum degree of any vertex in a tree with n vertices can be $(n - 1)$. Now, on summing over all possible values of k, we get the number of labelled trees $T(n)$ on n vertices as

$$T(n) = \sum_{k=1}^{n-1} T(n, k) = \sum_{k=1}^{n-1} \binom{n-2}{n-1}(n-1)^{n-k-1} = n^{n-2}$$

Thus, a labelled tree is a tree in which each vertex is given a *unique* label. Clearly, we can consider any tree presented in this chapter as labelled tree. On the other hand, a *recursive tree* is a labelled rooted tree, where the vertex labels respect the tree order (i.e., if $u < v$ for two vertices u and v, then the label of u is smaller than the label of v). For example, labels of the tree as shown in Figure 10.1 are v_1 and v_2, respectively.

FIGURE 10.1 A simple tree

However, it is sometimes useful to label the vertices or edges of a tree to indicate that the tree is being used for a particular purpose. This is especially true for many uses of trees in computer science. For example, in expression tree (discussed in Section 10.9), each label indicates the order of computation of sub-expression (i.e., priority of the operators).

10.2.3 Some Diagrams of Directed and Undirected Trees

Several diagrams of directed and undirected trees are presented below for better understanding of the above discussion (specifically the definitions of terminologies of tree).

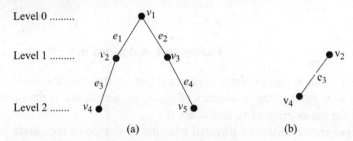

FIGURE 10.2 (a) An undirected tree with root at v_1, (b) one sub-tree of the tree as shown in (a)

From Figure 10.2(a), it is clear that the labels of the vertices of the tree are, respectively, v_1, v_2, v_3, v_4, v_5, i.e., the number of the vertices in the tree is 5. Also, the total number of edges is 4.

The *root* vertex v_1 of this tree is at level 0, and v_2 and v_3 are its children (at level 1). Hence, v_1 is the *parent* of v_2 and v_3, and v_2 and v_3 are *siblings* to each other, as their parent is same (i.e., v_1). Clearly, v_4 is the only child of v_2.

One *sub-tree* of this tree with parent v_2 is shown in Figure 10.2(b).

Again, in this tree, the degrees of v_1, v_2, v_3, v_4, and v_5 are, respectively, 2, 2, 2, 1, and 1 (since degree of a vertex is the total number of edges incident with it). Obviously, both v_4 and v_5 (at level 2) are the *leaf* (terminal) nodes of this tree, as these have no children (i.e., degree of each of these node is 1). On the other hand, v_1, v_2, and v_3 are called as *internal* (intermediate) nodes (as a degree of these nodes is greater than 1).

Now, consider any two vertices of this tree, say, v_1 (*source*) and v_4 (*destination*), it can easily be observed that there is only one path in between these two vertices which is v_1–v_2–v_4. In fact, no other alternate path is present here to move from v_1 to v_4. Further, the path distance to move to v_4 from v_1 is 2, as the *total* number of edges required to traverse here is 2. Also, the *height* of this tree is 2, as the *maximum* level of this tree is 2.

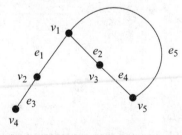

FIGURE 10.3 A graph with cycle

Certainly, Figure 10.3 may not be a tree, since one cycle is formed due to the presence of more than one path in between two vertices v_1 and v_5.

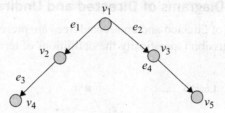

FIGURE 10.4 A directed tree

Figure 10.4 shows an example of *directed* tree. In this tree, the *in-degree* of v_1 is 0 and its *out-degree* is 2. Further, the in-degrees of v_2, v_3, v_4, and v_5 are 1, the out-degrees of v_2 and v_3 are 1, and the out-degrees of v_4 and v_5 are 0.

As per the definition of directed tree, the in-degree of root node is always 0, whereas the out-degree of every leaf node is 0. Accordingly, v_1 is the root of this tree (since its in-degree is 0) and v_4 and v_5 are its leaf nodes (since the out-degree of these is 0).

Comment In any type of tree (directed or undirected), there must exist a path between root and other vertices which is not more than 1.

The above comment is not valid, i.e., there may not exist a path between root and other vertices of a directed tree or there may exist more than one path between root and other nodes if any one of the following statements is not true.
(a) The in-degree of root must be zero.
(b) In directed tree, the in-degree of any node (except *root*) is at most 1 and the out-degree of every leaf node is 0.

The truth of the above statements is explained through some examples (presented below).

(i) Let us consider v_1 to be the root of the directed tree as shown in Figure 10.5(a). This tree is not a *valid* directed tree with *root* vertex at v_1, since the in-degree of v_1 (root) is 1, and there exists no path from v_1 to v_3 or v_5 to traverse. However, if we consider v_3 as the root of this tree, then it is a *correct* example of tree (since, in such case, all the properties of directed tree are satisfied here.

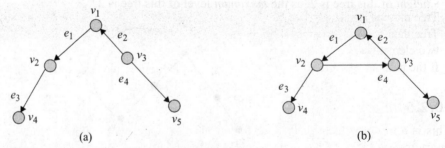

FIGURE 10.5 A directed tree considering root at v_1

Note The proper selection of root in a directed graph may treat the graph as tree.

(ii) In Figure 10.5(b), consider that the root is at v_1. The in-degree of root v_1 is 1, and there is a path: v_1–v_2–v_3–v_5, from v_1 (root) to v_5. But one *cycle* is formed through: v_1–v_2–v_3–v_1, and this is due to the in-degree of v_1 greater than 0. In this tree, there are *multiple* paths. Clearly, it also may not be a *correct* tree, and cannot be considered (in any way) a valid tree considering any suitable root, since *cycle* exists here.

(iii) Assume v_1 as the root of the tree as shown in Figure 10.6. It is clear that the in-degree of v_1 is 0, and there is a path from v_1 to v_5, i.e., v_1–v_3–v_5. But another alternate path: v_1–v_2–v_3–v_5 is also present here. Therefore, there exist multiple paths from v_1 to v_5, and this is because of the in-degree 2 of v_3 (i.e., greater than 1, which is not allowed).

Hence, it may not be an example of tree with root v_1, as there are more than one path between root to node v_3. However, no cycle is formed here.

Further, it cannot be treated (in any way) as a valid tree considering any suitable root, since the in-degree of any node is greater than '1'.

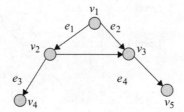

FIGURE 10.6 A directed diagram considering root at v_1

10.2.4 Review of the Basic Properties of a Tree

A tree is a *non-linear* data structure consisting of nodes (containing data) and edges (connections between nodes), such that

- One node, the root has no parent
- Every other node has exactly one parent node
- There is a unique path from the root to each node (i.e., the tree is connected and there are no cycles)
- Tree may be empty
- Tree data structure contains not only a set of elements but also connection (edge) between two elements
- If the number of vertices in the tree is n, then the number of edges must be $(n-1)$

10.2.5 *m*-ary Tree

This is a tree in which every node has at most m number of children. However, in strictly m-ary tree, every node has either m children or no children. Now, if the value of m is 2, then that tree is termed as *binary tree*. The binary tree is the simplest form of tree. Clearly, in a binary tree, each node may have at most two children, whereas every node in a strictly binary tree possesses either two or no children. Some *special* kinds of binary tree like complete, full, and skewed, etc., are discussed here.

A *complete binary tree* is one in which leaf nodes appear on at most two adjacent levels.

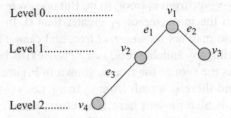

FIGURE 10.7 A complete binary tree

The example as shown in Figure 10.7 is a complete binary tree, as leaf nodes v_3 and v_4 appear at two adjacent levels 1 and 2. However, the above idea is true for *complete m-ary* tree also.

A *full binary tree* is one in which all leaf nodes appear at the same level and all the intermediate nodes have exactly two children.

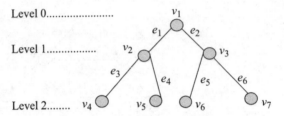

FIGURE 10.8 A full binary tree

The example (shown in Figure 10.8) is a full binary tree, as all the leaf nodes v_4, v_5, v_6, and v_7 are at the same level (i.e., at level 2), and all the intermediate nodes v_1, v_2, and v_3 have exactly two children.

One can think of a *full m-ary* tree in which every internal node has exactly m number of children and all leaf nodes appear at the same level.

Comment Some authors treat full *m*-ary and complete *m*-ary tree as same.

10.2.6 Why Are Skewed Trees Considered as Binary Trees?

A skewed tree is considered as a binary tree, since such a tree has only one side (or branch) either *left* or *right*. In fact, no other branch except these *two* is considered, i.e., at most two sides may be taken into account. So, such type of tree should be included into binary tree. However, they are not strictly binary tree.

Thus, it may be of two types: (a) *left-skewed* and (b) *right-skewed*. In the left-skewed, only left child of every node (except leaf) is present (i.e., every node has only left sub-tree but none of the nodes has any right child, i.e., no right sub-tree). On the other hand, in case of right-skewed, only right child of every node (except leaf) is present but none of the nodes has any left child. Examples are illustrated in Figure 10.9.

Equivalent Tree

Two trees are equivalent if and only if they are structurally equivalent and the data in the corresponding nodes of the two trees are same.

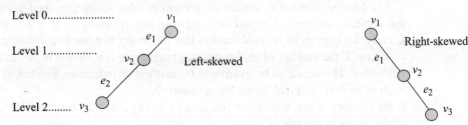

FIGURE 10.9 Two-skewed binary trees

10.3 SOME IMPORTANT RESULTS ON TREE

(i) The maximum number of nodes on level i in an m-ary tree is m^i. This is true only when the tree is full m-ary tree, assuming that the level starts from 0.

Correctness Since the m-ary tree is full, so each internal node has m children. Thus, at level 0, the number of nodes is 1 (which is the root). The number of nodes at level 1 is m (since the root has m children). In addition, every node of level 1 has m nodes, so the number of nodes at level 2 will be m^2.

Hence, by the method of induction, the number of nodes at ith level of a full m-ary tree will be m^i. Now, if the value of m is 2 (i.e., full binary tree), then it will be 2^i.

(ii) Maximum number of nodes with height h in an m-ary tree is $(m^{h+1} - 1)/(m - 1)$. This is true when the tree is a full m-ary tree. Also, it is assumed here that *level* starts from '0' (i.e., at level '0', the value of h is '0').

Correctness Since the tree is a full m-ary tree, each internal node has m children. So, at level 0, the number of nodes is 1 (which is root). The number of nodes at level 1 is m (since root has m children).

Now, suppose the tree has only *two* levels (i.e., h is 1), then the total number of nodes will be

$$(m + 1) = \frac{m^2 - 1}{m - 1} = \frac{m^{1+1} - 1}{m - 1}$$

Suppose the tree has three levels. Then, at level 2 (i.e., h is 2), the number of nodes is m^2 and the total number of nodes in the tree will be

$$m^2 + m + 1 = \frac{m^3 - 1}{m - 1} = \frac{m^{2+1} - 1}{m - 1}$$

Therefore, the total number of nodes in an m-ary tree with height h can be formulated as $(m^{h+1} - 1)/(m - 1)$ by the method of induction. Clearly, if it is a binary tree, then this value must be $(2^{h+1} - 1)$.

(iii) A full m-ary tree with i internal nodes consists of total $n = m * i + 1$ vertices. Since the tree is a full m-ary tree, so each internal node has m children. Further, the number of internal vertices in the tree is i. Accordingly, the total number of vertices will be $m * i + 1$ (this is also obtained by induction method).

On the other hand, if n vertices are present in a full m-ary tree, then the number of *internal* vertices $i = (n - 1)/m$ and the number of leaf vertices $L = n - i = n - (n - 1)/m$. Clearly, the number of internal vertices in a full m-ary tree must be divisible by m.

Again, if the number of leaf-nodes in a binary tree is n, then it must have $(n - 1)$ internal nodes. This can also be examined by method of induction. Further, it is obvious that each of $(n - 1)$ internal nodes has a degree 2.

(iv) If the number of leaf nodes in a full m-ary tree is p, then the total number of null pointers of the nodes in the tree is $m * p$.

In particular, a *null* pointer is a *link* which is not presently connecting to any node. However, every node in a full m-ary tree may have at most m pointers. Obviously, each internal node has exactly m children (i.e., no link of any internal node is unfilled). But only p nodes (which are leaves) do not have any children, i.e., their links are completely unfilled. Thus, the total number of null pointers of such tree is $m * p$, since each of p leaves has m null pointers.

On the other hand, if the tree is a *skewed binary tree* and it consists of n nodes, then the total number of null pointers in the nodes reduces to $(n - 1) + 2 = n + 1$ [as each of $(n-1)$ nodes has single null pointer and both the left and the right pointer of the last node (leaf) are null].

(v) A full m-ary tree with n nodes has height $h = \lfloor \log_m n \rfloor$. Thus, for a full binary tree, it is $\lfloor \log_2 n \rfloor$.

(vi) Let $v_1, v_2, v_3, \ldots, v_n$ denote n vertices in an n-vertex tree. Then the number of edges $(e) = n-1$. This can be easily understood by induction on n as follows.

If $n = 1$, then the tree has only one vertex. Since a tree cannot contain any cycle, so the number of edges e is zero (0). Now, if $n = 2$, then the number of vertices in such a tree is 2. Clearly, the number of edges e is 1, since it is a tree. Thus, for n number of vertices, the number of edges must be $n-1$.

(vii) An m-ary tree with p leaf nodes must have at least $\lceil \log_m p \rceil$ depth.

(viii) A left-skewed/right-skewed binary tree with n nodes has height $(n - 1)$, assuming that the height starts at 0.

In a skewed binary tree, each node has only one pre-decided side. So, if the number of nodes in such a tree is n, then the height of the tree will be $(n - 1)$.

(ix) In a binary tree, if the height is h ($h \geq 0$, i.e., h starts from 0), then the number of nodes n varies like: $(h+1) \leq n \leq 2^{h+1} - 1$, where $(h + 1)$ is the minimum number of nodes in the tree with height h (when the tree is a skewed tree) and $2^{h+1} - 1$ is the maximum number of nodes in the tree with height h (in case of a full binary tree).

(x) For n number of vertices (n is non-negative integer), then

(a) the total number of different binary trees, which can be constructed, is

$$b_n = \frac{1}{n + 1}({}^{2n}C_n) \qquad (10.1)$$

(b) the number of labelled trees on n nodes, which can be constructed, is n^{n-2}.

(xi) Each tree is a bipartite graph.

EXAMPLE 10.1 Prove that a tree with more than 1 vertex has at least 2 leaves.

Solution A tree T is given with more than 1 vertex. By definition, a tree is an acyclic, connected, and undirected graph.

Also, a tree with n vertices has exactly $(n-1)$ edges. Now, consider the sum of degrees of all vertices. It is obviously $2e = 2(n-1)$ [by *handshaking* theorem].

Let there be exactly k vertices with degree $= 1$, where $k \geq 2$, and hence all other $(n-k)$ vertices have degree ≥ 2.

In such case, the sum of degrees $\geq 2(n-k) + k = 2n - k$. But the sum of degrees of the tree is fixed at $2(n-1)$. So, considering the case of *equality*, we get $2n - k = 2(n-1)$, i.e., $-k = -2$, i.e., $k = 2$. Now, if we consider the case of the inequality, then we may write the sum of degrees $= (2n - k) + r$, where r is a positive integer.

This will give us $2n - k + r = 2(n-1)$, i.e., $-k + r = -2$, i.e., $k = r + 2$, i.e., $k > 2$. So, we get $k \geq 2$ on combining the above two results.

This proves that there must be at least 2 vertices of degree 1, i.e., at least 2 leaf nodes.

EXAMPLE 10.2 A tree has 2 vertices of degree 2, 1 vertex of degree 3, and 3 vertices of degree 4. How many vertices of degree 1 does it have if it has no vertex with degree greater than 4?

Solution Let x be the number of vertices each of which has degree 1. Now, as per the given question, the total number of vertices in the tree is $n = 2 + 1 + 3 + x = 6 + x$.

Therefore, the number of edges $(e) = n - 1 = (6 + x) - 1 = 5 + x$ [since a tree with n vertices has exactly $n - 1$ edges]. Obviously, the sum of the degrees of the vertices $= 2*2 + 1*3 + 3*4 + x*1 = 4 + 3 + 12 + x = 19 + x$. Further, we know that the sum of the degrees of the vertices $= 2$ (number of edges) [by *handshaking theorem* discussed in Chapter 9].

So, $19 + x = 2(5 + x)$, i.e., $19 + x = 10 + 2x$, therefore $x = 9$. Hence, total 9 vertices have degree 1.

EXAMPLE 10.3 An undirected tree has 1 vertex with degree 1, 2 vertices with degree 2 each, 3 vertices with degree 3 each, and n vertices with degree n each. Prove that no such tree exists.

Solution As per the question, total number of vertices

$$V = 1 + 2 + 3 + \cdots + n = \frac{n(n+1)}{2}$$

Clearly, the sum of degrees of all vertices $= 1^2 + 2^2 + 3^2 + \cdots + n^2 = \frac{n(n+1)(2n+1)}{6}$

So, the number of edges

$$V - 1 = \frac{n(n+1)}{2} - 1 \text{ [since a tree with } V \text{ vertices has } V - 1 \text{ edges]}$$

Further, the sum of degrees of all vertices $= 2*$(number of edges) [using *handshaking theorem*]. This gives

$$\frac{n(n+1)(2n+1)}{6} = n(n+1) - 2$$

$$\Rightarrow \quad n(n + 1)(2n + 1) = 6n(n + 1) - 12$$
$$\Rightarrow \quad n(n + 1)(2n - 5) = -12$$
$$\Rightarrow \quad n(n + 1)(2n - 5) + 12 = 0$$

Therefore, we may write the above equation as a function $f(n) = n(n + 1)(2n - 5) + 12$.
It is true that $f(n) > 0, \forall n > 0$.
So, there does not exist any $n > 0$ for which $f(n) = 0$.
$\Rightarrow \quad f(n) = 0$ has no positive integer solution.
$\Rightarrow \quad$ No such tree exists.

EXAMPLE 10.4 A tree has 3 vertices of degree 3 each. What is the number of leaves in this tree?

Solution Let n be the number of vertices in the tree. Now, there are 3 vertices each of degree 3.

Let there be x number of vertices of degree 1 each (since the leaf vertex has degree 1), and all the remaining vertices have degree ≥ 2 (but degree $\neq 3$).

So, the sum of degrees $\geq 3*3 + x + 2(n - x - 3)$,
i.e., the sum of degrees $\geq 9 + x + 2n - 2x - 6$, i.e., the sum of degrees $\geq 2n - x + 3$.

However, the sum of degrees is fixed at $2(n - 1)$ [since a tree with n vertices has exactly $(n-1)$ edges, and the sum of the degrees of the vertices is $2(n - 1)$ which is known from the *handshaking theorem*].

So, considering the case of *equality*, we get

$$2n - x + 3 = 2(n - 1)$$

or,

$$-x + 3 = -2$$

or,

$$x = 5 \tag{10.2}$$

Next, by considering the case of inequality, we may write

$$\text{the sum of degrees} = (2n - x + 3) + k$$

where k is a positive integer. This gives

$$2n - x + 3 + k = 2(n - 1)$$

or,

$$-x + 3 + k = -2$$

or,

$$x = 5 + k$$

$$\Rightarrow \quad x > 5 \tag{10.3}$$

Hence by combining the results from Eqs (10.2) and (10.3), we get $x \geq 5$, i.e., the number of leaves is greater than or equal to 5.

EXAMPLE 10.5 Does there exist a tree for the degree sequence (4, 4, 3, 2, 1, 1, 1, 1, 1)? If yes, then draw the tree—else explain why such a tree cannot exist?

Solution This tree has 2 vertices of degree 4. Let us try to work out the minimum number of leaf vertices (i.e., vertices of degree 1 for this tree).

Let n be the number of nodes in the tree. Now, out of n vertices, 2 vertices have degree 4.

Let there be x vertices with degree 1. Assume that each of the rest of vertices (i.e., $n - x - 2$ vertices) has degree ≥ 2 but not degree 4.

Now, the sum of degrees $\geq (2*4) + x*1 + 2*(n - 2 - x)$,
i.e., the sum of degrees $\geq 8 + x + 2n - 4 - 2x$, i.e., the sum of degrees $\geq 2n + 4 - x$.

However, it is known that the sum of degrees of a tree with n vertices $= 2(n - 1)$.
Therefore, taking the case of equality, we get

$$2n + 4 - x = 2(n - 1)$$

or,

$$4 - x = -2$$

or,

$$x = 6 \qquad (10.4)$$

Next, taking the case of inequality, we may write

$$\text{the sum of degrees} = (2n + 4 - x) + r$$

where r is a positive integer.

So, plugging into the result for the sum of degrees, we get

$$(2n + 4 - x) + r = 2(n - 1)$$

or,

$$4 - x + r = -2$$

or,

$$x = 6 + r$$

$\Rightarrow$

$$x > 6 \qquad (10.5)$$

Thus, from the above two cases: Eqs (10.4) and (10.5), we get $x \geq 6$, which means that the tree must have at least 6 leaf nodes, if it has exactly 2 vertices of degree 4. However, as per the given degree sequence, there are only 5 nodes with degree 1. Hence, no such tree exists.

EXAMPLE 10.6 If both G and G^c (complement of G) are trees, then show that the number of vertices $n = 4$ or 1.

Solution Let e and e' be the numbers of edges, respectively, in G and G^c.

Now, as per the given question, we may write: $e = e' = n - 1$, where n is the number of vertices in G.

Again, each of G and G^c gives a distinct tree, so *union* of both the trees must form a simple graph with maximum edges $n(n - 1)/2$. Further, this number equals to the sum of the edges of both the trees formed by G and G^c.

Clearly, $n(n - 1)/2 = 2(n - 1)$. This gives, $n = 4$ or 1. So, the question is true if and only if the number of vertices n is either 4 or 1.

10.4 SEQUENTIAL REPRESENTATION OF A BINARY TREE

Trees can be represented in two ways: (a) sequential (using array) and (b) linked list (using dynamic allocation of memory).

Suppose there are m number of nodes and the nodes are numbered as $1, 2, 3, \ldots n$, respectively. To represent the nodes in the form of a binary tree, an *array* with size n ($n \geq m$) is considered. Now, to accommodate the ith node of the tree into the array, the following steps are usually directed:

(i) parent (i) is at $\lfloor \frac{i}{2} \rfloor$, if $i \neq 1$. But i indicates the root node when $i = 1$. [It means that the parent of ith node will be at location $\lfloor \frac{i}{2} \rfloor$, if $i \neq 1$.]

(ii) lchild (i) is placed at $2i$, if $2i \leq n$. Clearly, if $2i > n$, then ith node has no left children. [Here, lchild (i) means left child of the ith node.]

(iii) rchild (i) is placed at $2i + 1$, if $(2i + 1) \leq n$. But the ith node has no right children if $(2i + 1) > n$. [Here, rchild (i) means right child of the ith node.]

This process continues until all the nodes are accommodated into the array, fulfilling the space (size) criteria of the array.

Comment If we need to construct a *binary search tree*, then some other criteria, such as if lchild (i) < parent (i), and so on, are essential to add along with the respective steps mentioned above. [We know that $\lceil x \rceil$ (i.e., *ceil(x)*) returns the *smallest* integer which is *greater* than or *equal* to x. For example, $\lceil 1.2 \rceil = 2$. Also, $\lfloor x \rfloor$ (i.e., *floor(x)*) returns the *largest* integer which is *less* than or *equal* to x. For example, $\lfloor 1.2 \rfloor = 1$. Further, $\lceil -x \rceil = -\lfloor x \rfloor$ and $\lfloor -x \rfloor = -\lceil x \rceil$.]

Clearly, a full binary tree with n vertices may be compactly stored in a one-dimensional array, say, Tree $[1 \cdots n]$. On the other hand, arrays with sizes: 2^{n-1} and $(2^n - 1)$ are essential, respectively, for a left-skewed and a right-skewed binary trees, each with n number of nodes. The correctness of the sizes can be easily verified on the two-skewed binary trees (shown in Figure 10.10), following the tree representation scheme discussed above.

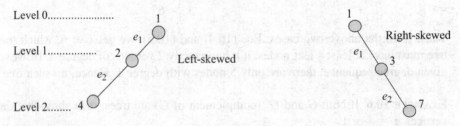

FIGURE 10.10

In the left-skewed binary tree (shown in Figure 10.10), the number of nodes n is 3, and these are placed, respectively, at the locations: 1, 2, and 4 of an array, following the scheme presented in this *section*. Hence, the array *size* should be at least 4, and it can easily be determined by 2^{n-1} (i.e., $2^{3-1} = 2^2 = 4$).

Again, in the *right-skewed* binary tree, the number of nodes n is also 3, and these are needed to place, respectively, at the *locations*: 1, 3, and 7 of an array, following the same tree *representation scheme*. Clearly, the size of the array should be at least 7, and it comes directly from the formulation: $2^n - 1$ (i.e., $2^3 - 1 = 8 - 1 = 7$).

10.5 OPERATIONS ON TREE

The following basic operations can be performed on a tree:

(a) Creation of a tree
(b) Insertion of a new node into a tree
(c) Deletion of an existing node from the tree
(d) Copying a tree into another tree
(e) Searching of a particular node in the tree on the basis of value, location, etc.
(f) Traversal of tree
(g) Counting number of nodes
(h) Finding height of the tree, etc.

10.5.1 Tree Traversal

Traversal is the process to traverse the containers of any data structure such as list. In particular, the goal of tree traversal is to traverse a tree in a systematic way so that each vertex is visited exactly once. Clearly, we can say that traversal involves the operation of traversing the nodes. However, traversing a node may or may not visit the node. Now, *visit* of node means reaching the node and then processing of its data, i.e., not simply crossing over the node. On the other hand, *traversing* of node implies reaching the node but may or may not *operate/process* the data of the node, i.e., without operating the data, it may leave the node.

Clearly, to visit a node, traversing is necessary. Further, during any traversal, the number of node traversals is more as compared to visits.

Three common ways of traversing a tree are normally considered. These are: (1) pre-order, (2) in-order, and (3) post-order. Generally, these are applied on a binary tree in which every node contains *three* parts/fields: *data* (D) to store data, *left link* (L) to connect its left-child, *right link* (R) to connect its right child. The *pre-order*, *in-order*, and *post-order* traversal techniques on a binary tree are discussed below.

1. *Pre-order* Any node is numbered (visited) before numbering the nodes of its left and right sub-trees but traversal always moves left as far as possible from any position. So, the traversing sequence followed is: *data* (D), *left* (L), and *right* (R). Now, considering this idea, all the vertices of the tree are visited. Further, the *pre-order* traversal is nothing but the *depth-first* tree traversal. [Algorithmic version of this traversal is presented in Section 10.7.]

2. *In-order* Any node is numbered only after numbering the nodes in its left sub-trees but not nodes in its right sub-trees (however, traversal always move left as far as possible from any position). Hence, the sequence: *left* (L), *data* (D), and *right* (R) is followed. On the basis of this conception, all the vertices of the tree are visited.

3. *Post-order* Any node is numbered only after numbering nodes in its left and in its right sub-trees (here also, traversing always moves left as far as possible). Thus, here the sequence: *left* (L), *right* (R), and *data* (D) is followed. Considering this view, all the vertices of the tree are visited exactly once.

> **Note** The idea on the *pre-*, *in-*, and *post*-order traversals discussed above is indeed based on *general tree* in which any node may have *n* number of nodes (value of *n* may be greater than 2), i.e., each node may have more than one child. However, we have considered here simply a binary tree for better understanding. A brief discussion on traversal of general tree is presented later in this chapter.

EXAMPLE 10.7 Consider the binary tree (shown in Figure 10.11) and determine the data sequences obtained from pre-order, in-order, and post-order traversal, respectively.

Solution

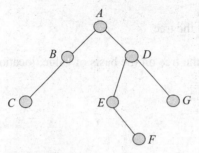

Pre-order: A B C D E F G
In-order: C B A E F D G
Post-order: C B F E G D A

FIGURE 10.11

EXAMPLE 10.8 Construct the pre-order, in-order, and post-order traversal data sequences of the binary tree (shown in Figure 10.12).

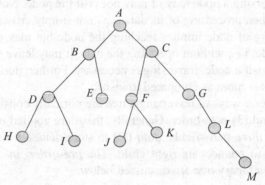

FIGURE 10.12

Pre-order: A B D H I E C F J K G L M
In-order: H D I B E A J F K C G L M
Post-order: H I D E B J K F M L G C A

EXAMPLE 10.9 Under what circumstances, the last visited node in pre-order and in-order is same?

Solution If the tree is a complete binary tree, then the last visited node in both cases is same. Let us consider a simple complete binary tree as shown in Figure 10.13.

In both the *pre-order* and *in-order* traversals, the last visited node is C, since the tree is a complete binary tree.

FIGURE 10.13

10.5.2 More Discussions on Tree Traversals

(a) A binary tree with n vertices has $(n - 1)$ edges, and each edge in any traversal (pre/in/post) is traversed twice plus one initial traversal of root. Therefore, the total number of node traversals performed by any traversal is $2(n - 1) + 1 = 2n - 1$. This can also be shown by induction method. However, in any traversal approach, total number of node visits is n.

(b) Each node in a binary tree is associated with three components: (i) data, (ii) left child, and (iii) right child, so the *maximum* number of traversals associated with a binary tree can be at most 3! = 6. These are such as: DLR (known as pre-order), LDR (known as in-order), LRD (known as post-order), DRL, RDL, and RLD.

(c) The observations found from the *pre-order* and the *post-order* traversals of tree are that in the pre-order traversal, the *root* of the tree is visited at first; whereas in post-order, it is visited at last.

Hence, to draw a unique binary tree, the values corresponding to any one of the *two* combinations: (a) *in-order* and *pre-order* or (b) *in-order* and *post-order* are required to be known.

(d) It is not possible to construct a unique binary tree if the data sequences corresponding to only pre-order and post-order traversals are given, as we cannot determine the exact location of the *root* of the tree. Clearly, the pre-order and the post-order sequences of a binary tree do not uniquely define a binary tree. So, in order to construct a unique binary tree, it requires an *in-order* sequence as one of the two given sequences.

(e) In a full binary tree, the last visited node in *in-order* and *pre-order* is always the same.

10.5.3 Construction of a Unique Binary Tree When the Pre-Order and the In-Order Traversal Sequences Are Given

Suppose the given values corresponding to the pre-order and the in-order traversals of a binary tree are:

Pre-order: A B C D E F G and *In-order*: C B A E F D G

The following illustration shows the construction procedure of a unique binary tree.

From the given pre-order sequence, the *root A* (visited at first) is identified. Now, consider the corresponding in-order traversal sequence and mark *A* in it. Then, consulting the given in-order values, find the nodes in the left and the right sub-trees of *A* as follows.

The nodes in the *left sub-tree* of root are present in the portion left to *A* (here, it is *C B*), whereas the nodes in the *right sub-tree* of root are in the portion right to *A* (here, it is *E F D G*). This is also shown in Figure 10.14.

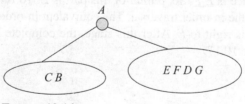

FIGURE 10.14

Now, for both the portions: *C B* and *E F D G*, consider the given *pre-order* as well as *in-order* traversals to continue the construction of the binary tree.

The *pre-order* traversal sequence of the portion *C B* can be viewed as <u>B C</u> (found from the *pre-order* sequence). Thus, the parent of this portion is *B*. Now, to find *B*'s *left* sub-tree and *right* sub-tree, consult again the given in-order traversal sequence which in fact results as <u>C B</u>. Therefore, from its in-order traversal portion, it is concluded that *C* is left to *B*. Hence, the present constructed tree is shown in Figure 10.15.

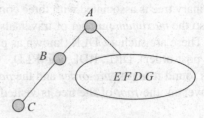

FIGURE 10.15

Next, start to construct the binary tree for the portion *E F D G*. For this portion also, consider the *pre-order* as well as the *in-order* traversals. The pre-order traversal sequence of the portion: *E F D G* must look as: $\underline{D\ E\ F\ G}$ (found from the pre-order traversal). Clearly, the parent of this portion is *D*. Now, to find *B*'s *left* sub-tree and *right* sub-tree, consult with the in-order traversal which results here as $\underline{E\ F\ D\ G}$. The in-order sequence $\underline{E\ F\ D\ G}$ infers that *E* and *F* are *left* to *D*, whereas *G* is *right* to *D*. In this stage, the ongoing constructed binary tree looks like Figure 10.16.

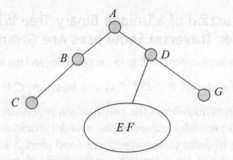

FIGURE 10.16

At present, construction of the binary tree for the only portion *E F* is left. To build it, consult again with the pre-order traversal for finding parent of this part. Obviously, for this portion, the pre-order sequence is $\underline{E\ F}$. So, *parent* of this part is *E*. To find its *left* sub-tree as well as *right* sub-tree, follow the in-order traversal. The equivalent in-order sequence of this portion is $\underline{E\ F}$, which means *F* is right to *E*. After this stage, the complete binary tree is constructed, and it looks like Figure 10.17.

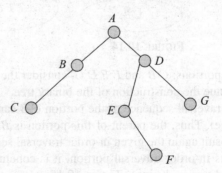

FIGURE 10.17

In the same fashion, the unique binary tree can be built step by step if the post-order as well as the in-order traversals are given. However, in such combination, from the post-order traversal sequence, the element visited at last is selected as parent of a portion. But to find, respectively, the *left* sub-tree and the *right* sub-tree of that portion, corresponding to the in-order traversal for that portion is needed to consider. This procedure continues until one complete binary tree is constructed for the given traversal sequences.

10.5.4 Algorithm to Construct a Unique Binary Tree Using the Pre-Order and the In-Order Sequences

Step 1 Select the first value from the pre-order sequence and consider it as the root node of the binary tree.

Step 2 To find the right and the left sub-trees of the root node, first, mark the selected root node in the in-order sequence. Then, all the nodes left to this root are placed in the left sub-tree, and all the nodes right to this root are placed in the right sub-tree. Selecting the nodes in the left and the right sub-trees, draw the current tree.

Now, apply the pre-order sequence on the left sub-tree to find the root of this sub-tree. Similarly, apply the pre-order sequence on the right sub-tree to find the root of this sub-tree.

Step 3 Repeat Step 2 until each sub-tree consists of a single node.

10.6 BINARY SEARCH TREE

Binary search tree (BST) is a binary tree which satisfies the following properties:

(a) Every node has at most two children.
(b) The value of left child of a parent is *less* than the value of its parent.
(c) The value of right child of a parent is greater than or equal to the value of its parent.

Some authors define binary search tree as follows:
Each node in a binary tree with a single key such that for any node x, and nodes in the left sub-tree of x have keys $\leq x$ and all nodes in the right sub-tree of x have keys $\geq x$. [In this book, the first definition is considered.]
One simple binary search tree is shown in Figure 10.18.

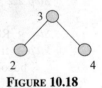

FIGURE 10.18 A simple BST

EXAMPLE 10.10 What is the maximum height of the binary search tree on the set of elements $S = \{1, 3, 7, 8, 9\}$?

Solution The maximum height will be 4, since the number of elements is 5 and if element 1 is placed at the root. Clearly, the constructed tree on the elements, taking 1 at root, will be a right-skewed binary tree.

EXAMPLE 10.11 Consider the problem of generating binary search trees, using a given set of numbers in the given order. The tree is to be constructed by inserting the numbers into the current partial tree such that at any node, the following condition is satisfied: value of left child < value of node < value of right child. Which of the following sequences of numbers will result in a tree i.e., strictly binary, i.e., every node has either two children or no children?

(a) 4 5 2 1 3   (b) 4 6 5 7 3 2   (c) 4 6 2 1 3 5   (d) 4 5 3 2 6

Solution The sequence 4 5 2 1 3 is correct, since one strict BST can be constructed from this sequence.

10.6.1 Linked List Representation of a Binary Tree

In *linked list* representation of a binary tree, one must use recursive data structure like *self-referential* structure (allowed in C, C++, etc.) with three fields to represent a node. These fields are commonly: (a) *data*, (b) *LPTR* (left pointer to point a left child node of same type), and (c) *RPTR* (right pointer to point a right child node of same type). The simple declaration of such structure can be written in C (language) as follows.

```
struct node {
            int data ;
            struct node *LPTR, *RPTR *right_ptr ;};
```

However, the type of data may be *real*, *char*, etc. Also, *to add* node except the root node into a binary tree, one may follow any one of the following criteria:
 (a) Insertion of any node after a node with value, say, x.
 (b) Insertion of any node to the left sub-tree of a node or to the right sub-tree of a node, etc.
 (c) Considering the properties of binary search tree (this technique is frequently used).

10.6.2 Construction of Binary Search Tree

Consider a BST T and T points to the *root* of the tree. Suppose x is the *item* to be inserted into the BST. The following steps are required to be considered to insert x into the BST.

Step 1 If T has no node, then create a new tree T with one node, and insert x into it, and then STOP.
Step 2 If T has node, then check the content of root T [denoted by data (T)] with x.
Step 2.1 If $x <$ data (T), then proceed to the left child of the current root.
 Currently, left child of T is considered as root, i.e., $T =$ left (T).
Step 2.2 If $x \geq$ data (T), then proceed to the right child of the current root. Currently, right child of T is considered as root, i.e., $T =$ right (T).
Step 3 Repeat step 2 until T becomes NULL.
Step 4 If T is NULL, attach the data x (storing into a new node) to the node from where T becomes NULL.

The high level description of the above outlines is given below, using *linked representation*.

```
      Algorithm BST( t, x)
/* 't' is a pointer variable of struct node type for creating
tree and 'x' is the data */
  begin
if (t= NULL) then
//no node is added to the tree till now allocate memory for 't'
    t→data:= x
    t→LPTR:= NULL
    t→RPTR:= NULL
  end
else
```

```
begin
  if ((x < (t→data)) then
    BST(t→LPTR, x)
/* i.e., if the value of 'x' is less than the data of current
node then recursively search the left sub-tree of the current
node to place this new node with data 'x' */
    else
      BST(t→RPTR, x)
/* otherwise recursively search the right sub-tree of the cur-
rent to place this new node with data 'x' */
  end
end
```

(*Problem*: Draw a BST for the given sequence: *E, A, S, Y, Q, U, K, S, T, I, O, N*.)

10.6.3 Useful Results from Binary Search Tree

1. When a BST consisting of numerical values is traversed in *in-order* way, then values are obtained in *ascending* order. On the other hand, if a BST contains expression (i.e., if it is an *expression tree*), then the pre-order traversal on that tree generates *pre-fix* expression of the corresponding *in-fix* expression. Similarly, the equivalent *post-fix* expression is obtained from the post-order traversal. [Example of *expression tree* is given later in this chapter.]
2. BST guarantees that the minimum key (value) is located at the left-most node of BST. Similarly, the *maximum* key is located at the right-most node of BST.
3. If n data items: $A_1, A_2, A_3, \ldots, A_n$ are already sorted like $A_1 < A_2 < A_3 < \cdots < A_n$ and it is assumed that the items are inserted (in the given order) into an empty BST, then *height* of that BST is $(n-1)$, as right-skewed binary tree is built following such sequence.

Remark In BST, time required to *insert* one element into tree with i nodes is $O(i)$ in worst case and $O(\log_2 i)$ in average/best case. Obviously, the worst case occurs if the tree is a *skewed* BST, but the average/best case is achieved when it is *full* or *complete* binary tree.

Now, in worst case, the total running time to insert n elements is expressed as follows:

$$O\left(\sum_{i=1}^{n} i\right) = O(1 + 2 + 3 + \cdots + n) = O(n^2) \cdots (2)$$

In particular, to get a sorted list of elements (in ascending order) from a BST with n nodes, the in-order traversal should be applied, and the traversal takes $O(n)$ time to visit n nodes. However, the total required time to get the *sorted* list in *worst case* will be measured as follows.

Total time = time to construct BST with n nodes in worst case + time to visit n nodes
= $O(n^2) + O(n) = O(n^2)$ [it is $O(n^2)$, since the BST may be skewed due to construction, i.e., if the elements come either in ascending/descending order].

Further, *in average case*, time to construct a BST with n nodes will be

$$O\left(\sum_{i=1}^{n} \log_2(i)\right) = O(\log_2(1) + \log_2(2) + \cdots + \log_2(n)) = O(\log_2(1 * 2 * 3 * \cdots * n))$$

$$\leq (\log_2 n + \log_2 n + \log_2 n + \cdots + \text{up to } n\text{th term})$$

$$\leq (n * \log_2 n) = O(n * \log_2 n)$$

10.7 RECURSIVE PROCEDURE FOR BINARY TREE TRAVERSAL

In this section, recursive procedure for only *pre-order* traversal is discussed. The rests are left to the reader.

Basic outlines are first mentioned, and then a *high-level* description is given.

Algorithm **PREORDER( )**

Step 1 Search the left sub-tree, if it exists.
Step 2 Search the right sub-tree, if it exists.
Step 3 Visit the root.

High level description of pre-order

```
        Procedure PREORDER( t )
/* 't' is a pointer variable of struct node type for creating
node of binary tree */
        begin
          if ( t = NULL ) then return
        else
            begin
              print(t→data)
              call PREORDER( t→LPTR )
              call PREORDER( t→RPTR )
            end
        end
```

Note The above procedure is indeed a recursive procedure, and to implement recursion, one stack (containing calling address and other relevant data) is to be used.

Further, in any recursive call, from any point (i.e., from any *address* of function call), next statement for that call is not started until its earlier operation (operations appearing before calling) is completed. When all the possible statements in the body of a function (called from a point/address) is over, then *control* (flow) automatically (even if return statement is not present in the body) moves back to the point from which it was called. Therefore, every recursion must have one *stopping* criteria.

Explanation of the pre-order traversal on the binary tree is shown in Figure 10.19.

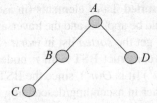

FIGURE 10.19 A binary tree

In any traversal, calling (pre/in/post) is started from the *root* node. In the *pre-order* traversal, whenever this traversal procedure is called from a node, immediately the content (data) is retrieved.

The *root* node of this tree (shown in Figure 10.19) is *A* (tree is not *null*, i.e., it has some nodes) and *t* is pointing to *A* (root node), so its (root's) content is retrieved and printed first.

From node *A*, *t* → LPTR (i.e., *A* → LPTR) is checked, and it is currently found that *t* → LPTR ≠ NULL. Therefore, from node *A*, next PREORDER( ) call is performed to its *left* sub-tree. Here, node *B* is found as left of *A*. So, the content of node (suppose it is *B*) is retrieved (and printed).

Then, next PREORDER( ) call is performed to the left sub-tree of node *B*, and node *C* is found as left of *B*. Accordingly, content of this node (suppose it is *C*) is retrieved (and printed).

But now node *C* has *no* left child. So, at this moment, as per PREORDER( ) call from *C*, it will check any *right* child (of *C*) is present or not. But *C* has currently *no* right child also. Accordingly, the call goes back to the *previous* call [i.e., from where control (call!) came]. Here, it is node *B*, and it is checked if *right* child (of *B*) is present or not (as its *left* sub-tree is traversed). But here it is not present. So, again control goes back to the *previous* call, and checks in such way until all the vertices in the tree are visited exactly *once*.

Similarly, we can write *recursive* procedure for the *in-order* and the *post-order* tree traversals and gets explanation.

Note Paying special attention, one can count the total number of nodes in a tree by incrementing the value of a counter variable, during any traversal (pre-order or in-order or post-order).

EXAMPLE 10.12 Design an algorithm to determine the height of a binary tree.

Solution Use modified form of the pre-order traversal given below:

```
Algorithm Height(node N)
   begin
      if (N==null) return 0;
   if((N–>left == NULL) and(N–>right== NULL))
      return 0;
   else
   return(1 + max(Height(N–>left),Height(N–>right));
   end
```

EXAMPLE 10.13 A BST contains the values 1, 2, 3, 4, 5, 6, 7, 8 and the tree is traversed in pre-order and the values are printed. Which of the following sequences is a valid output?

(a) 53124786 [As it is a pre-order traversal, so 5 is the root and 7 is the first value which is greater than 5 (so, 7 is the first right child of root 5 in its right sub-tree and the portion of the sequence after 7 are the contents of its right sub-tree) and the next value is 8. But as per property of BST, 6 should not come after 8 in the pre-order traversal.]
(b) 53126487 [Here, 5 is the root, 6 is the first value greater than 5 (so, 6 is the first right child of root 5 in its right sub-tree and the portion of the sequence after 6 are the contents of its right sub-tree). As per property of BST, now 4 may not come after 6.]
(c) 53241678 [Here, 5 is the root. Now, 1 may not come after 4.]
(d) 53124768 [Valid.]

Solution The validity can be checked in the following ways.

  (i) First, find the root. However for pre-order, first element is the root. Then we get one *sub-sequence* of elements with value *less* than the *root* (for its left sub-tree checking the

first value greater than the root from the sequence) and find the greatest value in this sub-sequence. Now, if any value comes to the right of this greatest value in this sub-sequence, which is less than any value placed to the left of this greatest value in this sub-sequence, then it is not a valid pre-order sequence.
(ii) No value less than the root may come after the value greater than the value of root.
(iii) In the right sub-tree also (after finding the right sub-tree), if any value comes to the right of the greatest value in a sub-sequence, which is less than any value placed to the left of this greatest value in this sub-sequence, then it is not a valid pre-order sequence.

10.7.1 Analysis of Time Complexities for Some Operations on Binary Tree

(i) If n nodes are present in a tree T, then running time for each traversal (like pre-order or post-order or in-order) is $O(n)$. It is, in fact, the time complexity to count the number of leaf nodes in a binary tree (since to count the leaf nodes, any traversal technique is needed to apply, which traverses all the nodes exactly once).
(ii) To search a particular node from a binary tree (not a BST) with n nodes, time complexity is $O(n)$ in *worst* case (because to find that node all the nodes in the tree may be traversed), and $O(1)$ in best case (as the target data may be found in the first searching). The searching is also performed by any basic traversal technique like *pre*, *in*, or *post*.
(iii) To search a particular node from a BST with n nodes, time complexity is $O(n)$ in *worst* case because the BST may also be left- or right-skewed due to construction; $O(\log_2 n)$ in *average case*, as the height of the BST is $\lfloor \log_2 n \rfloor$ when it is full or complete binary tree; and $O(1)$ in best case if the *required* data are found in the first searching. Here also, the searching can be done by any basic traversal technique like: *pre*, *in*, or *post*.

10.8 PREDECESSOR AND SUCCESSOR NODE

If node X has two sub-trees, its *predecessor* is the *maximum* value in its *left* sub-tree, and its *successor* is the *minimum* value in its *right* sub-tree.

If the tree is BST, then time taken to find predecessor and successor of a node is proportional, respectively, to the height (h) of the left and the right sub-trees of that node. However, to find the same, it is required to start from the *root*. So, exact time complexity must be proportional to $O(H)$, where H is the height of the BST. On the other hand, if the tree is not a BST, then it is $O(n)$; since almost all the n nodes (including the node X and the rest nodes) may be needed to visit to find the *predecessor/successor*.

10.9 EXPRESSION TREE

It is a binary tree through which *arithmetic expression* can be expressed in binary tree structure. Such tree satisfies the following rules:

(a) The intermediate nodes of the tree contain *operators* in the expression. Of course, *operators* are placed following their precedence rule, i.e., operators with *higher* precedence are placed in the *lowermost* sub-trees of the tree so that these operators are evaluated first. Here, the precedence rules of the arithmetic operators are same as mathematics.

(b) *Operands* or *constant* values in the expression are placed in the leaf nodes in the tree.

For example $a + b * c$

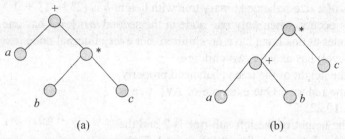

(a)   (b)

FIGURE 10.20 (a) Expression tree for $a + b * c$;   (b) Expression tree for $(a + b) * c$

In this expression, precedence of $*$ is more than that of $+$. Therefore, in between $+$ and $*$, operator $+$ is placed at the *root*. Further, in expression: $a + b*c$, since the *sub-expression* $b*c$ is right to $+$, so the sub-tree consisting of $b * c$ will be the right sub-tree of node $+$.

Further, operand a in this expression is left to $+$, so it is placed in the left sub-tree node $+$. Following this strategy (i.e., based on the precedence of the operators in the *in-fix* expression), the expression tree is constructed and shown in Figure 10.20(a).

However, if we consider the expression like $(a + b) * c$, then the sub-expression $(a + b)$ (i.e., sub-expression within parentheses) must have higher priority, and it will be considered as the lowermost sub-tree and evaluated first. The tree is shown in Figure 10.20(b).

 EXAMPLE 10.14 Find the pre-fix, in-fix, and post-fix form of the expression tree given in Figure 10.21.

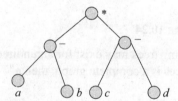

FIGURE 10.21

Solution Actually, by applying the pre-order, in-order, and post-order traversals on the given tree, we get the pre-fix expression as: $* - ab + cd$, the in-fix expression as: $(a - b) * (c + d)$, and finally the post-fix as: $ab - cd + *$. In particular, in finding the in-fix expression, we enclose each sub-expression of the same level of the expression tree by parentheses.

10.10 AVL TREE

In 1962, the two mathematicians: G. M. Adelson-Velskii and E. M. Landis created the *balanced* binary tree structure that is named after them as AVL tree.

An AVL tree (T, V_0) is a binary tree where for each $V \in T$, the height of $T(V_L)$ and $T(V_R)$ differ by at most one. Here, T denotes tree, V_0 is its root, V is any node of the tree, V_L and V_R denote the left sub-tree and the right sub-tree of the node, V.

It is also called as a size-balanced binary tree. The *maximum* number of nodes of a size-balanced binary tree with *height h* (assuming that h starts from 0) is $2^{h+1} - 1$, when every internal node has two children and all *leaves* appear at the same level. The *minimum* number of nodes of a size-balanced binary tree with height h is $(2^h - 1) + 1 = 2^h$. Minimum number of nodes occurs, when only one node at the second *last* level has one child only and the rest of the nodes at this level have no children, but every internal node (except the node at the second last level) has exactly *two* children.

In AVL tree, the height of the tree is balanced properly. So, searching time reduces. One example of AVL tree is shown in Figure 10.22.

For node A, the height of the left sub-tree is 2 and the height of the right sub-tree is 1. So, it differs by 1. Similarly, check for other nodes of this tree.

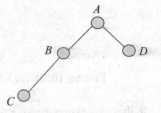

FIGURE 10.22 A sample AVL tree

10.11 SPANNING TREE

Let $G = (V, E)$ be an undirected, connected graph. A subgraph $T = (V, E')$ of G is a *spanning tree* of G iff it is a tree, i.e., the tree T of G consists of all the vertices of G but not necessary all the edges of G. Let us consider a graph G as shown in Figure 10.23.

FIGURE 10.23

The possible spanning trees of the above graph are shown in Figure 10.24.

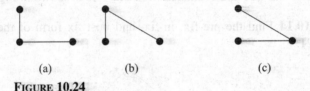

(a) (b) (c)

FIGURE 10.24

Obviously, many spanning trees may exist for an undirected connected graph. For example, if the graph G with n vertices is a complete graph, then n^{n-2} spanning trees in total can be generated from G.

Further, suppose a graph G is undirected connected but not complete graph, and the number of vertices (i.e., $|V|$) and edges in G are, respectively, n and m ($m > n$), then $^mC_{n-1} - k$ spanning trees may be constructed from G. Clearly, $^mC_{n-1}$ distinct graphs (each with $n - 1$ edges) from the original graph G can be constructed by selecting $(n - 1)$ edges out of m edges. However, out of this number, k is the number of graphs each with $(n - 1)$ edges but disconnected. Thus, $^mC_{n-1} - k$ graphs are considered as spanning trees, since each spanning tree must have $(V - 1) = (n - 1)$ edges and connected.

Spanning tree has many applications. For example, (i) they can be used to obtain an independent set of circuit equations for an electric network; (ii) the *optimality* principle of spanning tree (when edges weights are provided) can be used in computer network to link all the nodes; and (iii) the concept of spanning tree can be applied to connect n cities, considering minimum total *cost* or *length*.

Now, the graph searching techniques: BFS and DFS can be applied to find spanning trees from connected graph, and the trees generated from BFS and DFS are known respectively as BFST (breadth first Spanning tree) and DFST (depth first Spanning tree).

Note A spanning tree is a tree that reaches all nodes of a connected graph. Thus, a directed connected graph may or may not have spanning trees, depending on the reachability criteria among the vertices. But every undirected and connected graph must have spanning tree, and so we normally consider spanning tree of undirected graph instead of directed graph. Further, a *spanning forest* is a set of trees, one for each connected component of a graph.

10.11.1 Minimum Spanning Tree

A minimum spanning tree (MST) can be constructed from a weighted connected graph. Of course, a weighted connected graph may have many spanning trees, but the tree with minimum weight (cost/length) is called as *minimum spanning tree*. Further, a graph may have many MSTs with same cost also.

For example, an undirected graph G has n nodes. Its adjacency matrix is given by $n \times n$ matrix whose diagonal elements are 0s and non-diagonal elements are 1s. Then, G has *multiple* MSTs each of cost $(n - 1)$.

Generally, two methods are followed to find MST from the weighted connected graph. These are: (i) Prim's algorithm and (ii) Kruskal's algorithm. In particular, at each and every stage, both Prim's and Kruskal's algorithms add a *safe* edge (i.e., which does not add any cycle in the graph) of minimum possible weight to the spanning tree.

Prim's Algorithm

The algorithm begins by selecting an arbitrarily starting vertex. It then greedily grows the MST by choosing a new vertex as well as an edge. The procedure continues until all the vertices are selected but never forms a cycle during construction of the tree. Some important steps of Prim's approach are presented below.

Let $G = (V, E)$ be a weighted connected graph, V_T is the selected set of vertices and initially V_T contains Φ (null).

Step 1 Choose arbitrarily a vertex, say, v_1 (as starting vertex). Put v_1 into V_T.

Step 2 Choose one u vertex from $(V - V_T)$, which is incident to any vertex of V_T but the edge corresponding to u has a minimum weight. Also, inclusion of the new vertex does not make any cycle. However, if a cycle forms, discard it and choose the next vertex from V_T. Now, update V_T and V as follows:

$$V_T = V_T U\{u\} \quad \text{and} \quad V = V - \{u\}$$

where U is the union operator.

Step 3 Repeat Step 2 until V becomes empty.

Kruskal's Algorithm

Let $G = (V, E)$ be a weighted connected graph. We first arrange the edges in ascending order of their weights, and then select one by one until $(n - 1)$ edges are selected, where n is the number of vertices in G. However, the procedure never forms a cycle during building the tree. The algorithm is presented below highlighting some important steps.

Step 1 Arrange the edges in ascending order of their weights, and choose the edge (say e_1) with minimum weight. If more than one edge have the same minimum weight, then arbitrarily choose any one from them.

Step 2 Select the next minimum edge from the rest so that inclusion of it does not make any cycle.

Step 3 Repeat Step 2 until $(n-1)$ edges are selected (as the number of vertices in G is n).

EXAMPLE 10.15 Find an MST from the weighted graph G (shown in Figure 10.25) through Prim's algorithm, starting at vertex f.

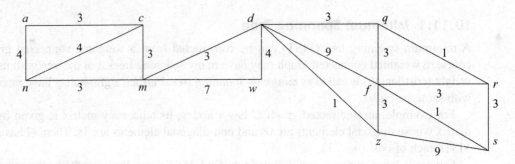

FIGURE 10.25

Solution We know that the MST of a graph G is an acyclic subset G' of the graph G such that:
(a) G' is a connected graph.
(b) The number of edges in G' is exactly $|V| - 1$, where $|V|$ is the number of vertices in G.
(c) The sum of the weights of the edges in G' is minimum.

Applying Prim's algorithm, the steps of computation of one possible MST are graphically shown in Figures 10.26–10.35. The highlighted edges demonstrate the step-by-step formation of the spanning tree, satisfying the property of edge minimality.

Prim's algorithm starts here at vertex f. Thus, V_T initially contains f, i.e., $V_T = \{f\}$, and $V = V - V_T = \{a, c, m, n, d, w, q, z, r, s\}$.

Step 1 As, vertex s is included to V_T, so $V_T = \{f, s\}$ and $V = \{a, c, m, n, d, w, q, z, r\}$.

FIGURE 10.26

Step 2 Here, vertex r is added to V_T. So, $V_T = \{f, s, r,\}$ and $V = \{a, c, m, n, d, w, q, z\}$.

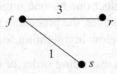

FIGURE 10.27

Step 3 At this stage, vertex q is included to V_T. Therefore, $V_T = \{f, s, r, q\}$ and $V = \{a, c, m, n, d\ w, z,\}$.

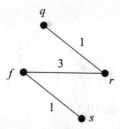

FIGURE 10.28

Step 4 Now, vertex z is added to V_T. Hence, $V_T = \{f, s, r, q, z,\}$ and $V = \{a, c, m, n, d, w\}$.

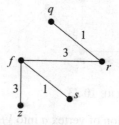

FIGURE 10.29

Step 5 Since vertex d is included to V_T, so $V_T = \{f, s, r, q, z, d\}$ and $V = \{a, c, m, n, w\}$.

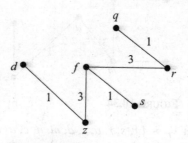

FIGURE 10.30

Step 6 Presently, vertex m is selected. So, $V_T = \{f, s, r, q, z, d, m\}$ and $V = \{a, c, n\ w\}$.

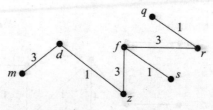

FIGURE 10.31

Step 7 Here, vertex r is included to V_T. So, $V_T = \{f, s, r, q, z, d, m, n\}$ and $V = \{a, c, w\}$.

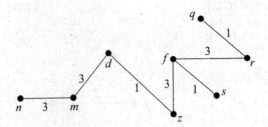

FIGURE 10.32

Step 8 Currently, c is added to V_T. Clearly, $V_T = \{f, s, r, q, z, d, m, n, c\}$ and $V = \{a, w\}$.

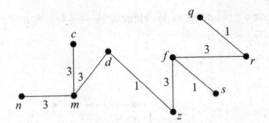

FIGURE 10.33

Step 9 Because of the inclusion of vertex a into V_T. We get $V_T = \{f, s, r, q, z, d, m, n, c, a\}$ and $V = \{W\}$.

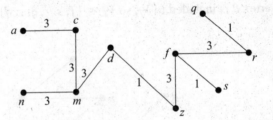

FIGURE 10.34

Step 10 Finally, we get $V_T = \{f, s, r, q, z, d, m, n, c, a, w\}$ and $V = \Phi$.

Thus, the edges in the spanning tree are: $(a, c) = 3$, $(c, m) = 3$, $(m, n) = 3$, $(m, d) = 3$, $(d, w) = 4$, $(d, z) = 1$, $(z, f) = 3$, $(f, s) = 1$, $(f, r) = 3$, $(r, q) = 1$.

Now, the total weight of the constructed MST is
$3 + 3 + 3 + 3 + 4 + 1 + 3 + 1 + 3 + 1 = 25$.

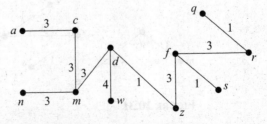

FIGURE 10.35

EXAMPLE 10.16 Find an MST from the weighted graph G (shown in Figure 10.36) using Kruskal's algorithm.

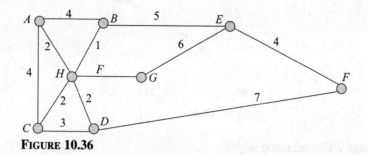
FIGURE 10.36

Solution The formation of an MST, using Kruskal's algorithm, is shown through Figures 10.37–10.43 by highlighting the edges as per ascending order.

Step 1 Initially, we add the edge BH to the ongoing constructed tree T, as it has the least weight.

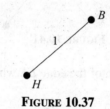
FIGURE 10.37

Step 2

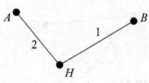
FIGURE 10.38

Step 3 Edges AH and CH are simultaneously included to T at Steps 2 and 3.

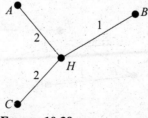
FIGURE 10.39

Step 4 Currently, edge *HD* is added to *T*.

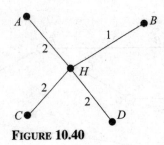

Figure 10.40

Step 5 New edge *HG* is attached with *T*.

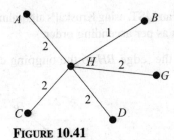

Figure 10.41

Step 6 This results the inclusion of the edge *EF* which is currently unconnected to *T*.

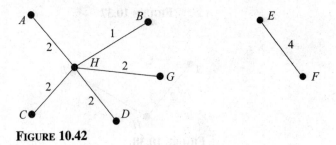

Figure 10.42

Step 7 However, at this stage, edge *BE* is attached with *T*, and it is, in fact, connecting edges *EF* as well as *BH* to return a complete tree.

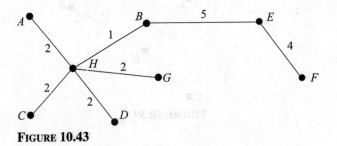

Figure 10.43

Thus, the selected edges in the tree are $(B, H) = 1, (A, H) = 2, (C, D) = 2, (D, H) = 2,$
$(H, G) = 2, (E, F) = 4,$ and $(B, E) = 5.$

Hence, the total weight of the resultant MST is $1 + 2 + 2 + 2 + 2 + 4 + 5 = 18.$

Note One basic difference between the two approaches is that the tree constructed by Prim's approach results always connected, which is not necessary for the tree obtained from Kruskal's method, i.e., forests may be found during construction of a tree by Kruskal's method, although one completely connected tree is finally built when the algorithm stops.

EXAMPLE 10.17 Find an MST for the graph G with the cost matrix as shown in Table 10.1. How many such trees are there?

TABLE 10.1

	A	B	C	D	E	F	G	H
A	0	12	0	14	11	0	17	8
B	12	0	9	0	12	15	10	9
C	0	9	0	18	14	31	0	9
D	14	0	18	0	0	6	23	14
E	11	12	14	0	0	15	16	0
F	0	15	31	6	15	0	8	16
G	17	10	0	23	16	8	0	22
H	8	9	9	14	0	16	22	0

Solution Kruskal's algorithm may be applied directly on the *cost matrix*, without necessarily drawing a diagram of the graph. We simply choose the *least* cost edges, making sure that no cycle is formed, and continue until *seven* edges have been chosen (since the number of vertices is 8). One MST is shown in Figure 10.44.

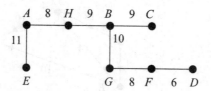

FIGURE 10.44

To construct the above MST, we first choose the edge FD with cost 6, and then the two edges GF and AH each with cost 8. Clearly, the next smallest cost edges are BH, CH, and BC, all with cost 9. However, if we select all these edges, then a cycle forms. Hence, only two of them may be used, and so edges BH and BC are selected here. Next, there is just one edge BG with cost 10, and one AE with cost 11.

However, it is obvious that we may generate three MSTs, as there are three possible choices of two edges from three edges with equal cost.

EXAMPLE 10.18 Table 10.2 of distances, in miles, between six villages, apply Kruskal's algorithm to find an MST.

TABLE 10.2

	A	B	C	D	E	F
A	0	5	6	12	4	7
B	5	0	11	3	2	5
C	6	11	0	8	6	6
D	12	3	8	0	7	9
E	4	2	6	7	0	8
F	7	5	6	9	8	0

Solution The smallest length is 2, and it is between B and E. The next smallest length is 3 between B and D. The next smallest length is 4 between A and E. At this stage, the next smallest length is 5 which is the distance between either A and B or between B and F. But we cannot choose the edge between A and B as this would make a cycle ABE. So, we have to choose the edge between B and F. There are three next shortest lengths of 6, between C and E, C and F, and A and C. We can freely choose any one of them, since inclusion of any one of them does not form a cycle. If we choose the edge AC, then one MST is obtained, as shown in Figure 10.45.

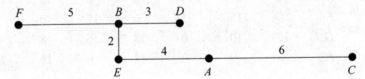

FIGURE 10.45

 EXAMPLE 10.19 Find the minimum cost spanning tree using Prim's algorithm from the graph given in Figure 10.46.

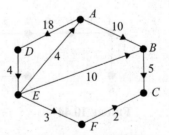

FIGURE 10.46

Solution Edges of one spanning tree of this digraph are $<A,B>, <B,C>, <A,D>, <D,E>,$ and $<E,F>$, when traversal starts at vertex A. Accordingly, the total cost is 40.

On the other hand, edges of the second spanning tree are $<E,F>, <F,C>, <E,A>,$ $<A,B>,$ and $<A,D>$. Here, traversal starts at vertex E. Now, the cost corresponding to this tree is 37.

Further, the edges of the third spanning tree can be seen as $<D,E>$, $<E,F>$, $<F,C>$, $<E,A>$, and $<A,B>$, when traversal starts at D. The total cost of this tree is 23.

Clearly, in all the above cases, Prim's algorithm is applied, and the trees are then obtained. In fact, the graph is a digraph, and so we are getting different trees by applying Prim's algorithm starting at different vertices. However, among these, the tree with cost 23 is the minimum cost spanning tree.

Note To find MST of a digraph G (if directions are not removed) by Prim's approach, it is essential to consider each vertex of G as *staring* vertex to generate several trees (if possible), and then choose the *minimum* cost tree among themselves.

EXAMPLE 10.20 The adjacency matrix of a graph is an $n \times n$ matrix in which all diagonal elements are 0s and all non-diagonal elements are 1s. What is the weight of any MST of the graph?

Solution From the matrix representation of the given graph, we may conclude that the given graph is a complete graph with no self-loop. Hence, the MST will always have weight $(n-1)$ (since each edge has weight equals to 1), and there exist several possible MSTs of same weight for this graph.

EXAMPLE 10.21 Consider a weighted complete graph G on the vertex set $\{v_1, v_2, v_3, \ldots, v_n\}$ such that the weight of the edge (v_i, v_j) is $2|i-j|$. What is the weight of any MST of this graph?

Solution Here, we have a weighted *complete* graph with n vertices. So, one possible minimum spanning tree T (by the simplest thinking) is given as $V_1-V_2-V_3-\cdots-V_n$.

Certainly, there exists a total $(n-1)$ edges in T, and for each edge (v_i, v_j) of T, we get its weight as: $2|i-j| = 2$ [such as for edge: (v_1, v_2), weight $= 2|1-2| = 2$; for edge: (v_2, v_3), weight $= 2|2-3| = 2$; and so on]. In fact, $|i-j|$ returns the absolute value of the subtraction: $(i-j)$.

Further, the $(n-1)$ edges of our simplest designed minimum spanning tree T are represented here by the vertices in sequence. Hence, W (weight of the MST) $= 2(n-1)$.

Theorem 10.1 If T is a spanning tree of G and e is an edge not in T, then inclusion of e to T makes a cycle.

Proof Let u and v be two vertices in T. Now, suppose e is an edge in G, directly connecting the vertices u and v. Also, assume that this edge is in G but not in T.

Now, since u and v are in T, so there must exist a path from u to v. Further, if we add the direct edge (u, v) of G to T, then there are two distinct paths between u and v, which means that there is at least one cycle containing u and v.

Theorem 10.2 The unique largest value edge of the weighted connected graph may not always be excluded from its MST.

Proof It is not necessarily true at all that the unique largest value edge of the weighted connected graph is always excluded from its MST. See the following graph (Figure 10.47).

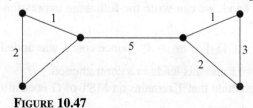

FIGURE 10.47

Clearly, the MST must include the maximum weight edge of the graph as shown in Figure 10.47. Otherwise, the tree will be disconnected. The MST corresponding to the above graph will be shown in Figure 10.48.

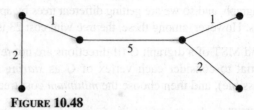

FIGURE 10.48

Theorem 10.3 The unique smallest value edge of the weighted connected graph is always included in its MST.

Proof Suppose that a weighted connected graph G with vertex set V and edge set E has a unique edge e with minimum weight, and it is the edge between two vertices, say, u and v. Let the weight of e be $w(e) = x$. Now, assume that T' is an MST of G, and it does not contain e.

Since T' is an MST of G and it does not contain e, so inclusion of e to T' makes a cycle. Next, let us remove an edge, say, f with weight $w(f)$ from this constructed cycle, and then it results a spanning tree, say, T, i.e., $T' = T + e - f$.

Clearly, the weight of T, i.e., $w(T) = w(T') + w(e) - w(f)$. Further, $w(e) < w(f)$, since e is the unique minimum weight edge in G.

Therefore, $w(T) < w(T')$, i.e., the spanning tree T', excluding the unique minimum edge e may not be an MST, and it is a contradiction.

EXAMPLE 10.22 Let G be an undirected graph with edge costs $C = [c_e]$ and T is an MST of G with respect to C. Prove or disprove that if we add 1 to all edge costs c_e then T is still an MST of G?

Solution We shall prove this by contradiction. Suppose that after adding 1 to all edge costs, the tree T is *not* an MST of the graph G with V vertices. This means that we need to replace a set of edges in T with some other set of edges of G to make it an MST. However, each spanning tree of the graph G must have $|V| - 1 = m$ edges.

Let S_1 be the subset of edges in T, which needs to be replaced by another set of edges S_2 from G, where $S_2 \subset T$ so as to convert T into an MST of the graph G.

This clearly means that $C[S_2] < C[S_1]$.

Further, let $C[S_1] - C[S_2] = k$, where $k > 0$ and k is a real number. Clearly, after adding 1 to all the edge costs, we have obtained the above expression. But S_1 and S_2 will have same number of edges m in each subset. In fact, we are replacing m number of edges of G in place of exactly m number of other edges (without introducing any cycles), so that the total number of edges remains fixed at $|V| - 1$.

On the other hand, we can write the following expression for the *case*: before adding 1 to each edge.

$C[S_1] - m - (C[S_2] - m) = k$ [since cost 1 was added later to each edge of G], i.e.,

$C[S_1] - C[S_2] = k$ and this leads to a contradiction.

Hence, we conclude that T remains an MST of G even after adding 1 to cost of each edge.

10.12 GENERAL TREE

A general tree is a tree in which each internal node may have any number of children.

10.12.1 Conversion of a General Tree to a Binary Tree

To convert a general tree to a binary tree, select the root of the general tree as the root of the binary tree. Now, for each parent, check the nodes of its (parent's) *next* level. The *left-most* node in the next level of the parent is selected as its left child. In the same level, the *immediate* right node of this *left* child is considered as the right child of the currently chosen left-child. Similarly, find the *right* child of the currently placed left child from the same level (if node exists in that level). The construction procedure is graphically shown in Figures 10.49 and 10.50.

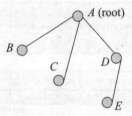

FIGURE 10.49 A general tree

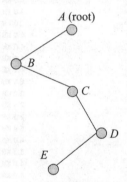

FIGURE 10.50 A binary tree equivalent to the general tree (shown in Figure 10.49)

Note Conversion from a general tree to a binary tree is important, as it is easier to represent and operate a binary tree instead of general tree.

10.12.2 Pre-Order Traversal for General Tree

Step 1 Visit root *r* and list (print) it.
Step 2 For each child of *r* from left to right, list the root of the first sub-tree then next sub-tree and so on, until we complete listing the roots of sub-trees at level 1.
Step 3 Repeat Step 2, until we arrive at the leaves of the given tree.
Step 4 Stop.

Similarly, we can traverse the general tree using in-order and post-order ways keeping their basic idea.

Note The binary search method can be easily explained in terms of tree, and the tree constructed in such representation is called as *search tree*.

10.13 SOME IMPORTANT APPLICATIONS OF TREE

Trees are fundamental data structures in computer science. Some important applications of tree are as follows.

(i) An *operating system* will maintain a directory/file hierarchy as a tree structure. *Files* are stored as *leaves*; *directories* are stored as internal (non-leaf) nodes. For example, take a look at the following screenshot taken from a computer running Mandrake Linux 9.2 operating system (Figure 10.51).

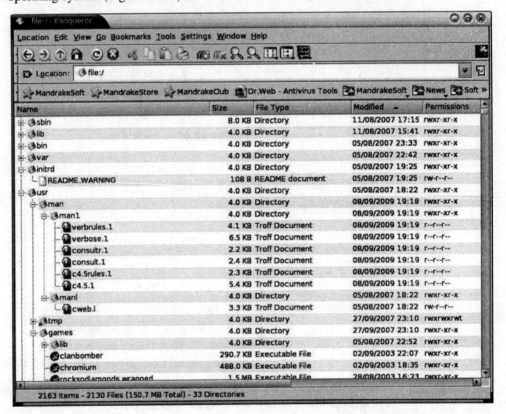

FIGURE 10.51

In this case, we can treat the node *usr* as the root of a sub-tree. The node *usr* has 3 child nodes, namely *man*, *tmp*, and *games* as visible in this image. The node *man* has 2 children, namely *man1* and *manl*. The node *man1* has 6 children, namely *verbrules.1*, *verbose.1*, *consultr.1*, *consult.1*, *c4.5rules.1*, *c4.5.1*. Each of these 6 children is leaf node, because none of them have any further children nodes. The node *manl* has just 1 child node, namely *cweb.l*, which is a leaf node, as it does not have any further children. Proceeding in the same way, it is clearly seen that the directory structure on the Mandrake Linux 9.2 operating system is implemented as a tree. In fact, the directory structure on any operating system used today follows a similar tree structure.

(ii) *Labelled trees* are of interest in several practical and theoretical areas of computer science. For example, some networks, such as Ethernet, are required to have one and only one path between every pair between every pair of terminal devices, and therefore are trees. Moreover, labelling the nodes is necessary because each device in such network is a distinct entity. Many algorithms for analysis of general networks require their spanning trees.

(iii) Implementation of dictionary in word processors. Usually, the dictionary of words contains millions of words, and it may not be practically feasible to load the entire data in the memory (RAM).

A common approach is to maintain pointers to the actual locations on hard disk, where the data are stored. Assume that the hard disk is not fragmented, so that all words in dictionary are stored in one straightline on the hard disk. In such case, we may implement a special kind of tree structure called as *B+* trees to ensure that only the relevant section of data is loaded into memory. A fragment of such *B+* tree is shown in Figure 10.52.

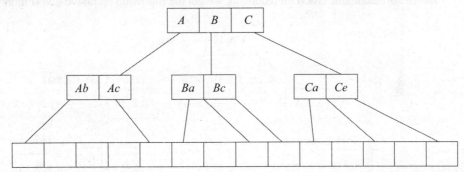

FIGURE 10.52

In Figure 10.31, the bottom-most straightline represents the hard disk on which the dictionary of words is stored. There are pointers to the beginning and at the end of locations which contain all words beginning with *Ab*, *Ac*, *Ba*, *Be*, etc. Finally, at the top, we have a root node containing the alphabets.

For example, suppose we want to search for a word *abstract* in the dictionary of English words, then the searching will be as follows. From the root node, take the path for the character *A*. Now, we have to choose between *Ab* and *Ac*. Clearly, we will choose *Ab*. Then, all words which begin with *Ab* will be loaded into RAM, after scanning only the relevant portion of hard disk. The word *abstract* will be searched only in that selected range of words. This prevents the need to unnecessarily load entire dictionary of words into the RAM.

(iv) Tree structure is preferred for faster and organized access of information in several cases such as in construction of *symbol table* and *parser* parts of compiler.

(v) Conversion of *in-fix* expression to *post-fix* expression (explained in Section 10.9).

(vi) The MST of a graph defines the cheapest subset of edges that keeps the graph in one-connected component. Telephone companies are particularly interested in MSTs, because the MST of a set of sites defines the wiring scheme that connects the sites using as little wire as possible. It is the mother of all network design problems.

The MSTs can be computed quickly and easily, and they create a sparse subgraph that reflects a lot about the original graph. They provide a way to identify clusters in sets of points. Deleting the long edges from MST leaves connected components that define natural clusters in the data set, as shown in the output figure above. They can be used to give approximate solutions to hard problems such as Steiner tree and travelling salesman.

Further, solving strategy of many problems can be easily expressed in terms of tree structure for easier understanding. For example, merge sort, quick sort, binary search, etc. Also, each recursive approach can be easily cleared through tree representation. One example is presented below.

EXAMPLE 10.23 Draw a recursive tree for merge sort of the list $a = \{179, 254, 285, 310, 351, 423, 450, 520, 652, 861\}$.

Solution The list elements can be represented as $a[0] = 179, a[1] = 254, a[2] = 285, a[3] = 310, a[4] = 351, a[5] = 423, a[6] = 450, a[7] = 520, a[8] = 652, a[9] = 861$. Here, location in the array a begins from 0 and ends at 9, since total 10 elements are present. Applying the *merge sort* technique based on recursion, we get the following recursive tree (Figure 10.53).

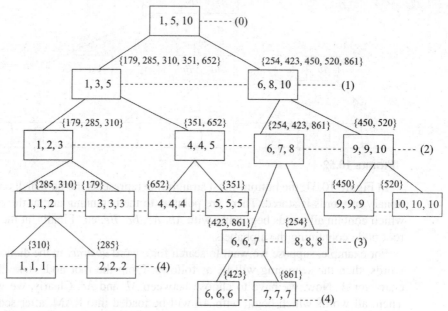

FIGURE 10.53

SUMMARY

A tree is a type of organizational structuring of classes of elements, databases, or directory files that has a type of generalization or hierarchical architecture. This type of structure looks like a tree with its different branches and nodes. Also, there is a unique path between any pair of nodes on the tree. The top of the structure is called the root node and each entry on a different branch is called a child node of the parent.

In this chapter, we begin with basic terminologies of tree. We then discuss m-ary tree along with some binary tree and some of its variations. Collection of useful results with proper explanations on tree is an important aspect of this chapter. We have provided the sequential and linked list representation of binary tree in respect of computer science. The chapter deals with the traversal techniques of binary tree, and shows how to create a unique binary tree as per the provision of appropriate traversal sequences. A comprehensive discussion on BST, AVL tree, and expression tree is also a major part of this chapter. A detailed study with various types of solved examples on minimum spanning tree is presented here.

The chapter ends with several applications of tree data structures.

EXERCISES

1. Find all the trees with 5 vertices.
2. How many different binary trees can be formed with 5 vertices? Draw them.
3. How many different labelled trees can be formed with 5 vertices? Draw them.
4. Draw 3 distinct rooted trees that have 4 vertices.
5. Given the tree with root at A:

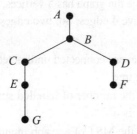

   (a) Find the parents of D and E.
   (b) Find the children of D, C, and E.
   (c) Find the siblings of B.
   (d) Find the leaves.
   (e) Find the internal vertices.
   (f) Draw the sub-tree with parent B.
   (g) Find the degree of each vertex.
6. Prove that a tree with more than 1 vertex has at least 2 leaves.
7. Prove that a tree with 3 vertices of degree 3 must have at least 5 leaves.
8. Determine the order in which the vertices of the given binary tree will be visited under
   (i) in-order
   (ii) pre-order
   (iii) post-order

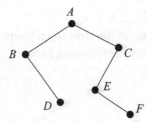

   Find the *height* of this tree. Also, find the *successor* and *predecessor* nodes of B and C.
9. Determine the order in which the vertices of the given binary tree will be visited under
   (i) in-order (ii) pre-order
   (iii) post-order.

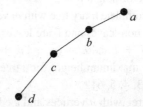

10. Construct a binary tree whose in-order and pre-order traversal sequences are given below.
    (a) (i) in-order: 5, 1, 3, 11, 6, 8, 2, 4, 7
        (ii) pre-order: 6, 1, 5, 11, 3, 4, 8, 7, 2
    (b) (i) in-order: d, g, b, e, i, h, j, a, c, f
        (ii) pre-order: a, b, d, g, e, h, i, j, c, f
11. Represent each of the expressions in a binary tree, and then find pre-order and post-order notations.
    (a) $(a + b) * (c - d)$
    (b) $[(a + b)/c] + d$
    (c) $((a - b)\uparrow 2)/(a + b)$
    (d) $((3 + x) - (4 + 5)) - x\uparrow y$
    (e) $(x + y)^6(x - y)$

12. Represent the expression tree for the expression: 3-4-6, making appropriate rule for such cases.
13. Compute the value of each of the following pre-fix expression:
    (a) $- *2/8, 4, 5$
    (b) $+ -*2, 3, 5/\uparrow 2, 3, 8$
14. Compute the value of each of the following post-fix expression:
    (a) $1, 2, 3 + -$
    (b) $1, 2, 3\uparrow 1, 2, 3 + +-$
15. Find the size of the array to represent appropriately the following tree.

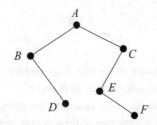

16. Show that the maximum number of vertices in a binary tree of height n is $2^{n+1} - 1$.
17. Let T be a complete n-ary tree with m vertices of which k are non-leaves and L are leaves. Prove that $m = nk + 1$ and $L = (n - 1)k + 1$.
18. What is the maximum height for a tree on $S = \{1, 2, 3, 4, 5, 6\}$?
19. Let T be a tree with n vertices and e edges. Derive a relation between n and e.
20. Prove that every tree with at least 2 vertices has a *pendant* vertex.
21. Prove that any connected graph with minimum number of edges is a tree.
22. Show that if a tree has two vertices of degree 3, then it must have at least 4 vertices of degree 1.
23. Construct an algorithm to determine if 2 trees are identical.
24. Draw AVL trees using the smallest possible number of vertices with height given like:
    (a) $h = 0$, (b) $h = 1$, (c) $h = 2$, and (d) $h = 3$.
25. Consider the following binary tree. Is it an AVL tree? If not, take the necessary steps to balance it.

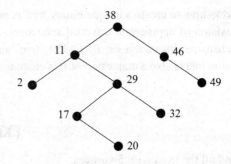

26. Draw all the spanning trees of the following graph.

[*Hints*: Since the graph has 5 vertices, a spanning tree will have 4 edges. So, two edges have to be removed.]

27. Prove that every connected undirected graph has a spanning tree.
28. Prove that the number of labelled spanning trees of K_n is n^{n-2}.
29. When will the MST of a graph include the maximum weight edge of the graph?
30. Prove that any maximum weight edge of a graph, which is a part of a cycle, is never included in any MST of the graph.
31. Show that any minimum weight edge of a graph is a part of some MST of the graph.
32. Let G be an undirected graph with edge costs $C = [c_e]$ and T is an MST of G with respect to C. Prove or disprove if we add 1 to all edge costs c_e then T is still an MST of G.
33. (a) Modify the original Kruskal's algorithm to find a maximal spanning tree from an undirected weighted graph G.
    (b) Find all conditions for which Prim's and Kruskal's algorithms, when run on an undirected graph, generate different MSTs.

34. Use Kruskal's algorithm to find an MST of the graph as shown below.

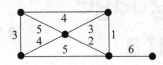

35. Construct a spanning tree of the graph as shown below.

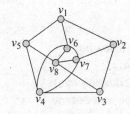

36. Given an MST of a graph G. How can you compute the second-best MST of G?

37. Let T be an MST of a graph $G(V, E)$. Let $c_1 \leq c_2 \leq c_3 \leq \cdots \leq c_n$ be the costs of edges in T. Let S be any arbitrary spanning tree of G with edge costs $d_1 \leq d_2 \leq d_3 \leq \cdots \leq d_n$. Show that $c_i \leq d_i$ for $1 \leq i \leq n$.

38. Prove that a graph having 2 distinct spanning trees must contain at least one cycle.

39. Construct an algorithm to determine the height of a general tree.

40. Construct asn algorithm to determine if 2 trees are identical. [*Hint*: Tree traversal]

41. Construct an algorithm to determine if a tree is a complete binary tree.

42. In an undirected graph, a Hamiltonian circuit is also the Euler circuit for the graph. Prove that deleting one edge of maximum weight gives a minimum spanning tree.

43. Prove or disprove that every tree is a 2-colour graph but the reverse is not true.

11 Formal Language and Automata

LEARNING OBJECTIVES

After reading this chapter you will be able to understand the following:
- The use of mathematics to communicate computer science
- Finite automata, acceptors, and transducers (Moore m/c and Mealy m/c)
- Deterministic and non-deterministic finite automata and their equivalence
- Reduction of the number of states in finite automata
- Regular expression and regular language
- Pumping lemma for regular language
- Grammar and its classifications
- Pushdown automata, turing machines, and cellular automata
- Introduction to computability, decidability, the halting problem, and complexity theory
- P, NP, and NP-completeness
- Fuzzy set and fuzzy logic
- Russell's paradox

11.1 INTRODUCTION

In theoretical computer science, automata theory is the study of abstract machines and their *capabilities* of problem solving. It is closely related to the formal languages theory, as the automata are often classified by the class of formal languages they are able to recognize. In fact, theoretical computer science is the collection of topics of computer science that focuses more on the abstract, logical, and mathematical aspects of computing, such as the theory of computation, analysis of algorithms, semantics of programming languages, and so on.

A formal language is a language in which interpretation of an expression is determined by precisely defined *rules* using some formal notations. Examples of formal languages are C, C++, Pascal, etc., whereas English, Hindi, Spanish, etc. are some examples of natural languages.

An *abstract machine*, also known as an *abstract computer*, is a theoretical (mathematical) model of a computer. A typical abstract machine consists of a definition in terms of input, output, and the set of allowable operations (called transitions) which are used to move from the current to the next. The input is read symbol by symbol, until it is consumed completely. Once the input is depleted, the automaton is said to have stopped.

Automata are abstract mathematical models of machines that perform computations on an input by moving through a series of *states* or *configurations*. If the computation of an automaton reaches an accepting configuration, it accepts that input. At each stage of the computation, a transition function determines the next configuration on the basis of a finite portion of the present configuration. *Turing machines* (TMs) are the most general automata. They consist of a finite set of states and an infinite tape which contains the input. The tape is used to read and write symbols during the computation. Automata play a major role in *compiler* design and *natural language processing*.

This chapter briefly describes some basic automata and their capabilities (i.e., what problems can be solved and what *cannot* be solved), formal languages and their grammar to gain insight of the design of *computer languages* and *their syntax*, brief discussion on different types of problems such as *P*, *NP*, and so on.

11.2 MATHEMATICAL PRELIMINARIES

The basic concepts of *symbols*, *alphabets*, *strings*, and *languages* are common to the descriptions of most of the automata.

11.2.1 Symbol

An arbitrary datum which has some meaning to or effect on the machine. Symbols are sometimes just called 'letters'. For example, $a, b, c, \ldots, z$, '1', $\ldots$, '9', etc., are normally considered as symbols.

11.2.2 Alphabet

An alphabet is a finite non-empty set of symbols. It is frequently denoted by Σ. For example, $\Sigma = \{0, 1, 2\}$.

11.2.3 String

A word or a string on Σ is a finite sequence of symbols. For example, $w = ab$ is a string derived from $\Sigma = \{a, b\}$. Strings are generally denoted by the symbols: u, v, w, x, y, z.

- *Empty string (null string)* is a string with no symbol, and it is normally denoted by the symbol Λ. Clearly, the length of Λ (denoted by $|\Lambda|$) is 0.
- *Substring* is a portion of a string. For example, let $w = abca$ is a string. Here, bc is a substring of $w = abca$.

Operations on String

1. *Reversal (transpose) of string* If w represents a string, then its reversal is denoted by w^R or w^T. Suppose $w = ab$ is a string on $\Sigma = \{a, b\}$, then its *reversal* or *transpose* is ba.
2. *Concatenation of strings* If w_1 and w_2 are two strings over Σ, then their concatenation is written as $w_1 \cdot w_2$ or $w_1 \circ w_2$, where both the symbols $\circ$ and $\cdot$ denote concatenation of strings.

For example, suppose $w_1 = ab$ is a string and $w_2 = aa$ is another string on $\Sigma = \{a, b\}$, then the concatenation of these two strings is $w = w_1 \cdot w_2 = ab \cdot aa = abaa$. Obviously, $w_1 \cdot w_2 \neq w_2 \cdot w_1$ (as $w_1 \cdot w_2 = abaa$ and $w_2 \cdot w_1 = aaab$).

11.2.4 Language

Language is normally denoted by the symbol L. A language may be thought of as a subset of all possible words. In other words, it is the collection of *sentences* (sentence means here meaningful sentence). In fact, a sentence consists of some words, and it must be meaningful following the set of *pre-defined rules*.

On the other hand, a word is formed from the selected set of symbols (called as *alphabet*), following some *rule* (expressed in terms of *regular expression*).

The term *the set of all possible words* may, in turn, be formally thought of as the set of all possible concatenations of strings, and it is termed as a *free monoid* (denoted by Σ^* which means it consists of Λ also). On the other hand, Σ^+ is the *positive closure* and it does not include Λ. The superscript symbol * is here called as the *Kleen Star*. For example, let $\Sigma = \{0, 1\}$. Now, Σ^* consists of all the strings (S) derived as follows: $S = \{\Lambda, \Sigma, \Sigma\Sigma, \Sigma\Sigma\Sigma, \ldots\} = \{\Lambda, 0, 1,$ 00, 01, 10, 11, 000, 001, 010, 011, 100, 101, 110, 111, $\ldots\}$.

Clearly, we may consider that this set of strings is indeed the *concatenation* of the strings over Σ. For example, if $w_1 = 0$ and $w_2 = 1$ are two strings in S, then their concatenations such as: $w_1 \cdot w_2 = 01$ and $w_2 \cdot w_1 = 10$ also belong to S, and so on.

Now, suppose we desire to construct one language L from this set of strings, satisfying the property that no string contains two consecutive 0s. Clearly, we need to extract here those strings from Σ^*, consisting of no two consecutive 0s (i.e., one subset from Σ^* is to be extracted to satisfy the desired property).

Hence, if w_1 and w_2 are two distinct strings both belonging to L, then their concatenation: $w_1 w_2$ returns a new string of L if the specified property holds.

Furthermore, dictionaries define the word language informally as a system suitable for the expression of certain *ideas*, *facts*, or *concepts* including a set of symbols and rules for their manipulation. Such languages are often called *natural language*. The syntax of a natural language is extremely complicated. A *formal language* consists of a set of rules (also called as *syntax*) by which these symbols can be combined into entities called sentences. In fact, formal languages are used to model languages to communicate with computer (i.e., computer languages).

Operations on Language

1. *Reversal of a language* If L is a language, then reversal of L is $L^R = \{x^R | x \in L\}$. For example, let $L = \{0, 01, 001\}$. Then $L^R = \{0, 10, 100\}$.
2. *Union operation* If L_1 is a language on Σ_1 and L_2 is another language on Σ_2, then their *union* forms another new language: $L = L_1 \cup L_2$ over $\Sigma = \Sigma_1 \cup \Sigma_2$, where $\cup$ is the union operator. Obviously, the new language consists of all the strings of L_1 and L_2.
3. *Intersection operation* If L_1 and L_2 are two languages over Σ_1 and Σ_2 respectively, then their *intersection* forms another new language $L = L_1 \cap L_2$ over $\Sigma = \Sigma_1 \cap \Sigma_2$, where $\cap$ is the intersection operator. It means that the new language consists of the strings common to both L_1 and L_2.
4. *Concatenation* Suppose L_1 and L_2 are the languages over Σ_1 and Σ_2, respectively, then their concatenation derives another new language $L = L_1 \circ L_2$ over $\Sigma = \Sigma_1 \cup \Sigma_2$, where

∘ is the concatenation operator. Each string in this new language consists of a string of L_1 followed by a string of L_2.

However, $L_1 \circ L_2 \neq L_2 \circ L_1$. Actually, in $L_1 \circ L_2$, string of L_1 comes before string of L_2, whereas in $L_2 \circ L_1$, string of L_2 comes before string of L_1.

Let us consider an example, suppose $L_1 = \{aa, b\}$ and $L_2 = \{c, d, e\}$, then $L_1 L_2$ will consist of the strings $\{aac, aad, aae, bc, bd, be\}$.

5. *Power of language* Power of language L is defined as L^n which means L is concatenated n times ($n \geq 0$) with itself. It is illustrated below.
   (i) $L^0 = \{\Lambda\}$, when $n = 0$.
   (ii) If $L = \{a^n b^n | n \geq 0\}$, then $L^2 = L.L = \{a^i b^i a^j b^j | 0 \leq i \leq n, 0 \leq j \leq n, n \geq 0\}$. In fact, the language $L = \{a^n b^n | n \geq 0\}$ consists of the set of strings like $\{\Lambda, ab, a^2b^2, a^3b^3, \ldots,\}$. On the other hand, in $L^2 = \{a^i b^i a^j b^j | 0 \leq i \geq n, 0 \leq j \leq n, n \geq 0\}$, the set of strings looks like $\{\Lambda, ab, abab, a^2b^2, ab a^2b^2, a^2b^2 ab, a^2b^2 a^2b^2, a^3b^3, \ldots\}$. Similarly, the languages $L^3, L^4, \ldots, L^n$ can be generated.

Clearly, the above two languages (L and L^2) are different, since each of them has a distinct property. In the first language L, equal number of a's is followed by equal number of b's, but no a appears after b. On the other hand, in the second language L^2, equal number of a's is followed by equal number of b's, and a may appear after the first substring: $a^i b^i$ but not after the second substring: $a^j b^j$.

Further, in context of language, we cannot say $\{a^n b^n a^n b^n | n \geq 0\} = a^{2n} b^{2n} | n \geq 0\}$.

11.3 AUTOMATA

The term *automata* is plural, whereas *automaton* is its singular form. A brief note on automata is given in the introduction section of this chapter. Now, we will present the details about automata and their capabilities. However, in this book, the terms automaton and m/c (machine) are used interchangeably.

In particular, each automaton can be described by the following basic *characteristics*:
(a) *State* At any moment, every automaton remains at a state. However, it starts its computation from a specific state called *start state*. State specifies the position of the machine at any instance during some computation. The state of system summarizes the information concerning past input that is needed to determine the behaviour of the system or subsequent input. Here, we may consider the following kinds of states:

(i) *Initial/start state* State from where m/c starts its operation. We normally represent such state by the symbol q_0.
(ii) *Final state* State which represents to accept an input, when the m/c moves to such state. We normally denote final state by the symbol q_f.
(iii) *Non-final state* When the m/c is in this state, the processed string is not accepted (i.e., it is rejected). The symbol q_i ($1 \leq i \leq n$) is generally used for non-final state.

However, if a m/c is either at final or non-final state and no input is left to process, then it implies that computation is over (i.e., the m/c stops/halts there). That is why these states are sometimes called as *halt state*. Clearly, halt state signifies that the processing is over, and the m/c has either moved to final or non-final state to stop its computation. In other words, it currently accepts the string (moving to final state) or rejects the string (moving to a non-final state), but can move to an appropriate final state once it gets enough information to make a decision. So, in both the cases, the m/c has no problem to stop its computation. Thus, both final and non-final states can represent as *halt* state.

(iv) *Dead/trapped state* It is used to indicate a *crash*, arising due to some abnormal situation in which the m/c cannot carry out its mission as expected. Accordingly, even if sufficient input is found later, the m/c cannot move out from this state (i.e., no further input can enable to escape from this state).

Hence, normal termination (i.e., halting) is not possible in such situation. Symbol h used in this chapter represents *dead/trapped* state.

(b) *Input* At every discrete instant, it takes input from a set of input symbols (Σ). In any kind of automaton, the input alphabet Σ is the finite set of input symbols, in which empty symbol is not included.

(c) *State relation* (*transition function or next state function*) The next state of an automaton at any instant of time is generally determined by the present state and the current input. It controls the next move for a particular input (i.e., to what state the automaton will move).

(d) *Output relation* Output is related to either state only or both to the input and the state.

11.3.1 Basic Categories of Automata

Four basic kinds of automata are generally considered in the theory of computation. The first kind of automaton is named as *finite automaton* (FA) (sometimes called as *acceptor*). The other three kinds are, respectively, *pushdown automaton* (PDA), *linear bounded automaton* (LBA), and *turing machine* (TM). However, the categorization is fully based on their capabilities.

Every kind of automaton possesses finite number of states. But the most *attractive* issue in designing automaton is to minimize the number of state, i.e., to design *minimum state automaton*.

State Transition Graph

We can graphically express automaton (i.e., the act of automaton), containing states (which are indeed nodes/vertices) and inputs (that are passed over directed edges), called *state transition graph*. Obviously, it is directed graph. Of course, the computation of automaton can also be described by tabular representation, called as *state table*.

11.3.2 Finite Automaton and Its Types

Finite automaton is a natural medium to describe behaviours of *reactive system*. A reactive system is a system that changes its *actions*, *outputs*, and *status* based on some information (input). To model a reactive system with FA, first identify the *states* and *operations* to be performed. The work of the computer devices and computer programs can generally be described by FA.

For example Consider the *very simplified version of login* process to a computer. It can be modelled through FA as follows.

Description of the login problem Initially, the computer waits for a user *name* to be typed in. This can be considered as a state of the system called as *start state*. When *name* is typed in, it checks whether the name is valid or not. If the user name does not match, then it remains at *start* state. On the other hand, if it matches (i.e., valid), then it moves to some other state; and asks for and waits for the *password*. If the typed password at this state is *valid*, it then moves to *final* state (i.e., the user has successfully entered into the system), otherwise moves back to *start* state. The process is shown diagrammatically in Figure 11.1.

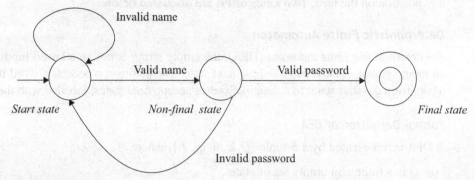

FIGURE 11.1

Clearly, we can say that the role of FA is normally to check the acceptance of strings, and that is why these are sometimes called as *acceptor* also. But there are some FA which are employed to produce some new output for the given input. *Moore* and *Mealy* machines are examples of such automata, and these are called as *finite state transducers*. Since, these are simply *converters*, so the number of final states in such FA is *zero*. Accordingly, we can conclude that the number of final states in FA may be zero or more. [*Moore* and *Mealy* machines are discussed later in this chapter.]

Now, we can *summarize* FA as follows: Finite automaton is the restricted model of an actual computer. It is in fact similar to a real computer with its 'central processor' of fixed capacity. It has no auxiliary memory to remember the past data, i.e., *it memorizes only the current state but not the path through which it has reached that state*. However, such m/c receives input string stored in INPUT tape by means of a movable reading *head*, and it is assumed that the size of the tape is very large. Here, the movement of head is unidirectional, i.e., the head can move in forward direction only. Further, the input in the INPUT tape is not erasable.

Thus, FA is, in other words, a *language recognition device*, i.e., it gives an indication of whether the input is acceptable or not. So, its power is limited. Transition diagram of FA is pictured in Figure 11.2.

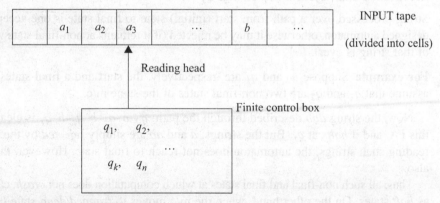

FIGURE 11.2 Model of FA

Here, $a_1, a_2, \ldots, b$ are the input symbols on tape, and $q_1, q_2, \ldots$, are the states of the FA. The *finite control box* with the help of the reading head can sense what symbol is written at any position on the tape. Two kinds of FA are discussed below.

Deterministic Finite Automaton

For deterministic finite automata (DFA), the empty string is not an allowed input. On reading an input symbol, it moves to a *unique* next state. The automaton accepts a word if there exists a path from q_0 (start state) to a state in F (set of accept/final states) labelled with the input word.

Formal Definition of DFA

A DFA is represented by a 5-tuple $\{Q, \Sigma, \delta, q_0, F\}$, where

(i) Q is a finite non-empty set of states.
(ii) Σ is a finite non-empty set of symbols, called as the *alphabet* of the language, that the automaton accepts.
(iii) δ is the transition function, i.e., defined as follows:

$$\delta : Q \times \Sigma \to Q$$

More clearly, the automata (machine) reads an input symbol (say a taken from Σ) at a state (say q_1 in Q), and moves to another state as next state. However, this new next state may be simply the present state itself.

(iv) q_0 is the *start* state, i.e., the state in which the automaton stays when no input has been processed yet (obviously, $q_0 \in Q$).
(v) F is a set of states of Q (i.e., $F \subseteq Q$), called *final/accept* states. Any final state is normally denoted by q_f, and it belongs to Q. Further, the states which are not final states, are called as *non-final* states. In fact, non-final states are normally called as intermediate states, and all the states except q_0 and q_f are considered as the *intermediate* states also.

Now, for a given input: $a \in \Sigma$, we may write the transition function δ as $q_i \times a \to q_j$ which can be viewed as $\delta(q_i, a) = q_j$ for all q_j and $q_j \in Q$, where q_j is the current state and q_j is the next state.

Acceptance and Rejection of Strings by FA

String processed over a path from start (initial) state to final state is one accepted string by a designed automaton, otherwise it may be rejected (if it remains at non-final state when processing of that string is over).

For example Suppose q_0 and q_f are, respectively, the start and a final states of a FA. Also, assume that q_1 and q_2 are two non-final states of the same m/c.

Now, the string *abb* described through the path: q_0–a–q_1–b–q_2–b–q_f is clearly accepted by this FA, and it *halts* at q_f. But the strings: a and ab are simply *rejected* by the same FA; since reading such strings, the automaton does not reach to final state. However, the FA halts here also.

Thus, all such non-final and final states at which computation does not *crash*, can be considered as *halt* states. On the other hand, when the m/c moves to *trapped/dead* state h for an input, it cannot escape from that state. Such input can be considered as rejected as it causes the automaton to *hang*. In other way, it implies that computation is allowed to terminate by *crashing* and further move (for any input) is not possible for such transition.

Hence, a string $x \in \Sigma^*$ is accepted by FA (say, M) if $\delta^*(q_0, x) \in M$, i.e., if the m/c moves to final state. On the other hand, a string is not accepted (i.e., rejected) if it remains at any non-final state after processing that string. Here, $\delta^*(q_0, x)$ means that, starting at q_0 to move to final (accept) state, the transition function is applied a number of times to process the string x.

Non-Deterministic Finite Automaton

Non-deterministic finite automaton (NDFA) allows a choice of zero or more next states for each input symbol (i.e., more outgoing arcs with the same input symbol). In other words, on reading an input symbol, it may move to a number of next states (i.e., next state is not unique). Further, the empty string is an allowed input in NDFA.

Clearly, states in such kind of an automaton can have multiple transitions for a symbol. However, the automaton accepts a word if there exists at least one path from q_0 (start state) to a state in F (set of accept/final states) labelled with the input word.

Formal Definition of NDFA

An NDFA is represented by a 5-tuple $\{Q, \Sigma, \delta, q_0, F\}$, where

(i) Q is a finite non-empty set of states.
(ii) Σ is a finite non-empty set of symbols, that we call the alphabet of the language the automaton accepts.
(iii) δ is the transition function that is defined as follows:

$$\delta : Q \times \Sigma^* \to 2^Q$$

More clearly, the automaton (machine) receives an input from Σ (this input may be Λ or single symbol or collection of symbols from Σ) at a state (say q_1 from Q), but may move one of multiple possible next states.
(iv) q_0 is the start state, i.e., the state in which the automaton stays when no input has been processed yet (obviously, $q_0 \in Q$).
(v) $F \subseteq Q$ is the set of final states.

Note The transition function $\delta : Q \times \Sigma^* \to 2^Q$ in NDFA can be written as $Q \times \Sigma^* \to P(Q)$, where $P(Q)$ is the *power* set of Q (i.e., the set of all subsets of Q). This transition can be explained as follows.

If the set Q of an NDFA consists of three states (say, $q_0, q_1,$ and q_2), then the set of next states from any state is selected from $P(Q)$, where $P(Q) = \{\Phi$ (*trapped/dead* state), $\{q_0\}, \{q_1\}, \{q_2\}, \{q_0, q_1\}, \{q_0, q_2\}, \{q_1, q_2\}, \{q_0, q_1, q_2\}\}$. Clearly, the number of subsets is $8 = 2^Q$, where Q in 2^Q represents the number of states in Q (here it is 3). So, it is obvious that, on reading an input from a state, the multiple next states will belong to $P(Q)$.

For example Suppose the NDFA, on receiving an input at q_0, moves to both the states q_1 and q_2. Here, q_1 and q_2 are denoted by the subset: $\{q_1, q_2\}$ of $P(Q)$.

11.3.3 Importance of NDFA

Designing NDFA is much simpler than DFA, since the automaton, from any state, can move to multiple number of states for an input. Further, it can move from one state to another state on reading Λ as input, which is not possible in DFA.

Hence, the number of states in NDFA is naturally *less* than that of DFA, which results in less space also. Obviously, non-deterministic finite automata reduce the complexity of the mathematical work in the theory of computation to establish many important properties such as *union* of two regular languages is regular and *concatenation* of two regular languages is also regular, and so on.

However, designing a real machine from NDFA is very difficult, since mapping from NDFA to a real machine is not a simple task. In fact, to model a real machine from NDFA, we first require to convert NDFA to DFA because '*For every NDFA there is an equivalent DFA*'. (Conversion is discussed later in this chapter.)

Note Languages that are accepted by FA are called *regular languages*. Also, there exists a strong equivalence: '*For every regular language, there is a finite state automaton, and vice versa*'.

11.3.4 Graphical Notations Used in Drawing Finite Automata

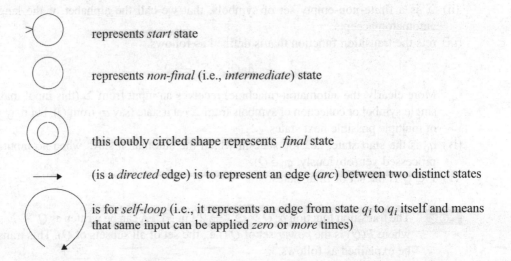

represents *start* state

represents *non-final* (i.e., *intermediate*) state

this doubly circled shape represents *final* state

(is a *directed* edge) is to represent an edge (*arc*) between two distinct states

is for *self-loop* (i.e., it represents an edge from state q_i to q_i itself and means that same input can be applied *zero* or *more* times)

Note While designing state diagram of an FA, from any state of that automaton, all the symbols of Σ should be considered as input. If transitions for one or more inputs are not explicitly defined from a state, it implicitly means that for such input, the particular automaton comes to *trapped/dead* state.

However, we are not bothered about dead state for all problems. For some problems, dead state needs to be considered; on the other hand, for some other problems, dead state may not arise. Hence, consideration of dead state is fully dependent on the nature of the problem.

11.3.5 Discussion on Designing of Some Basic FAs

EXAMPLE 11.1 Draw the state transition graphs and transition tables for the DFA which accepts sets of strings composed of zeros (0) and ones (1) over $\Sigma = \{0, 1\}$ which:

(a) Ends with the string 00.
(b) Possesses even numbers of 1s.

Solution

(a) *Problem analysis* The problem describes that every accepted string ends with 00. This suggests that the length of the valid string is ≥ 2. Further, any string ending with '1' or single '0' is rejected.

Designing procedure of the proposed automaton Here, we are interested in two consecutive 0s. Hence, in order to accept such string, the minimum 3 states (say, q_0, q_1, q_f) are essential, and the path of acceptance can be described as q_0–0–q_1–0–q_f (directed from q_0 to q_f), where q_0, q_1, and q_2 are assumed as, respectively, the start, the non-final, and the final states. It means that, for any valid string, the proposed automaton (say, M) starts from q_0, travels the path: q_0–0–q_1–0–q_f, and reaches to final state. Further, M must remain at q_f for input '0', since such string also ends with 00.

In particular, whenever a '1' is read and M is in one of states: q_0, q_1, and q_f; it returns to q_0, expecting to travel the path: q_0–0–q_1–0–q_f (to get valid string).

The working way is shown graphically in Figure 11.3. The corresponding state transition table is shown in Table 11.1.

State transition graph

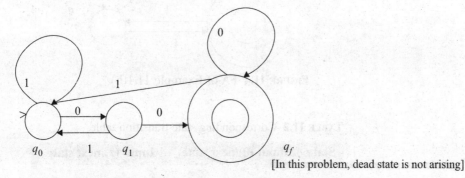

[In this problem, dead state is not arising]

FIGURE 11.3 FA of Example 11.1(a)

TABLE 11.1 Corresponding state transition table

State	Input (0)/next state	Input (1)/next state
>q_0	q_1	q_0
q_1	q_f	q_0
(q_f)	q_f	q_0

(b) *Problem analysis* The problem instructs that, in the accepted string, the number of 0s is immaterial (i.e., it may be zero, odd, or even) but the number of 1s must be even. This emphasizes that the proposed automaton (say M) starts from its start state and reaches to final state, traversing even number of 1s.

Designing procedure Since the valid strings consist of even number of 1s, so the strings with the following characteristics should be recognized by the proposed automaton.

(i) Strings with no (i.e., zero) 1 like Λ or: 0, 00, ..., are accepted, since these possess even number of 1s. However, to accept such strings, a single state (q_0) (which may be considered as both initial and final state) is sufficient.

(ii) Strings consisting of 1s and 0s but the number of 1s is multiple of 2 are also valid. Now, in order to accept such strings, two states are necessary. However, out of these two states, one is initial state as well as final state (q_0) (already considered and added). Thus, the rest should be non-final state (say q_1).

In brief, to process any string, M passes from state q_0 to q_1 or from back q_1 to q_0 when a '1' is read, but essentially ignores 0s always remaining in its current state when a '0' is read (since the number of 0s is immaterial).

Corresponding state transition graph and state table are given in Figure 11.4 and Table 11.2, respectively. Here, no dead state may arise.

State transition graph

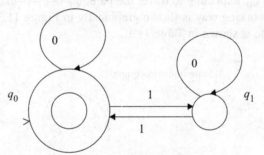

FIGURE 11.4 FA of Example 11.1(b)

TABLE 11.2 Corresponding state transition table

State	Input (0)/next state	Input (1)/next state
($>q_0$)	q_0	q_1
q_1	q_1	q_0

EXAMPLE 11.2 Draw the state graph for the DFA which accepts sets of strings of even numbers of 0s and odd numbers of 1s.

Solution *Problem analysis*: The problem summarizes that, in the valid string, the number of 0s and 1s should be even and odd, respectively. This implies that the proposed automaton starts from its initial state, and moves to its final state if and only if it traverses even number of 0s but odd number of 1s.

Designing procedure In designing such FA, the necessary states are as follows:

Step 1 Let us add start state (say, q_0) to the proposed FA (say, M). Now, q_0 may not be final state, since the number of 1s must be odd in the valid string. If q_0 becomes final state, then null string (which consists of zero number of 1s) is also accepted (i.e., a string with even number of 1s is accepted). Hence, starting at q_0, it is essential to add some new states to process valid strings.

Step 2 If input '1' is read from q_0, it should reach to its final state (say, q_1) (since the number of traversed 1s and 0s are now odd and even, respectively). However, M moves back to q_0 from q_1 on receiving input '1', since such string does not contain odd number of 1s. Thus, two states in total are described till now, and the latest included state q_1 is considered as final state.

Step 3 Let us assume that M is in q_0 and input is '0'. Obviously the next state may not be either q_0 or q_1. If it happens so, then specified conditions for valid string are violated. Hence, it is required to add one more distinct non-final state (say, q_2) for input '0' to move from q_0. Thus, the total number of states needed so far is 3 for the proposed automaton M.

Step 4 Now, if '1' is read from q_2, M should not return to q_0 or q_1 to maintain the acceptance property of string. Hence, one more non-final state (say, q_3) is essential to add into the automaton.

Thus, minimum four states are necessary, and it is sufficient also to accept the strings following the given properties. The work of the proposed automaton to accept the valid strings is presented with the help of a state diagram (shown in Figure 11.5).

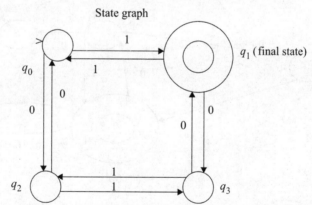

FIGURE 11.5

[In this problem, there may not be any *dead* state]

In this proposed automaton, q_0 is the start state, q_1 is the final state, and q_2 and q_3 are the non-final states. However, this graph can be redesigned (after a bit modification) for the problems such as *even* number of 0s and even number of 1s, and so on.

For example Automaton over $\{0, 1\}$ which accepts all strings containing even number of 0s and even number of 1s. In such case, initial state is obviously the final state too, because zero-size string has even (0) number of 0s and even (0) number of 1s.

EXAMPLE 11.3 Draw the state graph for the DFA which accepts all strings over $\Sigma = \{0, 1\}$ and end in 1, but do not contain the substring 00.

Solution Problem analysis: The problem describes that the accepted strings must satisfy the following conditions:

(a) Strings should end in a '1'.
(b) Strings should not contain 00 as substring (it is indeed a restriction).

This ensures that the proposed automaton (say, M) starts at its initial state to process a string, and reaches to final state if '1' received at the end but no two consecutive 0s over the path of traversal.

Designing procedure Let us add the start state (say, q_0) to the proposed automaton. As the accepted string ends in a '1', so start state q_0 may not be final state. [If so, then null string (which does not end with '1') is also accepted, i.e., acceptance property of string is violated in such case (and it is shown through Figure 11.6(b))]. Hence, some more states are essential to process the valid strings.

Now, if M reads '1' at q_0, it should reach to final state. Thus, the most recently added state is considered as final state (say, q_f), and M remains at q_f if a '1' is received here (as string ends with '1').

Further, for input '0' from states q_0 and q_f, respectively, M must not move to q_0. However, if it moves to q_0, then invalid string consisting of substring: 00 may also be accepted. So, one more non-final state (say, q_1) is necessary in M to resolve the occurred situation. Now, if '1' is encountered at q_1, it reaches to final state because of satisfying validity condition of string. Thus, the total three states (excluding dead state) are required for the proposed automaton.

The complete state diagram is given in Figure 11.6(a).

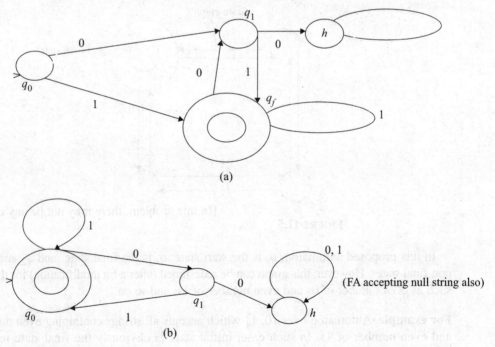

FIGURE 11.6

EXAMPLE 11.4 Draw the state graphs for the NDFA which accept the strings over $\Sigma = \{a, b\}$ if each string can be viewed as concatenation (or combination) of any number of substrings: *ab* or *aba*. Also, design a DFA for this problem.

Solution Any number of substring is meant here 0 (zero) or more substrings. Hence, null (empty) string also will be accepted in the automaton. Several possible NDFA are shown below.

Solution 1

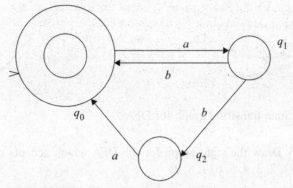

FIGURE 11.7

In this design, if the automaton receives b at q_1 (non-final state), it may move to either of the two different states q_0 (which is initial as well as final state) and q_2 (non-final state) to satisfy the characteristics of the given problem. The state graph is shown in Figure 11.7.

Solution 2

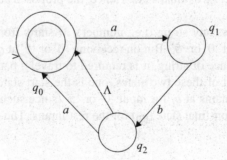

FIGURE 11.8

From the state graph (as shown in Figure 11.8), it is clear that the automaton moves to q_0 from q_2 on reading Λ (empty string).

To satisfy the properties of the given problem, the DFA [as shown in Figure 11.9(a)] needs four states, say, q_0, q_1, q_2, and q_3. In fact, all these states except q_1 must be the final state.

Further, in this design, it is true that every accepted string is viewed as concatenation of substrings: *ab* or *aba*, and the automaton moves to dead state for the string which cannot be viewed as the same one. Hence, if the automaton reads the inputs b, a, and b, respectively, from states q_0, q_1, and q_2, it must move to dead state (from which it cannot move to any state).

Comment If one designs an automaton [as shown in Figure 11.9(b)] for this problem, then some strings such as *abab* (i.e., combination of substring *ab*) may never be generated. But such types of strings are also desired to be accepted by the designed automaton.

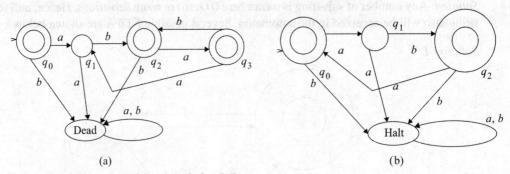

FIGURE 11.9 State transition graph for DFA

EXAMPLE 11.5 Draw the state graph for the DFA which accepts the numbers divisible by 5 over $\Sigma = \{0, 1, 2, 3, 4, 5, 6, 7, 8, 9\}$

Solution Problem analysis: Any decimal number is basically considered as a string of one or more digits taken from Σ. Further, it is true that a number which ends with either '0' or '5' is divisible by 5 (otherwise, not). This suggests that the proposed automaton starts from its initial state and reaches to final state if it traverses a path ending with either '0' or '5'.

Designing procedure In context to design an automaton (say, M) for this problem, describe first a state q_0 as start state. In particular, *null* string does not represent any number, so there is no chance to consider q_0 as final state. Hence, the proposed automaton needs at least one more state at this stage.

In order to process any string (i.e., number), M starts from q_0 and stays here for different digits (inputs) except '0' or '5'. But on receiving '0' or '5' at q_0, it will reach to final state, say, q_f (since for acceptance of string, it is required to travel a path ending with '0' or '5', starting from q_0). Hence, out of these two states, one is the start state and the other must be the final state. Further, M remains at q_f for input '0' or '5' (since such strings also end with '0' or '5') but moves back to non-final state (q_0) for the rest inputs. The complete state graph is shown in Figure 11.10.

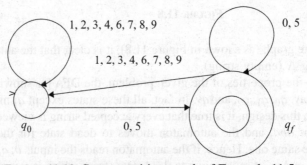

FIGURE 11.10 State transition graph of Example 11.5

EXAMPLE 11.6 Design a DFA which accepts strings over $\Sigma = \{0, 1\}$, containing substring as 011.

Solution Problem analysis and designing procedure: The condition to be satisfied for the valid string is the presence of at least one substring: 011. To satisfy this condition, the minimum number of states in the proposed automaton (say, M) should be 4, as the length of such a substring is 3.

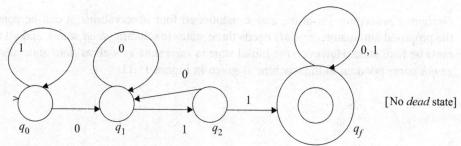

FIGURE 11.11 FA of Example 11.6

In order to accept a string, M starts from its initial state (say, q_0), and must traverse a path described by substring 011 to reach to final state (say, q_f).

Further, for any input (either '0' or '1') received at q_f, M moves back to q_f, since the string already consists of 011 as substring. But if it receives '0' at q_1 and q_2, respectively, it moves back to q_1, expecting to get valid string. On the other hand, it stays at q_0 for '1'.

The state graph of the designed automaton is shown in Figure 11.11.

EXAMPLE 11.7 Design a DFA which accepts strings over $\Sigma = \{0, 1\}$, ending with 011.

Solution Problem analysis and designing procedure: The condition to be satisfied here is that the accepted string ends with substring 011. Obviously, this problem is a bit different from the problem mentioned in Example 11.6. So, the corresponding automaton can be redesigned from the above automaton (as shown in Figure 11.11) also, after some *modification*.

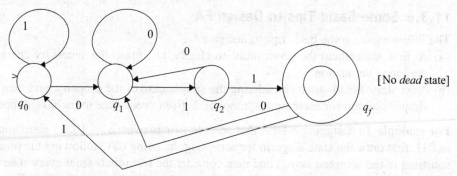

FIGURE 11.12 FA of Example 11.7

EXAMPLE 11.8 Design a DFA which accepts strings over $\Sigma = \{a, b\}$, not containing three consecutive b's.

Solution *Problem analysis*: The conclusions that can be drawn from this problem are as follows:

(a) The string with no b in the string is accepted.
(b) The string with single b at zero or more positions in the string is accepted.
(c) The string with two consecutive b's at zero or more positions in the string is accepted.
(d) But the string with three consecutive b's is not accepted and for such string the automaton moves to dead state (h).

Designing procedure From the above-mentioned four observations, it can be concluded that the proposed automaton (say, M) needs three states (excluding *dead* state), and all these states must be final state. However, the initial state is *start* state as well as *final* state also. The *state graph* corresponding to the machine is given in Figure 11.11.

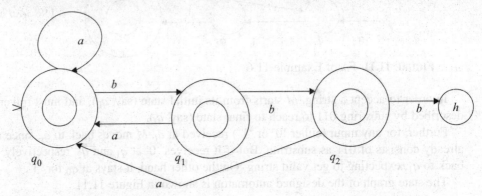

FIGURE 11.13 FA of Example 11.8

Comment In this automaton, it is shown that, for input a at q_1 and q_2, the automaton is moving to q_0 but not to the states q_1 and q_2, respectively, since there is possibility to move to *dead* state for three consecutive b's. Now, if it happens so, then the designed diagram may direct strings with no three consecutive b's to move to *dead* state (which should not be allowed).

11.3.6 Some Basic Tips to Design FA

The following are some basic tips to design FA:

(a) At first, understand the given problem clearly, i.e., grasp the necessary characteristics in designing automaton.
(b) Next, draw the diagram considering the simple case of the property, and then extend that simple diagram for each input symbol of Σ from every state to include all possibilities.

For example In designing a DFA that accepts strings over $\Sigma = \{0, 1\}$ containing substring as 011, first draw the state diagram for accepting the string 011 (following the property 011 as substring in the accepted string) and then consider the rest inputs from every state of the same diagram to extend it on the basis of analysis (see the state diagram shown in Figure 11.11 for this problem).

(c) To design DFA for accepting all the strings over $\{0, 1\}$ containing a '1' in the third position from the end, one may consider the following hints.

Here, the third position from the end of the string must be '1'. Obviously, the rest two symbols after the third position will be any of the possible combinations of '0' and '1', i.e., the accepted string may end with any one of the following substrings: (i) 100, (ii) 101, (iii) 110, and (iv) 111. Hence, the number of final states in total may be considered 4 (obtained by 2^2).

Thus, if Σ has only two symbols, and the *k*th position from the end of accepted string is a particular symbol of Σ, then the number of possible *final* states in the DFA can be formulated as 2^{k-1}. Similarly, if Σ consists of n symbols and the *k*th position from the end of accepted string is a particular symbol of Σ, then the number of possible final states in the DFA can be counted by n^{k-1}. However, some final states may be reduced by merging (if possible) under proper analysis.

(d) Some hints can be followed for another kind of problem. Suppose, it is needed to design a DFA of a problem in which the concept of *OR* logic is present, then there exists the possibility of selecting *two parallel* paths from a state (from which this *OR* situation is arising). But it is fully dependent on the *nature* of the problem.

For better understanding, some examples are explained below.

(i) Let us consider to design a DFA accepting the strings consisting of either *ab* or *bba* as substring.

Clearly, in this example, any of the two specified substrings is not the subset of other. So, while designing the state diagram for the simple case, two paths: one for *ab* and the other for *bba* will be considered. However, later on, both the paths may *meet* to a certain state. Further, on proper analysis, consider transitions for all the inputs from each state to draw the complete diagram.

(ii) Now, let us take another problem to design a DFA accepting the strings consisting of either *ab* or *aba* as substring.

Obviously, in this example, the larger substring *aba* includes the smaller substring *ab*. So, in such case, simply draw a DFA for the smaller substring, and then extend it on proper analysis.

(iii) If the problem is such that every string accepted by a DFA is viewed as *combination* of substrings *ab* or *aba*, then it is required to design DFA considering both the substrings (as shown in Figure 11.9(a) of Example 11.4). Here, if we design an automaton based on only one substring, then some valid strings may be rejected.

Hence, if the accepted string is dependent only on *criteria* 1 or *criteria* 2 (where criteria 1 or criteria 2 are substrings or some other instruction such as divisibility and so on), then both the criteria must be considered in designing DFA, otherwise some desired strings may not be accepted or some unwanted strings may be valid. One such example is shown in Figure 11.9(b).

(e) To design DFA such as to recognize strings over Σ, containing a particular symbol of Σ at least in r places out of last m places ($m > r$), we may consider the possible arrangements of the places for that particular symbol in the accepted strings as follows:

$$(n-1)!({}^mP_r + {}^mP_{r+1} + \cdots + {}^mP_m)$$

where n is the total number of distinct symbols in Σ, ${}^mP_r$ represents the possible strings with particular symbols in r places out of last m places; ${}^mP_{r+1}$, the possible strings with particular symbols in $r + 1$ places out of m; and so on. Finally, ^mP_m represents the possible strings with particular symbols in m places out of m. Further, we need to arrange the rest $(n - 1)$ symbols (i.e., except the particular symbol) among themselves, and it can be done in $(n - 1)!$ ways. For better understanding, we may take here one example such as *design a DFA to recognize strings over* $\Sigma = \{0, 1\}$, *which contain at least two zeros in its last four places*.

Clearly, in this example, $n = 2, r = 2, m = 4$, then possible combinations are $(2 - 1)!(^4P_2 + {}^4P_3 + {}^4P_4)$. Again, designing of DFA will be easier if we consider the tree structure.

Similarly, we may think about the case of *at most* logic.

Note *For every NDFA there is an equivalent DFA.* Now, if n number of states exist in an NDFA, then its corresponding DFA possesses 2^n states. In fact, for actual implementaion of FA, one must need DFA (not NDFA). Hence, it is required to eliminate the non-determinism case from NDFA.

11.3.7 Conversion Strategy from NDFA to DFA

The basic steps for such conversion are presented below.

Step 1 If an NDFA M has n states, than its equivalent DFA M_1 must have 2^n states with the same initial state of NDFA.

Step 2 The alphabet set Σ of M is also the same for M_1.

Step 3 The set of final states (F') of M_1 is constructed as follows: F' consists of subsets of Q, each of which includes *at least* one final state of M.

Step 4 The transition function δ' of M_1 is constructed, consulting δ of M. To represent the transitions/operations of M_1 in tabular form, the following ways may be applied.

First, consider the initial state as current state, and place the next states corresponding to the possible inputs of Σ (placed in different columns). Then, choose one state (as current state) from the next state columns if it is not already taken as current state. Now, if all the states in the next state columns are new, then select arbitrarily one state as current state. Construction of a table is stopped, when no new state appears in the subsequent columns.

EXAMPLE 11.9 Find a DFA equivalent to NDFA $M = (\{q_0, q_1, q_2\}, \{a, b\}, \delta, q_0, \{q_2\})$, δ is described below through the transition table (Table 11.3). Here, q_0 is initial state and q_2 is final state.

TABLE 11.3

State	Input (a)/next state	Input (b)/next state
($>q_0$)	q_0, q_1	q_2
q_1	q_0	q_1
q_2	—	q_0, q_1

Here, symbol '—' implies that from state q_2, on reading input '0', transition is *undefined*. In particular, it represents dead state.

Solution In this example, the NDFA M has three states $Q = \{q_0, q_1, q_2\}$. So, by applying Step 1 of the conversion procedure discussed above, the corresponding DFA M_1 will possess 2^3 states which are $\Phi(dead)$, $[q_0]$, $[q_1]$, $[q_2]$, $[q_0, q_1]$, $[q_0, q_2]$, $[q_1, q_2]$, $[q_0, q_1, q_2]$. It is basically power set of Q. Here, each state is enclosed by '[ ]' to represent a single state. For example, $[q_0, q_2]$ is a single state. Further, $[q_0]$ is the initial state of M_1, as it is the initial state of M.

Now, as per Step 2, $\Sigma = \{a, b\}$ of the given NDFA is also the Σ of the equivalent DFA.

Step 3 constructs $F' = \{[q_2], [q_0, q_2], [q_1, q_2], [q_0, q_1, q_2]\}$, since F' of DFA consists of subsets with final state q_2.

Finally, Step 4 gives the state table (as shown in Table 11.4) of the desired DFA equivalent to the given NDFA.

TABLE 11.4

State	Input (a)/next state	Input (b)/next state
[>q_0]	[q_0, q_1]	[q_2]
[q_2]	Φ	[q_0, q_1]
[q_0, q_1]	[q_0, q_1]	[q_2, q_1]
[q_2, q_1]	[q_0]	[q_0, q_1]

Note Let us consider the transition from state $[q_2, q_1]$ for input a. Clearly, we can explain it as $\delta'\{[q_2, q_1], a\} = \delta\{q_2, a\} \cup \delta\{q_1, a\} = \Phi \cup \{q_0\} = \{q_0\}$, since union of any set S_1 with Φ (null) is S_1 itself. Hence, in this transition, the automaton does not move to state Φ, as $\Phi \cup \{q_0\} = \{q_0\}$ is treated as single state.

11.3.8 Finite Automaton with Output

Moore machine An automaton in which the output depends only on its present state but not on the present input symbol is called *Moore machine*. A Moore m/c is a 6-tuple machine $M = (Q, \Sigma, \Delta, \delta, \lambda, q_0)$, where

(i) Q is a finite non-empty set of states.
(ii) Σ is the input alphabet.
(iii) Δ is the output alphabet (i.e., new symbol corresponding to input symbol).
(iv) δ is the transition function defined as $\delta: Q \times \Sigma \to Q$.
(v) λ is the output function defined as $\lambda(q_i) = x$, where $q_i \in Q$ and $x \in \Delta$.
(vi) q_0 is the start state of the machine, $q_0 \in Q$.

Thus, it is indeed one type of *output producer*. One example on input conversion along with the necessary operations of such a m/c is described in Table 11.5.

Now, consulting Table 11.5 we can say that the *input string* 011 is converted to 000. However, the produced string is actually obtained by applying the specific *output function* $\lambda(q_i) = x$, and output values x's corresponding to states q_i's are presented below with brief explanation.

TABLE 11.5 Transition table for Moore m/c

Present state	Next state (for input $a = 0$)	Next state (for input $a = 1$)	Output (λ)
>q_0	q_3	q_1	0
q_1	q_1	q_2	1
q_2	q_2	q_3	0
q_3	q_3	q_0	0

$\lambda(q_0) = 0$, since the start state of the machine is q_0. Next, on reading '0' at q_0, the m/c moves to q_3. Now, $\lambda(q_3) = 0$. Accordingly, output is 0. Then, on receiving '1' at q_3, the m/c moves to q_0, resulting output '0' [since $\lambda(q_0) = 0$]. Finally, for the current input '1' at q_0, the m/c moves to q_1, returning '0' as output.

Comment In case of Moore m/c, we should not instruct the m/c to receive *null* as input at any state. If so, then the m/c may fall in infinite loop. *For instance*, in the above example, if the proposed Moore m/c reads null input at q_1, it falls in infinite loop.

Mealy machine An automaton in which the output depends on its present state as well as the present input symbol is called *Mealy machine*. A Mealy m/c is a 6-tuple machine $M = (Q, \Sigma, \Delta, \delta, \lambda, q_0)$ where

(i) Q is a finite non-empty set of states.
(ii) Σ is the input alphabet.
(iii) Δ is the output alphabet (i.e., new symbol corresponding to input symbol).
(iv) δ is the transition function defined as $\delta: Q \times \Sigma \to Q$.
(v) λ is the output function defined as $\lambda(q_i, a) = x$, where $q_i \in Q, a \in \Sigma$ and $x \in \Delta$.
(vi) q_0 is the start state of the machine, $q_0 \in Q$.

For example The operations of one Mealy m/c for some kind of conversion are illustrated in Table 11.6.

TABLE 11.6 Transition table for Mealy m/c

Present state	Next state (for input $a = 0$)	Output (λ)	Next state (for input $a = 1$)	Output (λ)
>q_0	q_3	1	q_1	0
q_1	q_1	0	q_2	1
q_2	q_2	0	q_3	0
q_3	q_3	1	q_0	0

Now, consulting Table 11.6 it is clear that the *input string* 011 is converted to 111. However, the conversion takes as follows.

$\lambda(q_0, 0) = 1$, since the start state of the machine is q_0. On reading '0' at q_0, the m/c moves to q_3 producing output '1'. Now, $\lambda(q_3, 1) = 1$. This implies that on receiving '1' at q_3, the m/c gives output '1'. From Table 11.6, it is clear that the next state is q_0. Finally, on reading current input '1' from q_0, the m/c moves to q_1 resulting output '1'.

Note We may transform a Moore m/c to a corresponding Mealy m/c and the vice versa.

Transformation of Moore m/c to Mealy m/c

The conversion technique is explained here with example. The operation of a Moore m/c for some kind of conversion is described in Table 11.7.

TABLE 11.7 A transition table for a Moore m/c

Present state	Next state (for input $a = 0$)	Next state (for input $a = 1$)	Output (λ)
>q_0	q_3	q_1	0
q_1	q_1	q_2	1
q_2	q_2	q_3	0
q_3	q_3	q_0	0

Now, in order to convert the given Moore m/c into an *equivalent* Mealy m/c, we apply here the *transformation function*: $\lambda(\delta(q, a)) = x$ for every input state of the *present state* column of Table 11.7. Here, $q \in Q, a \in \Sigma, x \in \Delta$.

For instance, the first row of the Mealy table is formed as follows:

$$\lambda(\delta(q_0, 0)) = \lambda(q_3) = 0, \quad \lambda(\delta(q_0, 1)) = \lambda(q_1) = 1$$

[outputs are found from the last column of Moore table].

Clearly, the third row of the same table is again obtained as

$$\lambda(\delta(q_2, 0)) = \lambda(q_2) = 0, \quad \lambda(\delta(q_2, 1)) = \lambda(q_3) = 0 \text{ [by using Moore table]}.$$

In the same way, the second row of the same Mealy table is formed as follows:

$$\lambda(\delta(q_1, 0)) = \lambda(q_1) = 1, \quad \lambda(\delta(q_1, 1)) = \lambda(q_2) = 0$$

Finally, the *last* row is found as

$$\lambda(\delta(q_3, 0)) = \lambda(q_3) = 0, \quad \lambda(\delta(q_3, 1)) = \lambda(q_0) = 0$$

[consulting again the Moore table].

Hence, the corresponding Mealy m/c table (Table 11.8) will be as follows.

TABLE 11.8 Converted Mealy m/c table

Present state	Next state (for input $a = 0$)	Output (λ)	Next state (for input $a = 1$)	Output (λ)
>q_0	q_3	0	q_1	1
q_1	q_1	1	q_2	0
q_2	q_2	0	q_3	0
q_3	q_3	0	q_0	0

Transformation of Mealy m/c to Moore m/c

Such transformation is also explained with example. Let us consider a Mealy m/c for some kind of conversion which is shown in Table 11.9.

TABLE 11.9 A Mealy m/c table

Present state	Next state (for input $a = 0$)	Output (λ)	Next state (for input $a = 1$)	Output (λ)
$>q_1$	q_3	0	q_2	0
q_2	q_1	1	q_4	0
q_3	q_2	1	q_1	1
q_4	q_4	1	q_3	0

Now, in order to convert Mealy m/c to Moore m/c, we first extract each state (from the Mealy table) associated with different output values. Here, states q_2 and q_4 are identified as such states.

Next, split the states depending on their different output association. For example, state q_2 is associated with two different output values (0, 1). So, two *new states* formed from q_2 can be considered as q_{20} and q_{21} (actually, naming convention is taken here as per output values). Similarly, new states formed from q_4 can be considered as q_{40} and q_{41}.

After splitting the selected states, the current transformed table (Table 11.10) looks as follows.

TABLE 11.10 A converted table

Present state	Next state (for input $a = 0$)	Output (λ)	Next state (for input $a = 1$)	Output (λ)
$>q_1$	q_3	0	q_{20}	0
q_{20}	q_1	1	q_{40}	0
q_{21}	q_1	1	q_{40}	0
q_3	q_{21}	1	q_1	1
q_{40}	q_{41}	1	q_3	0
q_{41}	q_{41}	1	q_3	0

In the new table, no states such as q_2 and q_4 are present (since these are splitted). Thus, two separate rows for q_2 are currently considered. The same logic is applied for state q_4 also.

At this stage, the current table has no state associated with different output value. *However, if so found, extract them and split following the above-mentioned approach.* Clearly, splitting is stopped at present, and the corresponding *Moore table* is constructed and shown in Table 11.11.

As the output associated with q_1 is always 1 (it is, in fact, examined as [next state, output], here it is [q_1, 1]) in Table 11.10, so the value in the *output column* of Table 11.11 corresponding to q_1 must be 1.

TABLE 11.11 Converted Moore table

Present state	Next state (for input $a = 0$)	Next state (for input $a = 1$)	Output (λ)
$>q_1$	q_3	q_1	1
q_{20}	q_1	q_{40}	0
q_{21}	q_1	q_{40}	1
q_3	q_{21}	q_1	0
q_{40}	q_{41}	q_3	0
q_{41}	q_{41}	q_3	1

Note If the number of states in Mealy m/c is *n* and the number of output is *m*, then the corresponding Moore m/c has no more than $mn + 1$ states.

EXAMPLE 11.10 Design a Mealy m/c to convert binary number into its 1s complement form.

Solution The given problem instructs to convert any binary number into its 1s *complement* form. This is possible if each '0' of the given binary number is changed to '1', and vice versa.

In order to design such a m/c, one state is sufficient, and this is simply the *start* state q_0 Now, at this state, if input '0' is encountered, then its corresponding output will be '1' (which is represented as 0/1). On the other hand, if '1' is received, then the corresponding output will be '0' (which is represented as 1/0). An illustration is shown graphically in Figure 11.14.

Remark Finite automaton with output is a special kind of FA. Through such kind of automaton, acceptance of string (example such as string ending with 00) can also be checked, i.e., they may behave like normal FA. On the other hand, the normal FA cannot do conversion of string except acceptance of string.

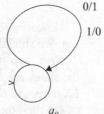

FIGURE 11.14

11.4 REGULAR EXPRESSION

Regular expression (RE) is a mathematical symbolism over Σ, which describes a set of strings accepted by an FA. A regular expression is, in fact, a *pattern checker*. It is sometimes called as *language recognizer*, since it helps to construct *sentence/statement* of a language identifying *token* such as data type, operand, operator, and so on.

REs are usually used to give a concise (compact) description of a set without listing all the elements in the set.

For example $S = \{\Lambda, a, aa, \ldots,\}$ can be concisely described as $(a)^*$. In fact, it generates the strings in *recursive* way.

Now, the language which can be described by regular expression is called *regular language*. In other words, a language *L* over Σ is a regular if there is some RE over Σ corresponding to that language. Again a language is *finite* if it contains finite number of strings, and so every finite language is regular.

Further, every RE is recognized by FA. So, there is an RE for every FA. However, more than one FA can be designed for the same problem, but there exists one and only one minimum state FA. Similarly, different REs (corresponding to the respective FAs) may be described for the same problem, but except one (i.e., corresponding to the minimum FA) all others are redundant. *The RE corresponding to the minimum FA is the unique one.*

11.4.1 Minimization of FA

A brief theoretical idea on minimization of FA and its importance is discussed below.

Let us first consider the case of *unreachable* states with the help of a diagram (as shown in Figure 11.15) in designing minimum finite automaton. It is obvious from the diagram that it is not possible to ever visit state 2, and states like this are called *unreachable*. So, we can simply remove them from the automaton without changing its behaviour. (This will be, indeed, the first step in our minimization algorithm.) In our case, after removing state 2, we get the automaton on the right.

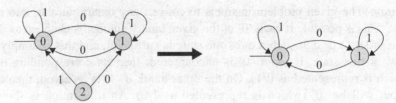

FIGURE 11.15

However, removing unreachable states is not sufficient in minimizing the states of automaton. We should merge some/all non-distinguishable states into a single state.

Now, two states p and q are *distinguishable* if the transitions for the same input from both the states are *final* and *non-final* states. For example, on reading input (say, a) at state p, the FA moves to *final* state, whereas for the same input, the FA moves to *non-final* state from q.

On the other hand, if the transitions of two states p and q for the same input (say, a) are either *final* or *non-final* (i.e., the next states are in the same group), then these states are *non-distinguishable* or *equivalent* state (also called as duplicate or redundant states).

Whatsoever, the objective of minimization of FA is to merge some/all non-distinguishable states into a single state. Consequently, such minimization results *less* space and *easy* processing of input. The minimization algorithm is presented below.

Minimization Algorithm

Given A finite automaton (A) with initial state q_0, final states F and transition function δ.

Step 1 Remove unreachable states.

Step 2 Mark the distinguishable pairs of states.

To achieve this task, we first mark all pairs p; q, where $p \in F$ and $q \notin F$ as distinguishable. Then we proceed as follows:

repeat
for all non-marked pairs p; q *do*
  for each letter a *do*
    if the pair $\delta(p; a)$, $\delta(q; a)$ is marked *then* mark p; q
until no new pairs are marked

Step 3 Construct the reduced automaton A'.

We first determine the equivalence classes of the indistinguishability relation. For each state q, the equivalence class of q consists of all states p for which the pair p; q is not marked in Step 2.

The states of A' are the equivalence classes. The initial state q'_0 (for A') is this equivalence class that contains q_0. The final states F' (for A') are these equivalence classes that consist of final states of A. The transition function δ' (for A') is defined as follows. To determine $\delta'(X; a)$, for some equivalence class X, pick any $q \in X$, and set $\delta'(X; a) = Y$, where Y is the equivalence class that contains $\delta(q; a)$.

EXAMPLE 11.11 We apply our algorithm to the automaton given in Figure 11.16.

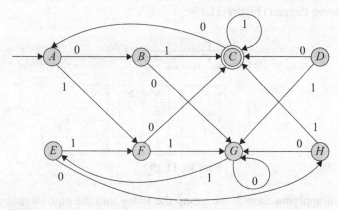

FIGURE 11.16

Solution By Step 1, we have one unreachable state, state *D*. We remove this state and proceed to Step 2. Step 2 then marks pairs of final and non-final states, getting the following tableau (Figure 11.17):

B						
C	X	X				
E			X			
F			X			
G			X			
H			X			
	A	B	C	E	F	G

FIGURE 11.17

In the first iteration we examine all unmarked pairs. For example, for pair *A*; *B*, we get $\delta(A; 1) = F$ and $\delta(B; 1) = C$, and the pair *C*; *F* is marked, so we mark *A*; *B* too. After doing it for all pairs, we get the following tableau (Figure 11.18).

B	X					
C	X	X				
E		X	X			
F	X	X	X	X		
G		X	X		X	
H	X		X	X	X	X
	A	B	C	E	F	G

FIGURE 11.18

In the next iteration, we examine the remaining pairs. For example, we will mark pair $A; G$ because $\delta(A; 0) = B$ and $\delta(G; 0) = G$, and the pair $B; G$ is marked. When we are done we get the following tableau (Figure 11.19):

B	X					
C	X	X				
E		X	X			
F	X	X	X	X		
G	X	X	X	X	X	
H	X		X	X	X	X
	A	B	C	E	F	G

FIGURE 11.19

Finally, on applying Step 3, we group the states into the equivalence classes. Since $A; E$ are equivalent and $B; H$ are equivalent, the classes are $\{A, E\}, (B, H\}, (C\}, \{F\}, \{G\}$. The minimal automaton A' is shown in Figure 11.20.

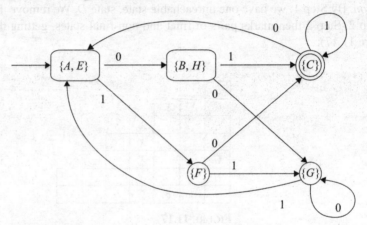

FIGURE 11.20

11.4.2 Brief Discussion to Derive REs

Normally, REs are denoted by the symbol R. Every single alphabet symbol is a regular expression, the empty string (Λ) itself is a regular expression. Thus, these are the *base* cases.

Now, if R and S are REs, then each of $R + S$, RS, R^*, and S^* forms a new RE. These are explained below.

1. Let $R_1 = (R + S) = (a|b) = (a + b) = (a \cup b)$ be an RE over Σ, where $R = a$ and $S = b$ are two REs. Here, all the symbols: $|, +, \cup$ express the same meaning. In fact, the semantic interpretation of $R + S$ is the union of all strings in R and S. Clearly, we can take here strings a, b but not like ab or ba, i.e., it generates only $\{a, b\}$.
2. Let R_1 and R_2 be two REs over Σ. The concatenation of R_1 followed by R_2 is written as $R_1 \circ R_2$ (or $R_1 R_2$), and it is a new RE over Σ. However, $R_1 \circ R_2 \neq R_2 \circ R_1$. Now, it is obvious that $R_1 R_2$ generates the set of strings xy, where x is in R and y is in S.

3. $R_1 = (a|b)^* = (R)^*$, where $R = (a|b)$, and R^* consists of all the strings which are zero or more concatenations of strings in R. Clearly, the set of generated strings from this RE includes $\{\Lambda, a, b, ab, ba, \ldots\}$.

However, the above set of strings is generated as follows:

$$\{\Lambda, (a|b), (a|b)\circ(a|b), (a|b)\circ(a|b)\circ(a|b), \ldots\}$$

where '∘' is the concatenation operator. Thus, $(a|b)^*$ denotes the set of all strings with no symbols other than a and b, including Λ (*null* string) also. Some more examples are discussed below.

$R = a|b^*$ results $\{\Lambda, a, b, bb, bbb, \ldots\}$.
$R = a^*$ gives $\{\Lambda, a, aa, aaa, \ldots\}$ i.e., zero or more occurrences of a.
$R = a^+$ gives $\{a, aa, aaa, \ldots\}$ i.e., one or more occurrences of a.

Clearly, a^+ can be expressed as aa^*.

11.4.3 Solved Problems on RE

EXAMPLE 11.12 Describe the following sets of strings by REs.

(a) $\{010\}$, (b) $\{0, 1\}$, (c) $\{\Lambda, ab\}$, (d) $\{1, 11, 111, \ldots\}$, (e) the set of all strings of 0s and 1s ending in 00, (f) the set of all strings of 0s and 1s beginning with '0' but ending with '1', (g) the set of all strings over $\{0, 1\}$ which has only one a, (h) the set of all strings over which has at most two 0s $\{0, 1\}$.

Solution

(a) $R = (010)$, as the set of strings contains only one string.
(b) $R = (0 + 1)$. Here, the set of strings contains only two simple strings '0' and '1'.
(c) $R = (\Lambda + ab)$.
(d) $R = 1(1)^* = 1^+$.
(e) Each string of this set ends with 00. But just before the last substring 00, any combination of '0' and '1' may appear, and such combination can be expressed as $(0 + 1)^*$. Hence, the set of strings can be described by the RE as $R = (0 + 1)^*00$.
(f) Each string in this set begins with '0' but ends with '1'. Further, the substring in between the first '0' and the last '1' of the valid string may possess any combination of '0' and '1'. Clearly, such combination can be expressed as $(0 + 1)^*$. Thus, the set of valid strings can be described as $R = 0(0 + 1)^*1$.
(g) The problem describes that each correct string contains exactly one a. However, in such a string, any number of b's may appear before the a and after the same a. Hence, it can be expressed as $R = b^*(a)b^*$.
(h) Each string of this set of strings has at most two 0s. So, the number of *zeros* in such a string may be 0, 1, or 2 (since number of 0s is at most 2). Also, any number of 1s may appear before and *after* every '0'. Thus, the above possibilities can be described by an RE as follows.

$$R = 1^* + 1^*01^* + 1^*01^*01^*$$

(as three different possibilities are essential here to consider, so concept of *OR* logic comes in picture. Clearly, *OR* is represented by '+' in RE.)

508 Discrete Mathematics

EXAMPLE 11.13 Describe the following sets of strings by REs:
(a) RE of odd length words over $\{a, b\}$.
(b) RE of even length word over $\{a, b\}$.
(c) RE of strings which end in '1' and does not contain the substring 00.
(d) RE for accepting the strings consisting of the substring 10 preceded by arbitrarily many 1s.
(e) RE of strings consisting of even number of 1s over $\{0, 1\}$.
(f) RE to generate the strings in the form $\{0^k : k$ is multiple of 2 or 3$\}$.

Solution

(a) $R = ((a + b)(a + b))^*(a + b)$. [The sub-expression $((a + b)(a + b))^*$ in R plays an important role to generate even length strings only. However, the reader may think to design here the RE as $R_1 = (a + b)^*(a + b)^*(a + b)$. But the sub-expression $(a + b)^*(a + b)^*$ of such an RE, where R_1 may generate odd length substring also; and then *concatenating* that odd length substring with the substring generated from $(a + b)$, new string of even length will be result in (which is not expected as per given instruction).]

> **Note** Hence, it can also be *concluded* that $R^*R^* \neq (RR)^*$.

(b) $R = ((a + b)(a + b))^*$.
(c) $R = (1 + 01)^+$.
(d) $R = (0^*1^+(10)0^*1^*)(1^+(10)0^*1^*)^* = R_1R_2$, where $R_1 = (0^*1^+(10)0^*1^*)$ and $R_2 = (1^+(10)0^*1^*)^*$. [The term *arbitrarily many* means here one or *more*. Now, no or more 0s expressed by 0^* may appear in the valid string before arbitrarily *many* 1s which has the expression 1^+. Further, it is instructed that every substring 10 precedes by arbitrarily many 1s. Clearly, all the possible strings consisting of single substring 10 and satisfying the given properties must be generated from R_1. Further, the valid strings, containing the subsequent substrings of 10 along with the prescribed properties, are generated by R_2. Thus, R_1R_2 together is the resultant RE for this problem.]
(e) $R = (0^*10^*10^*)^*$. [Each correct string consists of only even number of 1s but any number of 0s, and the 0s may be present at any position in the string.]

Comment If we design the above as $(0^*10^*1)^*$, then strings such as 110, 1100, ..., may *never* be generated. However, such strings are valid as per instruction. The rejection of the valid strings such as 110, 1100, ..., can easily be detected if we let $(0^*10^*1^*) = a$ and express $R = (0^*10^*1)^*$ as $R = (a)^* = \{\Lambda, a, aa, aaa, ...\}$.

(f) $R = (00)^* + (000)^*$. [Here, two distinct possibilities (i.e., the concept of *OR*) occurs. Now, we may think to design R as $(00)^*(000)^*$. Clearly, this RE generates string like 00000. But such string should not be accepted, as its length is 5 (which is not divisible by '2' or '3'). Hence, the second choice is not correct.]

EXAMPLE 11.14 Design RE for generating the set of all strings over $\{0, 1\}$ with three consecutive 0s.

Solution The RE for this problem is $R = (0 + 1)^*000(0 + 1)^*$.

EXAMPLE 11.15 Design RE for the set of strings defined as $\{a^n | n$ is divisible by 2 or 3, $n \geq 0\}$ over $\{a\}$.

Solution For satisfying the given criteria, we need an RE, $R = (aa)^* + (aaa)^*$.

EXAMPLE 11.16

(a) Design RE for generating the set of all strings over $\{0, 1\}$, containing exactly two 0s.

Solution Clearly, the RE corresponding to the given problem is $R = 1^*01^*01^*$.

(b) Design RE for the language $L = \{a^m b^n | m, n > 0\}$ over $\{a, b\}$.

Solution $RE = (aa^* bb^*) = (a^* ab^* b) = (a^+ b^+)$. [However, all these options are *equivalent*, since $RR^* = R^*R = R^+$ (as mentioned in identity (x) of Section 11.4.4). Hence, RE is here *unique* not *different*, only *representations* are different, i.e., it implies that all the representations have *identical* behaviour.]

> **Note** If L_1 and L_2 are regular languages over Σ, then $L_1 \cup L_2, L_1 \cap L_2, L_1 - L_2, L_1^c$ (*complement* of L_1) are also regular languages. These languages are basically new languages generated from the existing languages L_1 and L_2, applying the operations: $\cup, \cap, -, c$.

(c) Design RE for identifying valid identifier of C-programming language.

Solution In C language, the *valid* identifier starts with a letter followed by any combination of letters and digits. Hence, the desired RE $R =$ letter (letter|digit)*, where *letter* $= \{`a' - `z', `A' - `Z'\}$ and *digit* $= \{0, 1, 2, \ldots, 9\}$.

(d) Derive an RE to accept string over $\{0, 1\}$, each of which does not end with 0.

Solution The valid string here does not end with 0. It implies that the null string and the string which ends with 1 are the accepted strings. Such strings have the RE, $R = \Lambda + (1 + 0)^*1$.

11.4.4 The Identities on Regular Expresssion

If $P, Q,$ and R are three REs, then the following identities hold.

(i) $R + R = R$ but $RR \neq R$
(ii) $\Phi + R = R, \Phi R = \Phi$, where Φ is null set (which does not contain any string)
(iii) $P + Q = Q + P$
(iv) $P(Q + R) = PQ + PR$
(v) $(Q + R)P = QP + RP$
(vi) $(P + Q) + R = P + (Q + R)$
(vii) $P(QR) = (PQ)R$
(viii) $(R^*)^* = R^* = R^*R^* = R^*R^*R^* = \ldots$ [but $(RR)^* \neq R^*R^*$]
(ix) $\Lambda^* = \{\Lambda, \Lambda\Lambda, \ldots,\} = \Lambda$
(x) $RR^* = R^*R = R^+$ [but $RR^* \neq R$]

[$R^* = \{\Lambda, R, RR, \ldots,\}$. Hence, $RR^* = R\{\Lambda, R, RR, \ldots,\} = \{R, RR, \ldots,\} = R^+$]

(xi) $\Lambda + RR^* = \Lambda + R^+ = R^*$ [since $RR^* = R^+$]
(xii) $(P + Q)^* = (P^* Q^*)^* = (P^* + Q^*)^*$

To prove all the above-mentioned identities, we may choose suitable regular expression such as $P = a^*, Q = b^*$, and so on.

Note The presented identities are useful for simplifying complex REs.

EXAMPLE 11.17 Simplify $R = \Lambda + 1^*(011)^*(1^*(011)^*)^*$.

Solution Let $P = 1^*(011)^*$, where P is an RE.

Then, we may write

$$R = \Lambda + PP^* = P^* \qquad \text{[using identity (xi)]}$$

Now,

$$P^* = (1^*(011)^*)^* = (P_1^* Q_1^*)^*$$

Let $P_1 = (1)$ and $Q_1 = (011)$ be two REs.

$$= (P_1 + Q_1)^* \qquad \text{[using identity (xii)]}$$

$$= (1 + 011)^*$$

11.4.5 Rules for Constructing NDFA from Regular Expression

To construct NDFA from a given RE, we generally make use of the following rules (graphically shown below for each rule).

(i) *Parallel path* Suppose $(0 + 1)$ is one part of an RE. The NDFA corresponding to this part is shown in Figure 11.21(a).

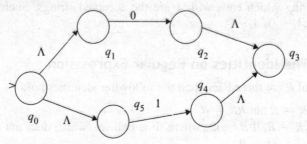

FIGURE 11.21(a) NDFA for $(0 + 1)$

(ii) *Serial path* Suppose (0.1) is part of an RE, then its corresponding NDFA will be as follows [shown in Figure 11.21(b)].

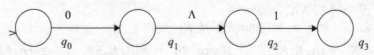

FIGURE 11.21(b) NDFA for (0.1)

(iii) *Closure* Suppose 0^* is a part of an RE, then its equivalent NDFA can be drawn as follows [shown in Figure 11.21(c)].

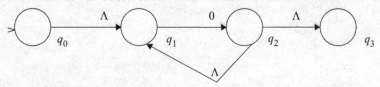

FIGURE 11.21(c) NDFA for 0^*

(iv) *Positive closure* Suppose 0^+ is a part of an RE, then its equivalent NDFA is shown in Figure 11.21(d).

FIGURE 11.21(d) NDFA for 0^+

[Considering the above rules, draw an NDFA for the RE, $R = a(a + b)^*$.]

EXAMPLE 11.18 Design a DFA for the RE $R = (1)^*$ over $\{1\}$.

Solution The set of generated strings is $\{\Lambda, 1, 11, 111, \ldots,\}$. Since Λ (*null* string) is also the accepted string, so initial state must be the final state of the DFA.

Further, string of any number '1' is accepted by this DFA. So, the DFA corresponding to the given RE is shown in Figure 11.22

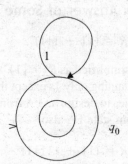

FIGURE 11.22 FA for 1^*

EXAMPLE 11.19

(a) Design a minimum state finite automata over $\{0, 1\}$ in which length of every string is divisible by '2' and '3'.
(b) Design a DFA which accepts all strings by the regular expression $(1 + 011)^*$.

Solution

(a) Clearly in this case, the string length has to be multiple of '6' (2 and 3). So, the regular expression in the simplest form is given by $((0 + 1)(0 + 1)(0 + 1)(0 + 1)(0 + 1)(0 + 1))^*$, and its equivalent minimum state automaton is shown in Figure 11.23(a).

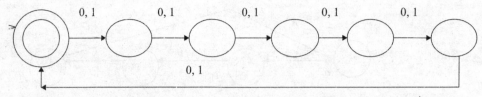

FIGURE 11.23(a) DFA for $((0 + 1)(0 + 1)(0 + 1)(0 + 1)(0 + 1)(0 + 1))^*$

(b) The RE $(1 + 011)^*$ can be expressed as $(1^*(011)^*)^*$ applying the above identities. Clearly, null string is also accepted. So, the initial state (say, q_0) of the proposed DFA must be also the final state. Put a *self-loop* for '1' at this initial state. This takes care of 1^* part. For the $(011)^*$ part, we need two more states. The designed automaton is shown in Figure 11.23(b).

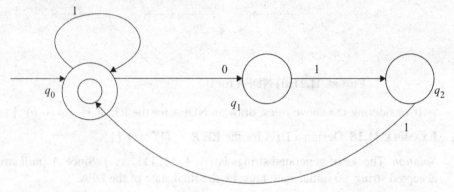

FIGURE 11.23(b) DFA for $(1 + 011)^*$

11.4.6 Tips to Get Quick Answer of Some Special Problems on FA and RE

EXAMPLE 11.20 Simplify RE, $R = (11 + 111)^*$.

Solution The given RE can be simplified as $R = \{1\}^* - 1$. It is clear that all the strings except the string '1' (i.e., all strings containing only 1s except the string with single '1') can be generated by this RE. Now, it is much easier to design an FA from the simplified version of the given RE, and that FA will be the minimum state FA also.

EXAMPLE 11.21 Simplify RE, $R = (11 + 11111)^*$.

Solution Analysis As per the given RE, we conclude that each string consists of only one symbol '1'. However, the only consideration for the accepted string is its length, and the length can be described by the expression $(2m + 5n)$, where $m \geq 0, n \geq 0$. Obviously, every even length string can be expressed by this expression. For example, 2 can be expressed by choosing $m = 1, n = 0$. Similarly, $4 = 2 \times 2 + 2 \times 0$, i.e., $m = 2, n = 0$, and so on.

Now, consider the case of odd length string. Let us assume any odd number $k \geq 5$. Clearly, we can express k in the form $k = 5 + 2n$, where $n \geq 0$. However, only two *odd* numbers 1 and 3 are left to be analysed. But it is true that none of these two can be expressed in the form $(2m + 5n)$. Thus, every odd length string except 1 and 3 can be expressed by $(2m + 5n)$. For example, 5 can be expressed by choosing $m = 0, n = 1$. Similarly, $7 = 5 \times 1 + 2 \times 1$, where $m = 1, n = 1$, and so on.

Thus, the given RE reduces to the form $1^* - \{1, 111\}$. It is easily seen that reducing the given expression to this form, makes it much easier to think about the design of automata.

EXAMPLE 11.22 If the number of states is n, then find the total number of strings with length up to $(n - 1)$ over alphabet consisting of m symbols.

Solution Obviously, the total number of string $N = m^0 + m^1 + m^2 + \cdots + m^{(n-1)}$. Also, the number of strings with length strictly n is m^n.

EXAMPLE 11.23 Find the minimum number of states required to construct an FA which accepts the strings of length n.

Solution The minimum number of states needed here is $(n + 1)$, as the given problem is concentrating on the length of string but not on other criteria.

EXAMPLE 11.24

(a) If length of string is divisible by m and n, then find the minimum number of states required to construct an FA.

Solution The minimum number of states in FA can be counted by finding LCM (m, n) (assuming Λ string is also accepted).

(b) If the number of 1s in string over $\{0, 1\}$ is divisible by m and n, then find the minimum number of states required to construct an FA.

Solution The minimum number of states in FA will be LCM (m, n) (assuming Λ string is also accepted).

Remark If the length of string is divisible by m or n, then also the minimum number of states in FA is counted by finding LCM (m, n) (assuming Λ string is also accepted). But in this case, the number of final states will be *more* as compared to (a) and (b), and it is around $(m + n - 1)$.

(c) Find *an* RE to represent the binary difference between two *positive* numbers.

Solution For any two integers, say, m and n, such that $m \geq 0$ and $n \geq 0$, the regular expression of the binary difference such as $|2^m - 2^n|$ is given by $1^* 0^+$. Actually, $|N|$ represents here the absolute value of number N.

EXAMPLE 11.25

(a) If L (length of string) mod $n = 0$, then find the minimum number of states required to construct an FA.

Solution The minimum number of states in the FA will be n (assuming Λ string is also accepted).

(b) If Σ consists of m number of symbols, then find the number of strings with length n.

Solution In such case, it requires $(n + 1)$ states to generate any string of length n over Σ. Thus, the total number of strings each with length n can be generated is m^n.

(c) Let L_1 and L_2 be two languages. If $L_1 \subseteq L_2$ and L_2 is not regular, then L_1 is not regular. Find the validity of the statement.

Solution False. Here, $L_1 \subseteq L_2$ and L_2 is not regular, i.e., there is no regular expression for the language L_2 Accordingly, L_2 cannot be recognized by any FA.

However, $L_1 \subseteq L_2$, i.e., some of the specific strings of L_2 belong to L_1. So, there may exist some *regular expression* for L_1.

An example to further justify it is given below:

L_1 = {set of all strings over Σ such that it has only 0s and no 1s at all} = 0^*.
L_2 = {set of all strings over Σ such that the number of 1s is either zero or a prime number}.

However, it is not possible to derive any RE for L_2. Further, L_2 restricts only on 1s but not on 0s, so L_2 obviously includes all strings of L_1 which has RE 0^*. Clearly, L_1 is regular but L_2 is NOT regular, although $L_1 \subseteq L_2$.

(d) Let L_1 be a language. If L_1 is a non-regular language, then L'_1 (complement of L_1) is non-regular. Find the validity of the statement.

Solution False. L_1 is a non-regular language, i.e., L_1 has no regular expression. Hence, L_1 cannot be accepted by any FA. Further, $L'_1 = \Sigma^* - L_1$, since L'_1 is the complement of L_1.

Now, L'_1 may have some regular expression, and hence it can be accepted by an FA. For example, let $\Sigma = \{0, 1\}$ and L_1 = {set of all strings such that string length is either an *odd* number or a *prime* number}.

Obviously, L'_1 = {set of all strings such that string length is an even number $\neq 2$} and it is a *regular* language.

11.4.7 Pumping Lemma for Regular Language

The logic of pumping lemma is a good example of the *pigeonhole* principle. This principle states that if n balls are to be put in m boxes, then at least one box will have more than one ball if $n > m$.

In context of language, a language is *regular* if it has a regular expression. Otherwise, it is not regular. However, just looking at the structure of any language, it often becomes very difficult to predict whether the language is regular or not. So, we try to prove in other way if such language is regular or not. Pumping lemma is such a way (*tool*) to check the regularity property of a language.

Statement of the Lemma Let $M = \{Q, \Sigma, \delta, q_o, F\}$ be an FA with n states. Let L be the regular language accepted by M and $w \in L$, where $|w| \geq m$. If $m \geq n$, then there exists strings x, y, z such that $w = xyz, y \neq \Lambda$ and also $x(y)^i z \in L$ for $i \geq 0$.

Proof Let $w = a_1 a_2 a_3 \cdots a_m, m \geq n$ [i.e., string w of length $m \geq n$ is considered; however, the number of available states here is n, it is less than or equal to m.]

Let us apply transition function δ on $w = a_1 a_2 a_3 \ldots a_m$, starting at *start* state, say, q_0. Accordingly, the number of required states for this string is $(m + 1)$ [since $\delta(q_0, a_1 a_2 a_3 \ldots a_m) = q_i, i = 1, 2, 3, \ldots, m$, i.e., $Q_1 = \{q_0, q_1, q_2, \ldots, q_m\}$].

However, there are only n distinct states in the FA, so at least two states of Q_1 must coincide to fulfil the available number of states in FA. Obviously, to satisfy this restriction, there must exist pairs of *repeated* states in FA, and let us consider the first pair of repeated states as q_j and q_k (i.e., $q_j = q_k$), where $0 \leq j < k \leq m$. Now, the string w can be decomposed into three *substrings* as follows:

$(a_1 a_2 a_3 \cdots a_j) (a_{j+1} \cdots a_k) (a_{k+1} \cdots a_m)$ as x, y, and z, respectively

This implies that the automaton M starts at state q_0. On reading the string x from q_0, it reaches to $q_j (= q_k)$. On receiving the string y from q_j, it reaches to $q_j (= q_k)$ (i.e., it makes a *loop* for string y). Further, on reading the string z from q_j, it reaches to final state q_k. It can be shown graphically in Figure 11.24.

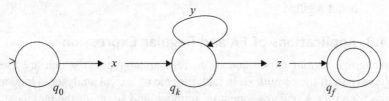

FIGURE 11.24 A simple diagram on pumping lemma

EXAMPLE 11.26 Prove that $L = \{a^n \mid n \text{ is } prime\}$ is not regular.

Solution First, assume that the given language is regular. Now, as it is regular, so it must satisfy the properties of pumping lemma. Let $w = a^p$ (a string of the form a^n) be a string in L, and suppose $|w| = p \geq n$, where n is the number of states of the FA by which string w of length p is supposed to be accepted.

Also, as per the statement of pumping lemma (since L is assumed regular), the string w can be decomposed into *three* substrings. Let us decompose it as $w = a^{m1} a^m a^{m2} = xyz$, where $x = a^{m1}, y = a^m, z = a^{m2}$ and $m_1 + m + m_2 = p$.

Further, $xy^i z$ should also belong to L, for $i \geq 0$.

Now,

$$|xy^i z| = |xyz| + |y^{(i-1)}| = p + (i-1)m$$

$$= p + mp \quad [\text{letting } i = (p+1), \text{ since we may select any value of } i]$$

$$= p(1+m) = k \text{ (say)}$$

Clearly, length of $xy^i z$ is not a *prime* number, as two *distinct* factors p and $(m+1)$ are present in k.

Thus, it proves that length of the selected string is not prime. But it should be prime as per the property of the language. Hence, a *contradiction* is evolved here. So, the given language may not be regular.

EXAMPLE 11.27 Prove that $L = \{0^i 1^i \mid i \geq 0\}$ is not regular.

Solution Let L be regular and $w = 0^i 1^i$ is a string belongs to L.

Now, $|w| = |0^i 1^i| = 2i$. Let us decompose $w = 0^i 1^m 1^p$, where $x = 0^i, y = 1^m$, and $z = 1^p$.

Further, $xy^k z$ should also belong to L (as per pumping lemma), and its length can be computed as follows.

$$|xy^k z| = |xyz| + |y^{k-1}| = 2i + m(k-1)$$

$$= 2i + m(1+i-1) \quad [\text{letting } k = (1+i), \text{ since we may select any value of } k]$$

$$= 2i + mi = m(2+i) = 2j \text{ (say)}$$

Therefore, $j = m(2 + i)/2$. However, this expression may result in fractional (non-integer) value of j, but it is not allowed as per the nature of the string of L. Hence, a *contradiction* occurs here. So, the given language is not regular.

Note Following the illustration of the above example, we can prove that $L = \{0^i 1^{2i} \mid i \geq 0\}$ is not regular.

11.4.8 Applications of FA and Regular Expression

FA recognize regular expressions (i.e., regular languages) which are used to design *lexical analyser* (part of the *compiler*). In fact, the role of lexical analyser is to identify types of *tokens* such as identifier, keywords, operator, number, and so on, of the language. Further, FA is used in construction of *cellular automata* which are used in *biological modelling*, *image processing*, *speech processing*, etc.

11.5 GRAMMAR

It is the study of the rules governing the use of a given natural language. Hence, this is a field of *linguistics*. A rule of a grammar can be described by collection of *strings* (words) or precisely by using some formal notations (called as mathematical way). Further, a grammar is said to be *formal grammar* if its rules are described only by formal notations.

In fact, a grammar describes the syntactical construction of a language. For example, English grammar refers to the rules of the *English* language itself. Accordingly, each language has its own distinct *grammar*. Let us try to understand the need of grammar through a simple sentence of English language.

For example '*Ram is a boy*' is a sentence in English language and it is correct also. But the question arises why is it correct? Why not '*boy a is Ram*' is correct?

The answer is that the sentence 'Ram is a boy', is constructed following the rules (syntax) of English grammar. But the next one is not satisfying any of the proper rules of grammar.

Some *common* rules for making correct English sentences can be written as: (1) <Noun> <Verb> <Article> <Noun>, (2) <Noun> <Verb> <Adverb>, (3) <Noun> <Verb> <Adjective> <Adverb>, . . . , i.e., numerous number of rules are designed to build correct sentence. Here, the notation (like < >) of *Backus–Naur form* (BNF) is used to describe the rules of grammar.

So, if a sentence in English follows any of the rules in its grammar, it is then said to be a *grammatically* correct sentence. Here, in the first sentence, 'Ram' is substituted as *Noun*, 'is' is substituted as *Verb*, 'a' as an *Article*, and 'boy' as *Noun*. Clearly, it is satisfying *Rule 1* specified above. But the second sentence 'boy a is Ram' does not match with any proper rule after word by word (i.e., token by token) substitution.

Thus, a grammar is also called as *language generator*. It is true that grammar of every language can be represented formally, and it must have a distinct *start* symbol. In order to generate a string from a grammar, we begin with its start symbol, and then successively apply the rules any number of *times*. Each string of a language can be generated in this manner.

Any particular sequence of legal choices taken during this *process* yields one particular string in the language, and it is called as *derivation* of *string*. In computer science, this process is also known as *parsing* (which is viewed in the form of tree called *parse tree*). However, the same approach is applied for every grammar such as *natural language*, *computer language*, and so on. In context of natural language, a formal string basically represents 'a sentence' (i.e., collection of strings). Each formal symbol of the formal string is a string (word) of the sentence of a natural language.

The *branch* of mathematics which is concerned with the properties of formal grammar and languages is called *formal language theory*.

11.5.1 Formal Definition of Grammar

In the classic formalization of generative grammar (first proposed by Noam Chomsky in 1950), a grammar G consists of the following components.

(i) A finite set of non-terminal symbols V_N (which can be expanded further). These are denoted normally by *capital* letters such as A, B, C, and so on.

(ii) A finite set of terminal symbols Σ (which cannot be expanded further) and $V_N \cap \Sigma = \Phi$. These are denoted normally by *small* letters such as a, b, c, and so on.

(iii) A special start symbol S (from where parsing/derivation process starts) and $S \in V_N$.

(iv) A finite set of production rules P, each has the form:

$$\text{LHS (left-hand side)} \rightarrow \text{RHS (right-hand side)}$$

and it is expressed as follows.

$$(V_N \cup \Sigma)^* \cup V_N \cup (V_N \cup \Sigma)^* \rightarrow (V_N \cup \Sigma)^*$$

where $\cup$ denotes set *union* and $*$ is the *Kleen star* operator. The expression implies that LHS of every rule must have *at least* one non-terminal symbol. However, RHS of each rule can be of any combination of non-terminal and terminal symbols (even, RHS may be *null* string). Clearly, production rule means rule which produce*s* or derives a set of words to make a meaningful sentence.

Thus, from the above discussion on grammar, we can write the *mathematical model* of grammar as $G = (V_N, \Sigma, P, S)$.

Recursive Rule

If the same non-terminal symbol is present at both LHS and RHS of a rule, then that rule is called as *recursive rule*. For example, $S \rightarrow aSb$ is a recursive rule.

Further, when multiple rules have the same LHS, then we can simply use the symbol '|' (i.e., *or* symbol) on the RHS to represent these multiple rules distinguishably. For example, we can write $S \rightarrow aSb, S \rightarrow c$ as $S \rightarrow aSb|c$.

Derivation/Generation of String

The language of a formal grammar $G = (V_N, \Sigma, P, S)$, denoted by $L(G)$, is defined as all those strings over Σ that can be generated by starting with the start symbol S and then applying the production rules in P until no more non-terminal symbols are present in the RHS of the current sentential form.

In particular, during the generation of string from the rules, at every step, RHS contains combination of terminal as well as non-terminal symbols which is called as *sentential* form. Finally, when the target string is derived, then the form is called as *sentence*.

For example Let us assume a grammar with $V_N = \{S\}$, $\Sigma = \{a, b\}$, the *start* symbol S and the following rules in P are:

1. $S \rightarrow aSb$
2. $S \rightarrow ba$

Now, we start with S, and choose a rule associated with S. If we select *rule 1*, we then obtain aSb on the RHS. It is the *sentential* form at this stage. However, if we choose here again rule 1 and replace S with aSb, then we get the *sentential* form $aaSbb$. In fact, this process can be repeated until all occurrences of S are removed, and only symbols from the alphabet remain (i.e., a and b). For example, if *rule 2* is applied in the *sentential* form $aaSbb$, then $aababb$ (as a *sentence*) is generated.

Thus, the entire derivation mentioned above can be written as

$$S \rightarrow aSb \rightarrow aaSbb \rightarrow aababb$$

Clearly, the language (L) of this grammar (G) generates the set of all the strings like $\{ba, abab, aababb, aaababbb, \ldots, \}$, using this process.

11.5.2 The Chomsky Hierarchy

Noam Chomsky first formalized generative grammar in 1956. He classified them into *four* types, now known as the *Chomsky hierarchy*. These are type 0 (phrase structured or unrestricted), type 1 (context sensitive), type 2 (context free), and type 3 (regular) grammar.

The difference between these types is that they possess increasingly strict production rules and can express fewer formal languages.

Type 0 (Phrase Structure or Unrestricted) Grammar

A type 0 grammar is phrase structure grammar without any restrictions. The production rules are of the form $\alpha \rightarrow \beta$, where

$$\alpha \in (V_N \cup \Sigma)^* \cup V_N \cup (V_N \cup \Sigma)^* \quad \text{and} \quad \beta \in (V_N \cup \Sigma)^*$$

In fact, only α contains *at least* one non-terminal symbol, no other restriction is present here.

Such type of grammar is recognized by *turing machine* (TM).

Type 1 (Context Sensitive) Grammar

A production rule of the form $\Phi A \psi \rightarrow \Phi \beta \psi$, where

$$A \in V_N, \quad (\Phi, \psi) \in (V_N \cup \Sigma)^*, \quad \beta \in (V_N \cup \Sigma)$$

(i.e., β may not be *null*) is type 1 production. Thus, type 1 grammar contains type 1 productions.

In other way, a type 1 grammar is *unrestricted* grammar in which every production has the form $\alpha \rightarrow \beta$ with $|\beta| \geq |\alpha|$, and *at least* one symbol of α is non-terminal symbol.

The language generated by such a grammar is called context sensitive language (CSL) and such language is recognized by *linear bounded automaton* (LBA).

In order to understand its rules, consider the following example.

$$G = (\{S, B\}, \{a, b, c\}, P, S),$$

where

$$V_N = \{S, B\}, \Sigma = \{a, b, c\}$$
$$P = \{S \rightarrow aBSc, S \rightarrow abc, Ba \rightarrow aB, Bb \rightarrow bb\}$$

In this example, $S \rightarrow aBSc$ and $S \rightarrow abc$ are of *type 1* production rules in which both the left and right *contexts* are Λ *(null)*. $Ba \rightarrow aB, Bb \rightarrow bb$ are of *type 1* productions in which the left *contexts* are Λ but the right *contexts* are, respectively, a and b. Obviously, this grammar generates the language $L = \{a^n b^n c^n | n \geq 1\}$.

Note Modifying the above rule set, we can generate language $L = \{a^n b^n a^n b^n | n \geq 1\}$.

Type 2 (Context Free) Grammar

It is a grammar in which the LHS of each production rule consists of only a single non-terminal symbol. Not all languages can be generated by context-free grammar, but those which are possible to be generated, are called context-free languages.

For example The language $L = \{a^n b^n c^n | n \geq 1\}$ is not a context-free language. But the language $L = \{a^n b^n | n \geq 1\}$ is context free. Now, in order to generate $L = \{a^n b^n | n \geq 1\}$, we can design a grammar $G = (V_N, \Sigma, P, S)$, where $V_N = \{S\}$, $\Sigma = \{a, b\}$, $P = \{S \rightarrow aSb, S \rightarrow ab\}$, S is the start symbol of this grammar. Derivation process of this language is explained below.

$$\begin{aligned}
\text{Derivation:} \quad & S \rightarrow aSb \rightarrow a(aSb)b \\
& \rightarrow aaSbb \rightarrow aa(aSb)bb \\
& \ldots \\
& \ldots \\
& \rightarrow a^n b^n \text{ (after repetition of } n\text{th step)}
\end{aligned}$$

Note PDA accepts this type of grammar. A context-free language can be recognized in $O(n^3)$ time. That is, for every context-free language, a machine can be built up that takes a string as input and determines in $O(n^3)$ time if the string is a member of the language, where n is the length of the string.

This is the most important and useful grammar in designing *parser* module of any compiler. Most of the compilers are designed by this type of grammar.

Type 3 (Regular) Grammar

In regular grammar, the LHS of each rule contains only one symbol and it is particularly a non-terminal symbol. Again, the RHS is also restricted. The RHS may be the *empty* string, or a single *terminal* symbol, or a single terminal symbol followed by a *non-terminal* symbol, or the reverse.

Forms of the rules are expressed like $A \rightarrow \Lambda, A \rightarrow a, A \rightarrow aB, A \rightarrow Ba$, where $A, B \in V_N$ and $a \in \Sigma$. Rule $A \rightarrow \Lambda$ is allowed if A does not appear on the RHS of any production.

In particular, the rules of the form $A \rightarrow aB$ are called as *right-linear* rules, whereas the rules of the form $A \rightarrow Ba$ are *left-linear*. However, one form can be converted to the other form. But in *parsing* point of view, left-linear rule should be converted to right-linear.

Sometimes, we may express the production rules for regular grammar as follows:

$$A \rightarrow wN \quad \text{or} \quad A \rightarrow w$$

where A and N are *non-terminals* and w is string of *terminal* symbols.

For example $A \rightarrow abcN$ is rule of regular grammar, where abc is string of *terminal* symbols. Of course, the rule $A \rightarrow abcN$ can be expressed as a collection of multiple rules, each of the form $A \rightarrow aB$ such as:

$$A \rightarrow aA_1, A_1 \rightarrow bA_2, A_2 \rightarrow cN$$

where A_1 and A_2 are the new non-terminal symbols required to be included into the set of existing non-terminal symbols.

Note If the regular grammar has the rules of the forms $A \rightarrow \Lambda, A \rightarrow a, A \rightarrow aB$, then such type of regular grammar is called as right-linear regular grammar. On the other hand, regular grammar with the rules of the forms $A \rightarrow \Lambda, A \rightarrow a, A \rightarrow Ba$ is left-linear. However, in a regular grammar, there should not be *mix-up* type of *left* and *right*. If so, transfer it into one kind (either *left* or *right*).

Further, all languages generated by a regular grammar can be recognized in *linear time* by an FA. Although, in practice, regular grammar are commonly expressed using *regular expressions*. But it is difficult to provide regular expressions for some regular languages. On the other hand, regular grammar makes easier to describe such regular languages. Such type of grammar is recognized by FA.

From Figure 11.25, it is clear that type 0 grammar includes type 1 grammar, where in turns type 1 includes type 2, and type 2 includes type 3 grammar. Hence, all the type 1, type 2, and type 3 grammar belong to type 0 grammar but the *reverse* is not true.

Several *examples* on grammar and languages generated by them are discussed below.

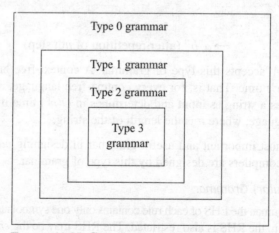

FIGURE 11.25 Grammar and their nature

EXAMPLE 11.28 Find the language generated by grammar $G = (V_N, \Sigma, P, S)$, where $V_N = \{S\}$, $\Sigma = \{a, b\}, P = \{S \rightarrow aS, S \rightarrow bS, S \rightarrow a, S \rightarrow b\}$, S is the start symbol of G.

Solution Clearly, the language generated by G is $L(G) = \{a, b\}^* - \{\Lambda\} = \{a, b\}^+$.

EXAMPLE 11.29 Find the language generated by the grammar with rules: $S \rightarrow aAa, A \rightarrow aAa$, $A \rightarrow a$, where S is the start symbol.

Solution $S \rightarrow aAa \rightarrow a(aAa)a \cdots \rightarrow a^{n-1}a\,a^{n-1}$ (substituting A by a at the nth derivation step).

$$a^{2(n-1)+1} = a^{2n-1}, \quad n \geq 2$$

where n is the number of derivation steps. Hence, the language

$$L(G) = \{a^{2n-1} | (n \geq 2)\}$$

EXAMPLE 11.30 Consider the grammar $G = (V_N, \Sigma, P, S)$, where $V_N = \{S\}$, $\Sigma = \{a, b\}$, $P = \{S \rightarrow aSb, S \rightarrow \Lambda\}$, S is the start symbol of G. Generate the language.

Solution After nth derivation step, the above grammar generates the language $L(G) = \{a^n b^n | n \geq 0\}$. Here, the value of n may be *zero*, as the grammar can derive *null* string Λ by applying the rule $S \rightarrow \Lambda$.

Comment The above grammar generates the language $\{a^n b^n | n \geq 0\}$, i.e., the derived language contains strings of *equal* number of a's followed by equal number of b's but no a appears after b. Hence, the strings such as aba^2b^2, aba^2b^2ab, ..., are not the correct strings of this language, since such strings contain a after b. However, in order to derive a language consisting of such strings, i.e., wherever a's appear in the valid strings, it must follow the same number of b's; we need a rule set such as $P_1 = \{S \rightarrow SS, S \rightarrow aSb, S \rightarrow \Lambda\}$. Thus, the language consisting of strings with equal number of a's followed by equal number of b's simply implies the second language presented here.

Further, let us consider a language consisting of strings, each with *equal* number of a's and b's. In such a language, the positions of a and b are *unpredictable*.

Clearly, both the rule sets $P = \{S \rightarrow aSb, S \rightarrow \Lambda\}$ and $P_1 = \{S \rightarrow SS, S \rightarrow aSb, S \rightarrow \Lambda\}$ are not able to generate string of equal number of a's and b's. To design rule set for such a language, we need to consider all possible combinations of a and b in rules such as

$$S \rightarrow aB|bA, A \rightarrow a|aS|bAA, B \rightarrow b|bS|aBB$$

EXAMPLE 11.31 Find the language generated by the grammar with rule $S \rightarrow SS$, where S is the start symbol.

Solution Obviously, $L(G) = \Phi$, since grammar G here consists of only one production rule $S \rightarrow SS$ with no terminal symbols on RHS (i.e., the rule cannot derive any terminal string).

EXAMPLE 11.32 Let L be the set of all palindromes over $\{a, b\}$. Construct G for L.

Solution Let S be the start symbol of G. The following patterns can be considered as palindrome over $\{a, b\}$.

(a) Λ (null) is a palindrome as its length is zero.

(b) a and b are palindromes (as these are strings with a single symbol only).

(c) axa, bxb are palindromes, where x is a string over $\{a, b\}$ and it is itself a palindrome.

Now, to derive Λ, a, and b, we need the simple rules $S \rightarrow \Lambda$, $S \rightarrow a$, and $S \rightarrow b$, respectively. On the other hand, the rules $S \rightarrow aSa$ and $S \rightarrow bSb$ are necessary for generating palindromes such as axa and bxb.

EXAMPLE 11.33 Construct grammar for

$$L(G) = \{WcW^T | W \in \{a, b\}^*\}$$

W^T is the reverse of W.

Solution Each of such string has two halves: one half (i.e., the first half) appears before terminal c, whereas the other half (i.e., the second half) appears after terminal symbol c.

Clearly, the second half is exactly the *reverse* of the first half. Keeping these points in mind, we must decide the production rules as follows: $S \to aSa$, $S \to bSb$. Now, applying rules $S \to aSa$, $S \to bSb$, the first half and its *reverse* (i.e., the second half) in WcW^T are easily derived.

Now, in order to generate a c in the *middle* of these two halves, rule $S \to c$ is *essential*. Thus, the complete rule set is $\{S \to aSa, S \to bSb, S \to c\}$.

The correctness of the designed rules can be verified to generate the string *abcba* as follows:

$$AS \to aSb \to a(bSb)a \quad \text{(replacing } S \text{ with } bSb\text{)}$$
$$\to abSba \to ab(c)ba \quad \text{(replacing } S \text{ with } c\text{)}$$

Here, $W = ab$ appears before c and $W^T = ba$ after c.

EXAMPLE 11.34 Construct the grammar to generate the strings that do not contain two consecutive 0s.

Solution Here, the *restriction* in the correct string is not the presence of two consecutive 0s. To satisfy such restriction, rules such as $S \to 0A$ and $A \to 1A$ are *essential*. In fact, these two rules *prevent* from generation of 0s at any two consecutive places in a *valid* string.

On the other hand, any number of 1s may be present at any *part* of the string. In order to fulfil such requirement, the rules $S \to 1S|1$ and $A \to 1A|1$ are needed.

Further, the simple string '0' is also accepted, and to derive it, we require the rule $S \to 0$.

Therefore, the *complete* rule set of this grammar will be $S \to 0A|1S|0|1, A \to 1A|1S|1$.

EXAMPLE 11.35 Construct the grammar to generate the strings of balanced parentheses, i.e., each left parenthesis has a matching *right* parenthesis.

Solution The rules of the grammar will be $S \to SS|(S)|\Lambda$.

For example The string (( ) ( )) has balanced parentheses, and it can be generated from the following derivations in sequence.

$$S \to (S) \to (SS) \to ((S)\ S) \to ((\Lambda)\ S) \to (( )\ (S)) \to (( )\ (\Lambda)) \to (( )\ ( ))$$

EXAMPLE 11.36 Construct the grammar for $L(G) = \{W = xaay | x, y \in \{a, b\}^*\}$

Solution Here, x and y are the substrings over $\{a, b\}$ with any combination of a and b. But in between x and y, one fixed substring aa must be present at every *correct* string of the language $L(G)$. In order to fulfil this property, the rule set (P) can be designed as follows: $S \to AaaA$, $A \to aA|bA|\Lambda$. Clearly, $A \to aA|bA|\Lambda$ is a subset of the rule set P, and it is *responsible* for generating the substrings x and y.

EXAMPLE 11.37 Construct the grammar for $L(G) = \{a^n ba^m | m, n \geq 1\}$.

Solution The valid string may be decomposed into three *substrings* (in sequence):
(a) The first substring contains any number of a's only.
(b) The second substring has a single b, and this b appears just after the first substring.
(c) The third substring also contains any number of a's only, but all these a's appear just after the b (i.e., the second substring).

Now, to derive the first substring, rules such as $S \to aA$, $A \to aA$ are *essential*. On the other hand, to generate the second kind of substring and then to proceed into third substring, rule $A \to bB$ is *sufficient*. Finally, rules $B \to aB$, $S \to a$ are *necessary* to derive the third string. Thus, the complete rule set of the grammar will be $P = \{S \to aA, A \to aA|bB, B \to aB|a\}$. The another set of rules for the same problem is $P = \{S \to AbA, A \to aA|a\}$.

EXAMPLE 11.38 Construct the grammar for $L(G) = \{a^j b^n c^n | j \geq 0, n \geq 1\}$.

Solution Clearly, every valid string in $L(G)$ can be divided into two parts (in sequence).
(a) The first part has substring a^j which can be generated by the rule $S \to aS$.
(b) The second part consists of $b^n c^n$, $n \geq 1$. This part can be easily derived from the rules $A \to bAc|bc$ (as $n \geq 1$).

Further, in a^j, $j \geq 0$, so the first part of the string can also be ignored from the string (i.e., accepted string may not have the *first* part at all). Thus, we can construct the complete rule set by *merging* the above subsets of rules as follows:

$$P = \{S \to aS|A, A \to bAc|bc\}$$

EXAMPLE 11.39 Construct the grammar for $L(G) = \{0^m 1^n | m + n \geq 1\}$.

Solution The basic properties of every correct string in $L(G)$ are
(a) All the 0s in the string must appear *before* all the 1s.
(b) Sum of the numbers of 0s and 1s is *greater* than or equal to 1.

Obviously, the second property may be examined and then satisfied as follows:
(i) The correct string may have one or more 0s but no 1s (as it fulfils the condition $m + n \geq 1$). Clearly, such strings can be generated by the rules $S \to 0S$ and $S \to 0$.
(ii) The generated string may have one or more 1s but no 0s (as it also fulfils the condition $m + n \geq 1$). Such strings can be generated through the rules $S \to 1$, $S \to 1A$, $A \to 1A$, and $A \to 1$, since no '0' appears after '1'.

Again, as per the first property of the string, the rules $S \to 0S$, $(S \to 1A, A \to 1A, A \to 1)$ must govern to appear all the 0s before all the 1s in the valid string.

Thus, the complete rule set of the grammar will be

$$P = \{S \to 0S|1A|0|1, A \to 1A|1\}$$

EXAMPLE 11.40
(a) Construct the grammar which generates equal number of a's and b's.

Solution The rule set of the grammar will be
$$P = \{S \to aB|bA, A \to a|aS|bAA, B \to b|bS|aBB\}$$

(b) Design a grammar to generate strings of a's and b's such that every string has $2k$ ($k = 0, 1, 2, 3, \ldots$) number of b's.

Solution The rule set to generate such strings is $\{S \to a|aS|SbSbS\,|\Lambda\}$.

(c) Design a context-free grammar (CFG) for the language $L = \{a^n b^{n-2}|n \geq 2\}$.

Solution Every string in L can be expressed as $a^2(a^{n-2}b^{n-2})$. Clearly, two *extra* a's in comparison to a number of b's are present in every valid string. Now, to generate these extra a's, rules such as $S \to aA$ and $A \to aB$ are needed.

Next, in order to derive the portion $a^{n-2}b^{n-2}$, the rules $B \to aBb, B \to \Lambda$ are *essential*. Hence, the complete rule set is $\{S \to aA, A \to aB, B \to aBb, B \to \Lambda\}$.

[The *another rule set* may be considered as $\{S \to aaA, A \to aAb, A \to \Lambda\}$.]

> **Note** The rule set of a grammar G for generating a language L will be designed in such a way that the *derivation* begins from *start* symbol of G and generates every accepted string (satisfying the specified properties for the string).

11.5.3 Derivation (Parsing)

Derivation or parsing is the process to generate (derive) the *target* string, using the rules of grammar. The derivation process of a string can be described in the form of a tree.

Derivation Tree (Parse Tree)

A derivation tree for a CFG, $G = (V_N, \Sigma, P, S)$ is a tree that satisfies the following properties:

(i) Every vertex has a label which is a non-terminal (variable) or terminal or Λ
(ii) The root has label S (*start* symbol of grammar)
(iii) The label of an internal vertex is a non-terminal
(iv) The label of leaf vertex is either terminal or Λ
(v) At every internal vertex, rule corresponding to the vertex label is applied, i.e., vertex label represented by the LHS of the rule is expanded

Why Derivation Tree for CFG Only?

Derivation tree is considered for CFG because a node of the tree may not represent more than one symbol altogether (taken from terminal or non-terminal). If it is a non-terminal, then it is clearly the LHS of any rule. Further, the LHS of each rule in CFG has only one symbol, and it is strictly non-terminal.

Example of Derivation Tree

Let us consider the rules of a grammar G be $S \to aS|ab$, and we need to derive the string $w = a^2b$. The derivation tree of this string is shown in Figure 11.26.

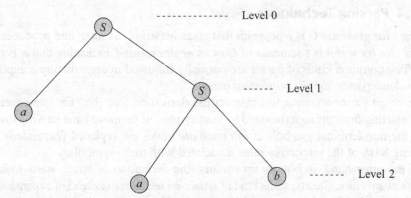

FIGURE 11.26 A simple derivation tree

Here, S is the start symbol of G, and it is placed at root. Terminal symbols a and b are placed at leaf nodes. Now, S at level 1 is an internal node, and the rule corresponding to S, i.e., $S \rightarrow ab$ is expanded.

Types of Derivation

Derivation may be of two types.

(a) *Leftmost derivation* A derivation $A * w$ is called a leftmost derivation if, at every step, a production is applied only to the leftmost non-terminal in the sentential form. Here, $*$ represents several number of derivation steps, and A and w are, respectively, the start symbol of G and the target string.

(b) *Rightmost derivation* A derivation $A * w$ is called a rightmost derivation if we apply a production only to the rightmost non-terminal at every step.

For example Consider the rules $S \rightarrow 0B|1A, A \rightarrow 0|0S|1AA, B \rightarrow 1|1S|0BB$. Now, for the target string 00110101, apply leftmost derivation and rightmost derivation.

Leftmost derivation $S \rightarrow 0B \rightarrow 00BB \rightarrow 00(1)B \cdots \rightarrow 00110101$. Here, at every step, only leftmost non-terminal is expanded to derive 00110101. Obviously, the rightmost derivation must never be applied at any moment of leftmost derivation.

Rightmost derivation $S \rightarrow 0B \rightarrow 00BB \rightarrow 00B(1S) \cdots \rightarrow 00110101$. At every step, rightmost non-terminal is certainly expanded to derive 00110101, i.e., leftmost derivation is never applied.

Note Derivation of a string from a grammar G is possible if and only if there is a corresponding derivation tree. In addition, during the formation of the tree, either leftmost or rightmost derivation is followed but not both (i.e., leftmost derivation at one step and rightmost derivation at any subsequent step or the vice versa).

However, if leftmost derivation is considered, then it simply implies that the parser scans the input from left to right. On the other hand, parser scans input from right to left in case of rightmost derivation. It is well known that most of the parsers follow *left* to *right* scanning.

11.5.4 Parsing Techniques

A *parser* for grammar *G* is a program that takes *w* (string) as input and produces output either a *parse tree* for *w* if it is a sentence of *G* or an *error message*, indicating that *w* is not a sentence of *G*. Two common kinds of parser are normally followed in constructing *compiler*. These are (i) *top-down* parser and (ii) *bottom-up* parser.

Top-down parser attempts to construct the derivation tree (i.e., the parse tree) of an input string, starting from the root (labelled by start symbol of grammar) and moving towards bottom. Here, the non-terminal symbols at *intermediate* nodes are replaced (expanded) by the corresponding RHS of the respective rules associated with these symbols.

On the other hand, in bottom-up parsing, the derivation of string starts from *bottom* and moves towards *root*. Clearly, some kind of *reduction* technique (instead of expansion) is followed. No matter, such derivation can also be represented by tree structure, and that is also derivation (parse) tree.

11.5.5 Ambiguous Grammar

A grammar is said to be ambiguous if more than one *distinct* (i.e., structurally different) parse tree can be generated for some string *w* (however, not *necessary* for each string of the language), following the same derivation at every step (either *leftmost* or *rightmost*).

EXAMPLE 11.41 $G = (\{S\}, \{a, b, *, +\}, P, S)$, $P = \{S \rightarrow S + S | S*S | a | b\}$. Prove that this grammar is an *ambiguous* grammar.

Solution Suppose we need to generate $w = a + a*b$, following the given grammar.

Here, we get more than one parse tree (which are shown in Figures 11.27(a) and (b), respectively).

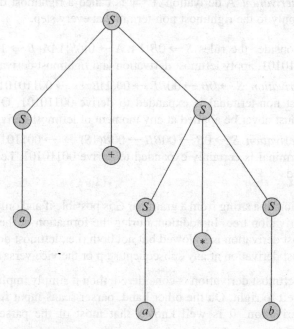

FIGURE 11.27(a) First parse tree corresponding to $a + a*b$

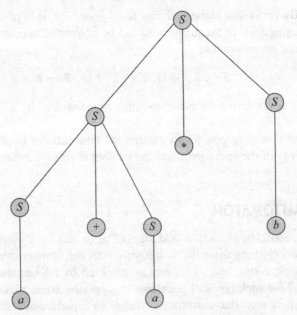

FIGURE 11.27(b) Second parse tree corresponding to $a + a*b$

Note If the grammar is an unambiguous grammar, then the rules of the grammar *never* generate multiple (distinct and *structurally* different) parse trees for the same string.

Demerits of Ambiguous Grammar

Because of the presence of ambiguous grammar in compiler, the parser can select any one, out of *multiple* possible parse trees. Accordingly, the undesired output may be computed for expression.

For instance, in Example 11.41, if the values of *a* and *b* are selected as 3 and 4 respectively and the parser selects the tree shown in Figure 11.27(a), then the *exact value* 15 for the expression 3 + 3*4 is computed (since the lower most subtree of this parse tree derives the sub-expression $a*b$ which is evaluated first).

On the other hand, if the parse tree shown in Figure 11.27(b) is selected by the parser, then it gives undesired output 36 for 3 + 3*4 (since the lower most subtree of this tree represents the *sub-expression* $a + b$ which will be evaluated first).

Note Ambiguity normally happens when the following occurs. Any non-terminal present on the LHS of a rule appears at least twice on the RHS of the *same* or *different* rule. Of course, a grammar may be ambiguous even if no such occurrence is present in the rules. Thus, an important *criterion* is that an ambiguous grammar derives *more* than one *distinct* trees for the same string.

Making Disambiguous Grammar

An ambiguous grammar may be made *unambiguous* by explicitly providing the *priority* and associativity of the operations or *transforming* the rules into another set.

For example By providing higher priority to '*' over '+', it is possible to make the above grammar as unambiguous, or the given rules can be converted into the new rules as follows to make the grammar unambiguous.

$$S \to S + T | T, T \to T * F | F, F \to a | b$$

This rule set may never derive the sub-expression associated with '+' after the sub-expression associated with '*'.

Clearly, conversion from one set to another set concentrates implicitly on the *priority* as well as *associativity* of the operations, and the number of rules increases due to transformation.

11.6 PUSHDOWN AUTOMATON

Problems with specifications such as odd number of 0s and even number of 1s in every valid string, length of every string divisible by a positive integer, number of a's is n-times that of b's, at least/atmost some count, and so on, can be resolved by FA because we are able to design REs (i.e., pattern) for such types of problems by applying some tricks. In fact, to solve such problem by FA, it is seen that automaton reaches to a particular state corresponding to that count (specification) and repeats that path if required.

On the other hand, problems such as *equal* number of a's and b's, n number of a's followed by n number of b's, checking *palindrome* and *well-formedness* of parentheses, and so on, cannot be resolved by simple FA, because to solve such problems the automaton requires some kind of memory to remember a count value or to check the matching of symbols in *reverse* order. In other words, it is required to memorize the path through which the machine is proceeding. Hence, we must need another special kind of automaton in which STACK (an auxiliary memory which follows *last in first out* principle) will be present, and such type of FA is termed as *pushdown automaton*.

Thus, STACK plays an important role to recognize some *non-regular* sets. However, regular set also must be recognized by PDA.

Concisely, PDA = FA + STACK. Hence, such machines are identical to DFAs (or NFAs), except that they additionally carry *memory* in form of a *stack*. However, the nature of the reading *head* is same as the FA. But the transition function (δ) of PDA depends on the symbol at the top of the stack, and specifies how the stack is to be changed at each transition. In particular, PDA *differ* from normal finite state machines (i.e., FA) in two ways.

1. They can use the top of the stack to decide which transition is to take.
2. They can manipulate the stack as part of performing a transition.

So, the *drawback* is that at any moment it can remember the stack-top symbol only. The general model of PDA is diagrammed in Figure 11.28.

Further, such kind of automata is equivalent to CFG. It means that, for every CFG, there exists a PDA such that the language generated by the grammar is identical to the language generated by the automaton; and for every PDA, there exists a CFG such that the language generated by the automaton is identical to the language generated by the grammar.

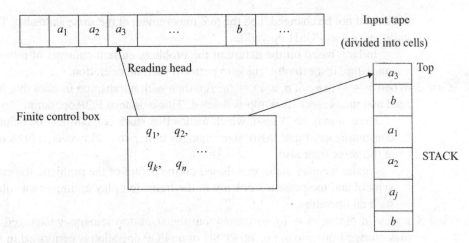

FIGURE 11.28 Model of PDA

11.6.1 Formal Definition of PDA

A PDA is formally defined as a 7-tuple machine $M = \{Q, \Sigma, \Gamma, \delta, q_o, Z_0, F\}$, where

(i) Q is a finite non-empty set of states.
(ii) Σ is a finite non-empty set of symbols, called alphabet of the language that the automaton accepts.
(iii) Γ is a finite non-empty set of stack symbols, called as *stack alphabet*.
(iv) δ is the transition function that is defined as follows: $\delta: Q \times (\Sigma \cup \{\Lambda\}) \times \Gamma \to Q \times \Gamma^*$.

This function expresses that the machine remains at a state (say, q_1) of Q, reads one input symbol (say, a) from $\Sigma \cup \{\Lambda\}$ and checks the current stack-top symbol. It then moves to another state as next state by either changing the stack-top symbol, or without making any changes to the stack-top symbol. However, the next state may be the present state q_1 itself.

(v) q_0 is the start state, i.e., the state at which no input has been processed yet (obviously, $q_0 \in Q$).
(vi) Z_0 is the initial stack symbol.
(vii) $F \subseteq Q$ is the set of final states.

[Here, the symbol Z_0 is not in Σ but it is in Γ. Thus, it is not *necessary* that all the symbols of Γ will be taken strictly from Σ.]

Let us consider the most general transition $\delta(p, u, v) = (q, \gamma)$, where p represents the present state of a PDA; u, the current input read by the automaton; v, the present stack-top content; q, the *next* state; and γ, the new stack-top content when any symbol is shifted into or deleted from stack. Further, γ may be simply v, when stack-top is not changed.

Clearly, v may never be *null* during any operation performed by such PDA, since Z_0 is assumed as the marker in its STACK.

Now, based on input symbol, topmost symbol in the stack and the present state, a number of choices are possible by a PDA. The three important choices are explained below.

Case 1 When $v \neq \gamma, \gamma \neq \Lambda$, and $u \neq \Lambda$ in the transition $\delta(p, u, v) = (q, \gamma)$, then it simply implies that stack-top is replaced by γ and state changes from p to q. Obviously, input u is shifted to stack-top of PDA. However, depending on the *nature* of the problem, the state need not be changed, i.e., the m/c may remain at the *same* state also. This is basically termed as PUSH operation.

In fact, based on the nature of the problem, current contents of *input* as well as *stack-top*, respectively, guide to perform such an operation.

Case 2 When $\gamma = \Lambda$, i.e., $\delta(p, u, v) = (q, \Lambda)$, then such a transition implies that input u is read and the present stack-top is deleted. This is indeed POP operation.

Here, u may be Λ also, which means that stack content may be deleted *without* consuming any input. Also, state changes from p to q. However, a PDA may remain at the *same* state also.

Again, we may claim that, based on the nature of the problem, the current *input* symbol and the present *stack-top* in the transition play an important role to decide such an operation.

Case 3 $v = \gamma$ in $\delta(p, u, v) = (q, \gamma)$ means that the stack-top remains unchanged, only state may change from p to q, i.e., no PUSH or no POP operation is performed in such a situation. Clearly, this is a new operation added to resolve some kinds of problems.

Note States except *initial/final* states used in the transitions generally avoid some situations. For example, let us consider two problems: (i) $L = \{a^n b^n | n \geq 0\}$ and (ii) $L = \{a^n b^n | n \geq 1\}$. Both the problems are almost same in nature except for the different values of n. The second problem ensures that the value of n must be *at least* 1, whereas it is 0 for the first problem. Obviously, state in the *transition* plays an important role to fulfil such restriction (and it is discussed in Example 11.42).

EXAMPLE 11.42 Design a PDA for accepting the language $L = \{a^n b^n | n \geq 1\}$.

Solution The basic idea to solve this problem is to shift (push) the input a's into stack (memory) of the PDA. But when a symbol b is read, state changes and *pop* (deletion of stack-top) is to be performed. If no input is left to read and the *stack-top* of the PDA reaches to Z_0 (initial stack-top symbol), then that string is accepted by the designed PDA.

Let $M = (Q, \Sigma, \Gamma, \delta, q_o, Z_0, F\}$, where $Q = \{q_0, q_1, q_f\}$, $\Sigma = \{a, b\}$, $\Gamma = \{a, Z_0\}$ (as the input symbol a only will be shifted to stack). Here, q_0 is the *start* state, and Z_0 is the initial stack symbol, $F = \{q_f\}$. The δ is defined as follows to solve the problem.

1. $\delta(q_0, a, Z_0) = (q_0, aZ_0)$ $\Rightarrow$ Push operation. aZ_0 implies here that the earlier content of stack-top was Z_0 but currently it is a.
2. $\delta(q_0, a, a) = (q_0, aa)$ $\Rightarrow$ Push operation. Here, aa implies that the earlier content of stack-top was a, and currently it is a too, i.e., the current a is just immediate above the earlier a. (Transition 2 will continue till input a is found.)
3. $\delta(q_0, b, a) = (q_1, \Lambda)$ $\Rightarrow$ Pop operation (i.e., when input is b and stack-top symbol is a, stack-top a is deleted). (Transition 3 continues till input b and stack-top a are found.)
4. $\delta(q_1, \Lambda, Z_0) = (q_f, Z_0)$ $\Rightarrow$ When no input is left to read (i.e., it is Λ) and stack-top is Z_0 (i.e., initial stack symbol), then the PDA moves to final state (q_f). Clearly, the processed string is accepted. In particular, this transition (specially the state in this transition) is very much vital to the m/c to move to final state.

Note If scanning of a string is over but the automaton does not move to the last transition (i.e., *Transition 4* mentioned above), then that string is not accepted by the automaton.

Further, someone may ask why the proposed PDA changes state in Transition 3. The answer is that if the m/c moves to *final* state from q_0 without changing the state (i.e., if it remains always at the same state q_0), then it means that the m/c moves to *final* state without consuming any input too. Clearly, no such input (i.e., *null* string) should be accepted as per instruction of the problem (since the value of n is greater than or equal to '1'). However, such a situation holds true for $n \geq 0$. Accordingly, this situation can be avoided if the m/c moves to final state from some other state rather than the *start* state q_0. That is why state is changed in *Transition 3*.

EXAMPLE 11.43 Design a PDA for accepting the language $L = \{W_c W^R | W \in \{a, b\}^*\}$.

Solution The basic idea to solve this problem is that such automaton reads the input string from left to right and must remember the *first* half (recognized by the symbol c), and then checks the *second* half of the string in *reverse* order against the *first* half of the string.

Let $M = (Q, \Sigma, \Gamma, \delta, q_o, Z_0, F)$, where $Q = \{q_0, q_1, q_f\}$, $\Sigma = \{a, b, c\}$, $\Gamma = \{a, b, Z_0\}$. Here, q_0 is the start state, Z_0 is the initial stack symbol, and $F = \{q_f\}$. The δ is defined as follows to solve the problem.

1. $\delta(q_0, a, Z_0) = (q_0, aZ_0)$ ⟹ Push operation. aZ_0 implies here that earlier stack-top content was Z_0 but currently it is a, i.e., Z_0 is just below a.
2. $\delta(q_0, b, Z_0) = (q_0, bZ_0)$ ⟹ Push operation. bZ_0 implies here that the earlier content of stack-top was Z_0 but currently it is b, i.e., Z_0 is just below b.
3. $\delta(q_0, a, a) = (q_0, aa)$ ⟹ Push operation. aa implies here that the just earlier content of stack-top was a but currently it is also a, i.e., the current a is just above the earlier a.
4. $\delta(q_0, b, a) = (q_0, ba)$ ⟹ Push operation. ba implies here that just earlier content of stack-top was a but currently it is b, i.e., b is just above the earlier a.
5. $\delta(q_0, a, b) = (q_0, ab)$ ⟹ Push operation. ab implies here that the immediate earlier content of b was a.
6. $\delta(q_0, b, b) = (q_0, bb)$ ⟹ Push operation. bb implies here that the immediate earlier content of stack-top was b but currently it is b too, i.e., the current b is *just* above the earlier b. [Transitions 1–6 continue until input c is read and no state is required to change.]
7. $\delta(q_0, c, b) = (q_1, b)$.
8. $\delta(q_0, c, a) = (q_1, a)$ [Both Transitions 7 and 8 imply that when the current input is c and the stack-top is either b or a, then state changes by maintaining the stack-top unchanged. However, if state is not changed, then the m/c accepts the *null* string also (as it is *palindrome*) from q_0 itself. But such a string should not be valid, as string contains at least c.]
9. $\delta(q_1, b, a) = (q_1, \Lambda)$.
10. $\delta(q_1, a, b) = (q_1, \Lambda)$ [By Transitions 9 and 10, pop operation is performed and this operation continues till input and stack-top symbols are reverse.]
11. $\delta(q_1, \Lambda, Z_0) = (q_f, Z_0)$, when no input is left to read (i.e., Λ) and stack-top is Z_0 (initial stack symbol), then the m/c moves to final state (q_f) and the processed string accepts it. Hence, this transition is very much vital to the m/c.

EXAMPLE 11.44 Design a PDA for accepting the string in which the number of a's is twice than that of b's. $\Sigma = \{a, b\}$.

Solution In particular to solve such a problem, we prefer to design a 2-stack PDA. Let $M = (Q, \Sigma, \Gamma, \delta, q_0,) <Z_0, Z_1>, F\}$, where $Q = \{q_0, q_1, q_f\}$, $\Sigma = \{a, b\}$, $\Gamma = \{c, Z_0, Z_1\}$. Here, Z_0 and Z_1 denote the initial symbols (*markers*) of stacks, say, S_0 and S_1, respectively. Further, q_0

represents the *start* state of m/c, whereas $F = \{q_f\}$ is the set of final states. Also, a new symbol c is used to push into stacks for both the inputs: a and b. Interestingly, in this approach, a single c is pushed into stack S_0 for an input a, whereas two consecutive c's are pushed into stack S_1 for an input b. The transition function: δ for such a PDA can be generalized as follows:

$$\delta(p, u, <v_1 Z_0, v_2 Z_1>) = (q, <v_3, v_{41}>).$$

In fact, the behaviour of the above transition is same as the behaviour of transition of 1-stack PDA. However, the notation $<>$ is used to specify the status of both the stacks. The necessary transitions to solve this problem are shown below.

1. $\delta(q_0, a, <Z_0, Z_1>) = (q_1, <cZ_0, Z_1>)$
2. $\delta(q_1, a, <c, Z_1>) = (q_1, <cc, Z_1>)$
3. $\delta(q_0, b, <Z_0, Z_1>) = (q_1, <Z_0), ccZ_1>)$
4. $\delta(q_1, b, <c, Z_1>) = (q_1, <c, ccZ_1)>)$
5. $\delta(q_1, b, <Z_0, c>) = (q_1, <Z_0, ccc>)$
6. $\delta(q_1, b, <c, c>) = (q_1, <Z_0, ccc>)$
7. $\delta(q_1, a, <Z_0, c>) = (q_1, <cZ_0, c>)$
8. $\delta(q_1, a, <c, c>) = (q_1, <cc, c>)$
9. $\delta(q_1, \Lambda, <c, c>) = (q_1, <\Lambda, \Lambda>) \Rightarrow$ pop symbol c from both the stacks
10. $\delta(q_1, \Lambda, <Z_0, Z_1>) = (q_f, <Z_0, Z_1>)$

11.6.2 Types of PDA

PDA may be of two types: *DPDA* (deterministic pushdown automaton) and *NPDA* (non-deterministic pushdown automaton).

Non-determinism occurs here if the m/c moves to multiple states from a state on reading the same input as well as the same stack symbol. For example, there exists a CFG for the problem WcW^R (where W is a string over Σ, W^R is the reverse of W, and $c \in \Sigma$), and we can easily design a *DPDA* for the same (since marker c is present in between the two *halves*). Similarly, for the problem WW^R, there is CFG also but we cannot design a DPDA for the same (as there is no symbol to decide deterministically when to *pop* the stack contents). Clearly, we must have to design here a NPDA which helps to solve this problem.

Therefore, *NPDA is more powerful than DPDA*. However, power of DFA and NDFA is same.

Note If we allow a finite automaton to access two stacks instead of just one, we obtain a more powerful device which is indeed equivalent in power to a *TM* (turing machine). An *LBA* (linear bounded automaton) is a device which accepts context sensitive grammar (CSG) is more *powerful* than a *PDA* but less so than a TM.

11.7 TURING MACHINE

TM is extremely basic abstract symbol-manipulating device which, despite its simplicity, can be adapted to simulate the logic of any computer algorithm (as we understand). Such a m/c was first described by Alan Turing in 1936.

A TM, in its simplest form, is composed of a *tape* (is basically *memory*) of infinite length. The tape is used to store data. There is a *head* (which is basically microprocessor) that points to the tape. The head can read the symbol, choose to write a new symbol in place, and then move *left* or *right*. Hence, it is a device that is the generalization of an FA from the perspective that it allows two-way *movement* of the head for *reading* as well as *writing* of symbols, and that is why it can solve any type of *feasible algorithm* (many of which *cannot* be solved by PDA and LBA).

It consists of number of *transitions* (i.e., a set of instructions) and finite number of *states*. The list of transitions says, given a current state and a symbol currently under the head, what should be written on the tape, what state the machine should go, and whether the head should move left or right.

Further, a TM which is able to simulate any other TM is called a *universal turing machine* (UTM). In particular, UTM consists of a single tape but its working style is close to 3-tape TM (i.e., a 3-tape TM is simulated on a single tape TM).

A basic model of TM is shown in Figure 11.29.

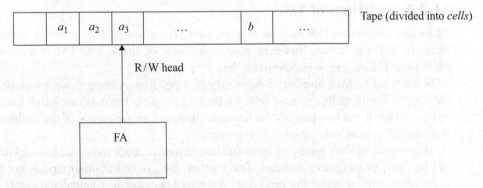

FIGURE 11.29 TM model

11.7.1 Formal Definition of TM

A TM is formally defined as a 7-tuple machine $M = (Q, \Sigma, \Gamma, \delta, q_o, b, F\}$, where

(i) Q is a finite non-empty set of states
(ii) Σ is a finite set of symbols, called alphabet of the language accepted by the automaton.
(iii) Γ is a finite non-empty set of *tape* symbols
(iv) b *(blank)* $\in \Gamma$ (but does not belong to Σ), [*blank* symbol is, in fact, used to mark the end of input]
(v) q_0 is the start state (obviously, $q_0 \in Q$)
(vi) $F \subseteq Q$ is the set of final states that the m/c accepts
(vii) δ is the transition function that is defined as follows:

$$\delta: Q \times \Gamma \rightarrow Q \times \Gamma \times \{L, R\}$$

Clearly, this function describes that the R/W head of the machine reads one input symbol from tape at a state, say, q_1 (from Q) and then the head moves towards *left* (L) or *right* (R) from its current position by *replacing* the present input symbol (if necessary). Obviously, the machine moves to another *next* state. However, the next state may be the present state itself. In particular, the transition (especially the current state and the current input in the transition) plays important role to satisfy the properties of the problem while solving.

Why Is Replacement of Input Symbol Required?

The head of a TM is allowed to move towards *left* from its current position. Thus, if an input already processed by the m/c is not replaced by some other symbol, then the *same* input may be processed further. Consequently, it may cause to fall the m/c in *infinite* loop. That is why, depending on the nature of the problem, the processed symbols may be needed to replace by some other symbols to avoid such a situation.

11.7.2 Improvement in TM

(i) The symbol in the tape is erasable.
(ii) The head (which is referred here as read and write head) can move into both directions left or right from its current position.
(iii) Any number of tapes and heads can be added.
(iv) Size of each tape is unbounded.

11.7.3 Variations of TMs

There are varieties of TMs such as *single-tape* TM, *multi-tape* TM, arbitrary numbers of R/W heads in TM, and so on. However, *time complexity* of single-tape TM is more than that of multi-tape TM but *space complexity* is less.

In multi-tape TM, a number of tapes may be employed. Among those, one tape must be an *input tape*. The input tape is used only for reading purpose. On the other hand, the requirement of output tape is not compulsory. Its necessity depends on the *nature* of the problem. Actually, it is needed to store data/results while processing.

The respective R/W heads for input and output tapes cannot move *backward*. However, they can be *static* or can move *forward* also. Further, the symbol in *input* tape is not replaced by some other symbol (since the head does not move backward), whereas the symbol on *output* tape may be replaced by some other symbol (if necessary).

Now, the other tapes (except *input* and *output* tapes) are called as INTERMEDIATE or MEMORY tapes that are used for processing purpose. Each MEMORY tape must have a unique left-end *marker*.

EXAMPLE 11.45 Design a single-tape TM that accepts language $L = \{a^n b^n | n \geq 1\}$.

Solution To design such a TM, first replace a by some symbol (say, x), move towards right until you get a b. Replace this by some other symbol (say, y) and move left until you get one x. When replaced symbol x for a is found first, move towards right for replacing the next a.

Repeat the above procedure until all the a's in valid string are replaced by x. However, if the string is not valid (i.e., the number of a's is less than that of b's or the *reverse*), then processing stops.

The procedure can be implemented through the transitions (operations) presented below. In other words, we can model the machine as follows:

$Q = \{q_0, q_1, q_2, q_3, q_f\}$, where q_0 and q_f are, respectively, the initial and the final states of the machine, $\Sigma = \{a, b\}$; $\Gamma = \{a, b, x, y, \#\}$, where '#' denotes *blank* symbol.
D (movement of head) $= \{S, L, R\}$, where S, L, and R stand, respectively, for *static* (no move), *left*, and *right* move.
Transitions The transitions are as follows:

1. $\delta(q_0, a) = (q_1, x, R)$ $\implies$ when a is read at q_0, state changes to q_1 and the R/W head moves right from its current position. Also, the symbol a is replaced by x.
2. $\delta(q_1, a) = (q_1, a, R)$ $\implies$ when a is read at q_1, the m/c stays at q_1. But the head moves right from its current position without replacing a. This operation continues till a is found.
3. $\delta(q_1, y) = (q_1, y, R)$ $\implies$ if y is read at q_1, the m/c stays again at q_1. Here too, the head moves again towards right from its current position, but y is not replaced. Of course, the

input y may be read at q_1 just after the first replacement of b with y. This operation is repeated till y is found. [Transitions 1, 2, and 3 play important role to search one b to be replaced.]

4. $\delta(q_1, b) = (q_2, y, L)$ ⇒ when b is received at q_1, state changes from q_1 to q_2. The head moves left from its current position, and b is replaced by y.
5. $\delta(q_2, y) = (q_2, y, L)$ ⇒ if y is found at q_2, it remains at q_2. But the head moves towards left from its current position without replacing y. This operation continues till y is found.
6. $\delta(q_2, a) = (q_2, a, L)$ ⇒ if a is read at q_2, the state and the input are not changed; only the head moves towards left from its current position. This operation is repeated till a is found. [Transitions 4, 5, and 6 are responsible to find the next a to be replaced.]
7. $\delta(q_2, x) = (q_0, x, R)$ ⇒ when x is read at q_2, state changes from q_2 to q_0 but input x is not replaced. The head moves right from its current position in order to replace the next a by applying *Transition* 1.
8. $\delta(q_0, y) = (q_3, y, R)$ ⇒ if y is read at q_0, state changes to q_3. But the head moves towards right from its current position without replacing the symbol. This occurs when no a is left to replace.
9. $\delta(q_3, y) = (q_3, y, R)$ ⇒ if y is again read at q_3, both the state and the input are not changed. Only the head moves right from its current position. Actually, it happens when no a is left to replace. This operation continues till y is found.
10. $\delta(q_3, \#) = (q_f, \#, S)$ ⇒ if $\#$ is read at q_3, the m/c reaches to final state without changing the symbol, and remains there.

 EXAMPLE 11.46 Design a 2-tape TM that accepts language $L = \{a^n b^n | n \geq 1\}$.

Solution

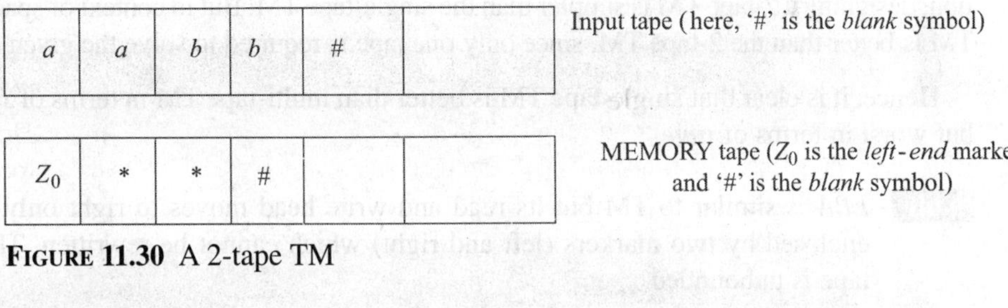

a	a	b	b	$\#$	

Input tape (here, '#' is the *blank* symbol)

Z_0	*	*	#		

MEMORY tape (Z_0 is the *left-end* marker and '#' is the *blank* symbol)

FIGURE 11.30 A 2-tape TM

Now, to solve the above problem, a *2-tape* TM can be designed as follows:

$Q = \{q_0, q_1, q_f\}$, where q_0 is the initial state and q_f is the final state of the machine,
$\Sigma = \{a, b\}$,
$\Gamma = \{a, b, *, \#\}$, where '#' denotes *blank* symbol and '*' is the *replaced symbol* for a. Two tapes of the proposed TM are shown in Figure 11.30.
D (movement of head) $= \{S, L, R\}$, where $S, L,$ and R stand, respectively, for *static* (no move), *left*, and *right* move.
Transitions Transition of 2-tape TM for solving the present problem can be generalized as follows:

$$\delta(q_i, <a_i, b_k>) = (q_j, <r_i>, <M_1, M_2>), \quad M_1, M_2 \in \{S, L, R\}$$

The *current* state and the *present* input symbol are important to satisfy the properties of the problem.

Explanation of such transition The above transition is described as follows: $q_i \in Q$ is the current state of the machine. $<a_i, b_k>$ means that two inputs: a_i from the first tape (here, it is input tape) and b_k from the second tape (here, it is MEMORY tape) are read by the m/c. $q_j \in Q$ represents the next state, and $<r_i>$ is the replaced symbol in place of b_k on the second tape.

Further, $<M_1, M_2>$ represents here basically the movements of the heads of the two tapes, respectively, the first tape and the second tape.

In order to solve this problem, for each a's on the first tape, the proposed TM first needs to write a symbol, say, '*', in place of a *blank* symbol (say, '#') in the second tape. In fact, this operation helps to match the number of b's with that of a's in the next part of processing the string. The necessary transitions (operations) are presented below.

Formal presentation of the transitions to solve the given problem:
1. $\delta(q_0, <a, Z_0>) = (q_0, <Z_0>, <S, R>)$ ⇒ when inputs a from the first tape and 'Z_0' from the second tape are read, state remains the same. The head of the first tape remains static, but the head of the second tape moves right from its current position without replacing its symbol in order to ensure the exact matching of the number of a's with the number of the replaced symbol.
2. $\delta(q_0, <a, \#>) = (q_0, <^*>, <R, R>)$ ⇒ symbol of the second tape is replaced by '*' and both the heads move right, but state remains the same.
3. $\delta(q_0, <b, \#>) = (q_1, <\#>, <S, L>)$
4. $\delta(q_1, <b, ^*>) = (q_1, <^*>, <R, L>)$
5. $\delta(q_1, <\#, Z_0>) = (q_f, <Z_0>, <S, S>)$

Comparison The problems described in Examples 11.45 and 11.46 are same. But the time consumed by the *single-tape* TM is more as compared to the designed 2-*tape* TM machine. In addition, designing a 2-tape TM is simpler than the single-tape TM. But in context of space, single-tape TM is better than the 2-tape TM, since only one tape is required to solve the given problem.

Hence, it is clear that single-tape TM is better than multi-tape TM in terms of *space* and *cost* but worst in terms of *time*.

> **Note** LBA is similar to TM but its read and write head moves to right only. The input is enclosed by two markers (left and right) which cannot be rewritten. The size of the tape is unbounded.

11.7.4 Halting Problem

Halting problem is the problem which means that the m/c does not halt (i.e., it is neither at a *final* state nor at a *non-final* state). More clearly, it goes to *dead/trapped* state. Halting problem of TM is indeed *unsolvable*.

11.7.5 Turing Acceptable Language

This is a language (L) if for string $w \in L$, the m/c accepts it (i.e., output is yes), but for string $w \notin L$, it does not take any decision (i.e., output is neither yes nor no). Turing acceptable language is called as *recursively enumerable language*.

11.7.6 Turing Decidable Language

This is a language if for string $w \in L$, the m/c accepts it (i.e., output is yes). On the other hand, it takes as decision 'no' (i.e., output is no) if $w \notin L$. Turing decidable language is called as *recursive language*. For any decidable problem, there exists some definite algorithm which always terminates (halts) with two inputs (yes/no). Otherwise, the problem is said to be undecidable.

Thus, *every recursive language is recursively enumerable language but the reverse is not true.*

11.7.7 Properties of Recursive and Recursively Enumerable Languages

The properties of recursive and recursively enumerable languages are as follows:
 (i) The *complement* of recursive language is recursive.
 (ii) The *union* of two recursive languages is recursive.
 (iii) The *union* of two recursively enumerable languages is recursively enumerable.
 (iv) If a language L and its *complement* L' are both recursively enumerable, then L is recursive.

11.7.8 Church Thesis

Turing m/cs are formal versions of algorithms (i.e., operations in an algorithm are represented by the transitions of TM), and no computational procedure will be considered an algorithm unless it can be presented as a TM. This is known as *Church thesis* or *Church-turing thesis*.

11.8 POST-CORRESPONDENCE PROBLEM

Post-correspondence problem (PCP) was first introduced by Emil Post in 1946. It has many applications in the theory of formal languages. The PCP determines whether there exist identical (common) solutions (obtained from two different sources) or not. Actually, the solutions are generated, following the same sequence.

More clearly, let us consider two lists $X = \{x_1, x_2, \ldots, x_n\}$, $Y = \{y_1, y_2 \ldots, y_n\}$ of *non-empty* strings over $\{a, b\}$. Now, we can check if it is possible to form string such that the string formed from X by a sequence and the string formed from Y by the same sequence are identical. Obviously, the length of the strings in both the cases is same. For example, let $X = \{b, bab^3, ba\}$ and $Y = \{b^3, ba, a\}$. Here, $x_1 = b$, $x_2 = bab^3$, $x_3 = ba$; $y_1 = b^3$, $y_2 = ba$, $y_3 = a$. Now, let us construct the strings as follows: $x_2 x_1 x_1 x_3 = bab^3 \, bbba$ (taken from X), $y_2 y_1 y_1 y_3 = ba \, b^3 \, b^3 \, a$ (taken from Y). So, the PCP here has post correspondence solution.

Note The PCP is undecidable, since such problem has no definite algorithm.

EXAMPLE 11.47 Prove that PCP with two lists $X = \{01, 1, 1\}$ and $Y = \{01^2, 10, 1^2\}$ has no PCP solution.

Solution Here, length of each string in X is less than the length of the correspondence string in Y. Hence, this problem may not have any PCP solution.

11.9 CLASSES OF PROBLEMS

Different types of problems are
  Decision problem A problem is decision problem if it takes an input and provides some output.
  Optimization problem It gives some *feasible* optimization result for some input values to the problem.

P-class problem A problem is P-class problem if it is solvable deterministically in polynomial time, i.e., the designed deterministic algorithm takes polynomial time $O(n^k)$, where n is the input size and k is constant.

NP-class problem A problem is *NP*-class problem if it is solvable non-deterministically (i.e., based on guess) in polynomial time, i.e., the designed non-deterministic algorithm takes polynomial time $O(n^k)$, where n is the input size and k is *constant*. In particular, such a problem can be *verified* in polynomial time. Obviously, no proper rule is followed to make the guess, since it is non-deterministic; and we cannot ensure that such an algorithm always returns the output, i.e., in one run the algorithm may provide the output but may not in the next run.

Comment Any problem in *P*-class is also in *NP*-class (since a *P*-class problem can be solved non-deterministically) but the reverse is not true. Thus, deterministic problems are just a special case of non-deterministic ones, i.e., we may conclude that $P \subseteq NP$.

NP-Hard and NP-Complete problems there are some problems for which it is very difficult to design polynomial time deterministic algorithm or there may not exist any appropriate polynomial time deterministic approach to solve. However, they may have non-polynomial time deterministic algorithm for such problem. Now, if we aim to design polynomial time algorithm to solve such problems, we must look for non-determinism approach. Such problems are then considered as *NP*-hard problems, for example travelling salesman, knapsack, Hamiltonian cycle, and graph colourability problems.

Now, the question is How to design a polynomial-time non-deterministic algorithm? Let us consider an example to understand it.

- Given a graph G, does G contain a Hamiltonian cycle?
- Hamiltonian cycle is a cycle passing through every vertex exactly once.

We may design non-deterministic algorithm for the above-mentioned problem as follows.

- Guess a permutation of all vertices.
- Check whether this permutation gives a cycle. If yes, then algorithm halts.

We have shown the relationship among *P*, *NP*, *NPC*, and *NP*-hard problems in Figure 11.31. In particular, some *optimization* and *decision* problems are very *hard*, and so they belong to *NP*-hard region. Further, all *NP*-hard problems are not NPC. Only the *decision NP*-hard problems are *NPC*, but some *optimization* problems may not be NPC. For example, we can find the shortest path between two vertices in a graph in polynomial time, but finding the largest path between two vertices is *NPC* problem. Other examples are *Hamiltonian cycle, graph colourability* problem, etc. However, some *optimization* problems may reduce to decision problems.

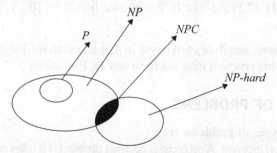

FIGURE 11.31 Relationship among *P*, *NP*, *NPC*, and *NP*-hard problems

If any NPC has a polynomial time solution, then every *NP* must have a polynomial time solution.

Lemma 11.1 Let P_1 and P_2 be two problems such that P_1 is NPC and P_1 is *reducible* to P_2, i.e., $P_1 \alpha P_2$, then P_2 is also *NPC*.

Lemma 11.2 If L_1 is a *decision* problem and L_2 is an *optimization* problem, and $L_1 \alpha L_2$, then L_2 is also *NPC*.

11.10 CELLULAR AUTOMATON

A regular lattice of cells in a *d*-dimensional space is set up for a system. A set of Boolean variables is attached to each site, giving the state of the cell. Again, at each time step, a rule describing the time evolution of the states is defined.

The rule is the same for all cells and it is applied to the cells in parallel (simultaneously). Normally, the new state is a function of the previous time step only, but sometimes it is necessary to take into account the states of (some) earlier time steps, too. In simulation, more memory is then needed for storing older states of the cells. Because the cellular automata (CA) rules are usually local, updating the state of a certain cell requires the knowledge of the states of the cells in the vicinity of it. This vicinity is called neighbourhood. Usually this neighbourhood is very restricted containing only a few cells (including, for example, 4 or 8 nearest neighbours in the two-dimensional case), otherwise the rule may become too complex. In practice, the simulation of a CA rule requires that the lattice cannot be infinite, certain boundary conditions have to be applied at the lattice edges. One possibility for managing the boundary cells is to apply a special rule for these sites, and this rule that is different from the rule applied to 'normal' cells in the lattice. In this method, it is also possible to use different rules at different boundaries. Another possibility is to define virtual cells beyond the boundaries, and the updating rule remains the same for all cells. In the used case of periodic boundary conditions, the one-dimensional lattice often forms a ring. The usual way to update the cells in a lattice is parallel (synchronous, simultaneous) updating. In some cases, other updating methods such as *random* or *sequential* are used.

Cellular automata are divided into two categories: *deterministic* and *probabilistic*. This subsection refers to a deterministic CA, and it bears the fact that the time evolution of the states of the CA is always the same if the starting configuration is the same. With probabilistic CA, the new value of a cell state in updating depends on the values of the cells in the neighbourhood and on a pre-determined probability to get a certain new value. Thus, starting from a particular configuration, a probabilistic CA may evolve into totally different configurations in separate simulation runs.

It seems that all one-dimensional CA (perhaps also others) fall into four distinct universality classes. These classes characterize the attractors in the evolution of CA configurations. Attractors are a family of certain states to which the trajectories in the configuration space evolve after many time steps having started from almost any initial state. Different classes can be characterized as follows:

Class 1 After a finite number of time steps, almost all initial states evolve to a unique homogeneous state, in which all sites have the same value (a 'limit point').

Class 2 A CA of this class generates in most cases separated simple structures from particular (typically short) initial site value sequences (the automaton serves as a 'filter'), the structures generated are either *stable* or *periodic*.

Class 3 Evolution from almost all initial states leads to chaotic, non-periodic patterns ('strange attractors').

Class 4 In most cases all sites attain value '0' after a finite number of time steps, but in few cases stable or periodic structures are formed, which persist for an infinite time-in some cases also propagating structures are formed.

Is CA Model Flawless?

The answer is 'it depends'. The behaviour of a system may depend only little on the details of the interactions between its elementary components. The complex behaviour of the macroscopic world coming from collective behaviour rather than from some specific characteristics of microscopic interactions may well justify the simplification of the microscopic laws connected to the phenomenon if these laws are not relevant at the macroscopic level of observation. In this kind of situations CA-type simplifications may be very desirable.

11.11 FUZZY SETS AND FUZZY LOGIC

A set is normally thought of as a collection of objects. It can be defined by enumerating the set

$$F = \{1, 2, 3, 4\ 5\}$$

or by indicating the set membership requirement

$$F = \{x | x \in Z^+ \text{ and } x \leq 5\}$$

A fuzzy set is a set, F, in which the set membership function, f, is a real-valued (as opposed to Boolean) function with output in the range [0, 1]. An element x is said to belong to F with probability $f(x)$ and simultaneously to be in $\neg(F)$ ('$\neg$' stands for *not*) with probability $1 - f(x)$. Actually, this is not a true probability, but rather the degree of truth associated with the statement that x is in the set. To show the difference, let us look at a fuzzy set operation. Suppose the membership value for Rakesh being tall is 0.7 and the value for his being thin is 0.4. The membership value for his being both is 0.4, the minimum of the two values. If these were really probabilities, we would look at the product of the two values.

Fuzzy sets have been used in many computer science and database areas. In the classification problem, all records in a database are assigned to one of the predefined classification areas. A common approach to solving the classification problem is to assign a set of membership function to each record for each class. The record is then assigned to the class that has the highest membership function. Similarly, fuzzy sets may be used to describe other data mining functions.

Let us consider a SALARY relation in traditional SQL. Suppose we want to retrieve the names of the employees with a salary more than Rs. 10000.00. Then, the corresponding SQL query will be as follows.

Select name from SALARY where salary > 10000.00

However, we may describe the above SQL query by an equivalent set member function as follows.

$$\{x | x \in R \text{ and } x.\text{Salary} > 10000.00\}$$

Here, x.Salary is the salary attribute within the tuple x. The membership function is Boolean here. Some queries, however, do not have a membership function, i.e., Boolean. For example, suppose that we wish to find the names of employees who are tall.

$\{x \mid x \in R \text{ and } x \text{ is tall}\}$, where SALARY relation has the following fields: (a) ID, (b) Name, (c) Address, and (d) Salary. The membership function is not Boolean, and thus the results of this query are *fuzzy*.

Fuzzy logic is reasoning with uncertainty. That is, instead of a two-valued logic (true or false), there are multiple values (true, false, may be). Fuzzy logic has been used in database systems to retrieve data with imprecise or missing values. In this case, the membership of records in the query result set is fuzzy. As with traditional Boolean logic, fuzzy logic uses operators such as ¬, ∧, and ∨. Assuming that x and y are fuzzy logic statements and that mem(x) defines the membership value, the following values are commonly used to define the results of these operations:

(a) mem(¬x) = 1 − mem(x)
(b) mem($x \wedge y$) = min (mem(x), mem(y))
(c) mem($x \vee y$) = max (mem(x), mem(y))

Fuzzy logic uses rules and membership functions to estimate a continuous function. Fuzzy logic is a valuable tool to develop control systems for such things as *evaluators*, *trains*, and *heating systems*. Most real-world classification problems are fuzzy.

11.12 RUSSELL'S PARADOX

It is also known as *Russell's antinomy*, Russell's paradox is the most famous of the logical or set-theoretical paradoxes. The paradox arises within naive set theory by considering the set of all sets that are not members of themselves. Such a set appears to be a member of itself if and only if it is not a member of itself, hence the paradox.

Some sets, such as the set of all teacups, are not members of themselves. Other sets, such as the set of all non-teacups, are members of themselves. Call the set of all sets that are not members of themselves R. If R is a member of itself, then by definition it must not be a member of itself. Similarly, if R is not a member of itself, then by definition it must be a member of itself.

11.12.1 History of the Paradox

Russell appears to have discovered his paradox in the late spring of 1901, while working on his *Principles of Mathematics* (1903). Cesare Burali-Forti, an assistant to Giuseppe Peano, had discovered a similar antinomy in 1897 when he noticed that since the set of *ordinals* is well-ordered, it too must have an ordinal. However, this ordinal must be both an element of the set of all ordinals and yet greater than every such element. Unlike Burali-Forti's paradox, Russell's paradox does not involve either *ordinals* or *cardinals*, relying instead only on the primitive notion of set.

Russell wrote to Gottlob Frege with news of his paradox on 16 June 1902. The paradox was of significance to Frege's logical work since, in effect, it showed that the axioms Frege was using to formalize his logic were inconsistent. Specifically, Frege's rule V, which states that two sets are equal if and only if their corresponding functions coincide in values for all possible arguments, requires that an expression such as $f(x)$ be considered both a function of the argument x and a function of the argument f. In effect, it was this ambiguity that allowed Russell to construct R in such a way that it could both be and not be a member of itself.

SUMMARY

In theoretical computer science, automata theory is the study of abstract machines and their *capabilities* of problem solving. It is closely related to formal languages theory, as the automata are often classified by the class of formal languages they are able to recognize.

Theoretical computer science is the collection of topics of computer science that focuses more on the abstract, logical, and mathematical aspects of computing, such as the theory of computation, analysis of algorithms, semantics of programming languages, and so on.

This chapter examines the preliminary requirements such as alphabet, string and its operations, sentence, language, and their operations to grasp the field *theory of computation*. This chapter also covers different types of automata, regular expression, context-free, and general phrase-structure languages along with their associated automata. We have shown the power of different types of computation and the theoretical limits of computers.

This chapter provides complexity theory with an introduction to some of the open classification problems relating to the classes *P* and *NP*. We finally end with a brief description on fuzzy set and cellular automata.

EXERCISES

1. Define length of a string. Find the length of the strings $u = abcag, w = gh$.
2. Perform all possible operations on the strings given in Exercise 1.
3. Define prefix and suffix of a string.
4. Define concatenations of languages L_1 and L_2. Also, define Kleen star.
5. Prove the equivalence of DFA and NDFA.
6. Design a DFA to recognize the language $L = \{a^n b \mid n \geq 0\}$.
7. Design a DFA that accepts the language $L = \{w \mid w \in \{a, b, c\}^*\}$ and w contains the pattern *abac*.
8. Design DFA and NDFA accepting all strings over $\{0, 1\}$, which end in 0 but do not contain 11 as substring.
9. Design an NDFA accepting the language $L = \{a^* \cup b^*\} a^*$.
10. Design a DFA accepting the positive integer divisible by 3.
11. Design a DFA accepting all strings over $\{a, b\}$, having at most one pair of a's or b's.
12. Design DFA over $\{a, b, c\}$ so that last two symbols are same.
13. Design DFA over $\{a, b, c\}$ so that every block of four characters contains at least two identical symbols.
14. Design DFA over $\{a, b\}$ containing exactly two a's but more than two b's.
15. Differentiate between Mealy m/c and Moore m/c. Describe the method of conversion from Mealy to Moore and Moore to Mealy.
16. Design RE for the following:
    (i) The set of all strings consisting of odd number of 0s and even number of 1s.
    (ii) $L = \{b^m a b^n \mid m > 0, n > 0\}$.
    (iii) The set of all strings over $\{a, b\}$ in which the number of occurrences of a is divisible by 3.
    (iv) Any string whose length is a multiple of 5 over $\{0, 1, 2\}$.
    (v) Any sentence that begins with a capital letter and ends with a full stop.
    (vi) Set of binary strings each with exactly three 1s.
    (vii) Obtain the regular expression for the language $L = \{a^{2n} b^{2m+1} \mid n, m \geq 0\}$.
17. Prove or disprove. If L_1 is regular and L_2 is non-regular language, then $L_1 \cup L_2$ is non-regular.
18. Prove or disprove. If L_1 is regular and L_2 is non-regular language, then $L_1 \cap L_2$ is regular.
19. By applying pumping lemma, prove or disprove $\{0^i 1^j \mid i, j \geq 0\}$ is regular.

20. By applying pumping lemma, prove or disprove $\{xx^R | x \text{ in } \{0, 1\}^*\}$ is regular.
21. By applying pumping lemma, prove or disprove $\{L = \text{odd length strings over } \{0, 1\} \text{ with middle symbol '0'}\}$ is regular.
22. By applying pumping lemma, prove whether the following languages are regular or not.
    (i) $L_1 = \{a^i b^j c^k : i, j, k \geq 0, \text{ and } i + j = k\}$.
    (ii) $L_2 = \{a^i b^j c^k : i, j, k \geq 0, \text{ and if } i = 1 \text{ then } j = k\}$.
23. What is CFG? Give an example.
24. What do you mean by right linear grammar? Give an example.
25. Construct grammar for the following:
    (i) For generating the set of strings over $\{a, b, c\}$ whose length is divisible by 3.
    (ii) Strings over $\{a, b\}$, such that every a is immediately followed by b.
    (iii) Strings over $\{a, b\}$, having unequal number of a's and b's.
    (iv) The set of all the strings over $\{a, b\}$ with exactly twice as many a's as b's.
    (v) The set of all the strings over $\{a, b\}$ such that the length of each string is odd.
    (vi) The set of all the strings over $\{a, b\}$ such that each string starts and ends with the same symbol.
26. Design a CFG for generating the language $L = \{a^i b^j c^k, j \neq i + k\}$.
27. (a) Convert the grammar for each of the 3 parts below into an NDFA.
    (b) Convert the NDFA you obtained in (a) to a DFA.
    (i) $S \rightarrow 0A|0C|1B, A \rightarrow 0A|1C|0B, B \rightarrow 0C|1B, C \rightarrow 0C|1C|\Lambda$
    (ii) $S \rightarrow 0A|0C, A \rightarrow 0B, B \rightarrow \Lambda, C \rightarrow 1D, D \rightarrow 0E|B, E \rightarrow 1F, F \rightarrow 0D$
    (iii) $S \rightarrow 0S|X, X \rightarrow 1Y, Y \rightarrow 0|S|1X$
28. For each of the following regular expressions:
    (a) Convert the regular expression into an NDFA.
    (b) Convert the NDFA you obtained in (a) to a DFA.
    (i) $(00 + 0)^* (00 + 1)^*$
    (ii) $(1 + 01 + 001)^* (\Lambda + 0 + 00)$
29. What do you mean by ambiguous language? Check the ambiguity of the grammar G with $P = \{S \rightarrow aAS, S \rightarrow a, A \rightarrow SS, A \rightarrow ba\}$, where S is the start symbol of G.
30. State the general form of transition function for a PDA.
31. Construct PDA for the following:
    (a) $L = \{0^{2i} 1^i | i, j \geq 0\}$, over $\{0, 1\}$
    (b) $L = \{0^m 1^n | n \neq m\}$, over $\{0, 1\}$
    (c) $L = \{0^m 1^n | n \leq m \leq 2n\}$, over $\{0, 1\}$
32. Construct a PDA M accepting the language $L = \{a^n b^m a^n | m, n \geq 1\}$.
33. Construct a PDA M accepting the strings over $\{0, 1\}$ in which equal number of 0s and 1s are present.
34. Give the general form of transition function for a TM, and justify that TM is more powerful than PDA.
35. Design a TM to accept the language $L = \{a^n | n \geq 0\}$.
36. Design a TM to accept the language $L = \{ww^R | w \in \{a, b\}^*, w^R \text{ is the reversal of } w\}$.
37. Design a TM to accept the language $L = \{a^n b^n c^n | n \geq 0\}$.
38. Design a TM to accept the language $L = \{a^n b^n | n \geq 0\}$.
39. Design a TM to add two positive integer numbers.
40. Design a TM to subtract two positive integer numbers.
41. Design a TM to multiply two positive integer numbers.
42. Design a TM to concatenate two strings.
43. State the implications of halting problem.
44. What do you mean by PCP? Explain with example.
45. State which of the sentences given below are correct, false, or unknown:
    (i) If a problem is in P, it must also be in NP.
    (ii) If a problem is in NP, it must also be in P.
    (iii) If a problem is NP-complete, it must also be in NP.
    (iv) If a problem is not in P, it must be NP-complete.

ANNEXURE

Designing DFA for the problems (with a small variation)

EXAMPLE A.1 Design DFA over $\Sigma = \{a, b\}$ for the following:

1. String with last two symbols are b
2. String with substring bb (i.e., string with two consecutive b's)
3. String in which each b appears by a pair (i.e., bb)
4. String which does not contain two consecutive b's

Solution

1. State transition graph

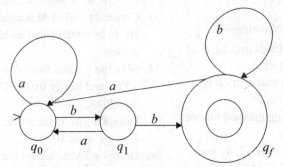

FIGURE A.1 DFA for the problem A.1 (1)

Figure A.1 shows that all possible combinations consisting of substring: bb at the end (even string consisting of bb and then bb at the end) are accepted. The RE corresponding to this problem is $(a + b)^* bb$.

2. State transition graph

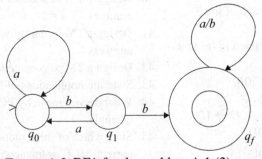

FIGURE A.2 DFA for the problem A.1 (2)

Figure A.2 shows that whenever the *first* substring bb is found in a string, then that string is accepted, and obviously any subsequent combination of a and b are certainly supposed to be accepted. The RE corresponding to this problem is $(a + b)^* bb(a + b)^*$.

3. State transition graph

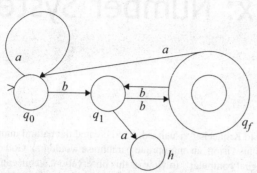

FIGURE A.3 DFA for the problem A.1 (3)

Figure A.3 suggests that a string is accepted if and only if b's appear in the string by a *pair*. For example, *bb*, *bbbb*, *bbabb* but not like *bbbabb*, etc. Therefore, if a is inputted at state q_1, then there is no chance to get a pair for the input b (taken immediately before the input a). That is why the FA then goes to *halt* state. The RE corresponding to this problem can be written as $(a^*bba^*)^+$.

4. State transition graph

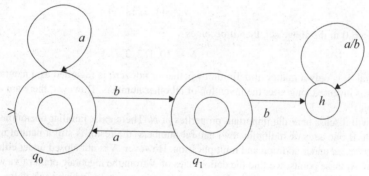

FIGURE A.4 DFA for the problem A.1 (4)

Figure A.4 ensures that the FA does not accept any string consisting of two consecutive b's.

Appendix: Number System

A German mathematician Leopold Kronecker visualized, 'God created the natural numbers, and all the rest is the work of man'. In response to his vision, an appropriate paraphrase would be, 'God created 0 and 1, and the rest is the performance of the digital computer'. In view of this observation, a conception of number structure should give due attention to the basic 'data pattern' of mathematics and computer science.

A.1 NUMBER STRUCTURE

Among the data patterns, the number structure/system is the most important constituent of computer. In this section, we introduce them, indicating their significance and the reasons towards their requirement in the process.

A.1.1 Natural Numbers

We start with *natural numbers N* (or *positive integers*), i.e., the set

$$N = \{1, 2, 3, \ldots\}$$

If we insert 0 in the above set, then it becomes

$$N = \{0, 1, 2, 3, \ldots\}$$

In general, it is only a matter of definition whether or not zero is regarded as a natural number, and it is true that it was invented long after the invention of all other numbers. However, there are authentic reasons for its inclusion.

We will discuss here the important properties of N. There exist familiar operations of addition and multiplication. If one adds or multiplies two natural numbers, the result is still a natural number, i.e., we can say that N is *closed* under addition and multiplication. However, N is not closed under either subtraction or multiplication. At these points, we find the deficiencies of N from the algebraic point of view. These lacking may be remedied in the larger systems Z (the integers) and Q (the rationals), which leads to the extension to Z and Q.

A.1.2 Integers

Next, we consider the system Z, the set of all integers, positive, negative, or zero

$$Z = \{\ldots, -2, -1, 0, 1, 2, \ldots\}$$

(the letter Z comes from the word 'Zahlen', which means 'the number' in German). The reason behind the extension from N to Z is to make subtraction always possible. This means that we may add, subtract, or multiply any two members of Z and the result belongs to Z (although we still cannot perform division in all cases).

A.1.3 Rational Numbers

The set of rational numbers, or fractions, is represented by Q. The system Q comprises all those numbers that can be written in the form a/b, with both a and b in Z (and $b \neq 0$).

By the extension, irrespective of addition, subtraction, and multiplication, we are able to divide as well, provided the divisor is non-zero. We generally avoid division by zero; but why? To search the background, assume that $x/y = z$ must be equivalent to $y \times z = x$. Actually, division is meant to be an 'inverse' to multiplication, implying that x/y should be that number (if any) which when multiplied by y, gives x. Suppose then that $x/0 = z$. This would therefore mean that $x = 0 \times z = 0$; so division by 0 could be performed only if x were itself equal to 0. But even in this case we could not obtain a sensible answer. Since $0 \times p = 0$ for any p at all, we could not decide which value to assign to 0/0, and it must therefore be regarded as meaningless.

A.1.4 Real Numbers

The real numbers are meant to represent all the points on a line, i.e., referred to 'the real line', which is shown in Figure A.1. Although the rational numbers appear to be very densely packed along the real line, there exist still many discontinuities (or gaps) in the sense that there are lengths which we can physically apprehend but which will correspond to no rational numbers.

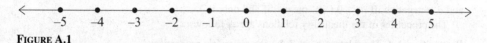

FIGURE A.1

For example The numbers $\sqrt{2}$ and π, respectively, represent the length of the diagonal of a square whose sides are of unit length and the length of the circumference of a circle of unit diameter, as shown in Figure A.2. Also, another example is e, the base of natural logarithms, whose geometrical intuition is not clear.

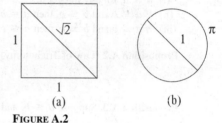

FIGURE A.2

Let us consider that R is the set of real numbers as consisting of all infinite recurring or non-recurring decimals, such as

$$512.237\,988\,032\ldots$$

Following this approach, we should ensure that certain pairs are also identified in which all these correspond to terminating decimals. For instance,

$$512.379\,999\,99\ldots = 512.380\,000\,00\ldots$$

A.1.5 Complex Numbers

The set C of complex numbers are represented by the set of all numbers of the form $a + ib$, where a and b are real numbers and $i^2 = -1$.

Complex numbers are represented in a computer as ordered pairs of real numbers. Thus, $a + ib$ is represented as (a, b), and we should ensure that addition and multiplication are performed according to the following rules:

$$(a, b) + (c, d) = (a + c, b + d)$$
$$(a, b) \times (c, d) = (ac - bd, ad + bc)$$

A.1.6 Computer Application of Numbers

With regard to computer implementation of numbers, the distinction between rationals and reals is not valid under the decimal representation. Since a computer is only a finite machine, it can hold only finitely many places of any real number expansion. Thus, one will normally have data types INTEGER and REAL and not RATIONAL (one should have RATIONAL in place of REAL, since one actually represents a rational approximation to the real). But real numbers play a vital role in any proper understanding of mathematical analysis, and are therefore indispensable for the theoretical development of the subject.

For example An illustration is given by the *intermediate value theorem*, which states that if f is a continuous function from R to R (considering positive and negative values), f takes the value 0 somewhere in between. Let us assume that $f(x) = x^2 - 2$. Then, $f(0) = -2$ and $f(2) = 2$, i.e., the hypotheses are satisfied. The function f cuts the x-axis at $x = \sqrt{2}$, lies in between $x = 0$ and $x = 2$. Thus, the theorem is true for $x = \sqrt{2}$. The same applies (with other functions f) to any real number.

A.2 INEQUALITIES

Let a and b be integers. We say a is *less than* b, which is written as $a < b$. If the difference $b - a$ is positive, i.e., if $b - a$ belongs to N.

The following notations should be described under inequalities:

$a > b$, means $b < a$ i.e., literally a is greater than b
$a \leq b$, means $a < b$ or $a = b$ i.e., literally a is less than or equal to b
$a \geq b$, means $b \leq a$ i.e., literally a is greater than or equal to b

The relations $<, >, \leq$, and $\geq$ are called *inequalities* in order to distinguish them from the relation $=$ of equality.

It may be noted that $a < b$ if and only if a lies to the left of b on the real number line (Figure A.1). For example, $2 < 5; -5 < -4; 3 \leq 3; 5 > -6; 6 \geq 0$; and $-8 \leq 0$. We also note that a is positive iff $a > 0$, and a is negative iff $a < 0$ ('iff' means 'if and only if').

The properties of the inequality relations are as follows:

Proposition A.1 The relation $\leq$ in Z has the following properties:

(i) $a \leq a$, for any integer a.
(ii) If $a \leq b$ and $b \leq a$, then $a = b$.
(iii) If $a \leq b$ and $b \leq c$, then $a \leq c$.

Proposition A.2 (Law of Trichotomy) For any integers a and b, exactly one of the following rules holds:

$$a < b, \quad a = b, \quad \text{or} \quad a > b$$

Proposition A.3 Suppose $a \leq b$, and let c be any integer. Then

(i) $a + c \leq b + c$
(ii) $ac \leq bc$ when $c > 0$; but $ac \geq bc$ when $c < 0$

Proof The proposition is true when $a = b$. Hence, we need to consider the case when $a < b$, i.e., when $b - a$ is positive.

(i) The following difference is positive:

$$(b + c) - (a + c) = b - a$$

Hence, $a + c < b + c$.

(ii) Let us assume that c is positive. By the property of the positive integers N (i.e., if a and b belong to N, $a + b$ and ab belong to N), the following product is also positive:

$$c(b - a) = bc - ac$$

Thus, $ac < bc$. Now, if c is negative, $-c$ is positive, and the following product is also positive:

$$(-c)(b - a) = ac - bc$$

Accordingly, $bc < ac$, whence $ac > bc$.

A.3 ABSOLUTE VALUE

The *absolute value* of an integer a, written as $|a|$, is defined by

$$|a| = \begin{cases} a, & \text{if } a \geq 0 \\ -a, & \text{if } a < 0 \end{cases}$$

From the geometric point of view, $|a|$ may be viewed as the distance between the points a and 0. Also, $|a-b| = |b-a|$ may be viewed as the distance between the points a and b.

For example

(i) $|-2| = 2, |5| = 5,$ and $|-11| = 11$
(ii) $|3 - 7| = |-4| = 4$ and $|7 - 3| = |4| = 4$
(iii) $|-2 - 7| = |-9| = 9$

Proposition A.4 Let a and b be any integers, then

(i) $|a| \geq 0$ and $|a| = 0$ iff $a = 0$
(ii) $-|a| \leq a \leq |a|$
(iii) $|ab| = |a\|b|$
(iv) $|a \pm b| \leq |a| \leq |b|$
(v) $\|a| - |b\| \leq |a \pm b|$

Proof The proof of property (iii) can be made by the following case studies:

Case I Let us assume that $a = 0$ or $b = 0$. Then $|a| = 0$ or $|b| = 0$ and so $|a\|b| = 0$. Also, $ab = 0$. Hence $|ab| = 0 = |a\|b|$.
Case II Let us assume that $a > 0$ and $b > 0$. Then $|a| = a$ and $|b| = b$. Hence $|ab| = ab = |a\|b|$.
Case III Let us assume that $a > 0$ and $b < 0$. Then $|a| = a$ and $|b| = -b$. Also, $ab < 0$. Hence $|ab| = -(ab) = a(-b) = |a\|b|$.
Case IV Let us assume that $a < 0$ and $b > 0$. Then $|a| = -a$ and $|b| = b$. Also, $ab < 0$. Hence $|ab| = -(ab) = (-a)b = |a\|b|$.
Case V Let us assume that $a < 0$ and $b < 0$. Then $|a| = -a$ and $|b| = -b$. Also, $ab > 0$. Hence $|ab| = ab = (-a)(-b) = |a\|b|$.

The proof of property (iv) is given as follows: Here, $ab \leq |ab| = |a\|b|$, and so $2ab \leq 2|a\|b|$. Hence $(a + b)^2 = a^2 + 2ab + b^2 \leq |a|^2 + 2|a\|b| + |b|^2 = (|a| + |b|)^2$. But, $\sqrt{(a+b)^2} = |a+b|$. Thus, the square root of the above expression yields

$$|a + b| \leq |a| + |b| \tag{A.1}$$

Also,

$$|a - b| = |a + (-b)| \leq |a| + |-b| = |a| + |b| \tag{A.2}$$

Combining Eqs (A.1) and (A.2),

$$|a \pm b| \leq |a| + |b|$$

A.4 DIVISION ALGORITHM

The fundamental property of arithmetic, enunciated below, is a result of long division.

Theorem A.1 (Division Algorithm) Let a and b be any integers with $b \neq 0$. Then there exists unique integers q and r such that $a = bq + r$ and $0 \leq r < |b|$.

Proof Let M be the set of non-negative integers of the form $a - xb$ for some integer x. If $x = -|a|b$, then $a - xb$ is non-negative; hence, M is non-empty. Assume that there exists a least element $r \in M$. Then, we have $r \geq 0$ and $r = a - qb$. For some integer q, we need only to show that $r < |b|$.

Let us assume that $r \geq |b|$. Let $r' = r - |b|$. Then, $r' \geq 0$ and also $r' < r$ because $b \neq 0$. Furthermore,

$$r' = r - |b| = a - qb - |b| = \begin{cases} a - (q + 1)b, & \text{if } b < 0 \\ a - (q - 1)b, & \text{if } b < 0 \end{cases}$$

In either case, $r' \in M$. This contradicts the fact that r is the least element of M. Accordingly, $r < |b|$. Thus, there exist q and r.

We now show that q and r are unique.

Let us assume that there exist integers q, r, q', and r' such that

$$a = bq + r \quad \text{and} \quad a = bq' + r' \quad \text{and} \quad 0 \leq r, r' < |b|$$

Then,
$$bq + r = bq' + r' \Rightarrow b(q-q') = r'-r$$

Thus, b divides $r' - r$. But $|r' - r| < |b|$, since $0 \le r, r' < |b|$. Accordingly, $r' - r = 0$. This implies that $q - q' = 0$ since $b \ne 0$. Consequently, $r' = r$ and $q' = q$, i.e., q and r are uniquely determined by a and b.

The number q in the above theorem is called the *quotient* and r is called the *remainder*.

A.4.1 Divisibility

Let a and b be any integers with $a \ne 0$. Let us assume an integer c such that, $ac = b$. We then say that a *divides* b or b is *divisible* by a and may be written as $a|b$. We may also say that b is a *multiple* of a or that a is a *factor* or *divisor* of b. If a does not divide b, we then write $a \nmid b$.

For example

(i) The division $2|4$ is true, since $2 \cdot 2 = 4$; and $-3|15$ since $(-3)(-5) = 15$.
(ii) The divisors
 (a) of 1 are ± 1
 (b) of 2 are $\pm 1, \pm 2$
 (c) of 4 are $\pm 1, \pm 2, \pm 4$
 (d) of 5 are $\pm 1, \pm 5$
 (e) and so on
(iii) If $a \ne 0$, then $a|0$, since $a \cdot 0 = 0$

We now discuss the theorem of divisibility.

Theorem A.2 Let us assume that a, b, and c are integers. Then,

(i) If $a|b$ and $b|c$, then $a|c$.
(ii) If $a|b$ then, for any integer x, $a|bx$.
(iii) If $a|b$ and $a|c$, then $a|(b + c)$ and $a|(b-c)$.
(iv) If $a|b$ and $b \ne 0$, then $a = \pm b$ or $|a| < |b|$.
(v) If $a|b$ and $b|a$ then $|a| = |b|$, i.e., $a = \pm b$.
(vi) If $a|1$, then $a = \pm 1$.

Proof

(i) If $a|b$ and $b|c$, then there exist integers x and y such that $ax = b$ and $by = c$. Replacing b by ax, we obtain $axy = c$. Hence, $a|c$.
(ii) If $a|b$, then there exists an integer c such that $ac = b$. Multiplying the equation by x, we obtain $acx = bx$. Hence, $a|bx$.
(iii) If $a|b$ and $a|c$, then there exist integers x and y such that $ax = b$ and $ay = c$, and adding them we obtain $ax + ay = b + c$ and $a(x + y) = b + c$. Hence, $a|(b + c)$. Again, subtracting them we obtain $ax - ay = b - c$ and $a(x - y) = b - c$. Hence, $a|(b - c)$.
(iv) If $a|b$, then there exists an integer c such that $ac = b$. Then $|b| = |ac| = |a||c|$.
 Now, if $|c| = 1$, then $|a| < |a||c| = |b|$, i.e., $|a| < |b|$. Again, if $|c| = 1$, then $c = \pm 1$, which implies that $a = \pm b$, as required.
(v) If $a|b$, then $a = \pm b$ or $|a| < |b|$. If $|a| < |b|$, $b \nmid a$. Hence, $a = \pm b$.
(vi) If $a|1$, then $a = \pm 1$ or $|a| < |1| = 1$. Also, $|a| > 1$. Hence, $a = \pm 1$.

A.5 PRIMES

One important concept based on divisibility is that of a prime number. A *prime* is an integer greater than 1, i.e., divisible only by 1 and itself. The study of prime numbers takes us back to ancient times.

Every positive integer greater than 1 is divisible by at least two integers, because a positive integer is divisible by 1 and by itself. Positive integers that have exactly two different positive integer factors are called *primes*.

A positive integer $p > 1$ is called prime if the only positive factors of p are 1 and p. A positive integer that is greater than 1 and is not prime is called as *composite*.

An important theorem, the Fundamental Theorem of Arithmetic, asserts that every positive integer can be expressed uniquely as the product of prime numbers. The length (or duration) to factor large integers into their prime factors plays an important role in cryptography.

Note The integer p is composite if and only if there exists an integer a such that $a|p$ and $a < a < p$.

For example

(i) The primes less than 50 are 2, 3, 5, 7, 11, 13, 17, 19, 23, 29, 31, 37, 41, 43, 47
(ii) Although 21, 24, and 1729 are not primes, yet each can be written as a product of primes:

$$21 = 3 \cdot 7, 24 = 2 \cdot 2 \cdot 2 \cdot 3 = 2^3 \cdot 3, \text{ and } 1729 = 7 \cdot 13 \cdot 19$$

EXAMPLE A.1 Show that 43 is a prime.

Solution Let us assume that $n = 43$. Since $6 < \sqrt{43} < 7$, and 2, 3, and 5 are the primes less than 6, so 43 is not divisible by 2, 3, and 5 and hence 43 must be a prime.

A.6 GREATEST COMMON DIVISOR

Let us assume that a and b are integers that are not equal to 0. An integer p is called a *common divisor* of a and b if p divides both a and b, i.e., if $p|a$ and $p|b$. It may be noted that 1 is a positive common divisor of a and b, and that any common divisor of a and b cannot be greater than $|a|$ or $|b|$. Thus, there exists the largest common divisor of a and b, which is denoted by gcd (a, b) and it is called the *greatest common divisor* (gcd) of a and b.

For example

(i) The common divisors of 12 and 18 are $\pm 1, \pm 2, \pm 3, \pm 6$. Thus, gcd $(12, 18) = 6$. Similarly, gcd $(12, -18) = 6$, gcd $(12, -16) = 4$, gcd $(29, 15) = 1$, and gcd $(14, 49) = 7$.
(ii) For any integer a, we have gcd $(1, a) = 1$.

A.6.1 Euclidean Algorithm

We will demonstrate here an efficient method for finding the gcd, called the *Euclidean algorithm* named by a Greek mathematician, Euclid.

Let us illustrate the method by evaluating gcd (540, 168).

First, divide 540, the larger of the two integers, by 168, the smaller, to obtain $540 = 168 \cdot 3 + 36$. Any divisor of 168 and 540 must also be a divisor of $540 - 168 \cdot 3 = 36$. Also, any divisor of 168 and 36 must also be a divisor of $540 = 168 \cdot 3 + 36$. Hence, the gcd of 540 and 168 is the same as the gcd of 168 and 36. This means that the evaluation of gcd (540, 168) is now being reduced to the evaluation of gcd (168, 36).

Next, divide 168 by 36 to obtain $168 = 36 \cdot 4 + 24$. Since any common divisor of 168 and 36 also divides $168 - 36 \cdot 4 = 24$ and any common divisor of 36 and 24 divides 168, it follows that gcd $(168, 36) = $ gcd$(36, 24)$.

The process is continued by dividing 36 by 24, to obtain $36 = 24 \cdot 1 + 12$ and thus, gcd $(36, 24) = 12$. Again, 12 divides 24, it follows that gcd $(24, 12) = 12$. Thus, gcd $(540, 168) = $ gcd $(168, 36) = $ gcd $(36, 24) = $ gcd $(24, 12) = 12$ and hence solved.

A.7 LEAST COMMON MULTIPLE

Let us assume that a and b are non-zero integers. It may be noted that $|ab|$ is a positive common multiple of a and b. Thus, there exists the smallest positive common multiple of a and b, denoted by lcm (a, b) and it is called the *least common multiple* (lcm) of a and b.

For example

(i) lcm $(2, 3) = 6$, lcm $(4, 6) = 12$, and lcm $(9, 10) = 90$.
(ii) For any positive integer a, we have lcm $(1, a) = a$.
(iii) For any prime p and any positive integer a, lcm $(p, a) = a$ or lcm $(p, a) = ap$, according to $p|a$ or $p \nmid a$.

Note Suppose a and b are non-zero integers, then lcm (a, b) can also be obtained from

$$\text{lcm }(a, b) = \frac{|ab|}{\gcd(a, b)}$$

A.8 FUNDAMENTAL THEOREM OF ARITHMETIC

In this section, we will discuss the fundamental theorem of arithmetic, for which it is necessary to have the exposure of relatively prime integers.

A.8.1 Relatively Prime Integers

Two integers a and b are said to be *relatively prime* or *coprime* if $\gcd(a, b) = 1$. Accordingly, if a and b are relatively prime, then there exists integers x and y such that $ax + by = 1$. Conversely, if $ax + by = 1$, then a and b are relatively prime.

For example
 (i) It may be seen that $\gcd(12, 35) = 1$, $\gcd(49, 18) = 1$, $\gcd(21, 64) = 1$ and $\gcd(-28, 45) = 1$.
 (ii) If p and q are distinct primes, then $\gcd(p, q) = 1$.

The relation of being relatively prime is particularly important because of the following theorem.

Theorem A.3 Let us assume that $\gcd(a, b) = 1$, and a and b both divide c. Then ab divides c.

Proof Since $\gcd(a, b) = 1$, there exist x and y such that $ax + by = 1$. Since $a|c$ and $b|c$, there exist m and n such that $c = ma$ and $c = nb$. Multiplying $ax + by = 1$ by c, we obtain $acx + bcy = c$ or $a(nb)x + b(ma)y = c$ or $ab(nx + my) = c$. Thus, ab divides c.

Corollary A.1 Let us assume that a prime p divides a product ab. Then, $p|a$ or $p|b$. In fact, the above corollary is the basis of the proof of the Fundamental Theorem of Arithmetic.

Theorem A.4 (Fundamental Theorem of Arithmetics) Every integer $n > 1$ can be expressed uniquely as a product of primes.

Proof First of all, we will show that products of primes exist. To perform this, let us use the method of induction.

Assume $n = 2$. Since 2 is a prime, n is a product of primes. Suppose $n > 2$, and the theorem holds for positive integers less than n. If n is prime, then n is a product of primes. If n is composite (i.e., if $n > 1$ is not prime, then n is composite), then $n = ab$, where $a, b < n$. By induction, a and b are products of primes. Hence, $n = ab$ is also a product of primes.

Next, we need to show that such a product is unique.

Suppose $n = p_1 p_2 \ldots p_k = q_1 q_2 \ldots q_r$, where the p's and q's are primes. Also, $p_1 | (q_1 q_2 \ldots q_r)$. Now, we arrange the q's so that $p_1 = q_1$. Then, $p_1 p_2 \ldots p_k = p_1 q_2 \ldots q_r$ and so $p_2 \ldots p_k = p_k = q_2 \ldots q_r$. Similarly, we can rearrange the remaining q's so that $p_2 = q_2$, and so on. Thus, n can be expressed uniquely as a product of primes.

EXAMPLE A.2 Find integers x and y such that $37x + 249y = 1$.

Solution To determine the values of x and y we first determine the gcd (37, 249), the procedure of which is illustrated below:

$$249 = 6.37 + 27$$
$$37 = 1.27 + 10$$
$$27 = 2.10 + 7$$
$$10 = 1.7 + 3$$
$$7 = 2.3 + 1$$
$$3 = 1.3 + 0$$

which shows that gcd (37, 249) = 1. Thus,

$$1 = 7 - 2 \cdot 3 = 7 - 2 \cdot (10 - 1 \cdot 7) = 3 \cdot 7 - 2 \cdot 10 = 3 \cdot (27 - 2 \cdot 10) - 2 \cdot 10 = 3 \cdot 27 - 8 \cdot 10$$
$$= 3 \cdot 27 - 8 \cdot (37 - 1 \cdot 27) = 11 \cdot 27 - 8 \cdot 37 = 11 \cdot (249 - 6 \cdot 37) - 8 \cdot 37$$
$$= 11 \cdot 249 - 74 \cdot 37 = 37 \cdot (-74) + 249 \cdot 11$$

which is of the form of $1 = 37x + 249y$, i.e., $x = -74$ and $y = 11$.

A.9 CONGRUENCE RELATION

Let m be a positive integer. We say that a is *congruent* to b modulo m, written as $a \equiv b$ (modulo m) or $a \equiv b \pmod{m}$ if m divides the difference $a - b$. The integer m is called the *modulus*. The negation of $a \equiv b \pmod{m}$ is written as $a \not\equiv b \pmod{m}$.

For example

(i) $67 \equiv 1 \pmod 6$ since 6 divides $67 - 1 = 66$
(ii) $72 \equiv -5 \pmod 7$ since 7 divides $72 - (-5) = 77$
(iii) $27 \not\equiv 8 \pmod 9$ since 9 does not divide $27 - 8 = 19$

Suppose m is positive, and a is any integer. By the division algorithm, there exist integers q and r with $0 \le r < m$ such that $a = mq + r$. Hence, $mq = a - r$ or $m|(a-r)$ or $a \equiv r \pmod m$.

Accordingly,

(i) Any integer a is congruent modulo m to a unique integer in the set $\{0, 1, 2, \ldots, m-1\}$. (The uniqueness yields from the fact that m cannot divide the difference of two such integers).
(ii) Any two integers a and b are congruent modulo m if and only if they have the same remainder when divided by m.

A.10 CONGRUENCE EQUATIONS

A *polynomial congruence equation* or, simply, a *congruence equation* (in one unknown x) is an equation of the form

$$a_n x^n + a_{n-1} x^{n-1} + \cdots + a_1 + a_0 \equiv 0 \pmod{m} \tag{A.3}$$

Such an equation is said to be of *degree n* if $a_n \not\equiv 0 \pmod m$.

Let us assume that $s \equiv t \pmod m$. Then, s is a solution of Eq. (A.3) if and only if t is a solution of Eq. (A.3). Thus, the *number of solutions* of Eq. (A.3) is defined to be the number of incongruent solutions or, equivalently, the number of solutions in the set $\{0, 1, 2, \ldots, m-1\}$.

The *complete set of solutions* of Eq. (A.3) is a maximum set of incongruent solutions, whereas the *general solution* of Eq. (A.3) is the set of all integral solutions of Eq. (A.3). The *general solution* of Eq. (A.3) can be found by adding all the multiple of the modulus m to any complete set of solutions.

EXAMPLE A.3 Consider the equations

(i) $x^2 + x + 1 \equiv 0 \pmod 4$
(ii) $x^2 + 3 \equiv 0 \pmod 6$
(iii) $x^2 - 1 \equiv 0 \pmod 8$

Find the solutions of the equations.

Solution

(i) The given equation possesses no solution, since 0, 1, 2, and 3 do not satisfy it.
(ii) There exists only one solution among $0, 1, \ldots, 5$ which is 3. Thus, the general solution consists of the integers $3 + 6k$ where $k \in Z$.
(iii) There are four solutions 1, 3, 5, and 7. This shows that a congruence equation of degree n can have more than n solutions.

Sometimes it is needed to seek a solution of simultaneous congruence equations like $x \equiv 2 \pmod 3$ and $x \equiv 4 \pmod 5$, which means that search a number which when divided by 3 and 5 leaves remainder 2 and 4, respectively. Observe that the moduli 3 and 5 are pairwise relatively prime. This idea is governed by a theorem known as Chinese Remainder Theorem, which is essentially a system of linear congruences, as discussed below. The theorem expresses that there exists a unique solution modulo $M = 3 \cdot 5 = 15$.

A.10.1 Chinese Remainder Theorem

Let $m_1, m_2, \ldots, m_k$ be pairwise relatively prime positive integers. Then the system of congruences

$$x \equiv r_1 \,(\text{mod}\, m_1), x \equiv r_2 \,(\text{mod}\, m_2), \ldots, x \equiv r_2 \,(\text{mod}\, m_2) \tag{A.4}$$

where r_i, given integers, possesses a unique solution modulo $M = m_1 m_2 \ldots m_k$.

Proof Consider the integer

$$x_0 = M_1 s_1 r_1 + M_2 s_2 r_2 + \cdots + M_k s_k r_k$$

where $M_i = M/m_i$ and s_i is the unique solution of $M_i x \equiv 1 (\text{mod}\, m_1)$. Assume j is given. Then, for $i \neq j$, we have m_j/M_j and hence

$$M_i s_i r_i \equiv 0 \,(\text{mod}\, M_j)$$

On the other hand, $M_j s_j \equiv 1 \,(mod\, m_j)$ and hence

$$M_j s_j r_j \equiv r_j \,(\text{mod}\, m_j)$$

Accordingly,

$$x_0 = 0 + \cdots + 0 + r_j + 0 + \cdots + 0 \equiv r_j \,(\text{mod}\, m_j)$$

In other words, x_0 is a solution of each of the equations in Eq. (A.4).

Now, it is to be shown that x_0 is the unique solution of the system (A.4) modulo M. Assume x_1 is another solution of all the equations in Eq. (A.4). Then

$$x_0 \equiv x_1 \,(\text{mod}\, m_1), x_0 \equiv x_1 (\text{mod}\, m_2), \ldots, x_0 \equiv x_1 \,(\text{mod}\, m_k)$$

Hence, $m_i/(x_0 - x_1)$, for each i. Since the m_i are relatively prime, $M = 1 \text{ cm } (m_1, m_2, \ldots, m_k)$ and so $M/(x_0 - x_1)$, i.e.,

$$x_0 \equiv x_1 \,(\text{mod}\, M)$$

Thus, the theorem is proved.

EXAMPLE A.4 Solve the following simultaneous congruence equations:

$$x \equiv 2 \,(\text{mod}\, 3), \quad x \equiv 3 \,(\text{mod}\, 5), \quad x \equiv 2 \,(\text{mod}\, 7)$$

Solution Let $m = 3 \cdot 5 \cdot 7 = 105$, $M_1 = m/3 = 35$, $M_2 = m/5 = 21$, and $M_3 = m/7 = 15$. We see that 2 is an inverse of $M_1 = 35$ modulo, since $35 \equiv 2 \,(\text{mod}\, 3)$; 1 is an inverse of $M_2 = 21$ modulo 5, because $21 \equiv 1 \,(\text{mod}\, 5)$; and 1 is an inverse of $M_3 = 15 \,(\text{mod}\, 7)$, because $15 \equiv 1 \,(\text{mod}\, 7)$. The solutions to this system are those x such that

$$x = a_1 M_1 y_1 + a_2 M_2 y_2 + a_3 M_3 y_3$$
$$= 2 \cdot 35 \cdot 2 + 3 \cdot 21 \cdot 1 + 2 \cdot 15 \cdot 1 = 233 \equiv 23 \,(\text{mod}\, 105)$$

It follows that 23 is the smallest positive integer that is a simultaneous solution. We conclude that 23 is the smallest positive integer that leaves a remainder of 2 when divided by 3, a remainder of 3 when divided by 5 and a remainder 2 when divided by 7. Thus, the value of x is the smallest positive integer 23.

Answers to Exercises

CHAPTER 1

1. Subsets of A: ϕ, $\{a\}$, $\{b\}$, $\{c\}$, $\{a,b\}$, $\{a,c\}$, $\{b,c\}$, $\{a,b,c\}$; number of subsets of A: 8 **2.** (i) $A = \{-2, -1, 0, 1, 2\}$ (ii) $B = \{1, 4, 9, 16\}$ (iii) $C = \{-2, 2\}$ (iv) $D = \{1, 2, 3, 4, 5, 6, 10, 12, 15, 20, 30, 60\}$ **3.** (i) $A = \{x : x$ is an even integer between 1 and 11$\}$ (ii) $B = \{x : x$ is an odd integer between 2 and 90$\}$ (iii) $C = \{x : x$ is an integer, $-5 < x < 2\}$ (iv) $D = \{x : x = n^2, n$ is a natural number $\leq 6\}$ **4.** Finite sets: (i) The set of students in a class. (ii) The set of months in a year, etc. Infinite sets: (i) $A = \{1, 1/2, 1/3, 1/4, 1/5, \ldots\}$ (ii) $B = \{0, 1, 2, \ldots\}$ etc. **5.** (i) $A \cup B = \{1, 2, 3, 4, 6, 9\}$ (ii) $B \cap C = \{4, 6\}$ (iii) $A \cap C = \{3, 4\}$ (iv) $A \cap (B \cap C) = \{4\}$ (v) $A \cap (B \cup C) = \{2, 3, 4\}$ **6.** (i) 330 (ii) 150 (iii) 80 **7.** (i) $\{(1, 4), (1, 5), (1, 7), (4, 4), (4, 5), (4, 7)\}$ (ii) $\{(1, 5), (4, 5)\}$ **8.** (i) $\{\phi, \{1\}, \{2\}, \{1, 2\}\}$ (ii) $\{\phi, \{\{a\}\}, \{\{b\}\}, \{\{a, b\}\}\}$ (iii) $\{\phi, \{a, b\}, \{c\}, \{(a, b), c\}\}$ (iv) $\{\phi, \{1\}, \{3\}, \{\{1, 2, 3\}\}, \{1, \{1, 2, 3\}\}, \{1, 3\}, \{\{1, 2, 3\}, 3\}, \{1, 3, \{1, 2, 3\}\}\}$ **9.** $R = \{(2, 4), (2, 6), (3, 3), (3, 6), (4, 4)\}$, Dom $(R) = \{2, 3, 4\}$ and Ran $(R) = \{3, 4, 6\}$ **10.** $R = \{(1, 1), (1, 2), (1, 3), (1, 4), (2, 2), (2, 3), (2, 4), (3, 3), (3, 4), (4, 4)\}$ the domain and range of $R = \{1, 2, 3, 4\}$ and both are equal to A **11.** $R' = \{(x, Z), (y, X), (y, Y), (z, X), (z, Y), (z, Z)\}$ $R \cup S = \{(x, X), (x, Y), (y, Z)\}$, $R \cap S = \{(x, Y), (y, Z)\}$, $R - S = \{(x, X)\}$ **13.** Symmetric closure $(R) = \{(4, 5), (5, 4), (5, 5), (5, 6), (6, 5), (6, 7), (7, 6), (7, 4), (4, 7), (7, 7)\}$ **14.** (i) *Composition of relation R*: Reflexive closure $(R) = \{(1, 1), (1, 2), (2, 2), (2, 3), (3, 1), (3, 3)\}$, Symmetric closure $(R) = \{(1, 2), (2, 1), (2, 3), (3, 1), (3, 2), (1, 3)\}$, Transitive closure $(R) = \{(1, 1), (1, 2), (1, 3), (2, 1), (2, 2), (2, 3), (3, 1), (3, 2), (3, 3)\}$ (ii) *Composition of relation R*: Reflexive closure $(R) = \{(1, 1), (1, 2), (2, 2), (2, 3), (3, 1), (3, 3)\}$, Symmetric closure $(R) = \{(1, 2), (1, 3), (2, 1), (2, 3), (3, 1), (3, 2)$, Transitive closure $(R) = \{(1, 1), (1, 2), (1, 3), (2, 1), (2, 2), (2, 3), (3, 1), (3, 2), (3, 3)\}$ (iii) *Graphical representation of R*:

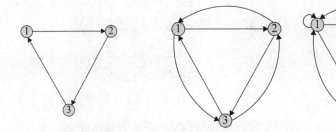
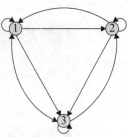

Graphical representation of R Reflexive closure   Symmetric closure   Transitive closure

15. $R^{-1} = \{(b, a), (c, a), (b, b), (c, b), (d, c), (d, d), (a, d), (b, d)\}$, the directed graph of R^{-1} is shown below. $R' = \{(a, a), (a, d), (b, d), (b, a), (c, a), (c, b), (c, c), (d, c)\}$, the directed graph of R' is shown below.

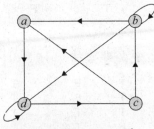
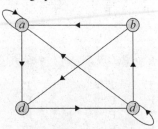

Directed graph of R^{-1} Directed graph of R'

16. The equivalence classes are concentric circles centred on the origin. **17.** (i) No (ii) No (iii) Yes **18.** (i) No (ii) No (iii) Yes **19.** Yes **20.** Yes **21.** (i) $f(3) = 6$ (ii) $f(12) = 29$ (iii) $f(-15) = -19$ (iv) $f^2(5) = 45$ **23.** f is an onto function **25.** $\Rightarrow f^{-1}(x) = 1/2(x+3)$ **26.** $(f \circ g)(x) = \sqrt{3x+1}$, $(g \circ f)(x) = 3\sqrt{x}+1$, $(f \circ g) \neq (g \circ f)$ **27.** $(f \circ g \circ h)(x) = (x^8+1)^{-3} - 4(x^8+1)^{-1}$, $(g \circ g)(x) = (1+x^2)^2/1 + (1+x^2)^2 (h \circ g \circ f)(x) = [(x^3-4x)^2+1]^{-4}$, $(g \circ h)(x) = (x^8+1)^{-1}$ **28.** $g(f(x)) = \begin{cases} 2(3x-2), x \leq 0 \\ (x^2+1), x > 0 \end{cases}$ **30.** $(g+f)(1) = s, (g+f)(2) = s, (g+f)(3) = r$ **32.** $f^{-1}(x) = (x+4)/3$ **33.** $f^{-1}(x) = (x-b)/a$

CHAPTER 2

1. 14 **2.** 456,976,000 **3.** 1296,360 **4.** 8 **5.** $n_1 + n_2 + n_3$ **6.** 15 **7.** 36 **8.** 5040 **9.** (i) 5040 (ii) 5040 (iii) 720 (iv) 1440 **10.** 4200 **12.** 604,800 **13.** (8! 8! 10!)/(4! 6!) **14.** (i) $^nC_2 - {}^nP_2 + 1$ (ii) $^nC_3 - {}^nP_3$ **15.** 485 **16.** $(n)/6 (n-4)(n-5)$ **17.** 1/5 **18.** 5/18 **19.** 1/2 **20.** $P(A) = 1/3$ or $1/4$ and $P(B) = 1/4$ or $1/3$ **21.** 1/2 **22.** 3/8 **23.** 24/29 **24.** 0.37 **25.** $a_r = (A_1 + A_2 r)3^r$ **26.** $a_r = A_1 3^r + A_2 4^r$ **27.** $a_r = A_1 + A_2 2^r$ **28.** $a_r^{(h)} = a_r^{(h)} = A_1(1/2)^r + A_2 2^r$; $a_r^{(p)} = -2/3(1/2)^r + 2/3(2)^r$ **29.** $a_r^{(h)} = A_1 2^r + A_2 5^r$; $a_r^{(p)} = 13(5)^r - 10(2)^r$ **30.** $a_r^{(p)} = (1/2)r^3 + r^2$ **31.** $a_r^{(p)} = (1 + r/6) r^2 2^r$ **33.** $a_r = C_1(-1)^r + C_2 \cdot 2^r - 1 - r/2 - r^2/2$ **34.** $a_r = A \cdot 1^r + r(r-1)(2r-1)/6$ **35.** $a_r = A_1 + A_2 r + (r^2 - 8r + 20)2^r$ **36.** $a_r = 3 - 5(r+1) + 4 \cdot 2^r$ **37.** $a_r = 1$ **38.** $a_r = 13(5)^r - 10(2)^r$ **39.** $A_0 r^2 3^r$; $(A_0 + A_1 r)r^2 3^r$; $(A_0 + A_1 r + A_2 r^2)3^r$ **40.** (i) 12 (ii) 18 (iii) 2

CHAPTER 3

1. (i) Proposition (ii) Not a proposition (iii) Not a proposition (iv) Proposition **2.** Conjunction: $p \wedge q$, Disjunction: $p \vee q$, Negation: $\sim p, \sim q, \sim (p \vee q), \sim p \wedge \sim q, \sim (\sim p)$ **3.** (i) It is not a warm day. (ii) It is false that it is a warm day or the temperature is 37°C. (iii) It is not true that it is a warm day and the temperature is 37°C. (iv) It is false that it is not a warm day. (v) It is a warm day or the temperature is 37°C. (vi) It is a warm day and the temperature is 37°C. (vii) It is neither a warm day nor the temperature is 37°C. (viii) It is false that it is not a warm day or the temperature is not 37°C. **4.** (i) $p \wedge \sim r$ (ii) $\sim p \wedge \sim q$ (iii) $\sim p \wedge \sim q$ (iv) $(p \vee q) \wedge \sim r$ (v) $\sim ((p \vee q) \wedge \sim r)$ (vi) $\sim (\sim r)$ (vii) $r \vee (p \wedge q)$ **5.** $(p \vee q) \rightarrow (r \leftrightarrow \sim s)$

6.

(i)

p	$\sim p$	$p \wedge \sim p$
T	F	F
F	T	F

(ii)

p	$\sim p$	$p \vee \sim p$
T	F	T
F	T	T

(ii)

p	$\sim p$	$\sim(\sim p)$
T	F	T
F	T	F

7.

(i)

p	q	$p \vee q$	$\sim(p \vee q)$
T	T	T	F
T	F	T	F
F	T	T	F
F	F	F	T

(ii)

p	q	$\sim q$	$p \vee \sim q$	$\sim(p \vee \sim q)$
T	T	F	T	F
T	F	T	T	F
F	T	F	F	T
F	F	T	T	F

(iii)

p	q	$p \wedge q$	$(p \wedge q) \vee (p \wedge q)$
T	T	T	T
T	F	F	F
F	T	F	F
F	F	F	F

(iv)

p	q	$p \vee q$	$\sim p$	$(p \vee q) \vee \sim p$
T	T	T	F	T
T	F	T	F	T
F	T	T	T	T
F	F	F	T	T

8. There exist people who are not intelligent. There exist students who are not graduate. **9.** (i) ~p: It is not cold (ii) $p \wedge q$: It is cold and raining (iii) $p \vee q$: It is cold or raining (iv) $p \vee \sim q$: It is cold or it is not raining **10.** $((p \Rightarrow q) \wedge (q \Rightarrow r)) \Rightarrow (p \Rightarrow r)$.

11.

p	$p \vee p$	$p \vee p \leftrightarrow p$
T	T	T
F	F	T

12.

p	q	$p \wedge q$	$\sim (p \wedge q)$	$(p \vee \sim (p \wedge q)$
F	F	F	T	T
F	T	F	T	T
T	F	F	T	T
T	T	T	F	T

13.

(i)

p	q	$\sim p$	$\sim q$	$p \wedge \sim q$	$\sim p \wedge q$	$(p \wedge \sim q) \vee (\sim p \wedge q)$
T	T	F	F	F	F	F
T	F	F	T	T	F	T
F	T	T	F	F	T	T
F	F	T	T	F	F	F

(ii)

p	q	$\sim p$	$\sim q$	$p \wedge q$	$\sim (p \wedge q)$	$\sim p \vee \sim q$	$\sim (p \vee q) \vee (\sim p \vee \sim q)$
T	T	F	F	T	F	F	F
T	F	F	T	F	T	T	T
F	T	T	F	F	T	T	T
F	F	T	T	F	T	T	T

14.

(i)

p	q	$p \to q$	$p \to (p \to q)$
T	T	T	T
T	F	F	F
F	T	T	T
F	F	T	T

(ii)

p	q	$q \to p$	$p \to (q \to p)$
T	T	T	T
T	F	T	T
F	T	F	T
F	F	T	T

(iii)

p	$\sim p$	$p \wedge \sim p$
T	F	F
F	T	F

15. The proposition r can be written in terms of p and q as $p \to (q \vee p)$. **17.** The argument is invalid.

18. (i)

p	q	r	$q \wedge r$	$p \vee (q \wedge r)$
T	T	F	F	T

(ii)

p	r	s	$r \wedge s$	$p \to (r \wedge s)$
T	F	F	F	F

$p \vee (q \wedge r)$ is true $p \to (r \wedge s)$ is false **19.** Contrapositive: 'If Indian team does not win then match is not in Kolkata, home town of Ganguly' Converse: 'If Indian team wins then match is in Kolkata, home town of Ganguly' Inverse: 'The match is not in Kolkata, home town of Ganguly, Indian team does not win' **21.** (i) False (ii) True (iii) True (iv) False **22.** (i) $(p \to q) \wedge (\sim p \wedge q) = \sim (p \wedge q) \vee (q \wedge \sim p)$ (ii) $\sim (p \to (q \wedge r)) = (p \wedge \sim q) \vee (p \wedge \sim r)$ **23.** (i) $\sim (p \to r) \wedge (p \leftrightarrow q) = (p \vee r) \wedge (\sim p \vee q) \wedge (\sim q \vee p)$ (ii) $(p \wedge q) \vee (\sim p \wedge q \wedge r) = (p \vee q) \wedge (p \vee r) \wedge (q \vee \sim p) \wedge q \wedge (q \vee r)$ **26.** It is a valid argument. **27.** (i) $(\forall x)(K(x) \to L(x))$ (ii) $(\exists x)(K(x) \wedge M(x))$ (iii) $(\exists x)(K(x) \wedge M(x) \to \sim L(x))$ (iv) $(\forall x)(K(x) \wedge L(x) \to M(x))$ **29.** The given argument is valid.

CHAPTER 4

2. (i) A is closed under multiplication. (ii) The set B is not closed under multiplication. (iii) C is not closed under multiplication. (iv) D is closed under multiplication. **3.** $(Z, +)$ is a group. **10.** The orders of the elements $a, a^2, a^3, a^4, a^5,$ and a^6 are 6, 3, 2, 3, 6 and 1, respectively.

12. $f + g = \begin{pmatrix} 1 & 2 & 3 \\ 1 & 3 & 2 \end{pmatrix}$ **13.** $f^2 = \begin{pmatrix} 1 & 2 & 3 & 4 & 5 & 6 & 7 & 8 \\ 1 & 2 & 3 & 4 & 5 & 6 & 7 & 8 \end{pmatrix}$; Order of f is 2 **14.** (i) f is even (ii) f is odd

15. p is odd

16. $\begin{pmatrix} 1 & 2 & 3 & 4 & 5 & 6 \\ 6 & 5 & 2 & 4 & 3 & 1 \end{pmatrix} = (1 \; 6)(2 \; 5 \; 3)$

17. The inverse is $\begin{pmatrix} 1 & 2 & 3 & 4 & 5 & 6 \\ 2 & 6 & 1 & 4 & 3 & 5 \end{pmatrix}$

19. The generators of G are a, a^3, a^5, a^7 and they are 4 in number.
28. The ring $(R, +, *)$ is commutative.

21.

o	(e, e)	(e, a)	(a, e)	(a, a)
(e, e)	(e, e)	(e, a)	(a, e)	(a, a)
(e, a)	(e, a)	(e, e)	(a, a)	(a, e)
(a, e)	(a, e)	(a, a)	(e, e)	(e, a)
(a, a)	(a, a)	(a, e)	(e, a)	(e, e)

CHAPTER 5

3. (i) $\begin{bmatrix} 5 & 1 \\ 1 & 26 \end{bmatrix}$ (ii) $\begin{bmatrix} 10 & -1 & 12 \\ -1 & 5 & -4 \\ 12 & -4 & 16 \end{bmatrix}$ **4.** $\begin{bmatrix} -13 & 52 \\ 104 & -117 \end{bmatrix}$ (ii) $\begin{bmatrix} 0 & 0 \\ 0 & 0 \end{bmatrix}$ **8.** Both (i) and (ii) are orthogonal matrices.

9. (i) $A^{-1} = \begin{bmatrix} 27 & -16 & 1 \\ 8 & -5 & 2 \\ -5 & 3 & -1 \end{bmatrix}$ (ii) B has no inverse. **12.** (i) Rank of the matrix is 1. (ii) Rank of the matrix is 2.

(iii) Rank of the matrix is 2. (iv) Rank of the matrix is 4. **13.** Case I: If $p = 2$, then $|A| = 1 \cdot 0 \cdot 8 \cdot (-)2 = 0$ and so, rank of $A = 3$ Case II: If $p = -6$, then the number of non-zero rows is 3 and so, rank of $A = 3$

14. (i) $\begin{bmatrix} 4 & 0 & 2 & 5 & -3 \\ 0 & 1 & -3 & 4 & 6 \\ 0 & 0 & 7 & -2 & 8 \end{bmatrix}$ (ii) $\begin{bmatrix} 0 & 0 & 5 & -4 & 7 \\ 1 & 2 & 3 & 4 & 5 \\ 0 & 0 & 0 & 0 & 0 \end{bmatrix}$ **15.** (i) $\begin{bmatrix} 1 & 0 & \frac{7}{9} \\ 0 & 1 & -\frac{26}{9} \\ 0 & 0 & 0 \end{bmatrix}$ (ii) $\begin{bmatrix} 1 & 3 & \frac{4}{11} & \frac{13}{11} \\ 0 & 1 & \frac{-5}{11} & \frac{3}{11} \\ 0 & 0 & 0 & 0 \\ 0 & 0 & 0 & 0 \end{bmatrix}$

16. (i) $P = \begin{bmatrix} 1 & 0 & 0 \\ -1 & 1 & 0 \\ -1 & 1 & 1 \end{bmatrix}$ and $Q = \begin{bmatrix} 1 & -1 & -1 \\ 0 & 1 & -1 \\ 0 & 0 & 1 \end{bmatrix}$ (ii) $P = \begin{bmatrix} 1 & 0 & 0 \\ -1 & 1 & 0 \\ -1 & 1 & 1 \end{bmatrix}$ and $Q = \begin{bmatrix} 1 & -3 & -9 & 7 \\ 0 & 1 & 1 & -2 \\ 0 & 0 & 1 & 0 \\ 0 & 0 & 0 & 1 \end{bmatrix}$

17. $P_{3 \times 3} = \begin{bmatrix} 1 & 0 & 0 \\ -\frac{1}{2} & \frac{1}{2} & 0 \\ -\frac{1}{4} & -\frac{1}{2} & \frac{1}{4} \end{bmatrix}$ and $Q_{3 \times 3} = \begin{bmatrix} 1 & 1 & 0 \\ 0 & 1 & -1 \\ 0 & 0 & 1 \end{bmatrix}$ Rank of $A = 2$

18. $AB = \begin{bmatrix} 0 & 1 & 1 \\ 0 & 1 & 0 \\ 1 & 1 & 1 \end{bmatrix}$; $BA = \begin{bmatrix} 1 & 1 & 1 \\ 1 & 0 & 0 \\ 0 & 0 & 1 \end{bmatrix}$; $A^2 = \begin{bmatrix} 1 & 0 & 0 \\ 1 & 1 & 0 \\ 1 & 0 & 1 \end{bmatrix}$

19. *Case I* If $\lambda \neq 8$, then the system has only trivial solution $x = 0, y = 0,$ and $z = 0$; *Case II* If $\lambda = 8$, $r(A) = r(A|B) = 2 < 3 = n$. System has non-trivial solutions. Thus, the infinite numbers of non-trivial solutions are

obtained for different values of k as $x = k, y = -4k, z = k$ **20.** (a) (i) If $\lambda = 3, \mu \neq 10$, then the system possesses no solution. (ii) If $\lambda \neq 3$, then μ may have any value and the system has unique solution. (iii) If $\lambda = 3, \mu = 10$, then the system possesses an infinite number of solutions. (b) (i) $\lambda = 5$ (ii) $\lambda \neq 5$ (iii) μ may have any value **21.** (i) The given equations are consistent and they have a unique solution given by $x = -\frac{16}{5}, y = \frac{29}{5}$, and $z = -\frac{4}{5}$ (ii) $x = \frac{7}{11}$, $y = \frac{3}{11}$, and $z = 0$ is a particular solution. (iii) The given system of equations is consistent and has infinite number of solutions. If $z = p$, where p is arbitrary, then the solutions are $x = 1, y = 3p - 2$, and $z = p$ (iv) The given set of equations is consistent and the solution is unique. The solutions are $x = 1, y = 2$, and $z = 3$ **22.** For a unique solution $\lambda \neq 5$ and μ may have any value. If $\lambda = 5$ and $\mu \neq 9$, the given system of equations possesses no solution. **23.** The system possesses no solution. **24.** (i) The eigenvalues are 3, 2, 5. The eigenvectors are
$X_1 = \begin{pmatrix} 1 \\ 0 \\ 0 \end{pmatrix}, X_2 = \begin{pmatrix} 1 \\ -1 \\ 1 \end{pmatrix}$ (ii) The eigenvalues are $-2, 3, 6$. The eigenvectors are $(-1, 0, 1), (1, -1, 1)$, and $(1, 2, 1)$.

(iii) The eigenvalues are 2, 3, 6. The eigenvectors are $X_1 = \begin{pmatrix} 1 \\ 0 \\ -1 \end{pmatrix}, X_2 = \begin{pmatrix} 1 \\ 1 \\ 1 \end{pmatrix}, X_3 = \begin{pmatrix} 3 \\ -2 \\ 1 \end{pmatrix}$ **25.** Sum of the eigenvalues = 5 and product of the eigenvalues = -21 **26.** Eigenvalues of $A = 2, 5, 4$, eigenvalues of $A^2 = 4, 25, 16$, eigenvalues of $A^{-1} = 1/2, 1/5, 1/4$ and eigen values of $A^{100} = 2^{100}, 5^{100}, 4^{100}$ **27.** The characteristic equation of the matrix A is $-A^3 + 6A^2 - 9A + 4 = 0$. **28.** The characteristic polynomial of the matrix A is $-\lambda^3 + 6\lambda^2 - 9\lambda + 4 = 0$

$A^{-1} = \frac{1}{4} \begin{bmatrix} 3 & 1 & -1 \\ 1 & 3 & 1 \\ -1 & 1 & 3 \end{bmatrix}$ **30.** For this particular problem, A^{-1} cannot be determined by using Cayley–Hamilton theorem.

CHAPTER 6

1. No **2.** Yes, $(P(S), \subseteq)$ is a partially ordered set. **3.** $R = \{(1, 1), (2, 2), (3, 3)\}$ on the set $A = \{1, 2, 3\}$ **5.** The relation R is not reflexive, not symmetric, antisymmetric, not transitive and not equivalence. Also, R is not a partial ordering relation. **6.** The relation R is not reflexive, not symmetric, antisymmetric, not transitive and not equivalence. Also, R is not a partial ordering relation.
8.

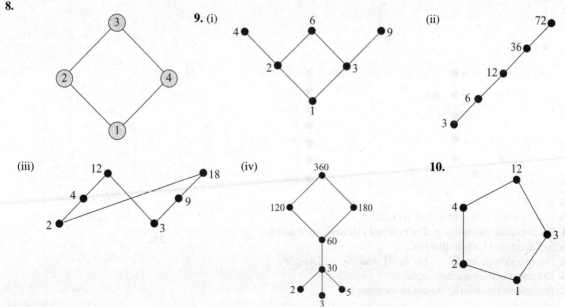

11. 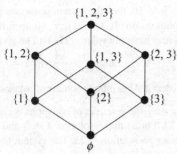 **12.** (i) 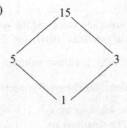 (ii) 16 | 8 | 4 | 2 | 1 (iii) 17 | 1

13. (i) Yes (ii) No (iii) Yes (iv) Yes (v) Yes **14.** The Hasse diagram of the poset D_{100} is described below.

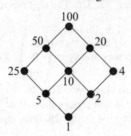

(i) The GLB (B) is 10 (ii) The LUB (B) is 20 (iii) The GLB (B) is 5 (iv) The LUB (B) is 100 **15.** (i) {0} and {1}
(ii) 4 and 6 **16.** (i) (0,0), (1,0), (1,1), (2,0), (2,1), (2, 2)} (ii) (Z^+, 'is a multiple of') (iii) (Z, ≤) (iv) ($S(Z)$, ⊇)
17. (i) (1, 1, 3) < (1, 3, 1) (ii) (0, 1, 2, 3) < (0, 1, 3, 2) (iii) (0, 1, 1, 5) < (0, 1, 3, 4) (iv) (0, 1, 1, 1, 0) < (1, 0, 1, 0, 1)
18. 0 < 0001 < 001 < 01 < 010 < 0101 < 011 < 11
19. **20.** (i) (ii)

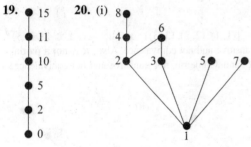

(iii) (iv) **21.**

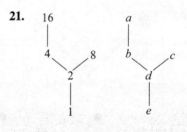

22. $n = 2$
23. (i) $a < e$ (ii) $c || b$ (iii) $b > d$ (iv) $a || d$
24. The maximal element is g. The minimal elements are a and b **25.**
26. (i) Lattice (ii) Lattice (iii) No
27. The two subsets are $S_1 = \{a, b, d\}$ and $S_2 = \{a, c, e\}$
28. The given lattices are isomorphic.
29. The lattices (i) and (ii) are not isomorphic.

CHAPTER 7

3. (i) A' (ii) $AC + B'C$ (iii) $AC' + B + D'$ (iv) $A' + B'C$ **4.** (i) y (ii) $ABC + ABD + BCD$ (iii) $BCD + A'BD'$
5. (i) $xy + x'z'$ (ii) $A'B + C'$ (iii) $a' + bc$ **6.** (i) y (ii) $(A + B)(A + C + D)(B + C')$ (iii) $(A + (A + D')(B' + D')$
10. $F = A + C$ **11.** $F = A' + (B \text{ XOR } C)$ **12.** $F = A \text{ XNOR } (B \text{ XOR } C)$ **13.** (i) $A + BC$ (ii) $Y + XZ$ (iii) Already simplified (iv) 1 (v) xy (vi) $x + y$ (vii) x (viii) $xy + x'z$ **14.** (i) $AB' + A'B$ (ii) X (iii) Already in DNF (iv) $A + B'$
15. XYZ **16.** (i) $F = w'x'z' + w'x'y' + x'y'z' + wxz + xyz + wyz$ (ii) $AB + BC$ **17.** $F = AB \text{ NOR } C$ **18.** $F = A (B + C + D)$ **19.** $F = A \text{ NAND } B \text{ NAND } C$ **20.** $F(A, B, C, D, E) = A + B + C(D \text{ NAND } E)$ **21.** $F(A, B, C, D, E) = A' + C + (B \text{ NOR } D)$ **22.** $F = B + C'$ **23.** $F = A + B + C'$ **24.** $F = A + C$ **25.** $F = A \text{ NOR } B$ **26.** $f = B' + (D \text{ NOR } E \text{ NOR } F)$ **27.** $f = B'$

CHAPTER 8

11. (a) $O(1)$ (b) $O(n)$ (c) $O(n^2)$ (d) $O(1)$ (e) $O(n!)$ (f) $O(n^2)$ (g) True for $n \geq 1$ if base of logarithm is 2 (h) True for $n \geq 1$ if base of logarithm is 2 **12.** (a) $T(n) = O(n)$ (b) $T(n) = O(n\log_2 n)$ (c) $T(n) = O(\sqrt{n})$ (d) $TT(n) = O(2^n)$ (e) $T(n) = O(n)$ (f) $T(n) = O(2^n)$ (g) $T(n) = O(n2^n)$ (h) $T(n) = \theta(n\sqrt{n}) = \theta(n^{3/2})$ (i) $T(n) = O(n\log_2(\log_5 n)) \approx O(n)$ (j) $T(n) = O((\log_2 n)^2)$ (k) $T(n) = \theta(\log_2 n)$ **16.** *Hint*: Think of binary search **17.** *Hint*: Think on the lines of binary search. Watch out for the 'interval' in which $A[low] > A[mid]$ or $A[mid] > A[high]$. The value of k is same as index of minimum element in the rotated array.

CHAPTER 9

2. (a) No such graph (b) No such graph (c) Graph exists (d) No such graph (e) No such graph (f) No such graph **3.** (a) 16 (b) 10 **4.** *Hint*: Suppose x vertices have degree 3 each and $(12-x)$ vertices have degree 6 each. **5.** *Hint*: Sum of all degrees is even. **6.** $(n^2 + n + 2)/2$ **9.** No **11.** (a) No **12.** *Hint*: Think in terms of number of edges in a complete graph. **15.** $1 + \text{floor}(2m/k_2) \leq V \leq \text{floor}(2m/k_1)$ **17.** (i) $2^k - 1$ (ii) Floor $(k/2)*\text{Ceil }(k/2)$ (iii) Yes **18.** The two distinct vertex sets are $\{a\}$ and $\{b, c, d, e, f, g\}$. **19.** A cycle of length 6 does exist in $K_{3,3}$. **22.** *Hint*: Think of a path between two disconnected vertices of $G_1 \cap G_2$. **23.** *Hint*: Think along the lines of definition of cut-vertex. **24.** *Hint*: There exists a path between any two vertices of an undirected tree. **25.** The algorithm will not work and counter examples exist. **26.** 8 path is $K \to C \to B \to L$ **27.** Two bridges exist—both edges are bridges. **32.** *Hint*: If same edge is contained multiple times, then vertices are also repeated. **33.** *Hint*: If same vertex is contained multiple times, it means that we hit the same vertex through multiple paths. **34.** $\left\lfloor \dfrac{n-1}{2} \right\rfloor$ **36.** Euler path exists, but no Euler circuit. Euler path is given by $v_1 \to v_2 \to v_3 \to v_4 \to v_5 \to v_6 \to v_2 \to v_5 \to v_3 \to v_6 \to v_1 \to v_4$. Removing the edge (v_1, v_4) will lead to a Euler circuit **41.** 8 (Just think of spanning tree, only $|V|-1$ edges are needed) **43.** *Hint*: Think of two different Hamiltonian paths passing through u and v. **45.** 5 **47.** *Hint*: Mathematical induction.

CHAPTER 10

6. *Hint*: Think of the sum of degrees. **7.** *Hint*: Think of the sum of degrees **29.** When the maximum weight edge is a bridge/cut-edge. **32.** Statement is true.

CHAPTER 11

3. If $w = xv$ for some v, then x is a prefix of v and for some x, v is the suffix of x **16.** (ii) b^+ab^+ (iii) $(b*ab*ab*ab*)*$ (iv) $(k_1 + k_2 + k_3 + \cdots + k_{243})*$ where each k_i is one possible string of length 5 over the given alphabet—there are a total of $3^5 = 243$ distinct strings of length 5 over the given alphabet (vi) $0*10*10*10*$ (vii) $(aa)*(bb)*b$ **19.** Regular **20.** Regular **21.** Not regular **22.** (i) Not regular (ii) Regular **25.** (i) $S \to k|kS$. There can be a total of 27 possible values of k—corresponding to each string of length 3 which can be formed by the given alphabet (ii) $S \to aA|bB|e, A \to b|bA|B, B \to aA|b|bB$ (v) $S \to a|b|aaS|abS|baS|bbS$ (vi) $S \to aAa|bAb, A \to a|b|aA|bA$

Bibliography

Agarwal, U., *Discrete Mathematical Structure*, Dhanpat Rai, New Delhi, 2008.
Alfred, V.A., J.E. Hopcroft, and J.D. Ullman, *Data Structures and Algorithms*, Addison-Wesley, 1983.
Ash and Ash, *Discrete Mathematics*, McGraw-Hill, New Delhi, 1989.
Bhisma Rao, G.S.S., *Discrete Structures and Graph Theory*, 3rd ed., Scitech, New Delhi, 2009.
Chandrasekharaiaha, D.S., *Mathematical Foundations of Computer Science*, Prism Books, New Delhi, 2006.
Chetwynd, A., and P. Diggle, *Discrete Mathematics*, Butterworth-Heinemann, Oxford, 1995.
Coremen, T.H., C.E. Leiserson, and R.L. Rivest, *Introduction to Algorithms*, PHI, New Delhi, 2008.
Diestal, R., *Graph Theory*, Springer-Verlag, New York, 2005.
Frank, H., *Graph Theory*, Narosa Publishing House, New Delhi, 1998.
Grimaldi, R.P., *Discrete and Combinatorial Mathematics: An Applied Introduction*, 4th ed., Pearson Education Asia, New Delhi, 2002.
Gupta, S.B., *Discrete Mathematics and Structures*, Laxmi Publication, New Delhi, 2006.
Hopcroft, J.E., and J.D. Ullman, *Introduction to Automata Theory, Languages and Computation*, Narosa Publishing House, New Delhi, 2002.
Iyengar, N.C.S.N., V.M. Chandrasekaran, K.A. Venkatesh, and P.S. Arunachalam, *Discrete Mathematics*, Vikas Publishing House Pvt. Ltd., New Delhi, 2003.
Johnsonbaugh, R., *Discrete Mathematics*, 7th ed., Pearson Education, New Delhi, 2008.
Krishnamurthy, V., *Combinatorics: Theory and Application*, East-West Press, New Delhi, 1986.
Levy, L.S., *Discrete Structures of Computer Science*, Wiley Eastern Limited, New Delhi, 1990.
Liptschutz, S., *Discrete Mathematics*, 3rd ed., McGraw-Hill, New Delhi, 2009.
Liu, C.L., *Elements of Discrete Mathematics*, 2nd ed., Tata McGraw-Hill Publication, New Delhi, 2000.
Malik, D.S., and M.K. Sen, *Discrete Mathematical Structures*, 1st ed., Cengage Learning, New Delhi, 2005.
Mishra K.L.P., and S. Chandrasekharan, *Theory of Computer Science*, PHI, New Delhi, 2001.
Mott, J.L., A. Kandel, and T.P. Baker, *Discrete Mathematics for Computer Scientists and Mathematicians*, Prentice Hall, New Delhi, 1986.
Narsingh Deo, *Graph Theory with Application to Engineering and Computer Science,* PHI, New Delhi, 1969.
Rathor, S.K., *Discrete Structures and Graph Theory*, EPH, New Delhi, 2004.
Rathore, S.K.S., and H. Chowdhury, *Discrete Structure and Graph Theory*, Everest Publishing House, Pune, 2001.
Sarkar, S.K., A *Textbook of Discrete Mathematics*, S. Chand & Company, New Delhi, 2004.
Shanker Rao, G., *Discrete Mathematical Structure*, New Age International, New Delhi, 2002.
Somasundaram, *Discrete Mathematical Structure*, 5th ed., PHI, New Delhi, 2005.
Tamilarasi, A., and A.M. Natarajan, *Discrete Mathematics and Its Application*, 2nd ed., Khanna Publishers, New Delhi, 2005.
Weiss, M.A., *Algorithms, Data Structures, and Problem Solving with C++*, Addison-Wesley, 2006.

Index

abstract data type, 350
abstract machine, 480
algebra of propositions, 130
algebra of sets, 14
algorithm, 36, 348
    criteria, 348
    matrix multiplication, 235
    Warshall, 36
alphabet, 481
alternating group, 194
ambiguous grammar, 526, 527
    demerit, 527
antichain, 285
anti-commute, 229
applications of graph, 433
arguments, 138
asymptote, 351
asymptotic notation, 354
    big omega, 357
    little oh, 358
    little omega, 358
    theta, 358
automata, 483
automorphism, 315
AVL tree, 461

basic principle of counting, 69
    multiplication, 70
biconditional statement, 129
binary operations, 162
    associative law, 164
    cancellation law, 167
    catenation, 177
    closure law, 163
    commutative law, 163
    identity element, 165
    inverse element, 165
    properties, 163
binary relation, 21
binary search tree, 455, 457

list representation, 457
    predecessor, 460
    successor node, 460
binomial distribution, 96
binomial theorem, 72
Boolean algebra, 320
Boolean expression, 325
    product of sums (POS)
        form, 325
    sum of products (SOP), 325
Boolean laws, 320
    associative law, 320
    boundedness law, 321
    commutative law, 320
    complement law, 321
    De Morgan's theorem, 321
    distributive law, 320
    duality, 320
    identity law, 320
    involution, 321
    redundance law, 321
    truth tables, 321
Boolean operations, 321
    logic gates, 321
    unique features, 325
bottom-up parsing, 526
bound variables, 150
bubble sorting, 379

Cauchy's, 170
Cayley–Hamilton theorem, 273
    inverse, 273
cellular automaton, 539
chain, 285
Chomsky hierarchy, 518
    context free, 519
    context sensitive, 518
    phrase structure, 518
    regular, 519
    type 0, 518

type 1, 518
type 2, 519
type 3, 519
unrestricted, 518
Church thesis, 537
classes of problems, 537
    NP-class, 538
    NP-complete problems, 538
    NP-hard, 538
    P-class, 538
closure operation, 26
    reflexive, 26
    symmetric, 27
    transitive, 34
colouring of graph, 424
combinations, 79
comparability of elements, 278
complexity, 350
component, 426
composition table, 174
compound propositions, 127
conditional probability, 87
conditional statement, 128
congruence relation, 182
conjunction, 123
connectives, 126
consistency, 151
    system specifications, 132
consistent enumeration, 395
contrapositive, 128
converse, 128
cosets, 196
    left, 196
    right, 196
counting method, 117
cover of an element, 281
    covering relation, 281
cut vertex, 427
cycle, 29
    disjoint, 191

database management system, 42
data structure, 348, 350
   array, 349
   categorization, 349
   operation, 348
   structure, 349
decoding and error
   correction, 207
   decoder, 203
   decoding function, 207
degree of membership, 18
De Morgan's laws, 134
De Morgan's theorem, 320
derivation, 517, 524
   tree, 524
different objects, 80
digraph, 29
   loop, 29
   nodes (or, vertices), 29
Dirichlet drawer principle, 118
   shoe box principle, 118
disambiguous, 527
discrete probability, 83
   axioms, 86
disjunction, 124
disjunctive syllogism, 143
dual poset, 278
eigenvalues, 267
eigenvectors, 267
   multiplicities of the elements, 17
   principal diagonal, 221
encoding function, 204
   encoder, 203
enumerable languages, 537
Euler graph, 418
   Fleury's algorithm, 419
event, 84
   compound, 84
   dependent, 85
   equally likely, 85
   exhaustive, 85
   independent, 87
   mutually exclusive, 85
   simple, 84
expression tree, 461

FA, 503, 516
   identities, 509
   minimization algorithm, 504
factorial notation, 71
Fibonacci numbers, 99

Fibonacci sequence, 103
field, 211
finite automaton with output, 499
   Mealy machine, 499
   Moore machine, 499
   the conversion, 501
   transformation of Mealy to Moore, 501
   transformation of Moore to Mealy, 501
finite automaton, 484, 486, 487
   deterministic, 486
   importance, 488
   Mealy, 485
   Moore, 585
   non-deterministic, 487
   types, 484
flow network, 428
   augmenting path, 429
   capacity, 429
   Ford–Fulkerson algorithm, 431
   free, 429
   network, 430
   positive, 429
   residual, 429, 430
   saturated, 429
free monoid, 482
function, 42
   absolute value, 58
   Ackermann, 60
   addition and multiplication, 45
   arrow diagram, 43
   associativity of composition, 50
   ceiling, 56
   classification, 46
   codomain, 42
   composition, 48
   constant, 48
   domain, 42
   floor, 56
   Hash, 54
   identity, 48
   image, 42, 54
   integer, 58
   inverse, 51
   invertible, 54
   mappings, 42
   one-to-one (injective), 46
   one-to-one and onto (bijective), 47
   onto (surjective), 47
   partial, 59

   primitive recursive, 60
   range, 42
   recursively defined, 55
   remainder, 58
   total, 59
fuzzy logic, 540
fuzzy set, 18, 540
   complement, 18
   intersection, 19
   operation, 18
   union, 19
general tree, 473
   application, 474
   pre-order, 473
generating function, 109
   binomial, 110
   exponential, 111
   Fibonacci sequence, 112
   initial condition, 113
   ordinary, 110
   partition, 116
generation, 518
generator, 185
   matrix, 205
grammar, 517
graph, 384, 404
   connected graph, 387
   connectivity, 387
   degree, 388
   degree sequence of graph, 389
   directed, 384
   multi-graph, 387
   multiple, 386
   parallel edges, 386
   path in a graph, 386
   pseudo-graph, 387
   self-edge, 385
   self-loop, 385
   simple graph, 387
   terminologies, 384
   types, 387
   undirected, 384
   vertex, 388
   weighted, 385
graph algorithms, 409
   breadth first search, 410
   depth first search, 410
greatest lower bound, 293
   infimum, 299
group, 168
   Abelian, 168
   commutative group, 168

cyclic, 185
dihedral, 194
finite group, 168
groupoid, 168
monoid, 168
multiplicative, 168
normal subgroup, 198
order, 168
products and quotients, 174
properties, 171
quotient, 182
subgroup, 183
submonoid, 178
symmetric, 194
group codes, 202
growth, 351

halting problem, 536
Hamiltonian graph, 419
Hasse diagram, 281
homomorphic image, 321
homomorphism, 315
 join, 315
 lattice, 315
 meet, 315
 order, 315
hypothetical syllogism, 143

ideal, 213
 left, 213
 right, 213
inclusion–exclusion principle, 61
 applications, 63
integral domain, 210
introduction to proofs, 150
 consistency, 151
 contradiction (reductio ad absurdum), 152
 contraposition, 152
 direct proof, 151
 mathematical induction, 153
 proof by cases, 157
 reversal law, 117, 252
isomorphic, 403
isomorphic ordered sets, 287
isomorphic posets, 287
isomorphism, 286, 321
isotonocity property, 302

join-irreducible, 313
Karnaugh map, 320, 332
Kleen star, 482

formal language, 482
 operations, 482
Lagrange's theorem, 200
language, 482, 503, 517
 computer, 517
 minimization, 503
 reversal, 482
language generator, 516
 natural, 517
lattice, 296
 bounded, 306
 complemented, 310
 complete, 306
 direct product, 304
 distributive, 306
 homomorphism, 321
 isomorphic, 313
 modular, 309
 non-distributive, 307
 sublattice, 303
law of contraposition (or Modus tollens), 142
law of detachment (or Modus Pones), 141
law of syllogism, 139
least upper bound, 292
 supremum, 292
left-linear, 520
lexicographic ordering, 288
linear, 350
linear algebraic equations, 260
 consistent, 262
 Gauss elimination, 265
 inconsistent, 262
 linear bounded automaton, 484
 linear homogenous, 261
 linear non-homogenous, 261
linear transformation, 269
 characteristic determinant, 269
 characteristic equation, 269
 characteristic polynomial, 269
 characteristic root, 269
 latent root, 269
linearly ordered set, 279
language recognizer, 503
 regular, 503
logic gate, 320
logical equivalence, 133
logical implication, 136
logical reasoning, 139
 fundamental principle, 139

matrix, 220
 addition and subtraction, 225
 adjoint, 244
 Boolean matrix or a zero-one, 256
 co-factor, 239
 column vector, 221
 comparable, 223
 complex, 234
 conjugate, 234
 conjugate transpose, 235
 determinant of matrices (square), 238
 diagonal, 222
 echelon form, 257
 elementary transformation, 258
 equal, 227
 Hermitian, 235
 idempotent, 243
 inverse, 244
 involutory, 243
 lower triangular, 224
 minor, 239
 multiplication, 231
 nilpotent, 244
 non-singular, 244
 normal or canonical form, 259
 null, 222
 operations, 225
 orthogonal, 241
 partition, 233
 positive integral power, 231
 post-factor (post-multiplier), 227
 pre-factor (pre-multiplier), 227
 rank, 257
 rectangular and square, 220
 row vector, 221
 scalar, 223
 scalar multiple, 226
 singular, 244
 skew-Hermitian, 235
 skew-symmetric, 234
 symmetric, 234
 transpose, 234
 transposed conjugate, 242
 unit or identity, 223
 unitary, 242
 upper triangular, 224
maxterm, 325
meet-irreducible, 314
minterm, 325
multinomial theorem, 73

multiplicative identity, 230
multiset, 17
  difference, 18
  intersection, 17
  sum, 17
  union, 17

NDFA from regular expression, 510
  pumping lemma, 514
NDFA, 488
  conversion, 498
negation, 125
non-linear, 350
normal form, 136
  canonical form, 136
  conjunctive normal form, 138
  disjunctive normal form, 137

operations, 404

paradox, 541
parity and generator matrices, 204
parse, 524
  tree, 524
parsing, 524
parsing techniques, 526
partially ordered set, 278
  greatest element, 291
  infimum, 293
  least element, 291
  lower bound, 292
  maximal element, 290
  minimal element, 290
  supremum, 293
  upper bound, 292
Pascal's triangle, 73
PDA, 532
  DPDA, 532
  NPDA, 532
permutation group, 187
  composition, 189
  cyclic, 191
  even, 193
  inverse, 190
  odd, 193
permutation identity, 188
permutations, 75
  circular, 78
  repetitions, 79
pigeonhole principle, 118
  generalized, 118
planar graph, 422
  properties of planar graph, 423

post-correspondence problem, 537
predicate calculus, 146
principle of counting, 69
  addition, 72
  multiplication, 70
principle of duality, 300
procedure, 361
product partial order, 281
product of sums, 320
propositional functions, 130
pushdown automaton, 484, 528
  types, 532
quantifier, 148
  existential quantifier, 149
  universal quantifier, 148
quotient structures, 176
random experiment, 84
  outcome, 84
  sample space, 84
random variable, 92
  expectation, 94
  probability distribution, 93
  standard deviation, 95
  variance, 95
recurrence relation, 99
  characteristic equation, 101
  characteristic root, 101
  E and Δ operator method, 104
  homogeneous equation, 104
  homogeneous solution, 101
  linear homogeneous, 100
  non-homogeneous equation, 100 104
  order and degree, 100
  particular solution, 104
  particular solution in tabular form, 105
  undetermined coefficients method, 104
recursion, 96
recursive, 97, 361, 537
  definition of a function, 100
  definition of a sequence, 96
  definition of a set, 98
  definition, 97
recursively, 537
recursive rule, 517
redundant digit, 205
regular expression, 503, 509, 516
  rules for, 510
  regular language, 514
  application, 516

relation, 20
  antisymmetric, 23
  associative, 24
  composition, 24
  connectivity, 29
  diagonal, 26
  equality, 26
  equivalence, 23
  inverse, 25
  matrix representation, 27
  model databases, 41
  n-ary, 40
  partial ordering, 40
  quaternary, 40
  reflexive, 22
  symmetric, 22
  ternary, 40
  transitive, 23
repetitions, 82
representation of graph, 405
  adjacency list, 406
  adjacency matrix, 405
  advantages, 408
  disadvantages, 408
  incidence matrix, 409
  linked list, 406, 408
  matrix, 405, 408
  representation, 409
right-linear, 519
rings, 208
  automorphism, 219
  commutative, 208
  endomorphism, 216
  epimorphism, 216
  homomorphism, 216
  identity, 208
  monomorphism, 216
  morphisms, 216
  quotient, 215
rotational symmetries, 195
rules of inference, 141
Russell's paradox, 541
  history, 541

searching, 370
  linear, 371
semigroup, 176
  commutative, 177
  congruence relation, 176
  free, 177
  products and quotients, 181
  quotient or factor, 186
  quotient structure, 176

subsemigroup, 178
sequence, 65
set, 2
   cardinality, 4
   Cartesian product, 5
   complement, 10
   computer representation, 19
   countable, 12
   denumerable, 13
   disjoint
   empty, 3
   intersection, 9
   non-denumerable, 13
   null, 3
   operation, 9
   ordered pairs, 5
   partially ordered power, 4
   relative complement, 11
   roster form, 2
   singleton
   symmetric difference, 11
   tabular form, 2
   uncountable, 12
   union, 9
   universal, 3
   void, 3
single-source shortest
   path problem, 412
   Dijkstra's algorithm, 412
sorting, 374
   merges sorting, 374
spanning tree, 462
   Kruskal's algorithm, 463
   minimum spanning tree, 463
   Prim's algorithm, 463
   spanning forest, 463
standard cases, 359
   average case, 359
   best case, 359
   worst case, 359
state, 483
   dead, 484
   final state, 483

   initial start, 483
   non-final, 483
   trapped, 484
state relation, 484
state table, 484
state transition graph, 484
statement, 122
   truth value, 122
string, 482, 517
   operations, 481
subgraph, 401
sub-matrix, 231
subring, 212
   improper or trivial, 212
   proper or non-trivial, 212
subset, 3
   proper subset, 3
sum of products, 320
summation, 65
switching network, 329
symbol, 481
tautologies and contradictions, 131
TM, 534
   variations, 534
top-down, 526
topological sorting, 284
totally ordered set, 279
trace, 221
transition function, 484
transposition, 192
tree, 444, 474
   average height, 440
   binary tree, 443
   child node, 439
   complete binary tree, 444
   depth, 440
   directed, 442
   equivalent tree, 444
   full binary, 444
   height, 440
   in-degree, 442
   internal, 441

   internal node, 439
   labelled tree, 440
   leaf node, 439
   left-skewed, 444
   level, 440
   m-ary tree, 443
   out-degree, 442
   path, 440
   right-skewed, 444
   root, 439
   siblings, 439, 441
   skewed trees, 444
   sub-tree, 440
   tree, 444
tree traversal, 451
   in-order, 451
   post-order, 451
   pre-order, 451
trial, 84
   Bernoulli, 91
   independent, 91
   repeated, 90
turing acceptable language, 536
turing decidable language, 537
turing machine, 532, 481, 484
   improvement, 534
type of derivation, 525
   leftmost, 525
   rightmost, 525
types of graph, 397
   bipartite graph, 399
   complete bipartite graph, 401
   complete graph, 397
   null graph, 397
   regular graph, 398

Venn diagrams, 8

well-formed formulae, 145
well-ordered set, 394